# Algorithmic Trading and Quantitative Strategies

# Algorithmic Trading and Quantitative Strategies

**Raja Velu**
Department of Finance
Whitman School of Management
Syracuse University

**Maxence Hardy**
eTrading Quantitative Research
J.P. Morgan

**Daniel Nehren**
Statistical Modeling & Development
Barclays

CRC Press is an imprint of the
Taylor & Francis Group, an **informa** business
A CHAPMAN & HALL BOOK

First edition published 2020
by CRC Press
6000 Broken Sound Parkway NW, Suite 300, Boca Raton, FL 33487-2742

and by CRC Press
2 Park Square, Milton Park, Abingdon, Oxon, OX14 4RN



CRC Press is an imprint of Taylor & Francis Group, LLC



***Library of Congress Control Number*: 2020932899**

ISBN: 9781498737166 (hbk)
ISBN: 9780429183942 (ebk)

Typeset in STIXGeneral
by Nova Techset Private Limited, Bengaluru & Chennai, India

# *Contents*

# Preface

Algorithms have been around since the day trading has started. But they have gained importance with the advent of computers and the automation of trading. Efficiency in execution has taken center stage and with that, speed and instantaneous processing of asset related information have become important. In this book, we will focus on the methodology rooted in financial theory and demonstrate how relevant data—both in the high frequency and in the low frequency spaces—can be meaningfully analyzed. The intention is to bring both the academics and the practitioners together. We strive to achieve what George Box (first author's teacher) once said:

> *"One important idea is that science is a means whereby learning is achieved, not by mere theoretical speculation on the one hand, nor by the undirected accumulation of practical facts on the other, but rather by a motivated iteration between theory and practice."*

We hope that we provide a framework for relevant inquiries on this topic. To quote Judea Pearl,

> *"You cannot answer a question that you cannot ask, and you cannot ask a question that you have no words for."*

The emphasis of this book, the readers will notice, is on data analysis with guidance from appropriate models. As C.R. Rao (first author's teacher) has aptly observed:

All knowledge is, in final analysis, history.
All Sciences are, in the abstract, mathematics.
All judgements are, in the rationale, statistics.

This book gives an inside look into the current world of Electronic Trading and Quantitative Strategies. We address actual challenges by presenting intuitive and innovative ideas on how to approach them in the future. The subject is then augmented through a more formal treatment of the necessary quantitative methods with a targeted review of relevant academic literature. This dual approach is also reflective of the dynamics, typical of quants working on a trading floor where commercial needs, such as time to market, often supersede consideration of rigorous models in favor of intuitively simple approaches.

Our unique approach in this book is to provide the reader with hands-on tools. This book will be accompanied by a collection of practical Jupyter Notebooks where select methods are applied to real data. This will allow the readers to go beyond theory into the actual implementation, while familiarizing them with the libraries available. Wherever possible the charts and tables in the book can be generated directly from these notebooks and the data sets provided bring further life to the treatment of the subject. We also add exercises to most of the chapters, so that the students can work through on their own. These exercises have been tested out by graduate students from Stanford and Singapore Management University. The notebooks as well as the data and the exercises are made available on: `https://github.com/NehrenD/algo_trading_and_quant_strategies`. This site will be updated on a periodic basis. While reading and working through this book, the reader should be able to gain insight into how the field of Electronic Trading and Quantitative Strategies, one of the most active and exciting spaces in the world of finance, has evolved.

This book is divided into five parts.

- Part I sets the stage. We narrate the history and evolution of Equity Trading and delve into a review of the current features of modern Market Structure. This gives the readers context on the business aspects of trading in order for them to understand why things work as they do. The next section will provide a brief high-level foundational overview of market microstructure which explains and models the dynamics of a trading venue heavily influenced by the core mechanism of how trading takes place: the price-time priority limit-order book with continuous double auction. This will set the stage for the introduction of a critical but elusive concept in trading: Liquidity.

- Part II provides an overview of discrete time series models applied to equity trading. We will address univariate and multivariate time series models of both mean and variance of asset returns, and other associated quantities such as volume. While somewhat less used today because of high frequency trading, these models are important as conceptual frameworks, and act as baselines for more advanced methods. We also cover some essential concepts in Point Processes as the actual trading data can come at irregular intervals. The last chapter of Part II will present more advanced topics like State-Space Models and modern Machine Learning methods.

- Part III dives into the broad topic of Quantitative Trading. Here we provide the reader with a toolkit to confidently approach the subject. Historical perspectives from Alpha generation to the art of backtesting are covered here. Since most quantitative strategies are portfolio-based, meaning that alphas are usually combined and optimized over a basket of securities, we will briefly introduce the topic of Active Portfolio Management and Mean-Variance Optimization, and the more advanced topic of Dynamic Portfolio Selection. We conclude this section discussing a somewhat recent topic: News and Sentiment Analytics. Our intent is to also remind the reader that the field is never "complete" as new approaches (such as this from behavioral finance) are embraced by practitioners once the data is available.

- Part IV covers Execution Algorithms, a sub-field of Quantitative Trading which has evolved separately from simple mechanical workflow tools, into a multi-billion dollar business. We begin by reviewing various approaches to modeling trade data, and then dive into the fundamental subject of Market Impact, a complex and least understood concept in finance. Having set the stage, the final section presents a review of the evolution and the current state of the art in Execution Algorithms.

- Finally, Part V deals with some technical aspects of developing both quantitative trading strategies and execution algorithms. Trading has become a highly technological process that requires the integration of numerous technologies and systems, ranging from market data feeds, to exchange connectivity, to low-latency networking and co-location, to back-office booking and reporting. Developing a modern and high performing trading platform requires thoughtful consideration and some compromise. In this part, we look at some important details in creating a full end-to-end technology stack for electronic trading. We also want to emphasize the critical but often ignored aspect of a successful trading business: Research Environment.

**Acknowledgments:** The ideas for this book were planted ten years ago, while the first author, Raja Velu, was visiting the Statistics Department at Stanford at the invitation of Professor T.W. Anderson. Professor Tze-Leung Lai, who was in charge of the Financial Mathematics program, suggested initiating a course on Algorithmic Trading. This course was developed by the first author and the other authors, Daniel Nehren and Maxence Hardy, offered guest lectures to bring the practitioner's view to the classroom. This book in large part is the result of that interaction. We want to gratefully acknowledge the opportunity given by the Stanford's Statistics department and by Professors Lai and Anderson.

We owe a personal debt to many people for their invaluable comments and intellectual contribution to many sources from which the material for this book is drawn. The critical reviews by Professors Guofu Zhou and Ruey Tsay at various stages of writing are gratefully acknowledged. Colleagues from Syracuse University, Jan Ondrich, Ravi Shukla, David Weinbaum, Lai Xu, Suhasini Subba Rao from Texas A&M, and Jeffrey Wurgler from New York University, all read through various versions of this book. Their comments have helped to improve its content and the presentation. Students who took the course at Stanford University, National University of Singapore and Singapore Management University and the teaching assistants were instrumental in shaping the structure of the book. In particular, we want to recognize the help of Balakumar Balasubramaniam, who offered extensive comments on an earlier version. On the intellectual side, we have drawn material from the classic books by Tsay (2010); Box, Jenkins, Reinsel, Ljung (2015) on the methodology; Campbell, Lo and MacKinlay (1996) on finance; Friedman, Hastie and Tibshirani (2009) on statistical learning. In the tone and substance at times, we could not say better than what is already said in these classics and so the readers may notice some similarities.

We have relied heavily on the able assistance of Caleb McWhorter, who has put the book together with all the demands of his own graduate work. We want to thank our doctoral students, Kris Herman and Zhaoque Zhou (Chosen) for their help at various stages of the book. The joint work with them was useful to draw upon for content. As the focus of this book is on the use of real data, we relied upon several sources for help. We want to thank Professor Ravi Jagannathan for sharing his thoughts and data on pairs trading in Chapter 5 and Kris Herman, whose notes on the Hawkes process are used in Chapter 8. The sentiment data used in Chapter 7 was provided by iSentium and thanks to Gautham Sastri. Scott Morris, William Dougan and Peter Layton at Blackthorne Inc, whose willingness to help on a short notice, on matters related to data and trading strategies is very much appreciated. We want to also acknowledge editorial help from Claire Harshberber and Alyson Nehren.

Generous support was provided by the Whitman School of Management and the Department of Finance for the production of the book. Raja Velu would like to thank former Dean Kenneth Kavajecz and Professors Ravi Shukla and Peter Koveos who serve(d) as department chairs and Professor Michel Benaroch, Associate Dean for Research, for their encouragement and support.

Last but not least the authors are grateful and humbled to have Adam Hogan provide the artwork for the book cover. The original piece specifically made for this book is a beautiful example of Algorithmic Art representing the trading intensity of the US stock universe on the various trading venues. We cannot think of a more fitting image for this book.

Finally, no words will suffice for the love and support of our families.

Raja Velu
Maxence Hardy
Daniel Nehren

# About the Authors

## *Raja Velu*

Raja Velu is a Professor of Finance and Business Analytics in the Whitman School of Management at Syracuse University. He obtained his Ph.D. in Business/Statistics from University of Wisconsin-Madison in 1983. He served as a marketing faculty at the University of Wisconsin-Whitewater from 1984 to 1998 before moving to Syracuse University. He was a Technical Architect at Yahoo! in the Sponsored Search Division and was a visiting scientist at IBM-Almaden, Microsoft Research, Google and JPMC. He has also held visiting positions at Stanford's Statistics Department from 2005 to 2016 and was a visiting faculty at the Indian School of Business, National University of Singapore and Singapore Management University. His current research includes Modeling Big Chronological Data and Forecasting in High-Dimensional settings. He has published in leading journals such as *Biometrika*, *Journal of Econometrics* and *Journal of Financial and Quantitative Analysis*.

## *Maxence Hardy*

Maxence Hardy is a Managing Director and the Head of eTrading Quantitative Research for Equities and Futures at J.P. Morgan, based in New York. Mr. Hardy is responsible for the development of the algorithmic trading strategies and models underpinning the agency electronic execution products for the Equities and Futures divisions globally. Prior to this role, he was the Asia Pacific Head of eTrading and Systematic Trading Quantitative Research for three years, as well as Asia Pacific Head of Product for agency electronic trading, based in Hong Kong. Mr. Hardy joined J.P. Morgan in 2010 from Societe Generale where he was part of the algo team developing execution solutions for Program Trading. Mr. Hardy holds a master's degree in quantitative finance from the University Paris IX Dauphine in France.

## *Daniel Nehren*

Daniel Nehren is a Managing Director and the Head of Statistical Modelling and Development for Equities at Barclays. Based in New York, Mr. Nehren is responsible for the development of algorithmic trading product and model-based business logic for the Equities division globally. Mr. Nehren joined Barclays in 2018 from Citadel, where he was the head of Equity Execution. Mr. Nehren has over 16 years' experience in the financial industry with a focus on global equity markets. Prior to Citadel, Mr. Nehren held roles at J.P. Morgan as the Global Head of Linear Quantitative Research, Deutsche Bank as the Director and Co-Head of Delta One Quantitative Products and Goldman Sachs as the Executive Director of Equity Strategy. Mr. Nehren holds a doctorate in electrical engineering from Politecnico Di Milano in Italy.

Dedicated to...

Yasodha, without her love and support, this is not possible.

R.V.

Melissa, for her patience every step of the way, and her never ending support.

M.H.

Alyson, the muse, the patient partner, the inspiration for this work and the next. And the box is still not full.

D.N.

# Part I

# Introduction to Trading

We provide a brief introduction to market microstructure and trading from a practitioner's point of view. The terms used in this part, all can be traced back to academic literature; but the discussion is kept simple and direct. The data, which is central to all the analyses and inferences, is then introduced. The complexity of using data that can arise at irregular intervals can be better understood with an example illustrated here. Finally, the last part of this chapter contains a brief academic review of market microstructure—a topic about the mechanics of trading and how the trading can be influenced by various market designs. This is an evolving field that is of much interest to all: regulators, practitioners and academics.

# 1

# *Trading Fundamentals*

## 1.1 A Brief History of Stock Trading

**Why We Trade:** Companies need capital to operate and expand their businesses. To raise capital, they can either borrow money then pay it back over time with interest, or they can sell a stake (equity) in the company to an investor. As part owner of the company, the investor would then receive a portion of the profits in the form of dividends. Equity and Debt, being scarce resources, have additional intrinsic value that change over time; their prices are influenced by factors related to the performance of the company, existing market conditions and in particular, the future outlook of the company, the sector, the demand and supply of capital and the economy as a whole. For instance, if interest rates charged to borrow capital change, this would affect the value of existing debt since its returns would be compared to the returns of similar products/companies that offer higher/lower rates of return. When we discuss trading in this book we refer to the act of buying and selling debt or equity (as well as other types of instruments) of various companies and institutions among investors who have different views of their intrinsic value.

These secondary market transactions via trading exchanges also serve the purpose of "price discovery" (O'Hara (2003) [275]). Buyers and sellers meet and agree on a price to exchange a security. When that transaction is made public, it in turn informs other potential buyers and sellers of the most recent market valuation of the security.

The evolution of the trading process over the last 200 years makes for an incredible tale of ingenuity, fierce competition and adept technology. To a large extent, it continues to be driven by the positive (and at times not so positive) forces of making profit, creating over time a highly complex, and amazingly efficient mechanism, for evaluating the real value of a company.

**The Origins of Equity Trading:** The tale begins on May 17, 1792 when a group of 24 brokers signed the Buttonwood Agreement. This bound the group to trade only with each other under specific rules. This agreement marked the birth of the New York Stock Exchange (NYSE). While the NYSE is not the oldest Stock Exchange in the world,[1] nor the oldest in the US,[2] it is without a question the most historically important and undisputed symbol of all financial markets. Thus in our opinion, it is the most suitable place to start our discussion. The NYSE soon after moved their operations to the nearby Tontine Coffee House and subsequently to various other locations around the Wall Street area before settling in the current location on the corner of Wall St. and Broad St. in 1865.

For the next almost 200 years, stock exchanges evolved in complexity and in scope. They, however, conceptually remained unchanged, functioning as physical locations where traders and stockbrokers met in person to buy and sell securities. Most of the exchanges settled on an interaction system called Open Outcry where new orders were communicated to the floor via hand signals and with a Market Maker facilitating the transactions often stepping in to provide short term liquidity. The advent of the telegraph and subsequently the telephone had dramatic effects in accelerating the trading process and the dissemination of information, while leaving the fundamental process of trading untouched.

**Electronification and the Start of Fragmentation:** Changes came in the late 1960s and early 1970s. In 1971, the NASDAQ Stock Exchange launched as a completely electronic system. Initially started as a quotation site, it soon turned into a full exchange, quickly becoming the second largest US exchange by market capitalization. In the meantime, another innovation was underway. In 1969, the Institutional Networks Corporation launched Instinet, a computerized link between banks, mutual fund companies, insurance companies so that they could trade with each other with immediacy, completely bypassing the NYSE. Instinet was the first example of an Electronic Communication Network (ECN), an alternative approach to trading that grew in popularity in the 80s and 90s with the launch of other notable venues like Archipelago and Island ECNs.

This evolution started a trend (Liquidity Fragmentation) in market structure that grew over time. Interest for a security is no longer centralized but rather distributed across multiple "liquidity pools." This decentralization of liquidity created significant challenges to the traditional approach of trading and accelerated the drive toward electronification.

**The Birth of High Frequency Trading and Algorithmic Trading:** The year 2001 brought another momentous change in the structure of the market. On April 9th, the Securities and Exchange Commission (SEC)[3] mandated that the minimum price increment on any exchange should change from 1/16th of a dollar ($\approx$ 6.25 cents) to

---

[1] This honor sits with the Amsterdam Stock Exchange dating back to 1602.

[2] The Philadelphia Stock Exchange has a 2 year head start having been established in 1790.

[3] SEC is an independent federal government agency responsible for protecting investors, and maintaining fair and orderly functioning of the securities market `http://www.sec.gov`

1 cent.[4] This seemingly minor rule change with the benign name of 'Decimalization' (moving from fractions to decimal increments) had a dramatic effect, causing the average spread to significantly drop and with that, the profits of market makers and broker dealers also declined. The reduction in profit forced many market making firms to exit the business which in turn reduced available market liquidity. To bring liquidity back, exchanges introduced the Maker-Taker fee model. This model compensated the traders providing liquidity (makers) in the form of rebates, while continuing to charge a fee to the consumer of liquidity (takers).[5]

The maker-taker model created unintentional consequences. If one could provide liquidity while limiting liquidity taking, one could make a small profit, due to the rebate with minimal risk and capital. This process needs to be fairly automated as the per trade profit would be minimal, requiring heavy trading to generate real revenue. Trading also needs to be very fast as position in the order book and speed of cancellation of orders are both critical to profitability. This led to the explosion of what we today call High Frequency Trading (HFT) and to the wild ultra-low latency technology arms race that has swept the industry over the past 15 years. HFT style trading, formerly called opportunistic market making, already existed but never as a significant portion of the market. At its peak it was estimated that more than 60% of all trading was generated by HFTs.

The decrease in average trading cost, as well as the secular trend of on-line investing led to a dramatic increase in trading volumes. On the other hand, the reduction in per-trade commission and profitability forced broker-dealers to begin automating some of their more mundane trading activities. Simple workflow strategies slowly evolved into a field that we now call Algorithmic Execution which is a main topic of this book.

**Dark Pools and Reg NMS:** In the meantime, the market structure continued to evolve and fragmentation continued to increase. In the late eighties and early nineties a new type of trading venue surfaced, with a somewhat different value proposition: Allowing traders to find a block of liquidity without having to display that information "out loud" (i.e., on exchange). This approach promised reduced risk of information leakage as these venues do not publish market data, only the notification of a trade is conveyed after the fact. These aptly but ominously named Dark Pools have become a staple in equity trading, and now represent an estimated 30-40% of all liquidity being traded in US Equities.

In 2005, the Regulation National Market System (Reg NMS) was introduced in an effort to address the increase in trading complexity and fragmentation. Additionally,

---

[4]The 1/16th price increment was a vestige of Spanish monetary standards of the 1600s when doubloons where divided in 2, 4, 8 parts.

[5]For some readers the concepts of providing and taking liquidity might be murky at best. Do not fret, this will all become clear when we discuss the trading process and introduce Market Microstructure.

it accelerated the transformation of market structure in the US. Through its two main rules, the intent of Reg NMS was to promote best price execution for investors by encouraging competition among individual exchanges and individual orders. The Access Rule promotes non-discriminatory access to quotes displayed by the various trading centers. It also had established a cap, limiting the fees that a trading center can charge for accessing a displayed quote. The Order Protection Rule requires trading centers to obtain the best possible price for investors wherever it is represented by an immediately accessible quote. The rule designates all registered exchanges as "protected" venues. It further mandates that apart from a few exceptions, all market participants transact first at a price equal or better to the best price available in these venues. This is known as the National Best Bid Offer (NBBO).

The Order Protection Rule in particular, had a significant effect on the US market structure. By making all liquidity in protected venues of equal status it significantly contributed to further fragmentation. In 2005 the NYSE market share in NYSE-listed stocks was still above 80%. By 2010, it had plunged to 25% and has not recovered since.

Fragmentation continued to increase with new prominent entrants like BATS Trading and Direct Edge. Speed and technology rapidly became major differentiating factors of success for market makers and other participants. The ability to process, analyze, and react to market data faster than competing participants, meant capturing fleeting opportunities and successfully avoiding adverse selection. Therefore, to gain or maintain an edge, market participants heavily invested in technology, in faster networks, and installed their servers in the same data centers, as the venues they transacted on (co-location).

**Conclusion:** This whirlwind tour of the history of trading was primarily to review the background forces that led to the dizzying complexity of modern market structure, that may be difficult to comprehend without the appropriate context. Although this overview was limited to the USA alone, it is not meant to imply that the rest of the world stayed still. Europe and Asia both progressed along similar lines, although at a slower and more compressed pace.

## 1.2 Market Structure and Trading Venues: A Review

### 1.2.1 Equity Markets Participants

With the context presented in the previous section, we now provide a brief review of the state of modern Market Structure. In order to get a sense of the underpinnings of equity markets, it is important to understand who are the various participants, why they trade and what their principal focus is.

**Long Only Asset Managers:** These are the traditional Mutual Fund providers like Vanguard, Fidelity, etc. A sizable portion of US households invest in mutual funds as part of their company's pensions funds, 401K's, and other retirement vehicles. Some of the largest providers are of considerable size and have accumulated multiple trillions of dollars under their management. They are called Long Only asset managers because they are restricted from short selling (where you borrow shares to sell in the market in order to benefit from a price drop). They can only profit through dividends and price appreciation. These firms need to trade frequently in order to re-balance their large portfolios, to manage in-flows, out-flows, and to achieve the fund's objective. The objective may be to track and hopefully beat a competitive benchmark. The time horizon these investors care about is in general long, from months to years. Historically, these participants were less concerned with transaction costs because their investment time scales are long and the returns they target dwarf the few dozen basis points normally incurred in transaction costs.[6] What they are particularly concerned about is information leakage. Many of these trades last from several days to weeks and the risk is that other participants in the market realize their intention and profit from the information to the detriment of the fund's investors.

**Long-Short Asset Managers and Hedge Funds:** These are large and small firms catering to institutional clients and wealthy investors. They often run multiple strategies, most often market-neutral long-short strategies, holding long and short positions in order to have reduced market exposure and thus hopefully perform well in both raising and falling markets. Investors of this type include firms like Bridgewater, Renaissance Technologies, Citadel, Point 72 (former SAC Capital), and many others. Their time horizon is varied, as with the strategies they employ, but in general they are shorter than their Long Only counterparts. They also tend to be very focused on minimizing transaction costs, because they often make many more smaller short term investments and the transaction costs can add up to be a significant fraction of their expected profits.

---

[6]In recent years though, competitive pressure as well as Best Execution regulatory obligations have brought a lot of focus on transaction costs minimization to the Long Only community as well.

**Broker-Dealers (a.k.a. Sell Side):** These firms reside between the "Buy Side" (generic term for asset management firms and hedge funds) and the various exchanges.[7] They can act either as Agent for the client (i.e., Broker) or provide liquidity as Principal (i.e., Dealer) from their own accounts. They historically also had large proprietary trading desks investing the firm's capital using strategies not dissimilar to the strategies that hedge funds use. Since the introduction of the Dodd-Frank Act,[8] the amount of principal risk that a firm can carry has dramatically decreased and banks had to shed their proprietary trading activities either shutting down the desks or spinning them off into independent hedge funds.

**HFTs, ELPs (Electronic Liquidity Providers), and DMMs (Designated Market Makers):** These participants generate returns by acting as facilitators between the above participants, providing liquidity and then unwinding it at a profit. They can also act as aggregators of retail liquidity, e.g., individual investors who trade using online providers like ETrade. Usually the most technology savvy operators in the market place, they leverage ultra-low-latency infrastructure in order to be extremely nimble. They get in and out of positions rapidly, taking advantage of tiny mis-pricing.

### 1.2.2 Watering Holes of Equity Markets

Now that we know who the players are, we briefly review various ways they access liquidity.

**Exchanges:** This is still the standard approach to trading and accounts for about 60–70% of all activity. This is where an investor will go for immediacy and a surer outcome. Because the full order-book of an exchange, arrivals/cancellations are all published, the trader knows exactly the liquidity that is available and can plan accordingly. This information, it should be kept in mind, is known to all participants, especially (due to their often technological advantages) the ones whose strategies focus on patterns of large directional trades. A trader that needs to buy or sell in large size will need to be careful about how much information their trades disseminate or they may pay dearly. The whole field of Algorithmic Execution evolved as an effort to trade large positions while minimizing market impact and informational leakage.

Apart from these concerns, interacting with exchanges is arguably the most basic task to trading. But it is not straightforward. At the time of this writing, a US equity trader can buy or sell stocks in 15 registered exchanges (13 actively trading).[9] As mentioned before when discussing Reg NMS, these exchanges are all "protected." Thus, the liquidity at the best price cannot be ignored. An exchange will need to

[7]Note that only member firms are allowed to trade on exchange and most asset management firms are non-members thus need an intermediary to trade on their behalf.

[8]`https://www.cftc.gov/LawRegulation/DoddFrankAct/index.htm`

[9]`https://www.sec.gov/fast-answers/divisionsmarketregmrexchangesshtml.html`

reroute to other exchanges where price is better (charging a fee for it). Smart Order Routers have evolved to manage this complexity.

Recent years have seen a consolidation of these exchanges in the hands of mainly three players: ICE, NASDAQ and CBOE. Here is a list of venues operated by each, respectively:

- NYSE, ARCA, MKT (former AMX, American Stock Exchange), NSX (former National Stock Exchange), CHX (former Chicago Stock Exchange)
- NASDAQ, PHLX (former Philadelphia Stock Exchange), ISE (former International Securities Exchange), BX (former Boston Stock Exchange)
- CBOE, BZX (former BATS), BYX (former BATS-Y), EDGA, EDGX

The newest exchange and as of now the only remaining independent exchange, is the Investors Exchange (IEX).[10] We will discuss more about this later.

An interesting observation is that this consolidation did not happen with a contemporaneous reduction in fragmentation. These venues continue to operate as separate pools of liquidity. While the exchange providers make valid arguments that they provide different business propositions, there is a growing concern in the industry that the revenue model for these exchanges is now largely centered around providing market data and charging for exchange connectivity fees.[11] Because the venue quotes are protected, any serious operator needs to connect with the exchanges and leverage their direct market data feeds in their trading applications. With more exchanges, the more connections and fees these exchange can charge. Recently regulators are starting to weigh in on this contentious topic.[12]

It is also interesting to note that most of these exchanges are hosted in one of four data centers located in the New Jersey countryside: Mahwah, Secaucus, Carteret, Weehawken. The distance between these data centers adds some latency in the dissemination of information across exchanges and thus creates latency arbitrage opportunities. Super fast HFTs co-locate in each data center and leverage the best available technology such as microwave and more recently laser technologies to connect them. These operators can see market data changes before others do. Then they either trade faster or cancel their own quotes to avoid adverse selection (a phenomenon known as liquidity fading).

All exchanges have almost exactly the same trading mechanism and from the trading perspective, behave exactly the same way during the continuous part of the trading day except for the opening and closing auctions. They provide visible (meaning that market data is disseminated) order books and operate on a price/time priority basis.[13]

---

[10]In 2019, a group of financial institutions filed an application for a members owned exchange, MEMX.

[11]`https://www.sifma.org/wp-content/uploads/2019/01/Expand-and-SIFMA-An-Analysis-of-Market-Data-Fees-08-2018.pdf`

[12]`https://www.sec.gov/tm/staff-guidance-sro-rule-filings-fees`

[13]Certain Futures and Options exchanges have a pro-rata matching mechanism which is discussed later.

Most of the exchanges use the maker-taker fee model that we discussed earlier but some: BYX, EDGA, NSX, BX, have adopted an 'inverted' fee model, where posting liquidity incurs a fee while a rebate is provided for taking liquidity. This change causes these venues to display a markedly different behavior. The cost for providing liquidity removes the incentive of rebate seeking HFTs; however these venues are used first when needing immediate liquidity as the taker is compensated. This in turn brings in passive liquidity providers who are willing to pay for trading passively at that price. This interplay adds a subtle and still not very well understood dynamic to an already complex market structure.

Finally, a few observations about IEX. It started as an Alternative Trading System (ATS) whose main innovation is a 38 mile coil of optical fiber placed in front of its trading engine. This introduces a 350 microsecond delay each way aptly named the "speed bump" that is meant to remove the speed advantages of HFTs. The intent is to reduce the efficacy of the more latency-sensitive tactics and thus provide a liquidity pool that is less "toxic." On June 17, 2016 IEX became a full fledged exchange after a very controversial process.[14,15] It was marketed as an exchange that would be different and would attract substantial liquidity due to its innovative speed bump. But, as of 2019, this potential remains still somewhat unrealized and IEX has not moved much from the 2% to 3% market share range of which 80% is still in the form of hidden liquidity. That being said IEX has established itself as a vocal critic[16] of the current state of affairs continuing to shed light on some potential conflicts of interest that have arisen in trading. At the time of this writing IEX does not charge any market data fees.[17]

**ATS/Dark Pools:** As discussed above, trading in large blocks in exchanges is not a simple matter and requires advanced algorithms for slicing the block orders and smart order routers for targeting the liquid exchanges. Even then the risk and the resulting costs due to information leakage can be significant. Dark Pools were invented to counterbalance this situation. One investor can have a large order "sitting" in a dark pool with no one knowing that it is there and would be able to find the other side without showing any signals of the order's presence. Dark pools do not display any order information and use the NBBO (National Best Bid/Offer) as the reference price. In almost all cases, to avoid accessing protected venues, these pools trade only at the inside market (at or within the bid ask spread). Although, in order to maximize the probability of finding liquidity most of the block interaction happens at the mid-point. Off-exchange trading has gained more and more traction in the last fifteen

---

[14] `https://www.sec.gov/comments/10-222/10-222.shtml`

[15] `https://www.bloomberg.com/news/articles/2016-06-14/sec-staff-recommends-approving-iex-application-wsj-reports`

[16] `https://iextrading.com/insights`

[17] `https://iextrading.com/trading/market-data/`

years and now accounts for 30–40% of all traded volume in certain markets like the US. ATS/Dark Pool volume makes up roughly 30% of this US equities off-exchange volume. Clearly, the growth of dark execution has spurred a lot of competition in the market. Additionally, the claim of reduced information leakage is probably overstated as it is still potentially possible to identify large blocks of liquidity by "pinging" the pool at minimum lot size. In order to counteract this effect orders are usually sent with a minimum fill quantity tag which allows the block to be transparent from small pinging.

As of 2019, there are 33 different equity ATSs![18] All these venues compete on pricing, availability of liquidity, system performance, and functionality such as handling of certain special order types, etc. Many of these ATS are run by the major investment banks and they have historically dominated this space. As far as overall liquidity goes, 10 dark pools account for about 75% of all ATS volume, and the top 5 make up roughly 50% of dark liquidity in the US (the UBS ATS and Credit Suisse Crossfinder are consistently at the top of the rankings[19]). Other venues were born out of the fundamental desire for investment firms to trade directly with each other, bypassing the intermediaries and thus reducing cost and information leakage. The main problem encountered by these buy-side to buy-side pools is that many of the trading strategies used by these firms tend to be highly correlated (e.g., two funds tracking the same benchmark) and thus the liquidity is often on the same side. Therefore as a result, these venues have to find different approaches to leverage sell side broker's liquidity to supplement their own such as access via conditional orders. BIDS and Liquidnet are the biggest ATS of this type. BIDS had significant growth in recent years with Liquidnet losing its initial dominance. This space is still very active with new venues coming up on a regular basis.

**Single Dealer Platform/Systematic Internalizers:** A large number of trading firms in recent years started providing direct access to their internal liquidity. Broker/Dealer and other institutional clients connect to a Single Dealer Platform (SDP) directly. These SDPs, also called Systematic Internalizers, send regular Indication Of Interest (IOI) that are bespoke to a particular connection and the broker can respond when there is a match. This approach to trading is also growing fast. Because SDPs are not regulated ATS, they can offer somewhat unique products. Brokers themselves are now starting to provide their own SDPs to expose their internal liquidity. A quickly evolving space that promises interesting innovations, but alas, it can also lead to further complication in an already crowded ecosystem. In the US, the other 70% of non-ATS off-exchange volume is comprised of Retail Wholesalers, Market Makers and Single Dealer Platforms, and Broker Dealers.

**Auctions:** Primary exchanges (exchanges where a particular instrument is listed) begin and end the day with a (primary) auction procedure, that leverages special order

---

[18] `http://www.finra.org/industry/equity-ats-firms`

[19] `https://otctransparency.finra.org/otctransparency/AtsData`

types to accumulate supply and demand and then run an algorithm that determines the price that would at best pair off the most volume. The Closing Auction is of particular importance because many funds set their Net Asset Value (NAV) using the official closing price. This generally leads traders to trade as close as possible to the closing time in order to optimize the dual objective of getting the best price but also not deviating too much from the close price.

The Auction also represents an opportunity for active and passive investors to exchange large amounts of shares (liquidity). Index constituents get updated on a regular basis (additions, deletions, weight increase/decrease), and as they get updated, passive investors need to update their holdings to reflect the optimal composition of the benchmark that they track. In order to minimize the tracking error risk, that update needs to happen close to the actual update of the underlying benchmark. Consequently, most passive indexers tend to rebalance their portfolios on the same day the underlying index constituents are updated, using the closing auction as a reference price, and this results in significant flows at the close auction. The explosive growth of ETFs (Exchange Traded Funds) and other passive funds have exacerbated this trend in recent years. At the time of this writing, about 10% of the total daily US volume in index names trade at the close. Recent months have brought a lot of movement in this area with brokers and SDPs trying to provide unique ways to expose internal liquidity marked for the closing auction.

**Beyond the US:** As previously mentioned the structure we presented above is not unique to the US. European and Asian exchanges that had operated as a single marketplace for longer than their US counterparts now offer a more diverse landscape. The evolution of ATS/Dark Pools and other MTFs (Multilateral Trading Facilities) have followed suit but in a more subdued manner. Trading in these markets has always been smaller and concentrated with fewer participants and there is not enough liquidity to support a large number of venues. Often new venues come on-line but are quickly absorbed by a competitor when they fail to move a significant portion of the traded volume. Additional complexity arises with regulatory environments in these various countries limiting cross-border trading. As of December 2018, fragmentation in European markets is still quite lower than in the US, with 59% traded on primary exchange, 22% on lit MTF (there are only 6, and 4 of them trade roughly 98% of the volume), 6.5% on dark MTF (10 different ones), and 6% traded in systematic internalizers. In APAC, with the exception of Australia, Hong Kong and Japan, most countries only have one primary exchange where all transactions take place. Even in the more developed market, Japan for instance, the Tokyo Stock Exchange still garners over 85% of the total volume traded.

**Summary:** Modern market structure may appear to be a jumbled mess. Yes, it is. Fully understanding the implications of these different methods of trading tied by regulation, competition, behavioral idiosyncrasies (at times due to participants lack of understanding and preconceived notions) is a daunting task. It is however an environment that all practitioners have to navigate, and it remains difficult to formulate these issues in mathematical models as they could be based on questionable heuristics to account for the residual complexity. But even with the dizzying complexity, modern market structure is a fascinating ecosystem. It is a continuously evolving system through the forces of ingenuity and competition. It provides tools and services to institutional investors who strive to reduce the cost of execution, an investment mandate. We hope the above treatment provides the reader with at least a foothold in the exploration of this amazing social and financial experimentation.

## 1.3 The Mechanics of Trading

In order to fully grasp the main topics of this book, one requires at least a good understanding of the mechanics of trading. This process is somewhat complicated and requires some technical details and terminology. In this section, we will strive to provide a brief but fairly complete overview of the fundamentals. This should suffice for our purposes. For a more complete and thorough treatment we refer the readers to the existing literature, most notably Harris (2003) [178].

### 1.3.1 How Double Auction Markets Work

The most common approach used by modern electronic exchanges can be termed, as time/price priority, continuous double auction trading system. The term double auction signifies that, unlike a common auction with one auctioneer dealing with potential buyers, in this case there are multiple buyers and multiple sellers participating in the process at the same time. These buyers and sellers interact with the exchange by sending instructions electronically, via a network protocol to a specialized software and hardware infrastructure called: The Matching Engine. It has two main components: The Limit Order Book and the Matching Algorithm.

**Limit Order Book (LOB):** It is a complex data structure that stores all non-executed orders with associated instructions. It is highly specialized so as to be extremely fast to insert/update/delete orders and then able to sort them and to retrieve aggregated information. For an active stock, the LOB can be updated and queried thousands of times every second, so it must be highly efficient and able to handle a high degree of concurrency to ensure that the state is always correct. The LOB is comprised of two copies of the core data structure, one for Buy orders and one for Sell orders often referred to as the two "sides" of the order book. This structure is the core abstraction for all electronic exchanges and so it is very important to understand it in detail.

The LOB supports three basic instructions: Insert, cancel, and amend, with insert initiating a new order, cancel removing an existing order from the market and amend modifying some of the parameters of the existing order. New orders must specify "order type" and associated parameters necessary to fully encapsulate the trader decision. We will review order types in more detail later, but we start with the two main types: The limit order and the market order. The main difference between the two order types is that a limit order has a price associated to it while a market order does not.

Accounting for limit and market orders, there are eight events at any given time, four on either side, that can alter the state of the order book:

- **Limit Order Submission:** A limit order is added to the queue at the specified price level.
- **Limit Order Cancellation:** An outstanding limit order is expired or canceled and is therefore removed from the Limit Order Book (LOB).
- **Limit Order Amendment:** An outstanding limit order is modified by the original sender (such as changing order size).
- **Execution:** Buy and sell orders at appropriate prices are paired by the matching algorithm (explained below) into a binding transaction and are removed from the LOB

**Matching Algorithm:** This software component is responsible for interpreting the various events to determine if any buy and sell orders can be matched in an execution. When multiple orders can be paired the algorithm uses the so-called price/time priority meaning that first the order with the most competitive prices are matched and when prices are equal the order that arrived prior is chosen. As we will see in later chapters this is only one of the possible algorithms used in practice but it is by far the most common. We will go into more details in the next sections.

The matching algorithm operates continuously throughout the trading hours. In order to ensure an orderly start and end, this continuous session is usually complemented by a couple of discrete auctions. The trading day generally starts with an open auction, then followed by the main continuous session, and ends with a closing auction. Some markets like Japan also have a lunch break which might be preceded by a morning closing auction and followed by an afternoon opening auction. We will now discuss these main market phases in chronological order.

### 1.3.2 The Open Auction

The Open Auction is only one type of call auction that is commonly held on exchanges. The term "call auction" explains the liquidity-aggregating nature of this event. Market participants are 'called' to submit their quotes to the market place in order to determine a matching price that will maximize the amount of shares that can be transacted. To facilitate timely and orderly cross, auctions have strict order submission rules, including specified timing for entries (see Table 1.1) and information dissemination to prevent wild price fluctuations and ensure that the process is efficient for price discovery.

Most exchanges publish order imbalance that exists among orders on the opening or closing books, along with the indicative price and volume. For instance, Nasdaq

Table 1.1: Nasdaq Opening Cross

| | |
|---|---|
| 4:00 a.m. EST | Extended hours trading and order entry begins. |
| 9:25 a.m. EST | Nasdaq enters quotes for participants with no open interest. |
| 9:28 a.m. EST | Dissemination of order imbalance information every 1 second. Market-on-open orders must be received prior to 9:28 a.m. |
| 9:30 a.m. EST | The opening cross occurs. |

publishes the following information[20] between 9:28 a.m. EST and 9:30 a.m. EST, every 1 second, on its market data feeds:

- **Current Reference Price:** Price within the Nasdaq Inside at which paired shares are maximized, the imbalance is minimized and the distance from the bid-ask mid-point is minimized, in that order.
- **Near Indicative Clearing Price:** The crossing price at which orders in the Nasdaq opening / closing book and continuous book would clear against each other.
- **Far Indicative Clearing Price:** The crossing price at which orders in the Nasdaq opening / closing book would clear against each other.
- **Number of Paired Shares:** The number of on-open or on-close shares that Nasdaq is able to pair off at the current reference price.
- **Imbalance Shares:** The number of opening or closing shares that would remain unexecuted at the current reference price.
- **Imbalance Side:** The side of the imbalance: B = buy-side imbalance; S = sell-side imbalance; N = no imbalance; O = no marketable on-open or on-close orders.

In a double auction setup, the existence of multiple buyers and sellers requires employing a matching algorithm to determine the actual opening price which we will illustrate with a practical example. Table 1.2 gives an example of order book submissions for a hypothetical stock, where orders are ranked based on their arrival time. Different exchanges around the world apply slightly different mechanisms to their auctions, but generally the following rules apply to match supply and demand:

- The crossing price must maximize the volume transacted.
- If several prices result in similar volume transacted, the crossing price is the one the closest from the last price.
- The crossing price is identical for all orders executed.
- If two orders are submitted at the same price, the order submitted first has priority.

[20]Source: Nasdaq Trader website.

Table 1.2: Pre-Open Order Book Submissions

| Timestamp | Seq. Number | Side | Quantity | Price |
|---|---|---|---|---|
| 9:01:21 | 1 | B | 1500 | 12.10 |
| 9:02:36 | 2 | S | 1750 | 12.12 |
| 9:05:17 | 3 | B | 4500 | 12.17 |
| 9:06:22 | 4 | S | 1750 | 12.22 |
| 9:06:59 | 5 | S | 2500 | 12.11 |
| 9:07:33 | 6 | B | 1200 | 12.23 |
| 9:07:42 | 7 | B | 500 | 12.33 |
| 9:08:18 | 8 | B | 500 | 12.25 |
| 9:09:54 | 9 | S | 1930 | 12.30 |
| 9:09:55 | 10 | B | 1000 | 12.21 |
| 9:10:04 | 11 | S | 3500 | 12.05 |
| 9:10:39 | 12 | B | 2000 | 12.34 |
| 9:11:13 | 13 | S | 4750 | 12.25 |
| 9:11:46 | 14 | B | 2750 | 12.19 |
| 9:12:21 | 15 | S | 10000 | 12.33 |
| 9:12:48 | 16 | B | 3000 | 12.28 |
| 9:13:12 | 17 | B | 5500 | 12.35 |
| 9:14:51 | 18 | S | 1800 | 12.18 |
| 9:15:02 | 19 | B | 800 | 12.17 |
| 9:15:37 | 20 | S | 1200 | 12.19 |
| 9:16:42 | 21 | S | 5000 | 12.16 |
| 9:17:11 | 22 | B | 12500 | 12.15 |
| 9:18:27 | 23 | S | 450 | 12.23 |
| 9:19:13 | 24 | S | 3500 | 12.20 |
| 9:19:54 | 25 | B | 1120 | 12.16 |

- It is possible for an order to be partially executed if the other side quantity is not sufficient.
- "At Market" orders are executed against each other at the determined crossing price, up to the available matching quantity on both sides, but generally do not participate in the price formation process.
- For the Open Auction, unmatched "At Market" orders are entered into the continuous session of LOB as limit orders at the crossing price.

The first step is to organize orders by limit price, segregating buys and sells as shown in Table 1.3. A buy order submitted with a limit price of 12.25 represents an intent to execute at any price lower or equal to 12.25. Similarly, a sell order submitted at 12.25 represents an intent to sell at any price higher or equal to 12.25. For each price level, we can then determine the cumulative buy interest and sell interest. The theoretical cross quantity at each price point is then simply the minimum of the cumulative buy interest and the cumulative sell interest as shown in Table 1.4. The

Table 1.3: Ranked Order Book Submissions

| Timestamp | Seq. Number | Buy Price | Buy Quantity | Sell Quantity | Sell Price |
|---|---|---|---|---|---|
| 9:13:12 | 17 | 12.35 | 5500 | | |
| 9:10:39 | 12 | 12.34 | 2000 | | |
| 9:07:42 | 7 | 12.33 | 500 | | |
| 9:12:21 | 15 | | | 10000 | 12.33 |
| 9:09:54 | 9 | | | 1930 | 12.30 |
| 9:12:48 | 16 | 12.28 | 3000 | | |
| 9:08:18 | 8 | 12.25 | 500 | | |
| 9:11:13 | 13 | | | 4750 | 12.25 |
| 9:07:33 | 6 | 12.23 | 1200 | | |
| 9:18:27 | 23 | | | 450 | 12.23 |
| 9:06:22 | 4 | | | 1750 | 12.22 |
| 9:09:55 | 10 | 12.21 | 1000 | | |
| 9:19:13 | 24 | | | 3500 | 12.20 |
| 9:11:46 | 14 | 12.19 | 2750 | | |
| 9:15:37 | 20 | | | 1200 | 12.19 |
| 9:14:51 | 18 | | | 1800 | 12.18 |
| 9:05:17 | 3 | 12.17 | 4500 | | |
| 9:15:02 | 19 | 12.17 | 800 | | |
| 9:16:42 | 21 | | | 5000 | 12.16 |
| 9:19:54 | 25 | 12.16 | 1120 | | |
| 9:17:11 | 22 | 12.15 | 12500 | | |
| 9:02:36 | 2 | | | 1750 | 12.12 |
| 9:06:59 | 5 | | | 2500 | 12.11 |
| 9:01:21 | 1 | 12.10 | 1500 | | |
| 9:10:04 | 11 | | | 3500 | 12.05 |

crossing price is determined as the price that would maximize the crossed quantity. In our example, the opening price will be 12.19, and the opening quantity will be 15,750 shares.

The list of buy orders executed during the auction is shown in Table 1.5 and the sell orders in Table 1.6. It is worth mentioning that the open auction tends to be considered as a major price discovery mechanism given the fact that it occurs after a period of market inactivity when market participants were unable to transact even if they have information. All new information accumulated overnight will be reflected in the first print of the day, matching buying and selling interests.

As market participants with better information are more likely to be participating in the open auction with more aggressive orders in order to extract liquidity (and, as such, setting the price), the price discovery mechanism is often considered to be quite volatile and more suited for short-term alpha investors. Similarly, the period immediately following the open auction also tends to be much more volatile than the rest of the day. As a result of which, most markets experience wider spreads while

Table 1.4: Cumulative Order Book Quantities

| Price | Sequence Number | Cumulative Buy Quantity | Buy Quantity | Sell Quantity | Cumulative Sell Quantity | Quantity Crossed at Price |
|---|---|---|---|---|---|---|
| 12.35 | 17 | 5500 | 5500 | | 38130 | 5500 |
| 12.34 | 12 | 7500 | 2000 | | 38130 | 7500 |
| 12.33 | 7 | 8000 | 500 | | 38130 | 8000 |
| 12.33 | 15 | 8000 | | 10000 | 38130 | 8000 |
| 12.30 | 9 | 8000 | | 1930 | 28130 | 8000 |
| 12.28 | 16 | 11000 | 3000 | | 26200 | 11000 |
| 12.25 | 8 | 11500 | 500 | | 26200 | 11500 |
| 12.25 | 13 | 11500 | | 4750 | 26200 | 11500 |
| 12.23 | 6 | 12700 | 1200 | | 21450 | 12700 |
| 12.23 | 23 | 12700 | | 450 | 21450 | 12700 |
| 12.22 | 4 | 12700 | | 1750 | 21000 | 12700 |
| 12.21 | 10 | 13700 | 1000 | | 19250 | 13700 |
| 12.20 | 24 | 13700 | | 3500 | 19250 | 13700 |
| 12.19* | 14 | 16450 | 2750 | | 15750 | 15750 |
| 12.19 | 20 | 16450 | | 1200 | 15750 | 15750 |
| 12.18 | 18 | 16450 | | 1800 | 14550 | 14550 |
| 12.17 | 3 | 20950 | 4500 | | 12750 | 12750 |
| 12.17 | 19 | 21750 | 800 | | 12750 | 12750 |
| 12.16 | 21 | 21750 | | 5000 | 12750 | 12750 |
| 12.16 | 25 | 22870 | 1120 | | 7750 | 7750 |
| 12.15 | 22 | 35370 | 12500 | | 7750 | 7750 |
| 12.12 | 2 | 35370 | | 1750 | 7750 | 7750 |
| 12.11 | 5 | 35370 | | 2500 | 6000 | 6000 |
| 12.10 | 1 | 36870 | 1500 | | 3500 | 3500 |
| 12.05 | 11 | 36870 | | 3500 | 3500 | 3500 |

market makers try to protect themselves against information asymmetry by quoting wider bids and offers. The increased volatility and wider spreads might discourage certain investors from participating in the market at the open auction and in the period immediately following the open. While this appears to be reasonable from a price risk perspective, it is worth mentioning that for many less liquid stocks (in particular small and mid cap stocks), the open auction can be a significant liquidity aggregation point that even surpasses the close auction. In Australia for instance, the bottom 50% less liquid stocks have more volume traded in the open auction than in the close auction. Similarly in Japan, the less liquid stocks have more volume traded in the open auction, but also in the afternoon open auction that follows the market lunch break.

From an execution standpoint, though, the usage of the open auction has to be considered carefully. While this represents a liquidity opportunity, the first print of the day can also have a significant anchoring effect on the stock price for the remainder of the day. So, participating in the open should be considered in light of the liquidity

Table 1.5: Crossed Buy Orders

| Price | Seq. Number | Buy Qty |
|---|---|---|
| 12.35 | 17 | 5500 |
| 12.34 | 12 | 2000 |
| 12.33 | 7 | 500 |
| 12.28 | 16 | 3000 |
| 12.25 | 8 | 500 |
| 12.23 | 6 | 1200 |
| 12.21 | 10 | 1000 |
| 12.19* | 14 | 2050 |

*Order number 14 was for 2750 shares but did not get fully executed as the bid quantity up to 12.19 exceeded the offered quantity at that price. The balance of order 14 will then be posted as a limit order in the continuous trading session

Table 1.6: Crossed Sell Orders

| Price | Seq. Number | Sell Qty |
|---|---|---|
| 12.19 | 20 | 1200 |
| 12.18 | 18 | 1800 |
| 12.16 | 21 | 5000 |
| 12.12 | 2 | 1750 |
| 12.11 | 5 | 2500 |
| 12.05 | 11 | 3500 |

demand of the order: Orders that are small enough can likely do without participating in the open auction and the period that continues immediately following, while large orders that try to extract significant liquidity from the market might benefit from participating in the open auction. The market intraday momentum study by Gao, Han, Li and Zhou (2018) [156] demonstrates how the first half-hour return on the market, as measured from the previous day's market close predicts the last half hour return.

### 1.3.3 Continuous Trading

This refers to the main market phase between the auctions. During this market session the state of the order book changes quite rapidly due to the multi-agent nature of financial markets and the prevalence of high frequency trading. Consequently, it is important to understand the dynamics of the LOB before implementing trading strategies. There exists quite a diversity of order types that are mostly relevant to the continuous trading session, but the two most basic ones are: Limit Orders and Market Orders, which we describe below.

A limit order has an associated side (Buy or Sell), a quantity and a price which represent the highest (lowest) price the trader is willing to buy (sell). As previously discussed, once a limit order is received by the exchange it is inserted in a data structure called a Limit Order Book (LOB) which contains two sub-structures, one per side. Orders are inserted in this structure in price priority, higher prices for buys, lower prices for sells, and for orders at the same price the orders are stored in the order in which they were received. That is what is meant by price/time priority.[21] If the price of a newly arrived order overlaps with the best price available on the opposite side, the order is executed either fully or up to the available quantity on the other side. These orders are said to be "matched" and again this matching happens in price and time priority meaning that the better prices (higher for buys, lower for sells) are executed first and orders that arrived beforehand at the same price level are executed first. Market orders on the other hand do not have a price associated with them and will immediately execute against the other side and will match with more and more aggressive prices until the full order is executed.

Orders on the buy side are called "bids" while those on the sell side are called "asks." The above events are illustrated in Figure 1.1 to Figure 1.3. When a market (or marketable) order is submitted, it decreases the number of outstanding orders at the opposite best price. For example, if a market bid order arrives, it will decrease the number of outstanding asks at the best price. All unexecuted limit orders can be canceled. When a cancellation occurs, it will decrease the number of outstanding orders at the specified price level.

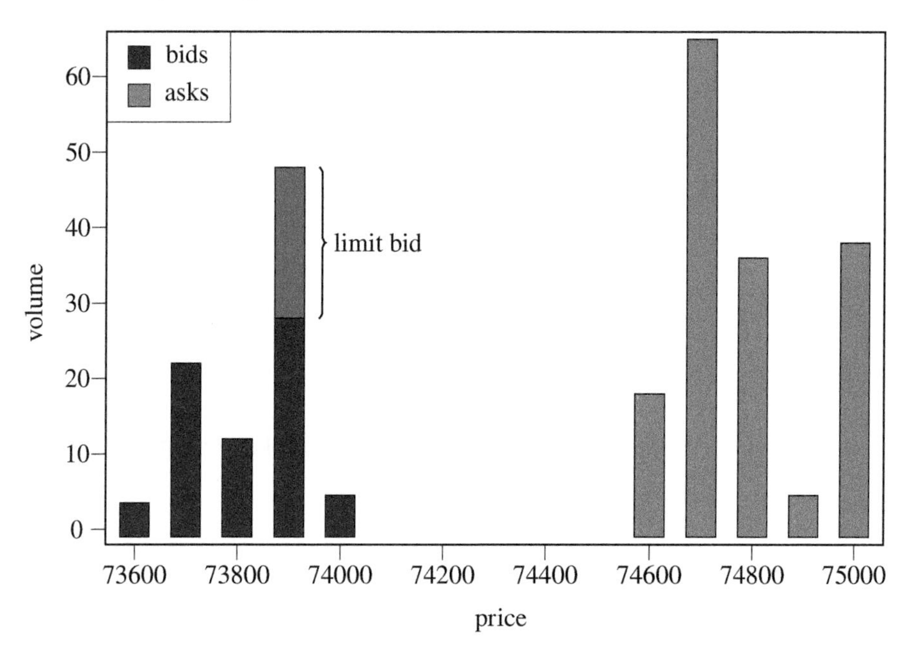


Figure 1.1: Limit Order Book—Limit Bid.

[21]Note: Not all exchanges are matching orders following a price/time priority algorithm; a key characteristic of the Futures market, for instance, is the existence of pro-rata markets for some fixed income contracts, where passive child orders receive fills from aggressive orders based on their size as a fraction of the total passive posted quantity.

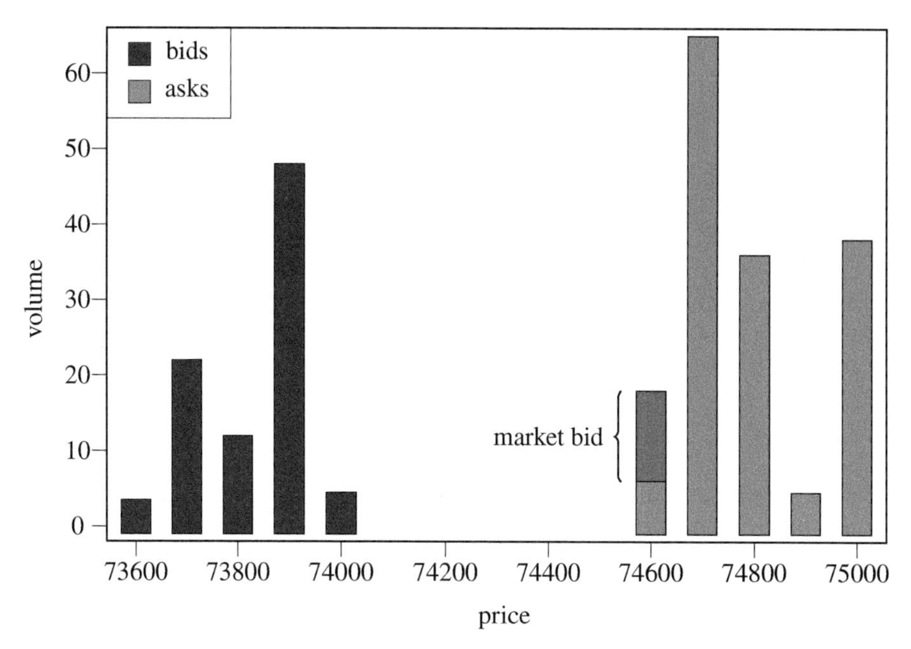


Figure 1.2: Limit Order Book—Marketable Bid.

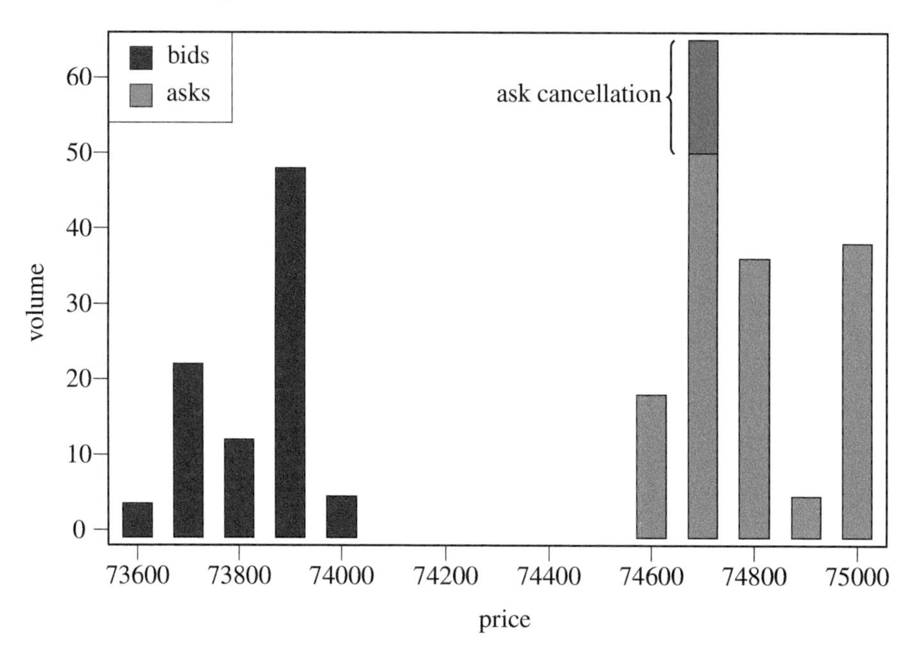


Figure 1.3: Limit Order Book—Ask Cancellation.

Limit orders make up a significant percentage (70%) of stock market trading activity. The main advantage of a limit order is that there is no price risk associated to it, that is, when the order is executed the limit price is the maximum (for a buy order) or minimum (for a sell order) price that will be achieved. But if the limit order is not marketable, the execution is not guaranteed and the time to get an order executed depends on various market factors. The trade-off between limit orders and marketable orders depends on the investor's need for immediate liquidity and the fill probability of limit orders. The limit price chosen (how deep in the order book is the order placed) as well as the amount of liquidity ahead of the submitted order (how many shares will need to trade before the order gets executed following, for instance, a price/time priority order matching of the exchange) affect both the order fill probability and its expected time to fill. These two metrics are of particular relevance for execution algorithms and will be studied in more depth later.

The execution of limit orders does affect how the quotes are posted and are updated. If the size of a market order exceeds the number of shares available at the top of book, it is usually split and is executed at consecutive order book levels until the order is filled. Market orders are usually restricted to be filled within a day and orders placed after the markets close might be entered the next day.[22]

**Order Types:** The diversity of order types is a key component of continuous double auction electronic markets. Order types allow participants to express precisely their intentions with regards to their interaction with the market via the limit order book. Over time, in an effort to cater to sophisticated electronic traders, exchanges around the world have raced to offer ever more complex order types. Here, we will just provide a brief description of some, besides the market and limit orders that were already mentioned:

- **Peg:** Specify a price level at which the order should be continuously and automatically repriced. For instance, an order pegged to the bid price will be automatically repriced as a higher price limit order, each time the market bid price ticks up. This order type is particularly used for mid-point executions in non-displayed markets. One would think that pegging order has the additional advantage of improving the queue priority of the order when it is repriced since this process is done directly by the exchange. That turns out not to be always true. Managing peg orders is the responsibility of a separate component at the exchange and its interaction speed with the order book is usually slower than that of ultra-low-latency operators.

- **Iceberg:** Limit order with a specified display quantity. In order to prevent information leakage to other market participants, a trader desiring to buy or sell a large quantity at a given price might elect to use an iceberg order with a small display size. For instance, for an order to buy 100,000 shares at $20 with a display size of 2,000 shares. Only 2,000 shares would be displayed in the order book. Once that quantity is executed, the order would automatically reload another

[22]Depending on the Time-in-Force selected.

2,000 shares at $20, and so on, until the full quantity is executed. Note that for iceberg orders, only the visible quantity has time priority and once that quantity has been executed the new tip of the iceberg will be placed at the back of the queue.

- **Hidden:** While they are available to trade, these orders are not directly visible to other market participants in the central limit order book.
- **Stop:** These orders are also not visible, but additionally are not immediately entered in the limit order book. They only become active once a certain price (known as the Stop Price) is reached or passed. They then enter the order book as either limit or market order depending on the user setup.
- **Trailing Stop:** These orders function like stop orders, but the stop price is set dynamic rather than static (for instance: $-3\%$ from previous close).
- **All-or-None:** Specifically, request a full execution of the order. If the order is for 1500 shares but only 1000 are being offered, it will not be executed until the full quantity is available.
- **On-Open:** Specifically, request an execution at the open price. It can be limit-on-open or market-on-open.
- **On-Close:** Specifically, request an execution at the close price. It can be limit-on-close or market-on-close.
- **Imbalance Only:** Provide liquidity intended to offset on-open/on-close order/imbalances during the opening/closing cross. These generally are limit orders.
- **D-Quote:** Special order type on the NYSE mainly used during the close auction period.
- **Funari:** Special order type on the Tokyo Stock Exchange which allows limit orders placed in the book during the continuous session to automatically enter the closing auction as market orders.

As described above, there exists a wide variety of orders types offered by different exchanges to facilitate various types of trading activities.

**Validity of Instructions:** In addition to conditions on price, it is possible to add conditions on the life duration of the order known as Time-in-Force (TIF). The most common types of TIF instructions include Day orders which are valid for the full duration of the trading session, Extended Day orders allow trading in extended hours, and Good-Till-Cancel (GTC) orders will be placed again on the exchange the next day with similar instructions if they were not completely filled. More sophisticated market participants aiming at achieving greater control over their executions tend to also favor Immediate-or-Cancel (IOC) and Fill-or-Kill (FOK) Time-in-Force instructions. An IOC order will get immediately canceled back to the sender after reaching the matching engine if it does not get an immediate fill, and in case of a partial fill, the unfilled portion will be canceled, thus preventing it from creating a new price level in the order book. In a Fill-or-Kill scenario, the order gets either filled in its entirety or does not get filled at all. This instruction is particularly popular with high frequency market makers and arbitrageurs for which partial fills might result in unwanted legging risk as discussed in Chapter 5 on pairs trading.

Finally, it is worth mentioning that some exchanges as well as alternative venues offer the ability of specifying minimum fill sizes. This means that a limit order which might be eligible for a fill due to an incoming order at the same price level, only receives a fill if the incoming order is larger than a pre-specified number of shares or notional value. This type of instruction is used by market participants as a way of minimizing the number of small fills which carry the risk of excessive information dissemination. This happens, in particular, in dark pools, where they can be used to detect the presence of larger limit orders that would be otherwise not visible to market participants.

### 1.3.4 The Closing Auction

The Closing Auction tends to be the most popular call auction for a variety of reasons. First, it is the last opportunity (unless one engages in the risky practice of off-hours trading) for market participants to transact in a relatively liquid environment, before being exposed to the overnight period (when new information accumulates, but trades cannot easily take place). Second, with the increase in passive investment strategies, providing investors with replication of a predetermined benchmark index, the closing auction has become a particularly relevant price setting event. For most passive funds, the net asset value (NAV) is based on close prices of the underlying assets. For those reasons, the closing auction has become extremely important to many investors. From an execution standpoint, it is a major liquidity event that must be handled carefully.

The mechanics of the closing auction are in most part similar to the ones described above for the open auction. The major differences across countries (and sometimes across exchanges within a country) are in the order submission times. Some countries, such as the US, have order submissions start and end before the continuous session is over (see Table 1.7 and Table 1.8), while some other markets have two non-overlapping continuous and close order submission sessions.

Table 1.7: Nasdaq Closing Cross

| | |
|---|---|
| 3:50 p.m. EST | Cutoff for amend/cancel of MOC/LOC orders |
| 3:55 p.m. EST | Dissemination of imbalance information begins |
| 3:55 p.m. EST | Cutoff for entry of MOC/LOC orders |
| 3:58 p.m. EST | Freeze period - Late LOC orders cannot be added<br>OI orders offsetting the imbalance are still accepted |
| 4:00 p.m. EST | The Closing Cross occurs |

Table 1.8: NYSE Closing Cross

| | |
|---|---|
| 3:50 p.m. EST | Cutoff for MOC/LOC order entry and modifications<br>Dissemination of imbalance information begins<br>Closing Offset orders can be entered until 4:00 p.m. |
| 3:55 p.m. EST | Dissemination of d-quote imbalance information |
| 3:58 p.m. EST | Cutoff for MOC/LOC cancellation for legitimate error |
| 3:59:50 p.m. EST | Cutoff for d-Quote order entry and modification |
| 4:00 p.m. EST | The Closing Auction starts |
| 4:02 p.m. EST | DMM can automatically process auctions not yet complete |

Table 1.9: London Stock Exchange Sessions Times

| | |
|---|---|
| 7:00-7:50 a.m. GMT | Pre-Trading |
| 7:50-8:00* a.m. GMT | Opening Auction Call |
| 8:00-12:00 p.m. GMT | Regular Trading |
| 12:00-12:02* p.m. GMT | Periodic Call Auction |
| 12:02-4:30 p.m. GMT | Regular Trading |
| 4:30-4:35* p.m. GMT | Closing Auction Call |
| 4:35-4:40 p.m. GMT | Closing Price Crossing |
| 4:40-5:15 p.m. GMT | Post-Close Trading |

*Each auction end time is subject to a random 30 second uncross period.
An additional intraday auction call takes place every 3rd Friday of each month for stocks underlying FTSE 100 index options, and the 3rd Friday of every quarter for stocks underlying FTSE 100/250 Index Futures to determine the EDSP (Exchange Delivery Settlement Price). The settlement price is determined as the index value derived from the individual constituents intraday auction taking place between 10 a.m. and 10:15 a.m. and during which the electronic continuous trading is suspended. Using a call auction ensures the settlement price for these contracts is more representative of a fair market price.

An aspect of the NYSE is the presence of floor brokers operating in an agency capacity for their customers. They play a particular role during the close auction thanks to their ability to handle discretionary electronic quote orders (known as "d-Quote" orders) that offer more flexibility than traditional market-on-close (MOC) and limit-on-close (LOC) orders. The main advantage of d-Quote orders is their ability to bypass the 3:45 p.m. cutoff and be submitted or canceled until 3:59:50 p.m. This allows large institutional investors to remain in control of their orders almost until the end of the continuous session, by delaying the decision of how much to allocate to the auction. They can therefore react to larger volume opportunities based on the published imbalance. They can also minimize the information leakage by not being part of the early published imbalance while there is still significant time in the continuous session for other participants to drive the price away. Since there is no restriction on the side of d-Quote orders submission, it is possible to see the total imbalance sign flip once the d-Quotes are added to the publication at 3:55 p.m., creating opportunities for other market participants to adjust their own close trading via d-Quotes or change their positioning in the continuous session.

## 1.4 Taxonomy of Data Used in Algorithmic Trading

Running a successful trading operation requires availability of different data sets. Data availability, storage, management, and cleaning are some of the most important aspects of a functioning trading business and the core of any research environment. The amount of data and the complexity of maintaining such a "Data Lake" can be daunting and very few excel at this aspect. In this section, we will review the most important data sets and their role in algorithmic trading research.

### 1.4.1 Reference Data

While often overlooked, or merely considered as an afterthought in the development of a research platform,[23] reliable reference data is the key foundation of a robust quantitative strategy development. The experienced practitioner may want to skip this section, however we encourage the neophyte to read through the tedious details to get a better grasp of the complexity at hand.

- **Trading Universe:** The first problem for the functioning of a trading operation is knowing what instruments will be required to be traded on a particular day. The trading universe is an evolving entity that changes daily to incorporate new listings (IPOs), de-listings, etc. To be able to just trade new instruments, there are several pieces of information that are required to have in multiple systems. Market Data must be made available, and some static data needs to be set or guessed to work with existing controls, parameters for the various analytics need to be made available or sensibly defaulted. For research, in particular quantitative strategies, knowing when a particular stock no longer trades is important to avoid issues like survivor bias.

- **Symbology Mapping:** ISIN, SEDOL, RIC, Bloomberg Tickers, . . . Quantitative strategies often leverage data from a variety of sources. Different providers key their data with different instrument identifiers depending on asset class or regional conventions, or sometimes use their own proprietary identifiers (e.g., Reuters Identification Code—RIC, Bloomberg Ticker). Therefore, symbology mapping is the first step in any data merging exercise. Such data is not static. One of the symbols can change on a given day and others remain unchanged for some time, complicating historical data merges.

[23]More details on research platforms are presented in Chapter 12.

It is important to note that such mapping needs to persist as point-in-time data and allow for historical "as of date" usage, requiring the implementation of a bi-temporal data structure. Over the course of time, some instruments undergo ticker changes (for example, from ABC to DEF on a later $T_0$) not necessarily without any particular change on the underlying asset. In such cases, market data recorded day-by-day in a trade database will change from being keyed on ABC to being keyed on DEF after the ticker change date $T_0$. This has implications for practitioners working on data sets to build and backtest quantitative strategies. The symbology mapping should allow for both backward and forward handling of the changes.

For instance, in the simple example mentioned below, in order to efficiently backtest strategies over a period of time spanning $T_0$, a robust mapping is needed, so that it will allow to seamlessly query the data for the underlying asset in a variety of scenarios such as:

- Signal generation: 30-day backward close time series as of date $T < T_0$:
  `select close from data where date in [T-30, T], sym = ABC`
- Signal generation: 30-day backward close time series as of date $T = T_0 + 10$:
  `select close from data where date in [T0-20, T0+10], sym = DEF`
- Position holding: 30-day forward close time series as of date $T = T_0 - 10$:
  `select close from data where date in [T0-10, T0+20], sym = ABC`

- **Ticker Changes:** For comparable reasons as the ones described above in the Symbology Mapping section, one needs to maintain a historical table of ticker changes allowing to seamlessly go up and down time series data.

- **Corporate Actions Calendars:** This category contains stock and cash dividends (both announcement date and execution date), stock splits, reverse splits, rights offer, mergers and acquisitions, spin off, free float or shares outstanding adjustments, quotation suspension, etc.

  Corporate actions impact the continuity of price and volume time series and, as such, must be recorded in order to produce adjusted time series. The most common events are dividend distributions. On the day the dividend is paid, the corresponding amount is removed from the stock price, creating a jump in the price time series. The announcement date might also coincide with the stock experiencing more volatility as investors react to the news. Consequently, recording these events proves to be valuable in the design of quantitative strategies as one can assess the effect of dividends announcement or payment on performance, and decide to either not hold a security that has an upcoming dividend announcement or, conversely, to build strategies that look to benefit from the added volatility.

  Another type of corporate events generating discontinuity in historical time series are stock splits or reverse splits and right offers. When the price of a stock becomes too low or too high, a company may seek to split it to bring the price back to a level that is more conducive to liquid trading on exchanges.[24] When a stock experiences a 2:1 split, everything else being equal, its price will be halved and hence, its volume will double. In order to prevent the time series from showing a discontinuity, all historical data will then need to be adjusted backward to reflect the split.

Mergers & Acquisitions and Spin-offs are also regular events in the lifecycle of corporations. Their history needs to be recorded in order to account for the resulting changes in valuation that might affect a given ticker(s). These situations can also be exploited by trading strategies known as Merger Arbitrage.

Stock quotations can be suspended as a cooling mechanism (often at the request of the underlying company) to prevent excess price volatility when significant information is about to be released to the market. Depending on the circumstance, the suspension can be temporary and intraday, or can last for extended periods of time if the market place allows it.[25] Suspensions result in gaps in data and are worth keeping track of, as they can impact strategies in backtesting (inability to enter or exit a position, uncertainty in the pricing of composite assets if a given stock has a significant weight in ETFs or Indexes, etc.). Some markets will also suspend trading if the price swings more than a predefined amount (limit up / limit down situations), either for a period of time or for the remainder of the trading session.

- **Static Data:** Country, sector, primary exchange, currency and quote factor. Static data is also relevant for the development of quantitative trading strategies. In particular, country, currency and sector are useful to group instruments based on their fundamental similarities. A well known example is the usage of sectors to group stocks in order to create pairs trading strategies. It is worth noting that there exist different types of sector classifications (e.g., GICS®from S&P, ICB®from FTSE) offering several levels of granularity,[26] and that different classifications might be better suited to different asset classes or countries. The constituents of the Japanese index TOPIX, for instance, are classified into 33 sectors that are thought to better reflect the fundamental structure of the Japanese economy and the existence of large diversified conglomerates.

  Maintaining a table of the quotation currency per instrument is also necessary in order to aggregate positions at a portfolio level. Some exchanges allow the quotation of prices in currencies different from the one of the country in which

[24] A very low price creates trading frictions as the minimum price increment might represent a large cost relative to the stock price. A very high price might also deter retail investors from investing into a security as it requires them to deploy too much capital per unit.

[25] For instance, it was the case for a large number of companies in China in 2016.

[26] The Global Industry Classification Standard (GICS®) structure consists of 11 sectors, 24 industry groups, 68 industries and 157 sub-industries. An example of this hierarchical structure would be: Industrials / Capital Goods / Machinery / Agricultural & Farm Machinery.

the exchange is located.[27] Additionally, thus, the Quote Factor associated with the quotation currency data needs to be stored. To account for the wide range of currency values and preserve pricing precision, market data providers may publish FX rates with a factor of 100 or 1000. Hence, to convert prices to USD one needs to multiply by the quote factor: USD price = local price · fx · quote factor. Similarly, some exchanges quote prices in cents, and the associated quotation currency is reflected with a small cap letter: GBP/GBp, ZAR/ZAr, ILS/ILs, etc.

- **Exchange Specific Data:** Despite the electronification of markets, individual exchanges present a variety of differences that need to be accounted for when designing trading strategies. The first group of information concerns the hours and dates of operation:

  - **Holiday Calendar:** As not all exchanges are closed on the same day, and trading days they are off do not always fully follow the country's public holidays, it is valuable to record them, in particular in the international context. Strategies trading simultaneously in several markets and leveraging their correlation, may not perform as expected if one of the markets is closed while others are open. Similarly, execution strategies in one market might be impacted by the absence of trading in another market (for instance, European equity markets volume tends to be 30% to 40% lower during US market holidays).
  - **Exchange Sessions Hours:** These seemingly trivial data points can get quite complex on a global scale. What are the different available sessions (Pre-Market session, Continuous core session, After-Hour session, etc.)? What are the auction times as well as their respective cutoff times for order submission? Is there a lunch break restricting intraday trading? And if so, are there auctions before and after the lunch break? The trading sessions in the futures markets can also be quite complex with multiple phases, breaks, as well as official settlement times that may differ from the closing time and have an effect on liquidity.

    In Indonesia, for instance, markets have different trading hours on Fridays. Monday through Thursday the Indonesia Stock Exchange (IDX) is open from 9:00 a.m. to 12:00 p.m., and then, from 1:30 p.m. to 4:00 p.m. On Fridays, however, the lunch break is one hour longer and stretches from 11:30 a.m. to 2:00 p.m. This weekday effect is particularly important to consider when building volume profiles as discussed in Chapter 5.

    Along with local times of operation, it is necessary to consider eventual Daylight Saving Time (DST) adjustments that might affect the relative trading hours of different markets (some countries do not have DST adjustment at all, while for countries that do have one, the dates at which it applies are

[27]For example, Jardine Matheson Holdings quotes in USD on the Singapore exchange while most of the other securities quote in Singapore Dollars.

not always coordinated). Usually, US DST starts about two weeks prior to its start in Europe, bringing the time difference between New York and London to four hours instead of five hours. This results in the volume spike in European equities associated to the US market open being one hour earlier, requiring adjustment of volume profiles used for trading executions.

Some exchanges may also adjust the length of trading hours during the course of the year. In Brazil for instance, the Bovespa continuous trading hours are 10:00 a.m. to 5:55 p.m. from November to March, but an hour shorter (10:00 a.m. to 4:55 p.m.) from April to October to be more consistent with US market hours. Finally, within one country there might also exist different trading hours by venues as it is the case in Japan where the Nagoya Stock Exchange closes 30 minutes after the major Tokyo Stock Exchange.

- **Disrupted Days:** Exchange outages or trading disruptions, as well as market data issues, need to be recorded so they can be filtered out when building or testing strategies as the difference in liquidity patterns or the lack of data quality may likely impact the overall outcome.

Additionally, exchanges also have specific rules governing the mechanics of trading, such as:

- **Tick Size:** The minimum eligible price increment. This can vary by instrument, but also change dynamically as a function of the price of the instrument (e.g., stocks under $1 can quote in increments of $0.0001, while above that price the minimum quote increment is $0.01).
- **Trade and Quote Lots:** Similar to tick sizes, certain exchanges restrict the minimum size increment for quotes or trades.
- **Limit-Up and Limit-Down Constraints:** A number of exchanges restrict the maximum daily fluctuations of securities. Usually, when securities reach these thresholds, they either pause trading or can only be traded at a better price than the limit-up limit-down threshold.
- **Short Sell Restrictions:** Some markets also impose execution level constraints on short sells (on top of potential locate requirements). For instance, while a long sell order can trade at any price, some exchanges restrict short sells not to trade at a price worse than the last price or not to create a new quote that would be lower than the lowest prevailing quote. These considerations are particularly important to keep in mind for researchers developing Long-Short strategies as this impacts the ability to source liquidity.

Because these values and their potential activation threshold can vary over time, one needs to maintain historic values as well, in order to run realistic historical backtests.

- **Market Data Condition Codes:** With ever-growing complexity in market microstructure, the dissemination of market data has grown complex as well. While it is possible to store daily data as a single entry per day and per instrument, investors building intraday strategies likely need tick by tick data of all the events occurring in the market place. To help classify these events, exchanges and market data providers attribute so-called condition codes to the trades and quotes they publish. These condition codes vary per exchange and per asset class, and each market event can be attributed to several codes at once. So, to aggregate intraday market data properly and efficiently, and decide which events to keep and which ones to exclude, it is necessary to build a mapping table of these condition codes and what they mean: Auction trade, lit or dark trade, canceled or corrected trade, regular trade, off-exchange trade reporting, block-size trade, trade originating from a multi-leg order such as an option spread trade, etc.

  For instance, in order to assess accessible liquidity for a trading algorithm, trades that are published for reporting purposes (e.g., negotiated transactions that happened off-exchange) must be excluded. These trades should also not be used to update some of the aggregated daily data used in the construction of trading strategies (daily volume, high, low, ... ). Execution algorithms also extensively leverage the distribution of intraday liquidity metrics to gauge their own participation in auctions and continuous sessions, or in lit versus dark venues, therefore requiring a precise classification of intraday market data.

- **Special Day Calendars:** Over the course of a year, some days present certain distinct liquidity characteristics that need to be accounted for in both execution strategies and in the alpha generation process. The diversity of events across markets and asset classes can be quite challenging to handle. Among the irregular events that affect liquidity in equity markets that need to be accounted for, we can mention the following non-exhaustive list for illustration purposes only: Half trading days preceding Christmas and following Thanksgiving in the US or on the Ramadan eve in Turkey, Taiwanese market opening on the weekend to make up for lost trading days during holiday periods, Korean market changing its trading hours on the day of the nationwide university entrance exam, Brazilian market opening late on the day following the Carnival, etc.

  There are also special days that are more regular and easier to handle. The last trading days of the months and quarters, for instance, tend to have additional trading activity as investors rebalance their portfolios. Similarly, options and futures expiry days (quarterly/monthly expiry, 'Triple Witching'[28] in the US, Special Quotations in Japan, etc.) tend to experience excess trading volume and different intraday patterns resulting from hedging activity and portfolio adjustments. Consequently, they need to be handled separately, in particular, when modeling trad-

[28]Triple witching days happen four times a year on the third Friday of March, June, September and December. On these days, the contracts for stock index futures, stock index options and stock options expire concurrently.

ing volume. As most execution strategies make use of relatively short interval volume metrics (e.g., 30-day or 60-day ADV), one single data point can impact the overall level inferred. Similarly autoregressive models of low order may underperform both on special days and on the days following them. As a result, modelers often remove special days and model normal days first. Then, special days are modeled separately, either independently or using the normal days as a baseline.

In order to reflect changes in the market and remain consistent with index inclusion rules, most indices need to undergo regular updates of the constituents and their respective weights. The significant indices do so at regular intervals (annually, semi-annually or quarterly) on a pre-announced date. At the close of business of that day, some stocks might be added to the index while others are removed, or the weight of each stock in the index might be increased or decreased. These events, known as index rebalances, are particularly relevant to passive investors who are tracking the index. In order to minimize the tracking error to the benchmark index, investors need to also adjust their holdings accordingly. Additionally, as most funds are benchmarked at the close price, there is an incentive for fund managers to try to rebalance their holdings at a price as close as possible to the official close price, on the day the index rebalance becomes effective. As a result, on these days, intraday volume distribution is significantly skewed toward the end of day and requires some adjustment in the execution strategies.

- **Futures-Specific Reference Data:** Futures contracts present particular characteristics requiring additional reference data to be collected. One of the core differences of futures contracts compared to regular stocks is the fact that instruments have an expiry date, after which the instrument ceases to exist. For the purpose of backtesting strategies, it is necessary to know which contract was live at any point in time through the use of an expiry calendar, but also which contract was the most liquid. For instance, equity index futures tend to be the most liquid, for the first contract available (also known as front month), while energy futures such as oil tend to be more liquid for the second contract. While this may appear to be trivial, when building a trading strategy and modeling price series, it is particularly important to know which contract carries the most significant price formation characteristics and what is the true liquidity available in order to properly estimate the market impact.

  The task of implementing a futures expiry calendar is further complicated by the fact there is no real standardized frequency that applies across markets. For instance, European equity index futures tend to expire monthly, while US index futures expire quarterly. Some contracts even follow an irregular cycle through the course of the year as it is the case for grain futures (e.g., wheat) that were originally created for hedging purposes and as a result have expiry months that follow the crop cycle.[29]

The fact futures contracts expire on a regular basis has further implications in terms of liquidity. If investors holding these contracts want to maintain their exposure for longer than the lifespan of the contract, they need to roll over their positions onto the next contract which might have a noticeably different liquidity level. For instance, the front contract of the S&P 500 (e.g., ESH8) trades roughly 1.5 million contracts per day in the weeks preceding expiry while the next month contract (ESM8) only trades about 30,000 contracts per day. However, this liquidity relationship will invert in the few days leading to the expiry of the front contract as most investors roll their positions, and the most liquid contract will become the back month contract (see Figure 1.4). As a result, when computing rolling-window metrics (such as average daily volume for instance), it is necessary to account for potential roll dates that may have happened during the time span. In the example above, a simple 60-day average daily volume on ESM8 taken in early April 2018 would capture a large number of days with very low volume (January-March) owing to the fact the most liquid contract at the time was the ESH8 contract, and would not accurately represent the volume activity of the S&P 500 futures contract. A more appropriate average volume metric to be used as a forward looking value for execution purposes would blend the volume time series of ESH8 prior to the roll date, and ESM8 after the roll date.[30] Additionally, in order to efficiently merge futures positions with other assets in an investment strategy, reference data relative to the quotation of these contracts, contract size (translation between the quotation points and the actual monetary value), currency, etc., must be stored.

Finally, futures markets are characterized by the existence of different market phases during the day, with significantly different liquidity characteristics. For instance, equity index futures are much more liquid during the hours when the corresponding equities markets are open. However, one can trade during the overnight session if they want to. The overnight session being much less liquid, the expected execution cost tends to be higher, and as such, the various market data metrics (volume profile, average spread, average bid-ask sizes, ...) should be computed separately for each market phase, which requires maintaining a table of the start and end times of each session for each contract.

---

[29]US Wheat Futures expire in March, May, July, September and December.

[30]It is worth noting that different contracts 'roll' at different speeds. While for monthly expiry contracts it is possible to see most of the open interest switch from the front month contract to the back month on the day prior to the expiry. For quarterly contracts it is not uncommon to see the roll happen over the course of a week or more, and the front month liquidity vanish several days ahead of the actual expiry. Careful modeling is recommended on a case by case basis.

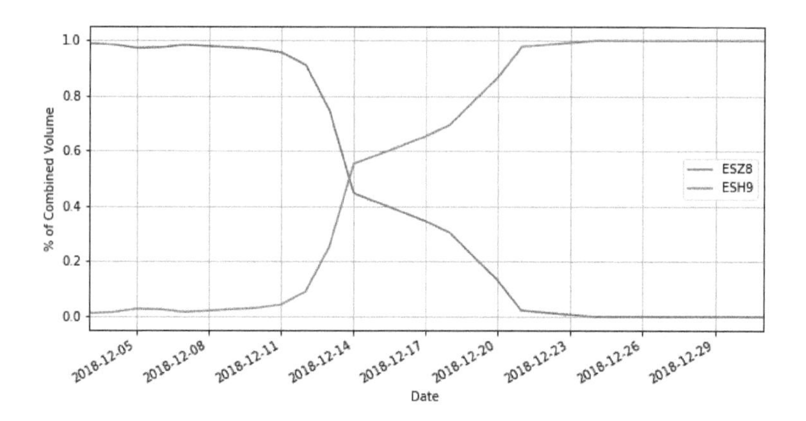


Figure 1.4: Futures Volume Rolling.

- **Options-Specific Reference Data (Options Chain):** Similar to futures contracts, options contracts present a certain number of specificities for which reference data need to be collected. On top of the similar feature of having a particular expiry date, options contracts are also defined by their strike price. The combination of expiries and strikes is known as the option chain for a given underlier. The ability to map equity tickers to option tickers and their respective strike and expiry dates allows for the design of more complex investment and hedging strategies. For instance, distance to strike, change in open interest of puts and calls, etc., can all be used as signals for the underlying security price. For an interesting article on how deviations in put-call parity contains information about future equity returns, refer to Cremers and Weinbaum (2010) [95].

- **Market-Moving News Releases:** Macro-economic announcements are known for their ability to move markets substantially. Consequently, it is necessary to maintain a calendar of dates and times of their occurrences in order to assess their impact on strategies and decide how best to react to them. The most common ones are central banks' announcements or meeting minutes releases about the major economies (FED/FOMC, ECB, BOE, BOJ, SNB), Non-Farm Payrolls, Purchasing Managers' Index, Manufacturing Index, Crude Oil Inventories, etc. While these news releases impact the broad market or some sectors, there are also stock specific releases that need to be tracked: Earning calendars, specialized sector events such as FDA results for the healthcare and biotech sectors, etc.

- **Related Tickers:** There is a wide range of tickers that are related to each other, often because they fundamentally represent the same underlying asset. Maintaining a proper reference allows to efficiently exploit opportunities in the market. Some non-exhaustive examples include: Primary tickers to composite tickers mapping (for markets with fragmented liquidity), dual listed/fungible securities in US and Canada, American Depository Receipt (ADR) or Global Depository Receipt (GDR), local and foreign boards in Thailand, etc.

- **Composite Assets:** Some instruments represent several underlying assets (ETFs, Indexes, Mutual Funds, ...). Their rise in popularity as investments over time makes them relevant for quantitative strategies. They can be used as efficient vehicles to achieve desired exposures (sector and country ETFs, thematic factor ETFs, ...), or as cheap hedging instruments, and they can provide arbitrage opportunities when they deviate from their Net Asset Value (NAV). In order to be leveraged in quantitative strategies, one needs to maintain a variety of information such as a time series of their constituents and the value of any cash component, the divisor used to translate the NAV into the quoted price, the constituent weights.
- **Latency Tables:** This last type of data would only be of interest for developing strategies and for research in the higher frequency trading space. For these, it might be relevant to know the distribution of latency between different data centers as they can be used for more efficient order routing as well as reordering data that may have been recorded in different locations.

While the above discussion provides a non-exhaustive list of issues on the reference data available to build a quantitative research platform, they highlight the challenges that must be taken into account when designing and implementing algorithmic trading strategies. Once in place, a proper set of reference data will allow the quantitative trader to systematically harness the actual content of various types of data (described later) without being caught off-guard by the minutiae of trading.

### 1.4.2 Market Data

While historically a large swath of modeling for developing strategies was carried out on daily or minute bar data sets, the past fifteen years have seen a significant rise in the usage of raw market data in an attempt to extract as much information as possible, and act on it before the opportunity (or market inefficiency) dissipates. Market data, itself, comes in various levels of granularity (and price!) and can be subscribed to either directly from exchanges (known as "direct feeds") or from data vendors aggregating and distributing it. The level of detail of the feed subscribed to is generally described as Level I, Level II or Level III market data.

**Level I Data: Trade and BBO Quotes:** Level I market data is the most basic form of tick-by-tick data. Historically, the Level I data feed would only refer to trade information (Price, Size, Time of each trade reported to the tape) but has grown over time into a term generally accepted to mean both trades and top of book quotes. While the trade feed updates with each trade printed on the tape, the quote feed tends to update much more frequently each time liquidity is added to, or removed from, the top of book (a rough estimate in liquid markets is the quotes present more updates than the trades by about an order of magnitude).

In order to build strategies, the timing of each event must be as precise as possible. While the time reported for trade and quotes is the matching engine time at the exchange, most databases also store a reception or record time to reflect the potential latency between the trade event and one being aware that it did happen and being able to start making decisions on it. Accounting for this real-world latency is a necessary step for researchers building strategies on raw market data for which opportunities may be very short lived and may be impossible to exploit by participants who are not fast enough.

The Level I data is enough to reconstruct the Best Bid and Offer (BBO) of the market. However, in fragmented markets, it is also useful to obtain an aggregated consolidated view of all available liquidity at a given price level across all exchanges. Market data aggregators usually provide this functionality for users who do not wish to reconstruct the full order book themselves.

Finally, it is worth noting that even Level I data contains significant additional information in the form of trade status (canceled, reported late, etc.) and, trade and quote qualifiers. These qualifiers provide granular details such as whether a trade was an odd lot, a normal trade, an auction trade, an Intermarket Sweep, an average price reporting, on which exchange it took place, etc. These details can be used to better analyze the sequence of events and decide if a given print should be used to update the last price and total volume traded at that point in time or not. For instance, if not processed appropriately, a trade reported to the tape out of sequence could result in a large jump in price because the market may have since moved, and consequently the trade status is an indication that the print should not be utilized as it is.

Capturing raw market data, whether to build a research database or to process it in real-time to make trading decisions, requires significant investments and expert knowledge to handle all the inherent complexity. Hence, one should carefully consider the trade-off between the expected value that can be extracted from the extra granularity, and the additional overhead compared to simpler solutions such as using binned data.

**Level II Data: Market Depth:** Level II market data contains the same information as Level I, but with the addition of quote depth data. The quote feed displays all lit limit order book updates (price changes, addition or removal of shares quoted) at any level in the book, and for all of the lit venues in fragmented markets. Given the volume of data generated and the decreasing actionable value of quote updates as their distance to top of book increases, some users limit themselves to the top five or ten levels of the order book when they collect and/or process the data.

**Level III Data: Full Order View:** Level III market data—also known as message data—provides the most granular view of the activity in the limit order book. Each order arriving is attributed a unique ID, which allows for its tracking over time, and is precisely identified when it is executed, canceled or amended. Similar to Level II, the data set contains intraday depth of book activity for all securities in an exchange. Once all messages from different exchanges are consolidated into one single data set ordered by timestamp, it is possible to build a full (with national depth) book at any moment intraday. For illustration purposes, we take the example of US Level III data and provide a short description below in Table 1.10.

Table 1.10: Level III Data

| **Variable:** | **Description** |
|---|---|
| Timestamp: | Number of milliseconds after the midnight. |
| Ticker: | Equity symbol (up to 8 characters) |
| Order: | Unique order ID. |
| T: | Message type. Allowed values:<br>• "B"—Add buy order<br>• "S"—Add sell order<br>• "E"—Execute outstanding order in part<br>• "C"—Cancel outstanding order in part<br>• "F"—Execute outstanding order in full<br>• "D"—Delete outstanding order in full<br>• "X"—Bulk volume for the cross event<br>• "T"—Execute non-displayed order |
| Shares: | Order quantity for the "B," "S," "E," "X," "C," "T" messages. Zero for "F" and "D" messages. |
| Price: | Order price, available for the "B," "S," "X" and "T" messages. Zero for cancellations and executions. The last 4 digits are decimal digits. The decimal portion is padded on the right with zeros. The decimal point is implied by position; it does not appear inside the price field. Divide by 10000 to convert into currency value. |
| MPID: | Market Participant ID associated with the transaction (4 characters) |
| MCID: | Market Center Code (originating exchange—1 character) |

While the display and issuance of a new ID to the modified order varies from exchange to exchange, a few special types of orders are worth mentioning:

1. Order subject to price sliding: The execution price could be one cent worse than the display price at NASDAQ; it is ranked at the locking price as a hidden order, and

is displayed at the price, one minimum price variation (normally 1 cent) inferior to the locking price. New order ID will be used if the order is replaced as a display order. At other exchanges the old order ID will be used.

2. Pegged order: Based on NBBO, not routable, new timestamp given upon re-pricing; display rules vary over exchanges.

3. Mid-point peg order: Non-displayed, can result in half-penny execution.

4. Reserve order: Displayed size is ranked as a displayed limit order and the reserve size is behind non-displayed orders and pegged orders in priority. The minimum display quantity is 100 and this amount is replenished from the reserve size when it falls below 100 shares. A new timestamp is created and the displayed size will be re-ranked upon replenishment.

5. Discretionary order: Displayed at one price while passively trading at a more aggressive discretionary price. The order becomes active when shares are available within the discretionary price range. The order is ranked last in priority. The execution price could be worse than the display price.

6. Intermarket sweep order: Order that can be executed without the need for checking the prevailing NBBO.

The richness of the data set allows sophisticated players, such as market makers, to know not only the depth of book at a given price and the depth profile of the book on both sides, but more importantly the relative position of an order from the top position of the side (buy or sell). Among the most common granular microstructure behaviors studied with Level III data, we can mention:

- The pattern of inter-arrival times of various events.

- Arrival and cancellation rates as a function of distance from nearest touch price.

- Arrival and cancellation rates as a function of other available information, such as in the queue on either side of the book, order book imbalance, etc.

Once modeled, these behaviors can, in turn, be employed to design more sophisticated strategies by focusing on trading related questions:

- What is the impact of market order on the limit order book?

- What are the chances for a limit order to move up the queue from a given entry position?

- What is the probability of earning the spread?

- What is the expected direction of the price movement over a short horizon?

As illustrated by the diversity and granularity of information available to traders, markets mechanics have grown ever more complex over time, and a great deal of peculiarities (in particular at the reference data level) need to be accounted for in order to design robust and well performing strategies. The proverbial devil is always in the details when it comes to quantitative trading but while it might be tempting to go straight to the most granular source of data, in practice—with the exception of truly high frequency strategies—most practitioners build their algorithmic trading strategies relying essentially on binned data. This simplifies the data collection and handling processes and also greatly reduces the dimension of the data sets so one can focus more on modeling rather than data wrangling. Most of the time series models and techniques described in subsequent chapters are suited to daily or binned data.

### 1.4.3 Market Data Derived Statistics

Most quantitative strategies, even the higher frequency strategies that leverage more and more granular data also leverage derived statistics from the binned data, such as daily data. Here we give a list of the most common ones used by practitioners and researchers alike.

#### Daily Statistics

The first group represents the overall trading activity in the instrument:

- **Open, High, Low, Close (OHLC) and Previous Close Price:** The OHLC provides a good indication of the trading activity as well as the intraday volatility experienced by the instrument. The distance traveled between the lowest and highest point of the day usually gives a better indication of market sentiment than the simple close-to-close return. Keeping the previous close value as part of the same time series is also a good way to improve computation efficiency by not having to make an additional database query to compute the daily return and overnight gap. The previous close needs, however, to be properly adjusted for corporate actions and dividends.
- **Last Trade before Close (Price/Size/Time):** It is useful in determining how much the close price may have jumped in the final moments of trading, and consequently, how stable it is as a reference value for the next day.
- **Volume:** It is another valuable source of trading activity indicator, in particular when the level jumps from the long term average. It is also worth collecting the volume breakdown between lit and dark venues, in particular for execution strategies.
- **Auctions Volume and Price:** Depending on the exchange, there can be multiple auctions in a day (Open, Close, Morning Close and Afternoon Open—for markets with a lunch break—as well as ad hoc liquidity or intraday auctions). Owing to their liquidity aggregation nature, they can be considered as a valuable price discovery event when significant volume prints occur.

- **VWAP:** Similar to the OHLC, the intraday VWAP price gives a good indication of the trading activity on the day. It is not uncommon to build trading strategies using VWAP time series instead of just close-to-close prices. The main advantage being that VWAP prices represent a value over the course of the day and, as such, for larger orders are easier to achieve through algorithmic execution than a single print.

- **Short Interest/Days-to-Cover/Utilization:** This data set is a good proxy for investor positioning. The short pressure might be an indication of upcoming short term moves: A large short interest usually indicates a bearish view from institutional investors. Similarly, the utilization level of available securities to borrow (in order to short) gives an indication of how much room is left for further shorting (when securities become "Hard to Borrow", the cost of shorting becomes significantly higher requiring short sellers to have strong enough beliefs in the short term price direction). Finally, days-to-cover data is also valuable to assess the magnitude of a potential short squeeze. If short sellers need to unwind their positions, it is useful to know how much volume this represents as a fraction of available daily liquidity. The larger the value, the larger the potential sudden upswing on heavily shorted securities.

- **Futures Data:** Futures markets provide additional insight into the activity of large investors through open interest data, that can be useful to develop alpha strategies. Additionally, financial futures offer arbitrage opportunities if their basis exhibits mispricing compared to one's dividend estimates. As such, recording the basis of futures contracts is worthwhile even for strategies that do not particularly target futures.

- **Index-Level Data:** This is also a valuable data set to collect as a source of relative measures for instrument specific features (Index OHLC, Volatility, ...). In particular for dispersion strategies, normalized features help identify individual instruments deviating from their benchmarks.

- **Options Data:** The derivative market is a good source of information about the positioning of traders through open interest and Greeks such as Gamma and Vega. How the broader market is pricing an instrument through implied volatility for instance is of interest.

- **Asset Class Specific:** There is a wealth of cross-asset information available when building strategies, in particular in Fixed Income, FX, and Credit markets. Among the basic ones, we would note:

  - Yield / benchmark rates (repo, 2y, 10y, 30y)
  - CDS Spreads
  - US Dollar Index

The second group of daily data represents granular intraday microstructure activity and is mostly of interest to intraday or execution trading strategies:

- **Number and Frequency of Trades:** A proxy for the activity level of an instrument, and how continuous it is. Instruments with a low number of trades are harder to execute and can be more volatile.
- **Number and Frequency of Quote Updates:** Similar proxy for the activity level.
- **Top of Book Size:** A proxy for liquidity of the instrument (larger top of book size makes it possible to trade larger order size quasi immediately, if needed).
- **Depth of Book (price and size):** Similar proxy for liquidity.
- **Spread Size (average, median, time weighted average):** This provides a proxy for cost of trading. A parametrized distribution of spread size can be used to identify intraday trading opportunities if they are cheap or expensive.
- **Trade Size (average, median):** Similar to spread size, trade sizes and their distribution are useful to identify intraday liquidity opportunities when examining the volume available in the order book.
- **Ticking Time (average, median):** The ticking time and its distribution is a representation of how often, one should expect changes in the order book first level. This is particularly helpful for execution algorithms for which the frequency of updates (adding/canceling child orders, reevaluating decisions, etc.) should be commensurate with the characteristics of the traded instrument.

  The daily distributions of these microstructure variables can be used as start of day estimates in trading algorithms and be updated intraday as additional data flows in through online Bayesian updates.

Finally, the last group of daily data can be derived from the previous two groups through aggregation but is usually stored pre-computed in order to save time during the research phase (e.g., X-day trailing data), or to be used as normalizing values (e.g., size as a percentage of ADV, spread in relation to long-term average, ...). Some common examples would be:

- $X$-day Average Daily Volume (ADV) / Average auction volume
- $X$-day volatility (close-to-close, open-to-close, etc.)
- Beta with respect to an index or a sector (plain beta, or asymmetric up-days/down-days beta)
- Correlation matrix

As a reminder to the reader, aggregated data need to fully support the peculiarities described in the Reference Data section (for instance: The existence of special event days which, if included, can significantly skew intraday distribution of values; mishandling of market asynchronicity resulting in inaccurate computations of key quantities such as beta or correlation, etc.).

**Binned Data**

The first natural extension to daily data sets is a discretization of the day into bins ranging from a few seconds to 30 minutes. The features collected are comparable to the ones relevant for daily data sets (period volume, open, high, low, close, VWAP, spread, etc.), but computed at higher frequency. It is worth mentioning that minute bar data actually underpins the vast majority of microstructure models used in the electronic execution space. Volume and spread profiles, for instance, are rarely built with a granularity finer than one minute to prevent introducing excess noise due purely to market frictions. The major advantage of binned-data is that discrete time series methods can be readily used.

These minute bar data sets are also quite popular for the backtesting of low-to-medium frequency trading strategies targeting intraday alpha (short duration market neutral long-short baskets, momentum and mean-reversion strategies, etc.). The main benefit they provide is a significant dimension reduction compared to raw market data (390 rows per stock per day in the US compared to millions for raw data for the liquid stocks), which allows researchers to perform rapid and efficient backtesting as they search for alpha. However, increasing data frequency from daily data to intraday minute bars also presents challenges. In particular, relationships that appear to be stable using close-to-close values become much noisier as granularity increases and signals become harder to extract (e.g., drop in correlation between assets, pricing inefficiency moves due to sudden liquidity demand, etc.). Similarly, for less liquid assets with a low trade frequency, there may not be any trading activity for shorter durations, resulting in empty bins.

### 1.4.4 Fundamental Data and Other Data Sets

A very large number of quantitative investment strategies are still based on fundamental data, and consequently countless research papers are available to the interested reader describing examples of their usage. Here we will only describe the main classes of existing fundamental data:

- **Key Ratios:** EPS (Earnings Per Share), P/E (Price-to-Earning), P/B (Price-to-Book Value), .... These metrics represent a normalized view of the financials of companies allowing for easier cross-sectional comparison of stocks and their ranking over time.

- **Analyst Recommendations:** Research analysts at Sell Side institutions spend a great deal of resources analyzing companies they cover, in order to provide investment recommendations usually in the form of a Buy/Hold/Sell rating accompanied by a price target. While individual recommendations might prove noisy, the aggregate values across a large number of institutions can be interpreted as a consensus valuation of a given stock, and changes in consensus can have a direct impact on price.

- **Earnings Data:** Similarly research analysts also provide quarterly earning estimates that can be used as an indication of the performance of a stock before the actual value gets published by the company. Here, too, consensus values tend to play a larger role. In particular, when the difference between the forecast consensus and the realized value is large (known as earning surprise), as the stock might then experience outsized returns in the following days. Thus, collecting analysts' forecasts as well as realized values can be a valuable source of information in the design of trading strategies.

- **Holders:** In some markets, large institutional investors are required to disclose their holdings on a regular basis. For instance, in the US, institutional investment managers with over $100 million in assets must report quarterly, their holdings, to the SEC using Form 13F. The forms are then publicly available via the SEC's EDGAR database. Additionally, shareholders might be required to disclose their holdings once they pass certain ownership thresholds.[31] Sudden changes in such ownership might indicate changes in sentiment by sophisticated investor and can have a significant impact on stock performance.

- **Insiders Purchase/Sale:** In some markets, company directors are required by law to disclose their holdings of the company stock as well as any increase or decrease of such holdings.[32] This is thought to be an indicator of future stock price moves from the group of people who have access to the best possible information about the company.

- **Credit Ratings:** Most companies issue both stocks and bonds to finance their operations. The credit ratings of bonds and their changes over time provide additional insight into the health of a company and are worth leveraging. In particular, credit downgrades resulting in higher funding costs in the future, generally have a negative impact on equity prices.

- **And much, much more:** Recent years have seen the emergence of a wide variety of alternative data sets that are available to researchers and practitioners alike.[33] While it is not possible to make a comprehensive list of all that are available, they can generally be classified based on their characteristics: Frequency of publication, structured or unstructured, and velocity of dissemination. The value of such data depends on the objectives and resources of the user (natural language processing or image recognition for unstructured data require significant time and efforts), but also—and maybe more importantly—on the uniqueness of the data

[31]In the US, Form 13D must be filed with the SEC within 10 days by anyone who acquires beneficial ownership of more than 5% of any class of publicly traded securities in a public company.

[32]In the US, officers and directors of publicly traded companies are required to disclose their initial holdings in the company by filing Form 3 with the SEC, as well as Form 4 within 2 days of any subsequent changes. The forms are then publicly available via the SEC's EDGAR database.

[33]For instance, `www.orbitalinsight.com` offers daily retail traffic analytics, derived from satellite imagery analysis monitoring over 260,000 parking lots, as well as estimates of oil inventories through satellite monitoring of oil storage facilities.

set. As more investors get access to it, the harder it becomes to extract meaningful alpha from it.

## 1.5 Market Microstructure: Economic Fundamentals of Trading

We first provide a brief review of market microstructure, an area of finance that studies how the supply and demand of liquidity results in actual transactions and modifies the subsequent state of the market, and then delve into some critical operational concepts. We draw upon key review papers by Madhavan (2000) [254], Biais, Glosten and Spatt (2005) [39] and O'Hara (2015) [276]. The core idea is that the efficient market hypothesis, which postulates that the equity price impounds all the information about the equity, may not hold due to market frictions. Algorithmic trading essentially exploits the speed with which investors acquire the information and how they use it along with the market frictions that arise mainly due to demand-supply imbalances. Taking an informational economics angle, Madhavan (2000) [254] provides a market microstructure analysis framework following three main categories.

- **Price Formation and Price Discovery:** How do prices impound information over time and how do the determinants of trading costs vary?

- **Market Design:** How do trading rules affect price formation?

  Market design generally refers to a set of rules that all players have to follow in the trading process. These include the choice of tick size, circuit breakers which can halt trading in the event of large price swings, the degree of anonymity and the transparency of the information to market participants, etc. Markets around the world and across asset classes can differ significantly in these types of rules, creating a diverse set of constraints and opportunity for algorithmic traders. Some early research on the effects of market design lead to following broad conclusions:

  - Centralized trading via one single market tends to result in more efficient price discovery with smaller bid-ask spreads. In the presence of multiple markets, the primary markets (such as NYSE, NASDAQ) remain the main sources of price discovery (see Hasbrouck (1995) [181]).

  - Despite market participants' preference for continuous, automated limit order book markets, theoretical models suggest that multilateral trading approaches such as single-price call auctions are the most efficient in processing diverse information (See Mendelson (1982) [263] and Ho et al. (1985) [197]).

- **Transparency:** How do the quantity, quality and speed of information provided to market participants affect the trading process?

  Transparency which is broadly classified into pre-trade (lit order book) and post-trade (trade reporting to the public, though at different time lags) is often a trade-off.

While in theory more transparency should lead to better price discovery, the wide disclosure of order book depth information can lead to thinner posted sizes and wider bid-ask spreads if participants fear revealing their intent and possibly their inventory levels, leading them to favor more off-exchange activity.

All the above points can be analyzed in light of the rapid growth of high frequency trading (HFT) described earlier in this chapter. O'Hara (2015) [276] presents issues related to microstructure in the context of HFT. While the basic tenet that traders may use private information or learn from market data such as orders, trade size, volume, duration between successive trades, etc., has remained the same, the trading is now mostly automated to follow some rules. These rules are based partly on prior information and partly on changing market conditions, monitored through order flows. With the high speed, adverse selection has taken a different role. Some traders may have access to market data milliseconds before others have it and this may allow them to capture short term price movements. This would expand the pool of informed traders. Here are some topics for research in microstructure:

- With a parent order sliced into several child orders that are sent to market for execution during the course of trading, it is difficult to discern who is the informed trader. Informed traders use sophisticated dynamic algorithms to interact with the market. Retail (known in the literature as 'uninformed') trades usually cross the spread.
- More work needs to be done to understand trading intensity in short intervals. Order imbalance is empirically shown to be unrelated to price levels.
- Informed traders may increasingly make use of hidden orders. How these orders enter and exit the markets require further studies.
- Traders respond to changing market conditions by revising their quoted prices. The quote volatility can provide valuable information about the perceived uncertainty in the market.

Although HFT has resulted in more efficient markets, with lower bid-ask spreads, the data related to trade sequences, patterns of cancellations across fragmented markets require new tools for analysis and for making actionable inference. In this review, we do not present any models explicitly and these will be covered throughout this book. Now keeping in line with the practical perspectives of this book, we highlight some key concepts.

### 1.5.1 Liquidity and Market Making

(a) **A Definition of Liquidity:** Financial markets are commonly described as carrying the function of efficiently directing the flow of savings and investments to the real economy to allow the production of goods and services. An important factor contributing to well-developed financial markets is in facilitating liquidity which enables investors to diversify their asset allocation and the transfers of securities at a reasonable transaction cost.

Properly describing "liquidity" often proves to be elusive as there is no commonly agreed upon definition. Black (1971) [42] proposes a relatively intuitive description of a liquid market: *"The market for a stock is liquid if the following conditions hold:*

- *There are always bid and ask prices for the investor who wants to buy or sell small amounts of stock immediately.*
- *The difference between the bid and ask prices (the spread) is always small.*
- *An investor who is buying or selling a large amount of stock, in the absence of special information, can expect to do so over a long period of time at a price not very different, on average, from the current market price.*
- *An investor can buy or sell a large block of stock immediately, but at a premium or discount that depends on the size of the block. The larger the block, the larger the premium or discount."*

In other words, Black defines a liquid market as a continuous market having the characteristics of relatively tight spread, with enough depth on each side of the limit order book to accommodate instantaneous trading of small orders, and which is resilient enough to allow large orders to be traded slowly without significant impact on the price of the asset. This general definition remains particularly well suited to today's modern electronic markets and can be used by practitioners to assess the difference in liquidity between various markets when choosing where to deploy a strategy.

**(b) Model for Market Friction:** We begin with a model to accommodate the friction in stock price ($P_t$). If $p_t^*$ is the (log) true value of the asset which can vary over time due to expected cash flows or due to variation in the discount rate. Given the publicly available information and the assumption of market efficiency, we have: $p_t = p_t^* + a_t$ and $p_t^* = p_{t-1}^* + \epsilon_t$. Therefore the observed return, $r_t = p_t - p_{t-1} = \epsilon_t + (a_t - a_{t-1})$ can exhibit some (negative) serial correlation, mainly a result of friction. The friction, '$a_t$' can be a function of a number of factors such as inventory costs, risk aversion, bid-ask spread, etc. When $a_t = cs_t$, where $s_t$ is the direction of the trade and '$c$' is the half-spread, the model is called a Roll model. This can explain the stickiness in returns in some cases.

**(c) Different Styles of Market Participants: Liquidity Takers and Liquidity Providers:** Liquid financial markets carry their primary economic function by facilitating savings and investment flows as well as allowing investors to exchange securities in the secondary markets. Economic models of financial markets attempt to classify market participants into different categories. These can be broadly delineated as:

**Informed Traders,** making trading decisions based on superior information that is not yet fully reflected in the asset price. That knowledge can be derived from either fundamental analysis or information not directly available nor known to other market participants.

**News Traders,** making trading decisions based on market news or announcements and trying to make profits by anticipating the market's response to a particular catalyst. The electronification of news dissemination offers new opportunities for developing quantitative trading strategies by leveraging text-mining tools such as natural language processing to interpret, and trade on, machine readable news before it is fully reflected in the market prices.

**Noise Traders** (as introduced by Kyle (1985) [234]), making trading decisions without particular information and at random times mainly for liquidity reasons. They can be seen as adding liquidity to the market through additional volume transacted, but only have a temporary effect on price formation. Their presence in the market allows informed traders not to be immediately detected when they start transacting, as market makers cannot normally distinguish the origin of the order flow between these two types of participants. In a market without noise traders, being fully efficient, at equilibrium each trade would be revealing information that would instantly be incorporated into prices, hereby removing any profit opportunities.

**Market Makers,** providing liquidity to the market with the intent of collecting profits originating from trading frictions in the market place (bid-ask spread). Risk-neutral market makers are exposed to adverse selection risk arising from the presence of informed traders in the marketplace, and therefore establish their trading decisions mostly based on their current inventory. As such, they are often considered in the literature to drive the determination of efficient prices by acting as the rational intermediaries.

Generalizing these concepts, market participants and trading strategies can be separated between liquidity providing and liquidity seeking. The former being essentially the domain of market makers whose level of activity, proxied by market depth, is proportional to the amount of noise trading and inversely proportional to the amount of informed trading (Kyle (1985) [234]). The latter being the domain of the variety of algorithmic trading users, described before (mutual funds, hedge funds, asset managers, etc.). Given the key role played by market makers in the liquidity of electronic markets, they have been the subject of a large corpus of academic research focusing on their activities.

(d) **The Objectives of the Modern Market Maker:** At present, market making can broadly be separated into two main categories based on the trading characteristics. The first one is the provision of large liquidity—known as blocks—to institutional investors, and has traditionally been in the realm of sell-side brokers acting as intermediaries and maintaining significant inventories. Such market makers usually transact through a non-continuous, negotiated, process based on their current inventory, as well as their assessment of the risk involved in liquidation of the position in the future. Larger or more volatile positions generally tend to come at a higher cost, reflecting the increased risk for the intermediary. But they provide the end investor with a certain price and an immediate execution bearing no timing risk that is associated with execution over time. These transactions, because they involve negotiations between two parties, still mostly happen in a manual fashion or over the phone and then get reported to an appropriate exchange for public dissemination.

The second category of market making involves the provision of quasi-continuous, immediately accessible quotes on an electronic venue. With the advent of electronic trading described in the previous section, market makers originally seated on the exchange floors have progressively been replaced by electronic liquidity providers (ELP). The ELP leverage fast technology to disseminate timely quotes across multiple exchanges and develop automated quantitative strategies to manage their inventory and the associated risk. As such, most ELP can be classified as high frequency traders. They derive their profits from three main sources: From liquidity rebates on exchanges that offer maker-taker fee structure, from spread earned when successfully buying on the bid and selling on the offer, and from short-term price moves favorable to their inventory.

Hendershott, Brogaard and Riordan (2014) [191] find that HFT activity tends to be concentrated in large liquid stocks and postulate that this can be attributed to a combination of larger profit opportunities emanating from trades happening more often, and from easier risk management due to larger liquidity that allows for easier exit of unfavorable positions at a reasonable cost.

(e) **Risk Management:** In the existing literature on informed trading, it is observed that liquidity supplying risk-neutral market makers are adversely selected by informed traders suddenly moving prices against them. For example, a market maker buy quote tends to be executed when large sellers are pushing the price down, resulting in even lower prices in the near term. This significant potential asymmetry of information at any point in time emphasizes the need for market makers to employ robust risk management techniques, particularly in the domain of inventory risk. For a market maker, risk management is generally accomplished by first adjusting market quotes upward or downward to increase the arrival rate of sellers or buyers and consequently adjusting the inventory in the desired direction. If biasing the quotes does not result in a successful inventory adjustment, the market makers generally employ limit orders to cross the spread.

Ho and Stoll (1981) [196] introduced a market making model in which the market maker's objective is to maximize profit while minimizing the probability of ruin by determining the optimal bid-ask spread to quote. The inventory held evolves through the arrival of bid and ask orders, where the arrival rate is taken to be a function of bid and ask prices. Their model also incorporates the relevant notion of the size dependence of spread on the market maker's time horizon. The longer the remaining time, the greater potential for adverse move risk for liquidity providers, and vice versa. This is consistent with observed spreads. In most markets, the spread is wider at the beginning of the day, narrowing toward the close. An additional reason annotated for the wider spread right after the beginning of the trading day is due to the existence of potentially significant information asymmetry accumulated overnight. As market makers are directly exposed to that information asymmetry, they tend to quote wider spreads while the price discovery process unfolds following the opening of continuous trading, and progressively tighten them as uncertainty about the fair price for the asset dissipates.

Understanding the dynamics of market makers inventory and risk management, and their effect on spreads, has direct implications for the practitioners who intend to deploy algorithmic trading strategies as the spread paid to enter and exit positions is a non-negligible source of cost that can erode the profitability of low alpha quantitative strategies.

Hendershott and Seasholes (2007) [195] confirms that market makers' inventories are negatively correlated with previous price changes and positively correlated with subsequent changes. This is consistent with market makers first acting as a dampener of buying or selling pressure by bearing the risk of temporarily holding inventory in return for earning the spread and thus potential price appreciation from market reversal. This model easily links liquidity provision and the dynamics of asset prices.

Finally, it has been observed that there is a positive correlation of market makers inventory with subsequent prices changes, inventories can complement past returns when predicting future returns. Since inventories are not publicly known, market participants use different proxies to infer their values throughout the day. Two commonly used proxies are trade imbalances (the net excess of buy or sell initiated trade volume) and spreads. Trade imbalance aims at classifying trades, either buy initiated or sell initiated, by comparing their price with the prevailing quote. Given that market makers try to minimize their directional risk, they can only accommodate a limited amount of non-diversified inventory over a finite period of time. As such, spread sizes, and more particularly their sudden variation, have also been used as proxies for detecting excess inventory forcing liquidity providers to adjust their positions.

# Part II

# Foundations: Basic Models and Empirics

In Part II, our goal is to provide a review of some basic and some advanced statistical methodologies that are useful for developing trading algorithms. We begin with time series models (in Chapter 2) for univariate data and provide a broad discussion of autoregressive, moving average models—from model identification, estimation, inference to model diagnostics. The stock price and return data exhibit some unique features and so we identify certain stylized facts regarding their behavior that have been empirically confirmed; this work will greatly help to discern any anomalies as and when they arise, as these anomalies generally indicate deviation from an efficient market hypothesis. Although for modeling price and return data, only lower order models are needed, increasingly other trading features such as volume and volatility are being used in developing trading strategies. In particular, predicting future volume flow to determine when to enter the market, requires the use of higher order models. Wherever possible we illustrate the methodology with examples and provide codes for computing, making sample data accessible to the reader. We also introduce novel methodologies that have some potential for developing competing trading models.

This chapter is followed by methodologies for multiple time series data in Chapter 3 which have applications from pairs trading to portfolio optimization. The last chapter (Chapter 4) in Part II contains advanced topics such as theory of point processes as trading takes place on a continual basis. Finally, this chapter also contains a very brief treatment of other modern topics such as machine learning.

A reader with a strong statistics background can afford to skip some sections but we recommend to peruse these chapters as they contain discussions on topics that need further research work. In presenting the methodologies here, we keep the discussion lucid but directly relevant to the main theme of this book—understanding the market behavior and developing effective trading strategies.

# 2

# *Univariate Time Series Models*

With the advent of electronic trading a vast amount of data on orders is now recorded and is available to traders to make real-time decisions. The Trades and Quotes (TAQ) data on time-stamped sequence of trades with the updates on the price and depth of the best bid and ask quotes was used by researchers and traders initially. This was later expanded to the so-called Level II data, with time-stamped sequence of trades and quotes for the top 5 levels of the order book. Both TAQ and Level II data still do not provide the complete story of the buy and sell sides. With Level III data now available, we can obtain time-stamped sequence of all events that occur (with the exception of the information on hidden orders), resulting in a quite voluminous, but irregularly spaced, data set. We take a broad view that for making strategic decisions to enter and exit the market, the trader needs to take an aggregate view of the trading activities, but that for the optimal execution of the strategies the trader needs to understand the market microstructure that relates to the actual process of trading. With the increased speed of decisions in an HFT world, the difference between the two steps, strategic and execution decisions, may be blurred but the aggregation process can help to look beyond trading frictions. In Section 2.1, we discuss the Trades and Quotes data and aggregation issues. In Section 2.2, we define trading algorithms that depend on prediction of short-term price movement.

The price movement, whether it is due to pure market frictions or due to information related to a stock, is captured through a drift term in the random walk model that needs to be carefully monitored. Because time series models for discrete time units are still widely used by traders in the form of aggregated price-bars, we focus on these methods in Section 2.5 to Section 2.11. We broadly classify these methods into modeling the mean (return) and into modeling the variance (volatility). As these techniques are well-documented and can be found elsewhere in the literature, we only present a brief but relevant overview. The established facts about the stock returns and variances are reviewed along with the methodology as these will provide expected benchmarks; successful algorithms as one would notice, generally exploit the deviations from these benchmarks.

## 2.1 Trades and Quotes Data and Their Aggregation: From Point Processes to Discrete Time Series

In its simplest form, the trading activities through an exchange that opens at 9:30 a.m. and closes at 4 p.m. can be described by a sequence of time stamps ("ticks") $t_0 < t_1 < \cdots < t_n$ and the "marks" $y_i$ at time $t_i$, in which $t_0$ denotes 9:30 a.m. and after, and $t_n$ denotes the time of the last trade that occurs before or at 4 p.m. The marks $y_i$ can be price, volume of an order placed on either the buy or sell side of the book and in general can represent the characteristics of the order book at the time of $i$th activity (see Figure 2.1). The events with the marks associated with the ticks can be described mathematically as a marked point process. But our goal here is to first show how these data points can be aggregated over regularly spaced time intervals and how methods for linear homogeneous time series can be used to analyze the aggregated data. Some tools that are relevant for the analysis of point processes will be presented in Chapter 4.

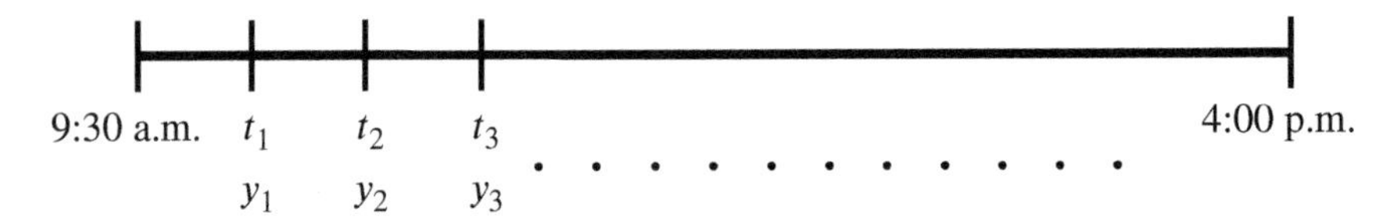


Figure 2.1: Trading Activities.

The typical Level III data for CISCO on a single day in 2011 is given below in Table 2.1. This outlay in Table 2.1 points to various issues that need to be addressed in processing this type of data. The order numbers are only unique during the lifetime of the order but can be reassigned at a later time. Several activities such as submission, cancellation or modification of an order, all can take place at the same time stamp. Hidden orders are revealed only when they are executed and as such do not have visible order numbers. Full cancellation quantity and price information should be traced back to the submission time. To summarize, at any given point in time using the data described in Table 2.1, the full limit order book that essentially records where each order is in the queue, can be constructed and thus providing the activity time and the associated marks. These and other characteristics of the limit order book (superbook—if we collate this information from all exchanges) will be taken up in a later chapter, but the focus here is on constructing discrete time series data from this point process data.

To illustrate the aggregation methods, we will initially focus on price of the stock, $P_t$, and the associated volume $V_t$. Two types of aggregation methods are proposed in the literature. One when the time span $T$ for the Exchange hours (9:30 a.m.–4:00 p.m.) is divided into '$K$' intervals (Clock Time) so that the regularly spaced intervals are of size $\Delta t = T/K$. The other method will do aggregation when there is a change in a marker, such as price (Volume Time). Here we focus on the first method. Various

Table 2.1: CISCO–Trade Data

| Timestamp | Order Number | Event | Quantity | Price | Exchange |
|---|---|---|---|---|---|
| ⋮ | ⋮ | ⋮ | ⋮ | ⋮ | ⋮ |
| 34561915 | 4254917 | D | 0 | 0 | J |
| 34561915 | 4253917 | B | 200 | 20.55 | J |
| ⋮ | ⋮ | ⋮ | ⋮ | ⋮ | ⋮ |
| 34561923 | 13056048 | S | 200 | 20.58 | Z |
| ⋮ | ⋮ | ⋮ | ⋮ | ⋮ | ⋮ |
| 34573369 | 13255731 | C | 98 | 0 | Q |
| ⋮ | ⋮ | ⋮ | ⋮ | ⋮ | ⋮ |
| 34577331 | 6225085 | E | 300 | 0 | K |
| ⋮ | ⋮ | ⋮ | ⋮ | ⋮ | ⋮ |
| 34577338 | 6225085 | F | 0 | 0 | K |
| ⋮ | ⋮ | ⋮ | ⋮ | ⋮ | ⋮ |
| 34573379 | 2030057 | S | 100 | 20.77 | K |
| ⋮ | ⋮ | ⋮ | ⋮ | ⋮ | ⋮ |
| 34573382 | NA | T | 1700 | 20.37 | P |
| ⋮ | ⋮ | ⋮ | ⋮ | ⋮ | ⋮ |

*Timestamp is the time since midnight; Event: B: Submission of Limit Order (LO) on Buy side; S: Submission of LO on Sell side; T: Full execution of a hidden LO; F: Full execution of a visible LO; E: Partial execution of a LO; D: Full cancellation of a LO; C: Partial cancellation of a LO; Exchange: Q-NASDAQ; J-EDGE-A; Z-BATS-Z; K-EDGE-K; P-ARCA; etc.

summary measures within each period $((i-1)\Delta t, i\Delta t)$ where $i = 0, 1, \dots, K$ can be computed;

- Number of transactions: '$n_i$'
- Average price: $\overline{P}_i = \sum P_{t_j}/n_i$
- Volume Weighted Average Price: $\overline{P}_{wi} = \sum V_{t_j} P_{t_j} / \sum V_{t_j}$
- Average duration between transactions: $\overline{d}_i = \sum_{j \in ((i-1)\Delta t, i\Delta t)} (t_j - t_{j-1})/n_i$
- Mid-Price: $(\text{Best bid}_i + \text{Best ask}_i)/2$

The typical daily data (also called price bars) that is available from popular online sources provide information on opening, closing, high and low prices along with the volume. This type of aggregation can also be done for shorter duration, such as one or five minutes, resulting in smoothed data that can cancel out short term market frictions and yet capture the price variations that may be due to information.

It is possible that for infrequently traded stocks, some time intervals may not contain any trading activity. For heavily traded stocks, the duration can be very small and thus may not provide any consequential information. We will discuss this further in Chapter 4. To begin with, we present the analysis for data aggregated over fixed time intervals. Depending on the size of the interval, the data can be called low frequency or high frequency but the time series methods that are discussed in this chapter apply to *all* aggregated data.

## 2.2 Trading Decisions as Short-Term Forecast Decisions

Most trading decisions generally involve the following considerations:

- How much to trade?
- When to trade, i.e., when to enter and when to exit?
- How to trade? How fast? What is the objective? What are the benchmarks?

Answers to these questions involve in some way predicting the market movement and in particular the future (short-term and long-term) price of assets. In the short term it is possible that prices have a tendency to 'trend' or exhibit 'momentum' but in the long run, prices have a tendency to 'mean-revert' as the information about the asset gets incorporated into the price.

Let $p_{it} = \ln P_{it}$ denote the price of the $i$th asset at time $t$ and let $p_t = (p_{1t}, p_{2t}, \dots, p_{nt})$ denote the price vector for '$n$' assets. Let $y_{it}$ denote a vector of characteristics, e.g., volume of transactions, price volatility, number of transactions and the intensity of trading as captured by the average duration between transactions, etc., of the $i$th asset at time $t$. These quantities are aggregated from high frequency data of the type given in Table 2.1 and illustrated in Figure 2.1 to be used as the characteristics associated with the price changes. In addition, we can also consider $r$ factors $f_t = (f_{1t}, f_{2t}, \dots, f_{rt})$ that may include market and industry factors as well as asset characteristics such as market capitalization, book-to-market ratio, etc. The trading rules can be broadly grouped as follows:

(A) **Statistical Arbitrage Rules:** $\mathrm{E}(p_{i,t+1} \mid p_{i,t}, p_{i,t-1}, \dots, y_{i,t}, y_{i,t-1}, \dots)$

- Predicting the price of $i$th stock at t+1 based on the past trading information; this is sometimes labelled as time series momentum.

(B) **Momentum:** $\mathrm{E}(p_{t+1} \mid p_t, p_{t-1}, \dots, y_t, y_{t-1}, \dots)$

- Predicting the cross-sectional momentum of a subset of stocks based on their past trading characteristics. This is not only useful for portfolio formation and rebalancing, but also for pairs trading which is based on tracking two or more stocks simultaneously so that their price divergence can be exploited.

(C) **Fair Value:** $\mathrm{E}(p_{t+1} \mid p_t, p_{t-1}, \dots, y_t, y_{t-1}, \dots, f_t, f_{t-1}, \dots)$

- Predicting the price using all relevant quantities. The factors normally include market and Fama-French factors. Many of these factors are at a more macro level than the time scale considered for the price prediction, but nevertheless could be useful, as explained in Chapter 3.

Thus the price (or alternatively, return) and volatility (or squared return) prediction can be formulated as a time series prediction problem and common tools such as autocorrelations and partial autocorrelations can be used to build autoregressive and ARCH models that are shown to have some predictive power.

Predicting stock market behavior has been studied extensively starting in the early 20th century. Cowles (1933, 1944) [91, 92] evaluated the performance of the stock market forecasters. Analyzing the performance of several financial services that made some 7500 recommendations on common stocks over a period, January 1, 1925 to July 1, 1952, the author points out that their record was worse than the performance of average common stock by 1.43% annually. Thus 'buy-and-hold' strategy out-performed the stock picking strategies of those days. In fact, a much harsher assessment is made: "A review of the various statistical tests, applied to the records for this period, of these 24 forecasters, indicates that the most successful records are little, if any, better than what might be expected to result from pure chance. There is some evidence, on the other hand, to indicate that the least successful records are worse than what could reasonably be attributed to chance." Many of these comments are still valid and have stood the test of the time. Yet, we will demonstrate in this book that there are opportunities for statistical arbitrage. The key being to have access to relevant information and using appropriate techniques to exploit the information in a timely manner.

## 2.3 Stochastic Processes: Some Properties

Formally, a discrete time series or stochastic process $Y_1, Y_2, \dots, Y_T$ is a sequence of random variables (r.v.'s) possessing a joint probability distribution. A particular sequence of observations of the stochastic process $\{Y_t, t = 1, \dots, T\}$ is known as a realization of the process. In general, determining the properties and identifying the probability structure which generated the observed time series are of interest. We do not attempt to study the joint distribution of $Y_1, Y_2, \dots, Y_T$ directly, as it is too complicated for large $T$, but we study the probabilistic mechanism which generates the process sequentially through time, and from this we want to derive the conditional distribution of future observations for purposes of prediction.

The means, variances, and covariances are useful summary descriptions of the stochastic process $\{Y_t, t = 1, \dots, T\}$, but it is not possible to estimate the unknown parameters (means, variances, and covariances) from a single realization of $T$ observations if these quantities vary with '$t$'. We impose some additional structure on the joint distribution of the process in order to substantially reduce the number of unknown parameters. The concept of stationarity of a process serves as a realistic assumption for many types of time series. Stationarity is motivated by the fact that for many time series in practice, segments of the series may behave similarly.

### Stationarity

A time series $\{Y_t\}$ is said to be stationary if for every integer $m$, the set of variables $Y_{t_1}, Y_{t_2}, \ldots, Y_{t_m}$ depends only on the distance between the times $t_1, t_2, \ldots, t_m$, rather than on their actual values. So a stationary process $\{Y_t\}$ tends to behave in a homogeneous manner as it moves through time. The means and variances of the $Y_t$ are the same for all $t$, that is, $\mathrm{E}(Y_t) = \mu$ and $\mathrm{Var}(Y_t) = \sigma^2$ are constant, for all $t$. So we may express the autocovariance function as

$$\gamma(s) = \mathrm{Cov}(Y_t, Y_{t-s}) = \mathrm{E}[(Y_t - \mu)(Y_{t-s} - \mu)], \text{ for all } s = 0, \pm 1, \ldots \tag{2.1}$$

Note that by this notation, $\gamma(0) = \mathrm{E}[(Y_t - \mu)^2] = \mathrm{Var}(Y_t)$. Also, the *autocorrelation function* of the stationary process may be expressed

$$\rho(s) = \mathrm{Corr}(Y_t, Y_{t-s}) = \frac{\mathrm{Cov}(Y_t, Y_{t-s})}{\left(\mathrm{Var}(Y_t)\,\mathrm{Var}(Y_{t-s})\right)^{1/2}} = \frac{\gamma(s)}{\gamma(0)}, \quad s = 0, \pm 1, \ldots \tag{2.2}$$

$\rho(s)$ will be referred to as the autocorrelation of the process at lag $s$. The autocorrelation function (ACF), $\rho(s)$ of a stationary process $\{Y_t\}$ is an important tool in describing the characteristics of the process, because it is a convenient summary of the correlation structure of the process over different time lags.

### Examples of Stationary Stochastic Processes

Now we present some select models, as examples, that tend to be the most useful in financial applications. The properties of these models would become relevant as we look at data.

**Example 2.1 (White Noise).** Let $\epsilon_0, \epsilon_1, \epsilon_2, \ldots$ be a sequence of independent random variables defined on the discrete time points $0, 1, 2, \ldots$, with mean $\mathrm{E}(\epsilon_t) = 0$ and variance $\mathrm{E}(\epsilon_t^2) = \sigma^2$, for all $t$. Set $Y_t = \mu + \epsilon_t$, $t = 0, 1, 2, \ldots$. Then $\mathrm{E}(Y_t) = \mu$ for all $t$, and since independence implies that the random variables are uncorrelated, we have $\mathrm{Cov}(Y_t, Y_{t-s}) = \mathrm{Var}(Y_t) = \sigma^2$ if $s = 0$ and $\mathrm{Cov}(Y_t, Y_{t-s}) = 0$ if $s \neq 0$. Thus the process is stationary. Such a process is referred to as a purely random process or a *white noise process*, and is the foundation for the construction of many other processes of interest. ◁

**Example 2.2 (Moving Average).** Let $\{\epsilon_t\}$ be independent r.v.'s as in Example 2.1, and define a new process $\{Y_t\}$ by

$$Y_t = \mu + \epsilon_t + \epsilon_{t-1}, \qquad t = 0, 1, 2, \ldots,$$

where $\mu$ is a constant. Then $\mathrm{E}(Y_t) = \mu$, for all $t$, and

$$\mathrm{Cov}(Y_t, Y_{t-s}) = \gamma(s) = \begin{cases} 2\sigma^2 & \text{if } s = 0 \\ \sigma^2 & \text{if } s = 1 \\ 0 & \text{if } s > 1 \end{cases}$$

which depends only on the time lag 1 and not on $t$. Hence the process $\{Y_t\}$ is stationary with ACF $\rho(s) = \gamma(s)/\gamma(0)$ such that $\rho(0) = 1$, $\rho(1) = 1/2$ and $\rho(s) = 0$ for $|s| > 1$.

Figure 2.2 (a) shows a time series plot of a computer simulated series of 100 values from a Gaussian white noise process $\{\epsilon_t\}$ with $\sigma^2 = 1$, while Figure 2.2 (b) shows a plot of the corresponding values of the series $Y_t = 1.0 + \epsilon_t + \epsilon_{t-1}$ using these same $\{\epsilon_t\}$. Note the relatively much smoother behavior over time of the data in (b) than in (a), due to the positive autocorrelation at lag one of the process in (b) which tends to make successive values of the series behave more similarly to each other. Similar to Example 2.2, it is easy to show that the process $\{Y_t\}$ defined by the relation $Y_t = \mu + \epsilon_t - \epsilon_{t-1}$ is stationary with autocorrelation function $\rho(0) = 1$, $\rho(1) = -1/2$, and $\rho(s) = 0$ for $|s| > 1$. A plot of a simulated series from this process is given in Figure 2.2 (d) which shows less smooth behavior over time relative to the behavior in (a) and (b) due to the addition of extra noise to the model. These figures form the basis of model building and the subsequent diagnostic checks on the fitness of models. ◁

## Examples of Nonstationary Stochastic Processes

Many time series that occur in finance do not exhibit stationary behavior. For example, stock price series show a trend or shifting mean level over time, reflecting information flow about the stock or investors' behavior. Series also exhibit seasonal behavior, either in the form of nonstationary or deterministic seasonal patterns, both of which need to be distinguished from purely nondeterministic stationary stochastic behavior. Other departures from stationarity include changes in variability of the process over time (nonconstant variance or volatility regimes), abrupt shifts in the mean level, and changes in the autocorrelation structure of the process over time which is a feature of the series which would typically be more difficult to detect. An important practical matter in the analysis of time series data involves methods of transforming a nonstationary time series into a stationary series, and accounting for and modeling of the nonstationary aspects of the original series. Many of the methods that are found useful in practice involve considering successive changes or differences over time in the original series to remove the nonstationary features of the series, especially changes in the mean level, or using linear regression techniques to help account for the nonstationary mean behavior. We will illustrate this with a few simple but commonly occurring examples.

**Example 2.3 (Random Walk with a Drift).** Let $\{\epsilon_t, t = 0, 1, 2, \dots\}$ be a sequence of independent random variables with mean 0 and variance $\sigma^2$, and define a new process $\{Y_t\}$ recursively by

$$Y_t = Y_{t-1} + \delta + \epsilon_t, \qquad t = 0, 1, 2, \dots, \quad Y_0 = 0,$$

where $\delta$ is a constant. By successive substitutions in the above relations for $Y_t$, we find that $\{Y_t\}$ can also be expressed as

$$Y_t = \delta\, t + \sum_{j=1}^{t} \epsilon_j = \delta\, t + \epsilon_t + \epsilon_{t-1} + \cdots + \epsilon_1, \qquad t = 1, 2, \dots .$$

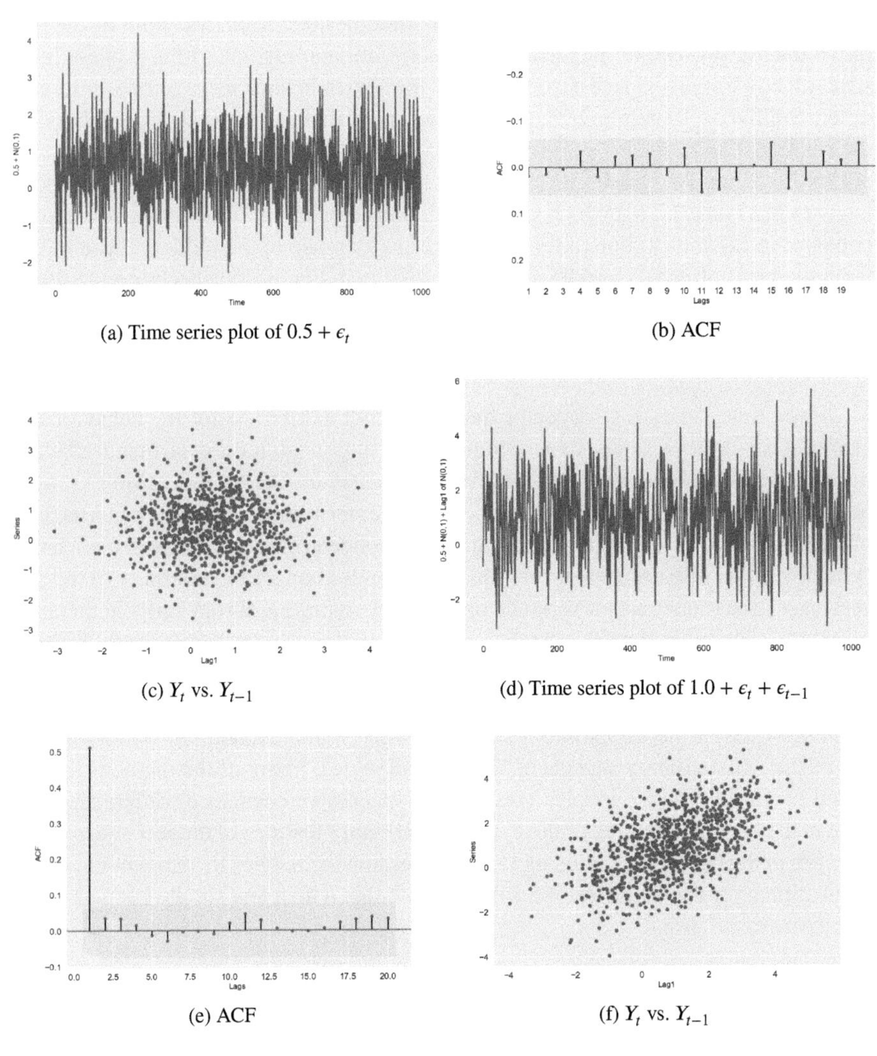


(a) Time series plot of $0.5 + \epsilon_t$

(b) ACF

(c) $Y_t$ vs. $Y_{t-1}$

(d) Time series plot of $1.0 + \epsilon_t + \epsilon_{t-1}$

(e) ACF

(f) $Y_t$ vs. $Y_{t-1}$

Figure 2.2: Processes & Diagnostics from $Y_t = 0.5 + \epsilon_t$ and from $Y_t = 1.0 + \epsilon_t + \epsilon_{t-1}$.

Hence we see that $\mathrm{E}(Y_t) = \delta t$ and $\mathrm{Var}(Y_t) = \mathrm{Var}(\sum_{j=1}^{t} \epsilon_j) = \sum_{j=1}^{t} \mathrm{Var}(\epsilon_j) = t\,\sigma^2$, so that the process $\{Y_t\}$ is not stationary and exhibits change in both mean and variance. Moreover, for any $s > 0$, we find that

$$\mathrm{Cov}(Y_t, Y_{t+s}) = \mathrm{Cov}\left(\sum_{j=1}^{t} \epsilon_j, \sum_{i=1}^{t+s} \epsilon_i\right) = \sum_{j=1}^{t} \mathrm{Var}(\epsilon_j) = t\,\sigma^2,$$

so that $\mathrm{Corr}(Y_t, Y_{t+s}) = t/\sqrt{t(t+s)} = 1/\sqrt{1+s/t}$. Note that this correlation is close to one for large $t$ and relatively small $s$, so that nearby values $Y_t$ and $Y_{t+s}$ of the process $\{Y_t\}$ are highly positively correlated. However, note that the time series of *first differences* of the process $\{Y_t\}$, defined by $Z_t = Y_t - Y_{t-1}$, forms a stationary process, because $Z_t = Y_t - Y_{t-1} = \delta + \epsilon_t$, a white noise series as in Example 2.1. Such a process $\{Y_t\}$ whose first differences form a white noise series, is called a random walk process with a drift. Time series plots of simulated data from random walk exhibit behavior with little tendency to vary around a fixed constant mean level. The parameter $\delta$ in the random walk model is referred to as the drift in the random walk process. Note the trend behavior in the process with $\delta > 0$. ◁

Of course, a more general class of models is obtained by supposing that a nonstationary process $\{Y_t\}$ is such that its first differences form a stationary process, but not necessarily white noise. It has been found from practical experience that this type of model is quite useful in producing adequate representations for actual observed financial series with short term drifts in many cases.

## 2.4 Some Descriptive Tools and Their Properties

Suppose we have a sample realization of $T$ observations $Y_1, Y_2, \ldots, Y_T$ from a stationary process $\{Y_t\}$. We are naturally interested in obtaining estimates of $\mu$, $\gamma(s)$, and $\rho(s)$ from the sample data. First, a natural estimator of $\mu$ is the sample mean $\overline{Y} = (1/T)\sum_{t=1}^{T} Y_t$. The most common estimator used for $\gamma(s)$ is the sample autocovariance function defined by

$$\hat{\gamma}(s) = \frac{1}{T}\sum_{t=1}^{T-s}\left(Y_t - \overline{Y}\right)\left(Y_{t+s} - \overline{Y}\right), \quad s = 0, 1, \ldots.$$

Usually one only computes $\hat{\gamma}(s)$ for lags $s$ much smaller than $T$, say, $0 \le s \le T/4$, since $\hat{\gamma}(s)$ for large lags are typically not very informative or of much interest. Note that $\hat{\gamma}(s)$ has the form of an ordinary sample covariance between the pairs of variables $(Y_t, Y_{t+s})$, except for the use of the common sample mean $\overline{Y}$ for both variables and the divisor $T$ in place of $T - s$. Note also that $\hat{\gamma}(0) = T^{-1}\sum_{t=1}^{T}(Y_t - \overline{Y})^2$ represents the sample variance of the data. The estimator of $\rho(s)$ follows directly from $\hat{\gamma}(s)$ and is given by

$$\hat{\rho}(s) = \frac{\hat{\gamma}(s)}{\hat{\gamma}(0)} = r(s) = \frac{\sum_{t=1}^{T-s}\left(Y_t - \overline{Y}\right)\left(Y_{t+s} - \overline{Y}\right)}{\sum_{t=1}^{T}(Y_t - \overline{Y})^2}, \quad s = 0, 1, \ldots, \tag{2.3}$$

which is called the sample ACF.

We now briefly discuss a few properties of the sampling distributions of the above estimators, with particular emphasis on the sample ACF $r(s)$. The mean of $\overline{Y}$ is easily seen to be $\mathrm{E}(\overline{Y}) = (1/T)\sum_{t=1}^{T}\mathrm{E}(Y_t) = \mu$, so that $\overline{Y}$ is an unbiased estimator of $\mu$. The variance of $\overline{Y}$ is

$$\mathrm{Var}(\overline{Y}) = \frac{\gamma(0)}{T}\left[1 + 2\sum_{u=1}^{T-1}\left(\frac{T-u}{T}\right)\rho(u)\right]. \tag{2.4}$$

Note that the variance of $\overline{Y}$ needs to account for the autocorrelations.

The sampling properties of the sample ACF $r(s)$ are quite complicated and exact results are not available except in a few special cases. [See Anderson and Walker (1964) [20].] In the white noise case where the $Y_t$ are independent r.v.'s, we then have $\rho(s) = 0$ for all $s \neq 0$ and it can be shown that $\mathrm{E}(r(s)) \approx 0$, $\mathrm{Var}(r(s)) \approx (T-s)/T^2$, and $\mathrm{Cov}(r(s), r(u)) \approx 0$. Hence we can use these results to check for "non-randomness" of a series by comparing the sample values $r(s)$ at various lags $s = 1, \ldots, k$ with the approximate 95% limits $\pm 2\sqrt{(T-s)/T^2} \approx 2/\sqrt{T}$ for the normal distribution to see whether the $r(s)$ generally fall within these limits. We will use this test repeatedly on the residuals as a diagnostic tool for model adequacy.

In the second case, suppose it is assumed that the theoretical ACF satisfies $\rho(s) = 0$ for all $|s| > q$. Then for any $s > q$,

$$\mathrm{Var}(r(s)) \approx \frac{1}{T}\sum_{v=-q}^{q}\rho(v)^2 = \frac{1}{T}\left[1 + 2\sum_{v=1}^{q}\rho(v)^2\right], \quad s > q \tag{2.5}$$

and

$$\mathrm{Cov}(r(s), r(u)) \approx \frac{1}{T}\sum_{v=-q}^{q}\rho(v)\,\rho(v-s+u), \quad s, u > q.$$

As one final special case, suppose the process has ACF of the simple form (which results from an autoregressive model of order one (AR(1)), with coefficient $\phi$) $\rho(s) = \phi^{|s|}$, $s = 0, \pm 1, \ldots$, with $|\phi| < 1$. Then

$$\mathrm{Var}(r(s)) \approx \frac{1}{T}\left[\frac{(1+\phi^2)(1-\phi^{2s})}{1-\phi^2} - 2s\phi^{2s}\right],$$

and in particular $\mathrm{Var}(r(1)) \approx \frac{1}{T}(1-\phi^2)$.

The plot of ACF for a white noise sequence for a large number of lags may have a few isolated lags lie outside the limits, $\pm\frac{2}{\sqrt{T}}$. But characteristic behavior of nonstationary series is that the sample ACF values $r(s)$ are slow to "dampen out" as the lag $s$ increases. We will use these facts repeatedly in our analysis.

## 2.5 Time Series Models for Aggregated Data: Modeling the Mean

In this section, we present a broad overview of the time series models for data aggregated over discrete time intervals. This data as mentioned in Section 2.1 is called price bars and the methodologies discussed in this chapter apply to any aggregated data for a fixed time unit of aggregation that is of equal length. As a close substitute for high frequency data, data aggregated over short time intervals such as two or five minutes can be used.

### Linear Models for Stationary Time Series

A stochastic process $\{Y_t\}$ will be called a *linear process* if it can be represented as

$$Y_t = \mu + \sum_{j=0}^{\infty} \psi_j \epsilon_{t-j}, \tag{2.6}$$

where the $\epsilon_t$ are independent with 0 mean and variance $\sigma_\epsilon^2$, and $\sum_{j=0}^{\infty} |\psi_j| < \infty$. This process may also be referred to as an infinite moving average process. With the backward shift operator $B$, defined by the property that $B^j Y_t = Y_{t-j}$, model (2.6) may be expressed as

$$Y_t = \mu + \sum_{j=0}^{\infty} \psi_j B^j \epsilon_t = \mu + \psi(B)\epsilon_t,$$

where $\psi(B) = \psi_0 + \psi_1 B + \psi_2 B^2 + \cdots$. Since the input $\{\epsilon_t\}$ is a stationary process, it follows that $\{Y_t\}$ in (2.6) forms a stationary process, with mean $\mathrm{E}(Y_t) = \mu$ and autocovariance function

$$\gamma(s) = \mathrm{Cov}(Y_t, Y_{t+s}) = \sigma_\epsilon^2 \sum_{j=0}^{\infty} \psi_j \psi_{j+s}, \tag{2.7}$$

because $\gamma_\epsilon(s) = \sigma_\epsilon^2$ if $s = 0$ and $\gamma_\epsilon(s) = 0$ if $s \neq 0$, so that $\gamma_\epsilon(j - k + s) = 0$ when $k \neq j + s$.

An important result called Wold's Theorem indicates the generality of the class of linear processes in stationary theory, that every purely non-deterministic stationary process may be expressed in an infinite moving average representation as in (2.6), although the $\epsilon_t$ need not be independent but merely uncorrelated.

The general linear representation model given by (2.6) although useful for studying the properties of time series models, is not directly useful in practice to represent a stationary process, since it requires the determination of an infinite number of unknown parameters $\psi_1, \psi_2, \ldots,$ from a finite set of data. Hence we now consider finite parameter models of the same type which will be useful representations and are sufficiently general to well approximate the model for any process of the form (2.6).

**Finite Moving Average Processes**

A direct way to obtain finite parameter models from the general form (2.6) is simply to restrict the $\psi_j$ to be zero beyond some lag $q$. Thus (with a change of notation), a stationary process $\{Y_t\}$ is said to be a *moving average process* of order $q$, which is denoted as MA($q$), if it satisfies

$$Y_t = \mu + \epsilon_t - \sum_{j=1}^{q} \theta_j \epsilon_{t-j}, \tag{2.8}$$

where the $\epsilon_t$ are independent with mean 0 and variance $\sigma^2$. Using the backward shift operator notation $B$, the MA($q$) model can be expressed as

$$Y_t = \mu + \theta(B)\, \epsilon_t,$$

where $\theta(B) = 1 - \sum_{j=1}^{q} \theta_j B^j$. An MA($q$) process is always stationary, by Wold's Theorem, because $\sum_{j=0}^{\infty} |\psi_j| = 1 + \sum_{j=1}^{q} |\theta_j|$ is always finite. The mean of the process is $\mu = \mathrm{E}(Y_t)$, and the autocovariance function is (from (2.7))

$$\gamma(s) = \mathrm{Cov}(Y_t, Y_{t+s}) = \sigma^2 \sum_{j=0}^{q-s} \theta_j \theta_{j+s}, \quad s = 0, 1, 2, \ldots, q,$$

and $\gamma(s) = 0$ for $s > q$, where $\theta_0$ is defined to be $-1$. So in particular, $\gamma(0) = \mathrm{Var}(Y_t) = \sigma^2(1 + \sum_{j=1}^{q} \theta_j^2)$. Hence the autocorrelation function of an MA($q$) process (2.8) is

$$\rho(s) = \frac{-\theta_s + \sum_{j=1}^{q-s} \theta_j \theta_{j+s}}{1 + \sum_{j=1}^{q} \theta_j^2}, \quad s = 0, 1, \ldots, q \tag{2.9}$$

and $\rho(s) = 0$ for $s > q$. The prominent feature of the ACF of an MA($q$) process is that it equals zero or "cuts off" after a finite number of lags, $q$. Thus in one sense the "memory" of an MA($q$) process is $q$ periods long. In practice, useful MA($q$) models are those for which the value of $q$ is typically quite small, such as $q = 1, 2$, or 3. For financial time series, usually $q = 1$ will suffice for modeling, which we will discuss below.

**Example 2.4 (Moving Average Model of Order 1 (MA(1))).** The simplest example of a moving average process is that of order one, MA(1), given by

$$Y_t = \mu + \epsilon_t - \theta \epsilon_{t-1} = \mu + (1 - \theta B)\, \epsilon_t.$$

Its autocovariances are $\gamma(0) = \sigma^2(1 + \theta^2), \gamma(1) = -\theta\,\sigma^2$, and $\gamma(s) = 0$ for $s > 1$, and hence the ACF is $\rho(1) = -\theta/(1 + \theta^2), \rho(s) = 0, s > 1$. It is easy to show that the largest possible value of $|\rho(1)| = |\theta|\,/(1 + \theta^2)$ for an MA(1) process is $\rho(1) = \pm 0.5$ which occurs when $\theta = \pm 1$. This illustrates that in general there are further restrictions on the autocorrelation function $\rho(s)$ of an MA($q$) process in addition to $\rho(s) = 0, s > q$. ◁

**Example 2.5 (Seasonal Moving Average).** For time series which may exhibit seasonal correlation, such as certain financial time series with week-day effect, the following type of special seasonal moving average (SMA) model is useful. Let $S$ denote the number of discrete time periods per seasonal period (for example, $S = 4$ for quarterly data of an annual seasonal nature), and define the process $\{W_t\}$ by

$$W_t = \mu + \epsilon_t - \tau_S \, \epsilon_{t-S}.$$

Then we find that $\gamma(0) = \sigma^2(1 + \tau_S^2)$, $\gamma(S) = -\tau_S/(1 + \tau_S^2)$, and $\rho(j) = 0$ for all $j \neq 0, \pm S$. Hence this process exhibits nonzero autocorrelation only at the seasonal lag $S$. This model is known as an SMA model of seasonal order 1 (and period $S$), and its generalizations are useful when modeling seasonal differences $W_t = Y_t - Y_{t-S}$ of an original seasonal time series $\{Y_t\}$. In the context of financial data, this could be used to model the so-called Monday and Friday effects on price and volatility. $\triangleleft$

We want to introduce another useful class of models, autoregressive models with the example below.

**Example 2.6 (Autoregressive Model of Order 1 (AR(1))).** Consider the process (2.6) with coefficients given by $\psi_j = \phi^j$, $j = 0, 1, \ldots$, for some $\phi$ with $|\phi| < 1$. Then the process $Y_t = \mu + \sum_{j=0}^{\infty} \phi^j \epsilon_{t-j}$ is stationary, because $\sum_{j=0}^{\infty} |\psi_j| = \sum_{j=0}^{\infty} |\phi|^j = 1/(1 - |\phi|)$ is absolutely convergent for $|\phi| < 1$. From (2.7), the autocovariance function of $\{Y_t\}$ is

$$\gamma(s) = \sigma_\epsilon^2 \sum_{j=0}^{\infty} \phi^j \phi^{j+s} = \sigma_\epsilon^2 \, \phi^s \sum_{j=0}^{\infty} \phi^{2j} = \frac{\sigma_\epsilon^2 \phi^s}{1 - \phi^2}, \qquad s \geq 0.$$

In particular, $\gamma(0) = \text{Var}(Y_t) = \sigma_\epsilon^2/(1 - \phi^2)$. Hence the autocorrelation function of $\{Y_t\}$ is $\rho(s) = \gamma(s)/\gamma(0) = \phi^s$, $s \geq 0$ which declines exponentially (geometrically) as the lag $s$ increases.

Note that the process $\{Y_t\}$ defined above actually satisfies the relation

$$Y_t = \phi \, Y_{t-1} + \delta + \epsilon_t,$$

$t = \ldots, -1, 0, 1, \ldots$, with $\delta = \mu(1 - \phi)$. This process is called a first-order autoregressive process, and is a special case of the general class of autoregressive processes which is discussed below. Also note that if the value of $\phi$ in the above is taken as $\phi = 1$, then the arguments for stationarity of $\{Y_t\}$ no longer hold, because $\sum_{j=0}^{\infty} |\psi_j| = 1 + 1 + \cdots$ does not converge. We see that in this case the process is actually a (nonstationary) random walk satisfying the relation $Y_t = Y_{t-1} + \epsilon_t$, a common representation of the behavior of the stock prices under the efficient market hypothesis. $\triangleleft$

**General Order Autoregressive Processes**

While stock price data exhibit simpler structures such as AR(1) in daily aggregation and returns exhibit MA(1) in smaller time unit aggregation such as five minutes, the volume of trading is shown to have more complex dependencies. The behavior of volume and other trade related data are best captured by higher order (autoregressive moving average) ARMA models.

Consider the AR($p$) process $\{Y_t\}$ defined by

$$Y_t = \phi_1 Y_{t-1} + \phi_2 Y_{t-2} + \cdots + \phi_p Y_{t-p} + \delta + \varepsilon_t \tag{2.10}$$

or $(1 - \phi_1 B - \phi_2 B^2 - \cdots - \phi_p B^p) Y_t = \delta + \varepsilon_t$. When $p = 1$ (Example 2.6), $Y_t = \phi Y_{t-1} + \delta + \varepsilon_t$, the ACF, $\rho(v) = \phi^{|v|}$ takes the simple form, as observed earlier.

**Result 1.** If the roots of $\phi(z) = 1 - \phi_1 z - \cdots - \phi_p z^p = 0$ are greater than one in absolute value, then $\{Y_t\}$ defined by (2.10) is a stationary AR($p$) process and has a one-sided infinite moving average representation as

$$Y_t = \mu + \sum_{j=0}^{\infty} \Psi_j \varepsilon_{t-j}, \tag{2.11}$$

where $\mu = \mathrm{E}(Y_t) = \delta/(1 - \phi_1 - \cdots - \phi_p)$, $\Psi(B) = \sum_{j=0}^{\infty} \Psi_j B^j$.

The result indicates that if a finite order AR process is modeled as an MA process, it may require a large number of lags. Now we introduce a property based on moments that is useful for estimation of AR models.

The autocovariance $\gamma(s)$ for the AR($p$) process satisfy the *Yule-Walker equations* given by

$$\gamma(s) = \mathrm{Cov}(Y_t, Y_{t-s}) = \phi_1 \gamma(s-1) + \phi_2 \gamma(s-2) + \cdots + \phi_p \gamma(s-p), \quad s = 1, 2, \ldots,$$

noting again that $\mathrm{Cov}(\varepsilon_t, Y_{t-s}) = 0$, for $s > 0$. Dividing by $\gamma(0)$, the ACF $\rho(s)$ also satisfies the relations

$$\rho(s) = \phi_1 \rho(s-1) + \phi_2 \rho(s-2) + \cdots + \phi_p \rho(s-p), \quad s = 1, 2, \ldots. \tag{2.12}$$

The Yule-Walker (YW) equations (2.12) for $s = 1, 2, \ldots, p$ are particularly useful for determining the autoregressive parameters $\phi_1, \ldots, \phi_p$ in terms of the autocorrelations. Note that these equations can be expressed in matrix form as $P_p \phi = \rho$, where

$$P_p = \begin{bmatrix} 1 & \rho(1) & \rho(2) & \cdots & \rho(p-1) \\ \rho(1) & 1 & \rho(1) & \cdots & \rho(p-2) \\ \vdots & & \ddots & & \vdots \\ \rho(p-1) & \cdots & & \rho(1) & 1 \end{bmatrix}, \quad \phi = \begin{bmatrix} \phi_1 \\ \phi_2 \\ \vdots \\ \phi_p \end{bmatrix}, \quad \rho = \begin{bmatrix} \rho(1) \\ \rho(2) \\ \vdots \\ \rho(p) \end{bmatrix}. \tag{2.13}$$

These equations can be used to solve for $\phi_1, \ldots, \phi_p$ in terms of $\rho(1), \ldots, \rho(p)$, with solution $\phi = P_p^{-1}\rho$. Note also, that the variance of the process, $\gamma(0) = \text{Var}(Y_t)$, and $\sigma^2 = \text{Var}(\varepsilon)$, are related by

$$\gamma(0) = \text{Cov}(Y_t, \phi_1 Y_{t-1} + \cdots + \phi_p Y_{t-p} + \delta + \varepsilon_t) = \phi_1\gamma(1) + \cdots + \phi_p\gamma(p) + \sigma^2, \quad (2.14)$$

because $\text{Cov}(Y_t, \varepsilon_t) = \sigma^2$. Hence $\sigma^2$ can be determined in terms of $\gamma(0)$ and the autocorrelations $\rho(s)$ from (2.14).

Also note that a measure of the strength of association or "predictability" of $Y_t$ based on its past values $Y_{t-1}, \ldots, Y_{t-p}$ can be given by the (squared) multiple correlation coefficients of $Y_t$ with $(Y_{t-1}, \ldots, Y_{t-p})$, which is $R^2 = (\gamma(0) - \sigma^2)/\gamma(0) = 1 - (\sigma^2/\gamma(0)) = \phi_1\rho(1) + \cdots + \phi_p\rho(p)$. Since $\gamma(0)$ represents the "total" variance of $Y_t$, and $\sigma^2$ may be interpreted as the "unexplained" or error variance (i.e., that portion of the variance not explained by the past values), the quantity $R^2$ has the usual interpretation as in linear regression of representing the proportion of the total variance of the variable $Y_t$ that is explained by (auto)regression on the past values $Y_{t-1}, \ldots, Y_{t-p}$.

**Partial Autocorrelation Function**

When an AR model is being fit to observed time series data, the sample version of the Yule-Walker equations (2.12), (2.13), in which the values $\rho(j)$ are replaced by the sample ACF values r(j), can be solved to obtain estimates $\hat{\phi}_j$ of the autoregressive parameters. However, initially it is not known which order $p$ is appropriate to fit to the data. The sample partial autocorrelation function (PACF) will be seen to be useful for determination of the appropriate order of an AR process as how the autocorrelation function (ACF) is useful for the determination of the appropriate order of an MA process. First, we discuss the concept of PACF in general.

Suppose $\{Y_t\}$ is a stationary process, not necessarily an AR process, with ACF $\rho(j)$. For general $k \geq 1$, consider the first $k$ Yule-Walker equations for the ACF associated with an AR($k$) process.

$$\rho(j) = \phi_1\rho(j-1) + \phi_2\rho(j-2) + \cdots + \phi_k\rho(j-k), \quad j = 1, 2, \ldots, k \qquad (2.15)$$

and let $\phi_{1k}, \phi_{2k}, \ldots, \phi_{kk}$ denote the solution for $\phi_1, \phi_2, \ldots, \phi_k$ to these equations. Given the ACF $\rho(j)$, (2.15) can be solved for each value $k = 1, 2, \ldots$, and the quantity $\phi_{kk}$, regarded as a function of the lag $k$, is called the (theoretical) *partial autocorrelation function* (PACF) of the process $\{Y_t\}$. The values $\phi_{1k}, \phi_{2k}, \ldots, \phi_{kk}$ which are the solution to (2.15) are the regression coefficients in the regression of $Y_t$ on $Y_{t-1}, \ldots, Y_{t-k}$; that is, values of coefficients $b_1, \ldots, b_k$ which minimize $\text{E}[(Y_t - b_0 - \sum_{i=1}^{k} b_i Y_{t-i})^2]$.

If $\{Y_t\}$ is truly an AR process of order $p$, the $\phi_{kk}$ will generally be nonzero for $k \leq p$, but $\phi_{kk}$ will always be zero for $k > p$. This is so since the ACF $\rho(j)$ actually satisfies the Yule-Walker equations (2.15) for $k = p$, and hence for any $k > p$ the solution to (2.15) must be $\phi_{kk} = \phi_1, \ldots, \phi_{pk} = \phi_p, \phi_{p+1,k} = 0, \ldots, \phi_{kk} = 0$, where $\phi_1, \ldots, \phi_p$ are the true AR coefficients. Thus, the PACF $\phi_{kk}$ of an AR($p$) process "cuts off" (is zero) after lag $p$, and this property serves to distinguish (identify) an AR($p$) process.

The quantity $\phi_{kk}$ defined above is called the *partial autocorrelation* at lag $k$, since it is actually equal to the partial correlation between the r.v.'s $Y_t$ and $Y_{t-k}$ adjusted for the intermediate variables $Y_{t-1}, Y_{t-2}, \ldots, Y_{t-k+1}$, and $\phi_{kk}$ measures the correlation between $Y_t$ and $Y_{t-k}$ after adjusting for the effects of $Y_{t-1}, Y_{t-2}, \ldots, Y_{t-k+1}$.

Based on a sample $Y_1, \ldots, Y_T$ from a process $\{Y_t\}$, the sample PACF value at lag $k$, $\hat{\phi}_{kk}$, is obtained as the solution to the sample Yule-Walker equations of order $k$,

$$r(j) = \hat{\phi}_{tk} r(j-1) + \cdots + \hat{\phi}_{kk} r(j-k), \quad j = 1, \ldots, k, \tag{2.16}$$

for each $k = 1, 2, \ldots$. These are the same form as the equations (2.15) but with sample ACF values $r(j)$ used in place of the theoretical ACF values $\rho(j)$. In matrix notation similar to that used in (2.13), the solution is $\hat{\phi} = R_k^{-1} r$. Now under the assumption that the process $\{Y_t\}$ is an AR of order $p$, the estimated PACF values $\hat{\phi}_{kk}$ for lags $k > p$ are approximately independently and normally distributed for large $T$, with mean $\mathrm{E}(\hat{\phi}_{kk}) = 0$ and St.Dev. $\hat{\phi}_{kk} = 1/\sqrt{T}$ for $k > p$. These facts concerning the sampling properties of the sample PACF $\hat{\phi}_{kk}$ can be used to assess whether the coefficients in estimated AR models may be treated as close to zero after some lag $p$ (e.g., if $|\hat{\phi}_{kk}| < 2/\sqrt{T}$ for $k > p$), and hence can be useful in the selection of an appropriate order $p$ for fitting an AR model to sample data.

**Linear Models for Nonstationary Time Series**

Often, in practice, series $\{Y_t\}$ will be encountered which are nonstationary. One type of nonstationary series that occurs commonly, are series that exhibit some homogeneous behavior over time in the sense that, except for local level and or local trend, one segment of the series may behave much like other parts of the series. In those cases, it may be found that the first difference of the series, $(1 - B)\, Y_t = Y_t - Y_{t-1}$, is a stationary series. For seasonal nonstationary time series that exhibit homogeneous behavior apart from a seasonal mean level or trend, with seasonal period $S$, the seasonal difference of the series, $(1 - B^s)\, Y_t = Y_t - Y_{t-s}$, may be stationary. More generally, a useful class of models for this type of homogeneous nonstationary time series $\{Y_t\}$ is obtained by assuming that the $d$th difference, $W_t = (1 - B)^d\, Y_t$, is a stationary series, and $W_t$ can be represented by an ARMA$(p, q)$ model. The most common case is $d = 1$, for financial time series.

**Autoregressive Integrated Moving Average Processes (ARIMA)**

We consider models for $\{Y_t\}$ such that $W_t = (1 - B)^d\, Y_t$ is a stationary ARMA$(p, q)$ process. The process $\{Y_t\}$ is then said to be an *autoregressive integrated moving average process* of order $(p, d, q)$, denoted as ARIMA$(p, d, q)$. Since the $d$th differences $W_t$ form an ARMA$(p, q)$ process, they satisfy $\phi(B)\, W_t = \delta + \theta(B)\, \varepsilon_t$. So the ARIMA process $Y_t$ is generated by

$$\phi(B)(1 - B)^d\, Y_t = \delta + \theta(B)\varepsilon_t, \tag{2.17}$$

where $\phi(B) = 1 - \phi_1 B - \cdots - \phi_p B^p$ and $\theta(B) = 1 - \theta_1 B - \cdots - \theta_q B^q$.

The use of the term "integrated" came as follows. When $d = 1$ so that the process $W_t = Y_t - Y_{t-1} = (1 - B)Y_t$ is a stationary ARMA$(p, q)$ process. Then the stationary series $W_t$ must be summed or "integrated" to obtain the process $\{Y_t\}$; that is,

$$Y_t = (1 - B)^{-1} W_t = (1 + B + B^2 + \cdots) W_t = \sum_{j=0}^{\infty} W_{t-j}.$$

More generally, with $d > 1$, the stationary process $W_t = (1 - B)^d Y_t$ must be summed $d$ times to obtain the process $Y_t$. Note that the random walk process $Y_t = Y_{t-1} + \delta + \varepsilon_t$ is the simplest example of an integrated process, and is also the most commonly observed in the context of financial time series. Because the first differences $W_t = Y_t - Y_{t-1} = \delta + \varepsilon_t$ form a (stationary) white noise process, i.e., the process $\{Y_t\}$ is ARIMA(0,1,0). As noted earlier, this is called a random walk model with a drift.

Differencing is appropriate to apply to processes which contain a stochastic trend component (not a deterministic trend component), and stochastic trend behavior is often more likely to be present in financial time series than deterministic trend behavior. Some observations are in order. First, if a series does not require differencing, over differencing will lead to unnecessarily more complicated models. Second, the differencing can sometimes wipe out all the information leaving series with white noise behavior. Some authors have suggested fractional differencing; that is, $0 < d < 1$, but there is no theoretic motivation for an appropriate choice of the value of '$d$'. Third, there are identifiability issues in ARMA models; it is possible to model the same data with various ARMA structures, but we will follow the principle of parsimony in modeling.

Two useful results that we present below illustrate how the general class of (ARIMA) models (2.17) can arise in practice.

**Result 2 (Aggregation).** Consider a process $Y_t$, composed of a trend component $U_t$ and a noise component $N_t$, $Y_t = U_t + N_t$. We assume that the (random) trend component follows a random walk model (possibly with a drift), $(1 - B)\,U_t = \delta + a_t$, and we assume that $N_t$ is a stationary AR(1) process $(1 - \phi B)\,N_t = b_t$, where $\{a_t\}$ and $\{b_t\}$ are independent white noise processes with zero means and variances $\sigma_a^2$ and $\sigma_b^2$, respectively. Then, applying the first differencing operator, we have $(1 - B)\,Y_t = (1 - B)\,U_t + (1 - B)\,N_t = \delta + a_t + (1 - B)\,N_t$, and hence

$$\begin{aligned} (1 - \phi B)(1 - B)\,Y_t &= (1 - \phi B)\,\delta + (1 - \phi B)\,a_t + (1 - \phi B)(1 - B)\,N_t \\ &= (1 - \phi)\,\delta + (1 - \phi B)\,a_t + (1 - B)\,b_t. \end{aligned} \tag{2.18}$$

It can be shown that $Z_t = (1 - \phi B)\,a_t + (1 - B)\,b_t$ is the sum of two independent MA(1) processes and thus an MA(1) process, so that $Y_t$ is an ARIMA(1,1,1) process. Thus ARIMA $(p, d, q)$ can arise naturally when several stochastic elements with different behaviors are added together. These stochastic elements may come from sources that can be justified by apriori theories.

This general observation concerning the sum of two ARMA processes gives rise to the following result.

**Result 3 (Aggregation).** If $X_t$ is an ARMA$(p_1, q_1)$ process and $Y_t$ is an ARMA$(p_2, q_2)$ process, with $X_t$ and $Y_t$ independent processes, then $Z_t = X_t + Y_t$ follows an ARMA$(p, q)$ model, where $p \leq p_1 + p_2$ and $q \leq \max(p_1 + q_2, p_2 + q_1)$.

These types of models have potential applications in studying aggregate market behavior of several series. For example in addition to modeling the individual stock behavior in an industry, we can study a model for the entire industry as well. Also where the observed series may be viewed as the sum of the true process of interest plus observational error or noise (as in the classical "signal-plus-noise" model), and in the use of structural component models where an observed time series is represented as the sum of unobservable components that correspond to factors such as trend, seasonality, stationary variations, and so on. These are modeled more compactly via Kalman filter which is presented in Chapter 4.

## 2.6 Key Steps for Model Building

In previous sections, theoretical linear models for stochastic processes and properties implied by these models were discussed. We now consider the problem of identifying appropriate models and fitting them to time series data. The following four steps advocated by George Box highlight the basic stages in the proposed model building and testing procedure:

1. Model Specification or Identification—specify or identify specific models to be entertained as appropriate, based on preliminary data exploration and examination of certain statistical features, such as features of the sample ACF and PACF for time series data.

2. Parameter Estimation—for the specified model or models, estimate parameters in the tentatively entertained models efficiently.

3. Model Checking—perform diagnostic checks on the adequacy of the estimated model, usually by examination of various features of the residuals from the fitted model.

4. Model Validation—confirm that the model is appropriate for out-of-sample prediction. When multiple models are tried out on a single set of data, it may result in data snooping bias. By keeping the data for building the model and the data for validating the model separate, we avoid the snooping bias. It should be kept in mind that overfit (estimated) models do not perform well in validation.

If the estimated model is assessed as being adequate, it can then be used for forecasting or other purposes of interest. Otherwise, one would return to step one and specify an alternate model or models for further consideration. We now discuss briefly some aspects of the model specification procedures in some detail, and point out the key features.

## Model Specification

Given the time series data $Y_1, \ldots, Y_T$, basic time series plotting of the series is a fundamental step in the initial model building, along with other data plots that might seem informative. We will then consider the sample ACF and PACF of the original series $Y_t$, and of certain transformed series, such as first and possibly second differences, seasonal differences, logarithms or other instantaneous transformation of the original series, residuals from regression to remove deterministic seasonal component or linear trend, and so on. That is, if the original time series appears to be nonstationary we consider differencing or using residuals from regression methods, of the original series which will yield a *stationary* series. More formal procedures to 'test' for certain (unit-root) type nonstationarity can also be considered, and will be discussed later in the context of estimation for AR models. While dealing with financial time series the following considerations are usually made: For example, $r_t$ = return = $\ln(P_t) - \ln(P_{t-1})$, where $P_t$ is the price of the stock, the differencing of the log price is naturally considered and for volume $V_t$, because of its size it is usually logged $v_t = \ln(V_t)$ to avoid heteroscedasticity.

The sample ACF and PACF of the stationary series are then compared against features of the theoretical ACF and PACF of various types of ARMA models to find an appropriate model for the observed series on the basis of close correspondence in features between sample and theoretical correlation functions. Of course, the sample ACF values that are defined in (2.3) are only *estimates*, and hence are subject to sampling errors. Thus, we must recognize that the sample ACF of an observed series will never correspond in all exact details to an underlying theoretical ACF, and we need to look for correspondence in broad features. To properly interpret the sample ACF $[\hat{\rho}(j)]$, we will use the sampling properties of the estimates $\hat{\rho}(j)$. In particular, under the assumption that $\rho(j) = 0$ for all $j > q$, we have $\mathrm{E}[\hat{\rho}(j)] = 0$ and $\text{St.Dev.}[\hat{\rho}(j)] = \frac{1}{\sqrt{T}}[1 + 2\sum_{i=1}^{q} \hat{\rho}(i)^2]^{1/2}$ for $j > q$ and the $\hat{\rho}(j)$ are approximately normally distributed, for moderate and large $T$, a result given in (2.5). Similarly, it is known that if the process $\{Y_t\}$ is an AR of order $p$, then the estimated PACF for lags $p+1$ and higher are approximately independently and normally distributed, with $\mathrm{E}[\hat{\phi}_{kk}] = 0$ and $\text{St.Dev.}[\hat{\phi}_{kk}] = 1/\sqrt{T}$ for $k > p$. These facts may be used to assess whether the last coefficient in an estimated AR model is essentially zero. We summarize some general patterns of the sample ACF and PACF, which might be useful in the initial model identification step.

1. The sample ACF of a unit-root nonstationary process will not generally dampen out sufficiently fast as the lag increases.

2. The sample ACF $[\hat{\rho}(j)]$ of a typical stationary process should dampen out sufficiently fast. The general behavior of the (stationary) MA($q$), AR($p$), and ARMA($p, q$) processes is:

    - MA($q$) process has ACF $[\rho(j)]$ that will 'cut off' after $q$ lags; that is, $\rho(j) = 0$ for $j > q$. For example, the MA(1) process has $\rho(j) = 0$ for $j > 1$.

- AR($p$) process has ACF[$\rho(j)$] which can have exponential decay or dampened sinusoidal behavior, or a mixture of both, so this will tend to be the behavior of the sample ACF. For example, the AR(1) process has $\rho(j) = \phi\rho(j-1)$, or $\rho(j) = \phi^j$, for all $j \geq 1$.
- ARMA($p, q$) process has ACF [$\rho(j)$] which can have an irregular pattern for the first $q$ lags and then similar behavior to a corresponding AR($p$) process for lags greater than $q$. For example, the ARMA(1,1) has $\rho(j) = \phi\rho(j-1)$, or $\rho(j) = \phi^{j-1}\rho(1)$, for $j \geq 2$, which is similar behavior to the AR(1) after lag 1.

**3.** The sample PACF $\hat{\phi}_{kk}$ is useful to identify the order $p$ of a finite order AR($p$) process, because for an AR($p$) process the theoretical PACF $\phi_{kk}$ 'cuts off' after lag $p$; that is, $\phi_{kk} = 0$ for all $k > p$. For the sample PACF $\hat{\phi}_{kk}$ of an AR($p$) process, we have E[$\hat{\phi}_{kk}$] = 0 and St.Dev.[$\hat{\phi}_{kk}$] = $1/\sqrt{T}$, for $k > p$.

Although the sample autocorrelation and partial autocorrelation functions are very useful in model identification, in some cases involving mixed ARMA models they will not provide unambiguous results. There has been considerable interest in developing additional tools for use at the model identification stage. Interested readers can refer to standard texts in time series for more in-depth discussion such as Shumway and Stoffer (2011) [302].

## Estimation of Model Parameters

As there are excellent sources available on this topic, our discussion will be quite brief.

### Method of Moments

Preliminary (initial) estimates of the parameters (2.17) in the model may be obtained by the *method of moments* procedure, which is based on the sample ACF $\hat{\rho}(j), j = 1, \ldots, p+q$. The method of moments estimates have the appeal of being relatively easy to compute, but are not statistically efficient (except for a pure AR); they can usefully serve as initial values to obtain more efficient estimates.

Assuming that $W_t = (1-B)^d\, Y_t$ follows an ARMA($p, q$) model, we know that the ACF $\rho(j)$ of $W_t$ satisfies the "generalized" Yule-Walker equations in (2.16). Hence we can obtain initial estimates for the AR parameters $\phi_1, \ldots, \phi_p$ by solving the sample version of these equations for $j = q+1, \ldots, q+p$,

$$\hat{\rho}(j) - \sum_{i=1}^{p} \hat{\phi}_i \hat{\rho}(j-i) = 0, \quad j = q+1, \ldots, q+p. \tag{2.19}$$

Note that these equations could also be useful in identifying the orders $(p, q)$ of an ARMA model. Since the mean $\mu_w = \mathrm{E}(W_t)$ of the stationary process $W_t$ in (2.17) is $\mu_w = \delta/(1-\phi_1-\cdots-\phi_p)$, we estimate $\mu_w$ by sample mean $\hat{\mu}_w = \overline{W}$ and hence the constant term $\delta$ by $\hat{\delta}_0 = (1-\hat{\phi}_1-\cdots-\hat{\phi}_p)\,\hat{\mu}_w$.

Initial estimates of the MA parameters could then be based on the autocovariances of the 'derived' MA(q) process, $W_t^* = \phi(B)W_t = \theta_0 + \theta(B)\varepsilon_t$. The autocovariances of $W_t^*$ are related to those of $W_t$ by

$$\gamma(s) = \mathrm{Cov}(W_t^*, W_{t+s}^*) = \sum_{i=0}^{p}\sum_{j=0}^{p} \phi_i\phi_j\gamma(s+i-j), \quad s = 0, 1, \ldots \tag{2.20}$$

with $\phi_0 = -1$, where $\gamma(j) = \mathrm{Cov}(W_t, W_{t+j})$. But since $W_t^* = \theta_0 + \theta(B)\varepsilon_t$ also satisfies the MA($q$) model

$$\gamma_*(s) = \sigma_\varepsilon^2(-\theta_s + \theta_1\theta_{s+1} + \cdots + \theta_{q-s}\theta_q), \quad s = 1, 2, \ldots, q, \tag{2.21}$$

with $\gamma_*(0) = \sigma_\varepsilon^2(1 + \theta_1^2 + \cdots + \theta_q^2)$. Hence, to obtain initial estimates of $\theta_1, \ldots, \theta_q$ and $\sigma_\varepsilon^2$ we first form the estimated autocovariances $\hat{\gamma}_*(s)$ of $W_t^*$ based on (2.20) using the initial $\hat{\phi}_i$ and the sample autocovariance $\hat{\gamma}(j)$ of $W_t$. We then substitute the $\hat{\gamma}_*(s)$ for $\gamma_*(s)$ in the equations (2.21), and solve for parameters $\theta_1, \ldots, \theta_q$ and $\sigma_\varepsilon^2$. The resulting equations are nonlinear in the $\theta_i$ and hence must be solved by an iterative numerical procedure (except for the MA(1) case where an explicit solution is available). We now illustrate this procedure with examples of a few simple models.

**Example 2.7 (AR(1) Model).** In this case, the only equation needed for the AR parameter $\phi_1$ is the first-order Yule-Walker equation, which gives the estimate $\hat{\phi}_1 = \hat{\rho}(1)$, and then $\gamma(0) = (1 - \phi_1^2)\,\gamma(0) \equiv \sigma_\varepsilon^2$, so that $\hat{\sigma}_\varepsilon^2 = (1 - \hat{\phi}_1^2)\,\hat{\gamma}(0)$. ◁

**Example 2.8 (ARMA(1,1) Model).** The equation for $\phi_1$ comes from $\rho(2) = \phi_1\rho(1)$, so that the moment estimate is $\hat{\phi}_1 = \hat{\rho}(2)/\hat{\rho}(1)$. Then we can form $\hat{\gamma}(0) = (1 + \hat{\phi}_1^2)\,\hat{\gamma}(0) - 2\hat{\phi}_1\,\hat{\gamma}(1)$, $\hat{\gamma}(1) = (1 + \hat{\phi}_1^2)\,\hat{\gamma}(1) - \hat{\phi}_1\,\hat{\gamma}(0) - (-\hat{\phi}_1\,\hat{\gamma}(2))$, and solve the equations, $\hat{\gamma}(0) = \sigma_\varepsilon^2(1 + \theta_1^2), \hat{\gamma}(1) = -\sigma_\varepsilon^2\theta_1$. We have $\hat{\rho}(1) \equiv \hat{\gamma}(1)/\hat{\gamma}(0) = -\hat{\theta}_1/(1 + \hat{\theta}_1^2)$, which leads to the solution to a quadratic equation as $\hat{\theta}_1 = \left(-1 + \sqrt{1 - 4\hat{\rho}(1)^2}\right)/(2\hat{\rho}(1))$ (provided $|\hat{\rho}(1)| < 0.5$, in which case we get an invertible value with $|\hat{\theta}_1| < 1$, and then $\hat{\sigma}_\varepsilon^2 = \hat{\gamma}(0)/(1 + \hat{\theta}_1^2)$. Alternatively, consider directly the sample versions of the first two autocovariance equations for the ARMA(1,1) process, $\hat{\gamma}(0) - \hat{\phi}_1\hat{\gamma}(1) = \sigma_\varepsilon^2[1 - \theta_1(\hat{\phi}_1 - \theta_1)]$ and $\hat{\gamma}(1) - \hat{\phi}_1\hat{\gamma}(0) = -\sigma_\varepsilon^2\theta_1$.

We can eliminate $\sigma_\varepsilon^2$ from the above two equations and obtain the initial estimate $\hat{\theta}_1$ by solving a quadratic equation. Then the initial estimate of $\sigma_\varepsilon^2$ is $\hat{\sigma}_\varepsilon^2 = [\hat{\gamma}(0) - \hat{\phi}_1\hat{\gamma}(1)]/[1 - \hat{\theta}_1(\hat{\phi}_1 - \hat{\theta}_1)]$. This same procedure would specialize to the MA(1) model, with the simplification that the first step of estimation of $\phi_1$ is eliminated and we use simply $\hat{\gamma}(0) = \sigma_\varepsilon^2(1 + \theta_1^2)$ and $\hat{\gamma}(1) = -\sigma_\varepsilon^2\theta_1$. ◁

**Example 2.9 (AR($p$) Model).** For the AR($p$) model, the method of moments estimates of the $\phi_1$ parameters come directly from solving the system of the first $p$ sample Yule-Walker equations in (2.12) for the series $W_t$. ◁

The method of moments can be used for preliminary parameter estimation and also for model specification. Now we will focus on more efficient estimation methods, and the general approach will be the use of Gaussian maximum likelihood (ML) methods, and the closely related method of least squares estimation. This discussion is made brief as there are excellent books such as Shumway and Stoffer (2011) [302] on these topics. For ease of presentation, we will first examine ML estimation for pure autoregressive models, and extend to the general ARMA model in later sections because certain estimation features are relatively simple for the AR model compared to the general ARMA model. It must be noted that the least-squares estimator does not guarantee that the roots of the characteristic function lie outside the unit circle, but the Yule-Walker estimator does.

**Conditional Likelihood Estimation**

We consider parameter estimation for the AR($p$) process with mean $\mu$, or constant term $\delta$.

$$Y_t = \delta + \sum_{i=1}^{p} \phi_i Y_{t-i} + \varepsilon_t = \mu + \sum_{i=1}^{p} \phi_i (Y_{t-i} - \mu) + \varepsilon_t, \tag{2.22}$$

where $\delta = \mu(1 - \phi_1 - \cdots - \phi_p)$, and we assume the $\varepsilon_t$ are i.i.d. normal $N(0, \sigma^2)$. Given a sample realization of $T$ observations $Y_1, Y_2, \ldots, Y_T$, we consider the conditional likelihood function (i.e., the conditional p.d.f of $Y_{p+1}, \ldots, Y_T$, given the first $p$ observations $Y_1, \ldots, Y_p$. The *conditional maximum likelihood estimates* (CMLE) of $\mu$, $\phi$ and $\sigma^2$ are values of the parameters that maximize the conditional likelihood function, where $\phi$ denotes $(\phi_1, \ldots, \phi_p)$.

First, consider the AR(1) case. Then the conditional (on knowing $y_1$) sum of squares function to be minimized is

$$S_*(\mu, \phi) = \sum_{t=2}^{T} \left(Y_t - \mu - \phi(Y_{t-1} - \mu)\right)^2 = \sum_{t=2}^{T} (Y_t - \delta - \phi Y_{t-1})^2.$$

Taking $\overline{Y}_{(0)} = \frac{1}{T-1} \sum_{t=2}^{T} Y_t$ and $\overline{Y}_{(1)} = \frac{1}{T-1} \sum_{t=2}^{T} Y_{t-1}$, the least squares estimates (LSE) of $\phi$ and $\delta = \mu(1 - \phi)$ are given by

$$\hat{\phi} = \frac{\sum_{t=2}^{T} (Y_t - \overline{Y}_{(0)})(Y_{t-1} - \overline{Y}_{(1)})}{\sum_{t=2}^{T} (Y_{t-1} - \overline{Y}_{(1)}^2)}, \quad \hat{\delta} = \overline{Y}_{(0)} - \hat{\phi}\overline{Y}_{(1)}.$$

Now since $\overline{Y}_{(1)} \approx \overline{Y}_{(0)} \approx \overline{Y}$, where $\overline{Y} = (1/T) \sum_{t=1}^{T} Y_t$ is the overall sample mean, we have approximately that $\hat{\mu} \approx \overline{Y}$, which is an intuitively appealing estimator. Then using $\hat{\mu} \approx \overline{Y}$, we obtain

$$\hat{\phi} = \frac{\sum_{t=2}^{T} (Y_t - \overline{Y})(Y_{t-1} - \overline{Y})}{\sum_{t=2}^{T} (Y_{t-1} - \overline{Y})^2}, \quad \hat{\delta} = (1 - \hat{\phi})\overline{Y}. \tag{2.23}$$

This estimator $\hat{\phi}$ is exactly the estimator one would obtain if the AR equation (adjusted for the overall sample mean $\overline{Y}$) $Y_t - \overline{Y} = \phi(Y_{t-1} - \overline{Y}) + \varepsilon_t, t = 2, \ldots, T$, is treated as an ordinary regression with $Y_{t-1} - \overline{Y}$ as the "independent" variable. A further approximation that can be noted from the Yule-Walker equations is $\hat{\phi} = \hat{\rho}(1)$, which is also an appealing estimator since $\rho(1) = \phi$ for the AR(1) process.

Higher order AR($p$) processes may also be estimated by conditional least squares in a straightforward manner. The estimates are simply obtained by ordinary least squares regression of the "dependent" variable $Y_t$ on the "independent" variables $Y_{t-1}, \dots, Y_{t-p}$ and a constant term. Except for the approximation $\hat{\mu} = \overline{Y}$, this approach will produce the conditional least squares estimates. A second approximate method consists of solving the first $p$ sample Yule-Walker equations for $\hat{\phi}_1, \dots, \hat{\phi}_p$. In matrix notation these sample Yule-Walker equations are $R_p\hat{\phi} = r$ with solution $\hat{\phi} = R_p^{-1}r$, which is the Yule-Walker estimator of $\phi$.

For moderately large sample size $T$, both conditional least squares estimation ("regression" methods) and Yule-Walker estimation will give approximately the same estimated values of $\phi_1, \dots, \phi_p$. This is because the sample sums of squares and cross-products that are involved in both estimation methods differ only by "end-effects" involving the treatment of the first and/or last few observations in the series. Generally, least squares (LS) estimators are preferred over Yule-Walker (YW) estimators since they tend to have better "finite-sample" properties, especially for AR models that are nearly nonstationary (i.e., have roots of $\phi(B)$ close to the unit circle). For inferences on the AR coefficients, it is observed that $\sqrt{T}\,(\hat{\phi} - \phi)$ converges as $T \to \infty$ to the multivariate normal distribution $N(0, \sigma^2\,\Gamma_p^{-1})$, a multivariate normal with mean 0 and covariance matrix $\sigma^2\,\Gamma_p^{-1}$, where $\Gamma_p \equiv \gamma(0)P_p$, where $P_p$ is defined in (2.13). Thus, for large $T$, we can use the approximation that $\hat{\phi}$ is distributed as $N(\phi, (\sigma^2/T)\Gamma_p^{-1})$. For a simple example, in the AR(1) case, the LS or Yule-Walker estimator $\hat{\phi} = r(1) = \hat{\rho}(1)$ has an approximate normal distribution with mean $\phi = \rho(1)$ and variance $\text{Var}(\hat{\phi}) = (\sigma^2/T)[\gamma(0)]^{-1} = (1/T)(\sigma^2/\gamma(0)) = (1/T)(1-\phi^2)$, since $\gamma(0) = \sigma^2/(1-\phi^2)$ for an AR(1) process, and $\text{Var}(\hat{\phi})$ is estimated by $(1-\hat{\phi}^2)/T$.

The maximum likelihood estimation is based on the augmented conditional likelihood function, augmented with the distribution of the initial values $Y_1, \dots, Y_p$. The procedure is quite cumbersome, but with large sample, $T$, the distribution of initial values does not make much of a difference in the estimation of the unknown parameters. Maximum likelihood and least squares estimation are much more computationally difficult for a mixed ARMA($p, q$) model than for a pure AR($p$) model because the log-likelihood and the sum of squares are more complicated (non-quadratic) functions of the unknown parameters. Interested readers can refer to standard textbooks in this area, but with financial series which tend to be fairly long, the discussion presented here will suffice.

## Model Selection Criteria

One approach to selecting among competing models is the use of information criteria such as AIC proposed by Akaike (1974) [5] or the BIC of Schwarz (1978) [299].

They are

$$\begin{aligned} \text{AIC}_{p,q} &= \log \hat{\sigma}_{\varepsilon}^2 + n^* \left( \frac{2}{T} \right) \\ \text{BIC}_{p,q} &= \log \hat{\sigma}_{\varepsilon}^2 + n^* \left( \frac{\log T}{T} \right), \end{aligned} \tag{2.24}$$

where $\hat{\sigma}_{\varepsilon}^2$ is the estimated model error variance, and $n^* = p+q+1$ denote the number of parameters estimated in the model, including the constant term. The first term in the above equations results from the maximized likelihood and can be made smaller with more parameters, while the second term acts as a "penalty factor" for the parameters used in the model. In this approach, models which yield a minimum value for the criterion are to be preferred, as these criteria penalize for overfitting.

## Model Diagnostics

As several models can be entertained from specification tools such as ACF and PACF, one quick way to check if all the informative dependencies are captured is to use the model residuals and examine if these residuals reflect the features that are assumed. One key feature is the assumption of independence that can be tested through ACF. Another feature is the constancy in the variance which again can be examined by the plot of squared residuals versus time scale and the time dependence by the ACF of the squared residuals. Thus residuals or functions of the residuals play an important role in model diagnostics. We recommend that descriptive tools should be complemented with visualization tools.

## Model Validation

The debate on the evaluation of a model for its consistency with the data at hand or its forecasting power is a long-standing one. The risk of overfitting a model purely by data exploration will be exposed through poor performance in forecasting. There is a chance to develop models that fit the features of the data that may be unique but these features may not be present in the future. We will follow in general the principle of parsimony; that is to develop models with fewer parameters that can adequately represent the data. These models should be robust to extreme values and should be stable over time.

Ideally, model validation requires two random samples from the same population; a training set and a test set. Because of the chronological nature of the time series, usually the training set precedes in time the test set. Thus if $y_t, t = 1, 2, \dots, n_1, n_1 + 1, \dots, n_1 + n_2$ and if the model is built on '$n_1$' observations and the last '$n_2$' observations are used for model validation, it is usual to compare the prediction mean square error (PMSE),

$$\text{PMSE} = \sum_{t=1}^{n_2} \frac{(y_{n_1+t} - \hat{y}_{n_1+t})^2}{n_2} \tag{2.25}$$

with the MSE ($\hat{\sigma}_{\varepsilon}^2$) of the fitted model. While other forms of cross-validation methods are applicable to non-dependent data, the method suggested here is usually applied for chronological data.

## 2.7 Testing for Nonstationary (Unit Root) in ARIMA Models: To Difference or Not To

The decision concerning the need for differencing is based, informally, on features of the time series plot of $Y_t$ and of its sample autocorrelation function (e.g., failure of the $\hat{\rho}(k)$ to dampen out quickly). This can be formally evaluated further based on the estimates of model parameters. However, it must be kept in mind that distribution theory for estimates of parameters is quite different between stationary and nonstationary models. Much of the notes below follow from Box, Jenkins, Reinsel, Ljung (2015) [53]

Consider for the simple AR(1) model $Y_t = \phi Y_{t-1} + \varepsilon_t$, $t = 1, 2, \ldots, T$, $Y_0 = 0$, testing that $|\phi| = 1$. The least squares (LS) estimator of $\phi$ can be written as,

$$\hat{\phi} = \frac{\sum_{t=2}^{T} Y_{t-1} Y_t}{\sum_{t=2}^{T} Y_{t-1}^2} = \phi + \frac{\sum_{t=2}^{T} Y_{t-1} \varepsilon_t}{\sum_{t=2}^{T} Y_{t-1}^2}. \tag{2.26}$$

If the process is stationary, $|\phi| < 1$ as it has been noted earlier, $T^{1/2}(\hat{\phi} - \phi)$ has an approximate normal distribution with zero mean and variance $(1 - \phi^2)$. However, when $\phi = 1$, so $Y_t = \sum_{j=0}^{t-1} \varepsilon_{t-j} + Y_0$, with $\text{Var}(Y_t) = \text{E}(Y_t^2) = t\sigma^2$, it can be shown that $T(\hat{\phi} - 1) = T^{-1} \sum_{t=2}^{T} Y_{t-1}\varepsilon_t / T^{-2} \sum_{t=2}^{T} Y_{t-1}^2$, the estimator $\hat{\phi}$ approaches its true value $\phi = 1$ with increasing sample size $T$ at a faster rate than in the stationary case. Both the numerator and the denominator of the above expression have non-normal limiting distributions.

We present the commonly used "Studentized" statistic, for testing $H_0 : |\phi| = 1$ vs. $H_a : |\phi| < 1$,

$$\hat{\tau} = \frac{\hat{\phi} - 1}{s_\varepsilon \left( \sum_{t=2}^{T} Y_{t-1}^2 \right)^{-1/2}}, \tag{2.27}$$

where $s_\varepsilon^2 = (T-2)^{-1}(\sum_{t=2}^{T} Y_t^2 - \hat{\phi} \sum_{t=2}^{T} Y_{t-1} Y_t)$ is the residual mean square, has been considered. The limiting distribution of the statistic $\hat{\tau}$ has been derived, and tables of the percentiles of this distribution under $\phi = 1$ are given in Fuller (1996) [153]. The test rejects $\phi = 1$ when $\hat{\tau}$ is "too negative."

For higher order AR($p+1$) model $Y_t = \sum_{j=1}^{p+1} \varphi_j Y_{t-j} + \varepsilon_t$, it is seen that the model can be expressed in an equivalent form as $W_t = (\rho - 1) Y_{t-1} + \sum_{j=1}^{p} \phi_j W_{t-j} + \varepsilon_t$, where $W_t = Y_t - Y_{t-1}$, $\rho - 1 = -\varphi(1) = \sum_{j=1}^{p+1} \varphi_j - 1$, and $\phi_j = -\sum_{i=j+1}^{p+1} \varphi_i$. Hence, the existence of a unit root in $\varphi(B)$ is equivalent to $\rho = \sum_{j=1}^{p+1} \varphi_j = 1$. Thus from the regression estimates of the model,

$$W_t = (\rho - 1) Y_{t-1} + \sum_{j=1}^{p} \phi_j W_{t-j} + \varepsilon_t, \tag{2.28}$$

it has been shown by Fuller (1996) [153, Chapter 10] that $(\hat{\rho}-1)/[s_\varepsilon(\sum_{t=p+2}^{T} Y_{t-1}^2)^{-1/2}]$ has the same limiting distribution as the statistic $\hat{\tau}$ in (2.27) for the AR(1) case.

The above methods extend to the case where a constant term is included in the least squares regression estimation, with the statistic analogous to $\hat{\tau}$ denoted as $\hat{\tau}_\mu$. Thus, for example, in the AR(1) model $Y_t = \phi Y_{t-1} + \theta_0 + \varepsilon_t$, one obtains the least squares estimate,

$$\hat{\phi}_\mu = \frac{\sum_{t=2}^{T}(Y_{t-1} - \overline{Y}_{(1)})(Y_t - \overline{Y}_{(0)})}{\sum_{t=2}^{T}(Y_{t-1} - \overline{Y}_{(1)})^2}, \tag{2.29}$$

where $\overline{Y}_{(i)} = (T-1)^{-1}\sum_{t=2}^{T} Y_{t-i}$, $i = 0, 1$. The corresponding test statistic for $\phi = 1$ in the AR(1) case is

$$\hat{\tau}_\mu = \frac{\hat{\phi}_\mu - 1}{s_\varepsilon\left(\sum_{t=2}^{T}(Y_{t-1} - \overline{Y}_{(1)})^2\right)^{-1/2}} \tag{2.30}$$

and percentiles of the distribution of $\hat{\tau}_\mu$ and when $\phi = 1$ are given in Table 2.2.

Table 2.2: Approximate Percentiles

| | Probability of a Smaller Value | | | | | | | |
|---|---|---|---|---|---|---|---|---|
| | 0.01 | 0.025 | 0.05 | 0.10 | 0.90 | 0.95 | 0.975 | 0.99 |
| $\hat{\tau}$ | −2.58 | −2.23 | −1.95 | −1.62 | 0.89 | 1.28 | 1.62 | 2.00 |
| $\hat{\tau}_\mu$ | −3.43 | −3.12 | −2.86 | −2.57 | −0.44 | −0.07 | 0.23 | 0.60 |

**Example 2.10 (To Difference or Not).** To illustrate here, we consider briefly the quarterly series of US Commercial Paper interest rates for 1953–1970 in Figure 2.3. The series is identified as ARIMA(1,1,0), and unconditional least squares estimation gave

$$W_t = (1 - B)\, Y_t = 0.4326\, W_{t-1}.$$

When conditional LS estimation of an AR(2) model for $Y_t$, as represented in (2.28), was performed, the results were $W_t = -0.0647 Y_{t-1} + 0.478 W_{t-1} + 0.277$, and the estimate $\hat{\rho} - 1 = -0.0147$ had estimated standard error of 0.0317. Clearly, the test statistic $\hat{\tau} = -2.04$ is not significant, and there is no cause to doubt the need for first differencing of the series $Y_t$ (i.e., that there is a unit root for $Y_t$). The use of the unit root test statistics will become apparent when we discuss pairs trading based on co-integration (see Chapter 5). A precondition for forming co-integrated pairs is that each series must be non-stationary on its own. ◁

## 2.8 Forecasting for ARIMA Processes

Given a realization of ARIMA$(p, d, q)$ process through time t, $\{Y_s, s \leq t\}$, we consider the problem of forecasting of future values $Y_{t+l}, l = 1, 2, \ldots$. For this purpose,

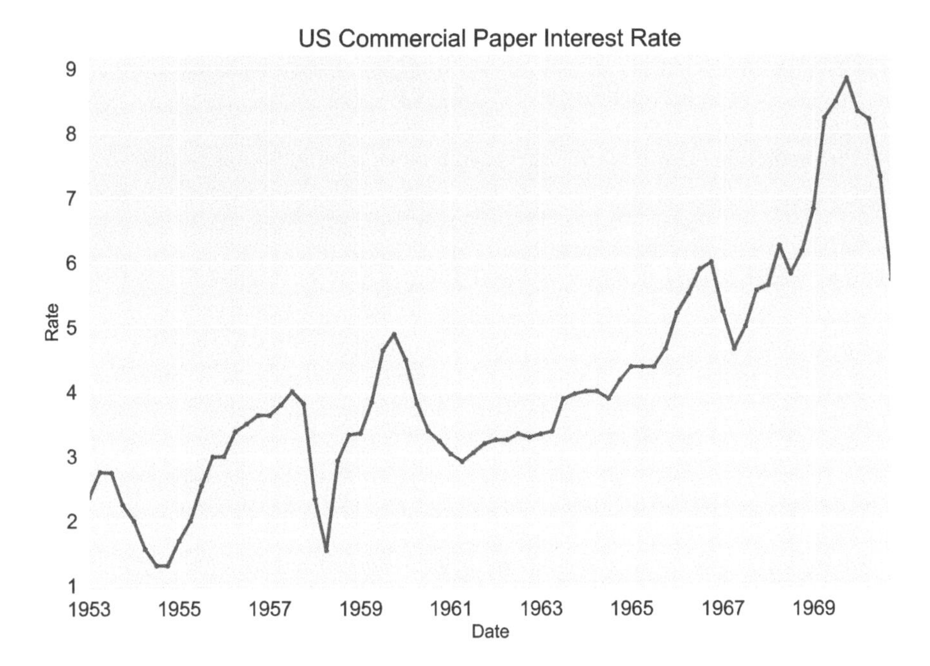


Figure 2.3: US Commercial Paper Interest Rate.

it is assumed that the model for $\{Y_t\}$ is known exactly including the values of the model parameters. Although, in practice, the model parameters are estimated from available sample data, errors due to estimation of parameters will not have too much effect on forecast properties for sufficiently large sample sizes. Also, the practical problem of future values based on the finite past sample data $Y_1, \ldots, Y_T$ will be considered briefly, but results based on the infinite past data assumption usually provide an adequate approximation to the finite past data prediction problem.

**Minimum Mean Squared Error Prediction**

Before presenting details of forecasting for ARIMA processes, we review some basic principles concerning prediction in the regression setting. Let $Y$ be a r.v. and let $X = (X_1, \ldots, X_p)'$ be an $p$-dimensional random vector related to $Y$, and consider the problem of predicting (estimating) the unknown value of $Y$ by some function of $X$, say $\hat{Y} = g(x)$. The mean squared error (MSE) of a predictor $\hat{Y}$ is $\mathrm{E}[(Y - \hat{Y})^2]$. We refer to the minimum mean squared error (minimum MSE) predictor of $Y$ as that function $\hat{Y} = g^*(X)$ such that, among all possible functions of $X$, $\hat{Y}$ minimizes the MSE $\mathrm{E}[(Y - \hat{Y})^2]$. Then it is well-known that the *minimum mean squared error* predictor is given by $\hat{Y} = \mathrm{E}(Y \mid X)$, the conditional expectation of $Y$ given $X$, with prediction error $e = Y - \hat{Y} = Y - \mathrm{E}(Y \mid X)$. One can also restrict prediction to consider only linear functions of $X$, and hence consider the minimum MSE *linear* predictor as the linear function $\hat{Y}^* = a + b'X$ which minimizes the prediction MSE among all linear functions of $X$. It is well-known that the minimum MSE linear predictor is

$$\hat{Y}^* = \mu_y + \Sigma_{yx}\,\Sigma_{xx}^{-1}(X - \mu_x) \tag{2.31}$$

with the prediction error $e^* = Y - \hat{Y}^*$ having mean zero and prediction MSE (variance)

$$\mathrm{Var}(e^*) = \mathrm{Var}(Y - \hat{Y}^*) = \sigma_y^2 - \Sigma_{yx}\,\Sigma_{xx}^{-1}\,\Sigma_{xy}, \tag{2.32}$$

where $\mu_y = \mathrm{E}(Y)$, $\mu_x = \mathrm{E}(X)$, $\sigma_y^2 = \mathrm{Var}(Y)$, $\Sigma_{yx} = \mathrm{Cov}(Y, X)$, and $\Sigma_{xx} = \mathrm{Cov}(X)$. Note $\Sigma_{yx}$ is a $1{\times}p$ row vector and $\Sigma_{xx}$ is a $p{\times}p$ variance-covariance matrix. Moreover, if the prediction error $e^* = Y - \hat{Y}^*$ of the best linear predictor is such that $\mathrm{E}(e^* \mid X) = 0$ (e.g., if $e^*$ is independent of $X$), then $\hat{Y}^*$ is also the minimum MSE predictor, i.e., $\hat{Y}^* = \hat{Y} = \mathrm{E}(Y \mid X)$ is a linear function of $X$ and the prediction error $e = e^* = Y - \hat{Y}$ has variance $\mathrm{Var}(e^*)$ as given above in (2.32).

**Forecasting for ARIMA Processes and Properties of Forecast Errors**

For forecasting in the time series setting, we assume that the process $\{Y_t\}$ follows an ARIMA$(p, d, q)$ model, $\phi(B)(1 - B)^d\, Y_t = \theta(B)\varepsilon_t$, and we assume the white noise series $\varepsilon_t$ are mutually independent random variables. We are interested in forecasting the future value $Y_{t+l}$ based on observations $Y_t, Y_{t-1}, \cdots$. From the result in the previous section, the minimum MSE forecast of $Y_{t+l}$ based on $Y_t, Y_{t-1}, \ldots$, which we will denote as $\hat{Y}_t(l)$, is such that $\hat{Y}_t(l) = \mathrm{E}(Y_{t+l} \mid Y_t, Y_{t-1}, \ldots)$. The prediction $\hat{Y}_t(l)$ is called the lead $l$ or $l$-step ahead forecast of $Y_{t+l}$, $l$ is the lead time, and $t$ is the forecast origin.

To obtain a representation for $\hat{Y}_t(l)$, recall the ARIMA process has the "infinite" MA form $Y_t = \psi(B)\varepsilon_t = \sum_{i=0}^{\infty} \psi_i \varepsilon_{t-i}$, and hence a future value $Y_{t+l}$ at time $t + l$, relative to the current time or "forecast origin" t, can be expressed as

$$Y_{t+l} = \sum_{i=0}^{\infty} \psi_i \varepsilon_{t+l-i} = \sum_{i=0}^{l-1} \psi_i \varepsilon_{t+l-i} + \sum_{i=l}^{\infty} \psi_i \varepsilon_{t+l-i}. \tag{2.33}$$

The information contained in the past history of the $Y_t's$, $\{Y_s, s \le t\}$, is the same as that contained in the past random shocks $\varepsilon_t$'s (because the $Y_t$'s are generated by the $\varepsilon_t$'s). Also, $\varepsilon_{t+h}$, for $h > 0$ is independent of present and past values $Y_t, Y_{t-1}, \ldots$, so that $\mathrm{E}(\varepsilon_{t+h} \mid Y_t, Y_{t-1}, \ldots) = 0, h > 0$. Thus

$$\hat{Y}_t(l) = \mathrm{E}(Y_{t+l} \mid Y_t, Y_{t-1}, \ldots) = \mathrm{E}(Y_{t+l} \mid \varepsilon_t, \varepsilon_{t-1}, \ldots) = \sum_{i=l}^{\infty} \psi_i \varepsilon_{t+l-i}, \tag{2.34}$$

using the additional property that $\mathrm{E}(\varepsilon_{t+l-i} \mid \varepsilon_t, \varepsilon_{t-1}, \ldots) = \varepsilon_{t+l-i}$ if $i \ge l$. The $l$-step ahead prediction error is given by $e_t(l) = Y_{t+l} - \hat{Y}_t(l) = \sum_{i=0}^{l-1} \psi_i \varepsilon_{t+l-i}$. So we have $\mathrm{E}[e_t(l)] = 0$ and the mean squared error or variance of the $l$-step prediction error is

$$\sigma^2(l) = \mathrm{Var}(e_t(l)) = \mathrm{E}[e_t^2(l)] = \mathrm{Var}\left(\sum_{i=0}^{l-1} \psi_i \varepsilon_{t+l-i}\right) = \sigma^2 \sum_{i=0}^{l-1} \psi_i^2, \tag{2.35}$$

with St.Dev. $(e_t(l)) = \sigma(l) = \sigma(1+\psi_1^2+\cdots+\psi_{l-1}^2)^{1/2}$. Note, in particular, that for $l = 1$ step we have $e_t(1) = Y_{t+1} - \hat{Y}(1) = \varepsilon_{t+1}$ with error variance $\sigma^2 = \text{Var}(\varepsilon_{t+1})$, so that the white noise series $\varepsilon_t$ can also be interpreted as the one-step ahead forecast errors for the process. Under the normality assumption that the $\varepsilon_t$ are normally distributed as i.i.d. $N(0, \sigma^2)$, the $l$-step ahead forecast errors $e_t(l) = Y_{t+l} - \hat{Y}_t(l) = \sum_{i=0}^{l-1} \psi_i \varepsilon_{t+l-i}$ will also be normally distributed with mean 0 and variance $\sigma^2(l) = \sigma^2 \sum_{i=0}^{l-1} \psi_i^2$. Hence, the conditional distribution of $Y_{t+l}$, given $Y_t, Y_{t-1}, \ldots$, is normal with mean $\hat{Y}_t(l)$ and variance $\sigma^2(l)$.

For forecasting purposes, the variance $\sigma^2(l)$ of the $l$-step forecast error provides a measure of the degree of accuracy of the point forecast $\hat{Y}_t(l)$. Along with the point forecast $\hat{Y}_t(l)$, we need to provide prediction intervals for the future observations $Y_{t+l}$ so that the degree of uncertainty of the point forecasts is properly reflected. From the preceding paragraph, it follows that the standardized variable $Z_t(l) = e_t(l)/\sigma(l) = (Y_{t+l} - \hat{Y}_t(l))/\sigma(l)$ is distributed as standard normal $N(0, 1)$, which is also the conditional distribution of this variable, given $Y_t, Y_{t-1}, \ldots$. Then, for example, from $0.90 = R[-1.65 < Z_t(l) < 1.65] = P[\hat{Y}_t(l) - 1.65\,\sigma(l) < Y_{t+l} < \hat{Y}_t(l) + 1.65\,\sigma(l)]$, it follows that a 90% prediction interval (interval forecast) for $Y_{t+l}$ is given by $\hat{Y}_t(l) \pm 1.65\sigma(l)$, where $\sigma(l) = \sigma(\sum_{i=0}^{l-1} \Psi_i^2)^{1/2}$. Prediction intervals for other desired levels of confidence, $100(1 - \alpha)\%$, can be constructed similarly. Of course, in practice, estimates of $\sigma^2$ and of the weights $\psi_i$ based on the estimates $\hat{\phi}_i$ and $\hat{\theta}_i$ of parameters in the ARIMA model obtained from the sample data will actually be used in estimating the standard deviation $\sigma(l)$ of the forecast errors and in forming the prediction intervals.

**Example 2.11 (IMA(1,1) Model).** Let $Y_t - Y_{t-1} = \delta + \varepsilon_t - \theta\varepsilon_{t-1}$. Then, from (2.33)–(2.34), the forecasts are $\hat{Y}_t(1) = Y_t + \delta - \theta\varepsilon_t$, $\hat{Y}_t(2) = Y_t(1) + \delta = Y_t - \theta\varepsilon_t + 2\delta$, and in general,

$$\hat{Y}_t(l) = \hat{Y}_t(l-1) + \delta = Y_t - \theta\varepsilon_t + \delta(l) = \hat{Y}_t(1) + \delta(l-1), \quad l = 1, 2, \ldots$$

So the "eventual" forecast function $\hat{Y}_t(l)$ is a straight line with a deterministic (fixed) slope $\delta$, and an "adaptive" intercept $\hat{Y}_t(1) - \delta = Y_t - \theta\varepsilon_t$ which is random and is adaptive to the process values which occur through time t. The variance of the $l$-step ahead forecast error $e_t(l)$ is $\sigma^2(l) = \sigma^2(1 + (l-1)(1-\theta)^2), l \geq 1$. Hence, we see that the values $\sigma^2(l)$ increase without bound as the lead time $l$ increases, which is a general feature for (nonstationary) ARIMA processes with $d > 0$. Also notice from the infinite AR form of the model (with $\delta = 0$) with AR coefficients $\pi_i = (1-\theta)\theta^{i-1}$, $i \geq 1$, $Y_{t+1} = (1-\theta)\sum_{i=1}^{\infty} \theta^{i-1} Y_{t+1-i} + \varepsilon_{t+1}$, we have the representation $\hat{Y}_t(1) = (1-\theta)\sum_{i=1}^{\infty} \theta^{i-1} Y_{t+1-i}$. This provides the exponential smoothing, or exponentially weighted moving average (EWMA), form of the one-step forecast $\hat{Y}_t(1)$ in the IMA(1,1). This model with a short term drift '$\delta$' is of immense interest in modeling the stock prices. ◁

## 2.9 Stylized Models for Asset Returns

It is generally known that the equity market is extremely efficient in quickly absorbing information about stocks: When new information arrives, it gets incorporated in the price without delay. Thus the efficient market hypothesis is widely accepted by financial economists. It would imply that neither technical analysts that study the past prices in an attempt to predict future prices nor fundamental analysts of financial information related to company earnings and asset values would carry any advantage over the returns obtained from a randomly selected portfolio of individual stocks.

The efficient market hypothesis is usually associated with the idea of the "random walk," which implies that all future price changes represent random departures from past prices. As stated in Malkiel (2012) [258]: "The logic of the random walk is that if the flow of information is unimpeded and information is immediately reflected in stock prices, then tomorrow's price change will reflect only tomorrow's news and will be independent of the price changes today. But news by definition is unpredictable, and, thus resulting price changes must be unpredictable and random."

The random walk (RW) model (Example 2.3) without the drift term can be stated as:

$$p_t = p_{t-1} + \varepsilon_t, \tag{2.36}$$

where $\varepsilon_t \sim N(0, \sigma^2)$ i.i.d. and $p_t = \ln(P_t)$. Note as observed earlier, this model is a particular case of AR(1) model if the constant term is assumed to be zero and the slope, $\phi$, is assumed to be one. Thus, the RW model is a non-stationary model and considering $\varepsilon_t = p_t - p_{t-1} = r_t$ is the differencing of the series, $p_t$, makes the series $\varepsilon_t$ stationary. Note $r_t \approx \frac{P_t - P_{t-1}}{P_{t-1}}$, the returns are purely random and are unpredictable. For any chronological data the decision concerning the need for differencing is based, informally, on the features of time series plot of $p_t$, its sample autocorrelation function; that is, its failure to dampen out sufficiently quickly.

**Testing for a RW Model:** Consider the AR(1) model $p_t = \phi p_{t-1} + \varepsilon_t$, $t = 1, \ldots, T$ and $p_0$ is fixed; we want to test $\phi = 1$. The details of this test are given in Section 2.7, along with the table of the percentiles of the distribution of the test statistics. While the results for testing for non-stationarity for higher order autoregressive models were discussed earlier, the price series once differenced does not generally exhibit autocorrelation structure that requires modeling beyond AR(1).

It will become obvious in subsequent discussions, that in order to exploit any pattern in the returns, one has to look beyond linear time dependencies. We observe a series of established stylized facts regarding stock prices (returns).

**Stylized Fact 1:** Absence of Autocorrelation in Returns: Linear autocorrelations are often insignificant for a lower frequency data.

To illustrate the main ideas, we consider the exchange rate between USD and GBP; these are daily closing rates from January 4, 1999 to September 16, 2013.

The time series plot of the data in Figure 2.4 clearly indicates that the price is non-stationary and it is confirmed by the never dying autocorrelation function in (Figure 2.4). The LS estimates $\hat{\phi} = 0.9996$, $s_e = 0.00958$ and $\sum_{t=2}^{T} p_{t-1}^2 = 10431$, provide a value of $\hat{\tau} = -0.001$ from (2.27) which confirms that the series is a random walk. The differenced series, $r_t = p_t - p_{t-1}$ is plotted in Figure 2.4 and its autocorrelation function is given in Figure 2.4. Thus it would appear that return series is a white noise process.

Many industry analysts tend to look at not just the closing price but also the price bars that consist of open, high, low and close prices in discrete time intervals. Select averages over these four quantities are monitored to study the stock momentum. Two such quantities discussed in the literature are pivot, the average of high, low and close prices; and mid-range, the average of high and low prices. The plot of the pivot for the exchange rate is overlaid on the closing price in Figure 2.5. It is not easy to distinguish between the two series from this plot, but an important property of the (aggregated) pivot series is worth noting here. The autocorrelation of the returns dated on pivot series at lag 1 is quite significant (Figure 2.6). This happens mainly due to 'aggregation' of data as observed by Working (1960) [326] and by Daniels (1966) [101]. Working (1960) [326] shows that if the average is computed over '$m$' terms in each time segment (in this illustration, a day) the first order correlation is $\frac{(m^2-1)}{2(2m^2+1)}$. When $m = 3$, we can expect the lag 1 autocorrelation to be about 0.21. Daniels (1966) [101] observes that we could expect lag 1 autocorrelation in the order of 0.4 for the mid-range series. Thus, the main point of this discussion is that the aggregation of prices (which is the basis of several low-frequency analyses) induces the autocorrelation by construction, but it does not mean that there is market inefficiency.

An assumption of the random walk model that is often made is that the errors (returns) follow a normal distribution. But in practice, the errors (or returns) tend to have heavier tails than the normal. We suggest using the quantile-quantile plot which is quite sensitive to even modest deviations from normality and also a test based on the properties of normal distribution. An omnibus test based on the skewness (= 0 for normal) and the kurtosis (= 3 for normal) is the Jarque-Bera test:

$$\text{JB} = n\left(\frac{\widehat{\text{SK}}^2}{6} + \frac{(\hat{K}-3)^2}{24}\right) \sim \chi_2^2, \tag{2.37}$$

where SK(ewness) $= E(\frac{X-\mu}{\sigma})^3$, K(urtosis) $= E(\frac{X-\mu}{\sigma})^4$. The corresponding sample numbers are substituted in JB (see Jarque and Bera (1980) [213]). Alternatively, the probability (Q–Q) plot of $r_t = \ln(p_t) - \ln(p_{t-1})$ as given in Figure 2.4 clearly indicates that the distribution has somewhat heavy tails. (For this data $\widehat{SK} = -0.31$ and $\hat{K} = 3.67$ resulting in JB = 129.7 with $n = 3236$, thus rejecting the normal distribution.)

**Stylized Fact 2:** Heavy Tails: The returns are likely to display a heavy tailed distribution such as a power-law or Pareto.

**Stylized Fact 3:** Asymmetry: When prices fall, they tend to fall faster than when they tend to rise, thus drawdowns tend to be larger than upward rises.

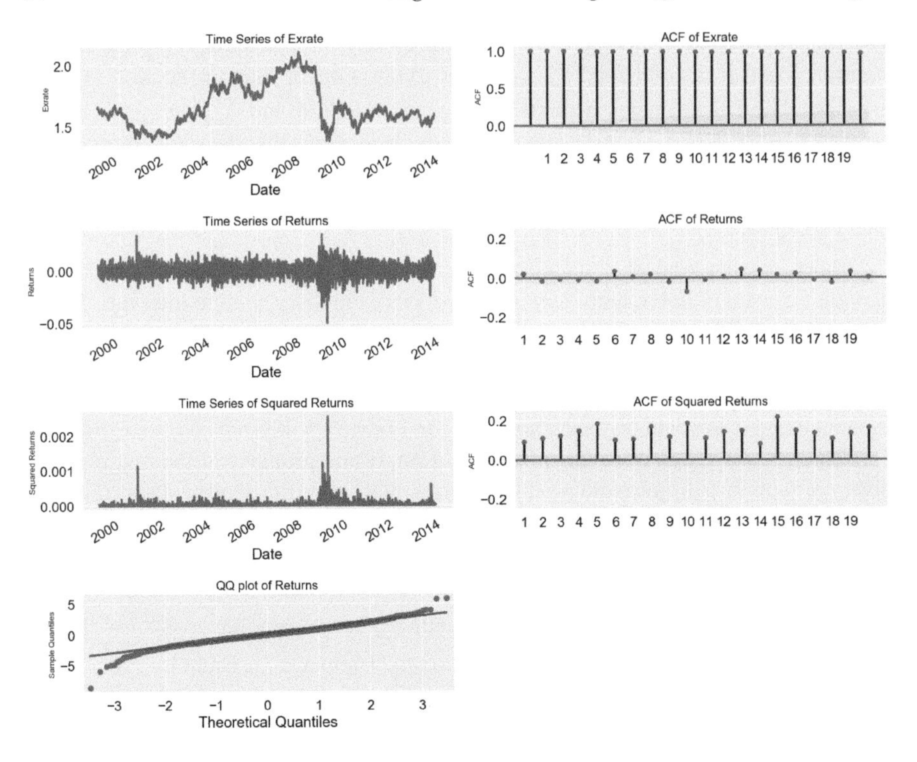


Figure 2.4: Exchange Rate Related Plots.

**Stylized Fact 4:** Positive Excess Kurtosis: Returns generally exhibit high values for the fourth moment due to different phases of high and low trading activity.

These stylized facts are particularly important for risk management of strategies, and how they should be optimally incorporated in trading decisions is still a subject of much research which will be discussed later. The thicker tail distribution warrants that the chance of extreme values is more likely to occur than what is predicted by a normal distribution and the asymmetry indicates that there is a difference between positive and negative sides of variance (volatility).

## 2.10 Time Series Models for Aggregated Data: Modeling the Variance

In the ARMA$(p, q)$ model $\phi(B)Y_t = \delta + \theta(B)\,\varepsilon_t$ for series $\{Y_t\}$, when the errors $\varepsilon_t$ are *independent* r.v.'s (with the usual assumptions of zero mean and constant variance $\sigma^2$), an implication is that the *conditional* variance of $\varepsilon_t$ given the past information

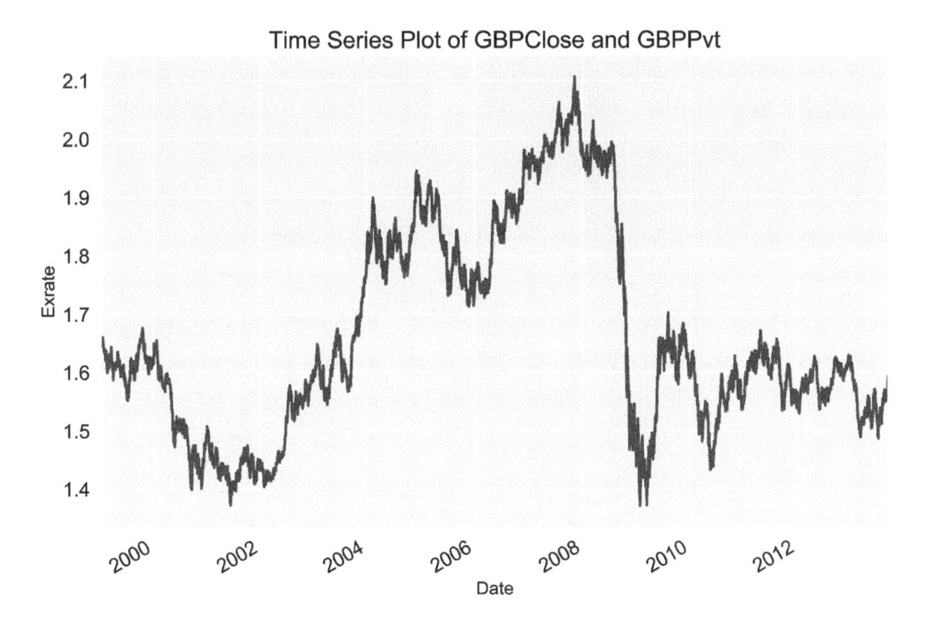


Figure 2.5: Times Series Plot of GBPClose, GBPPvt.

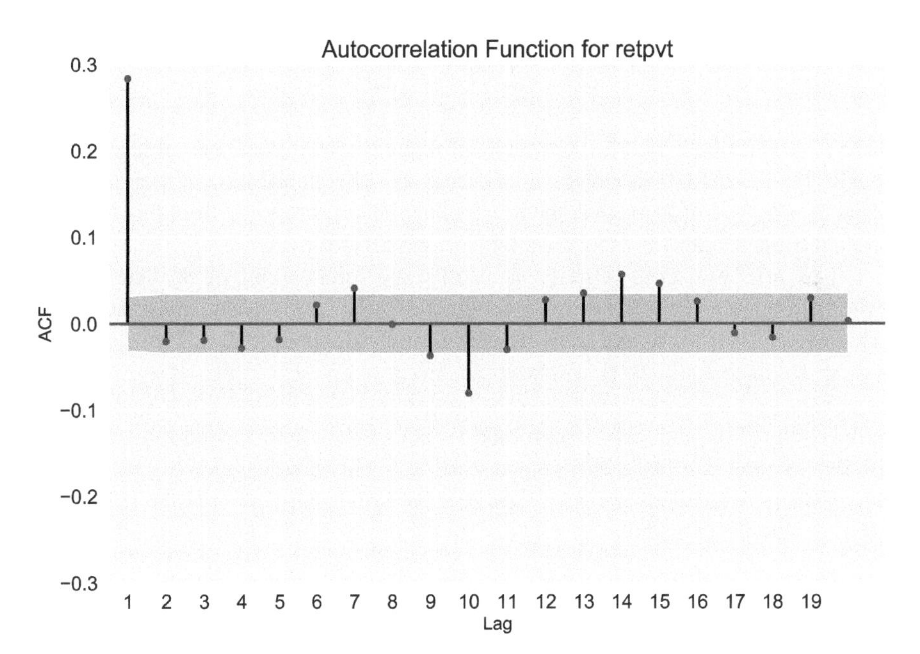


Figure 2.6: Autocorrelation Function for RetPvt.

is a constant not depending on the past. This, in turn, implies the same feature for $l$-step ahead forecast errors $e_t(l) = \sum_{i=0}^{l-1} \psi_i \varepsilon_{t+l-i}$. However, in some settings, particularly for financial data, the variability of errors may exhibit dependence on the past variability. Modeling the variance, which is essential for studying the risk-return relationship is an important topic in finance.

**Autoregressive Conditional Heteroscedastic (ARCH) Models**

The autoregressive conditional heteroscedastic (ARCH) models were originally proposed by Engle (1982) [120], Engle and Kraft (1983) [124], to allow for the conditional error variance in an ARMA process to depend on the past squared innovations, rather than be a constant as in ARMA models with independent errors. For example, an AR($p$) process with ARCH($q$) model errors is given as $Y_t = \sum_{i=1}^{p} \phi_i Y_{t-i} + \delta + \varepsilon_t$, where, $\mathrm{E}(\varepsilon_t \mid \varepsilon_{t-1}, \varepsilon_{t-2}, \dots) = 0$ and

$$h_t = \mathrm{Var}(\varepsilon_t \mid \varepsilon_{t-1}, \varepsilon_{t-2}, \dots) = \omega_0 + \sum_{i=1}^{q} \omega_i \varepsilon_{t-i}^2 \tag{2.38}$$

with $\omega_0 > 0$ and $\omega_i \geq 0$ for $i = 1, \dots, q$. In the generalized ARCH (GARCH) model, introduced by Bollerslev (1986) [46], it is assumed that

$$h_t = \omega_0 + \sum_{i=1}^{q} \omega_i \varepsilon_{t-i}^2 + \sum_{i=1}^{k} \beta_i h_{t-i}, \tag{2.39}$$

where $\beta_i \geq 0$ for all $i = 1, \dots, k$. Much subsequent research on and applications of the ARCH and GARCH models have occurred since the publication of their research papers. It is not possible to cover the range of applications resulted from this research here, but these results are widely available to the curious reader.

Let us briefly discuss some basic results and implications of the ARCH and GARCH models. The errors $\varepsilon_t$ in the model have zero mean, since $\mathrm{E}(\varepsilon_t) = \mathrm{E}[\mathrm{E}(\varepsilon_t \mid \varepsilon_{t-1} \dots)] = 0$, and they are serially uncorrelated; that is, $\mathrm{E}(\varepsilon_t \varepsilon_{t-j}) = 0$ for $j \neq 0$ (since for $j > 0$, for example, $\mathrm{E}(\varepsilon_t \varepsilon_{t-j}) = \mathrm{E}[\mathrm{E}(\varepsilon_t \varepsilon_{t-j} \mid \varepsilon_{t-1} \dots)] = \mathrm{E}[\varepsilon_{t-j}\, \mathrm{E}(\varepsilon_t \mid \varepsilon_{t-1} \dots)] = 0$). But the $\varepsilon_t$ are not mutually independent r.v.'s since they are inter-related through their conditional variances (i.e., their second moment). We will also assume that the $\varepsilon_t$ have equal unconditional variances, $\mathrm{Var}(\varepsilon_t) = \sigma^2$, for all $t$, so they are weakly stationary. Consider then, the simple case of the first-order ARCH or ARCH(1) model,

$$h_t = \omega_0 + \omega_1 \varepsilon_{t-1}^2. \tag{2.40}$$

Assuming that $\omega_1 < 1$ ($\omega_0$ and $\omega_1 \geq 0$ is always assumed), the *unconditional* variance of $\varepsilon_t$ is $\sigma^2 = \mathrm{Var}(\varepsilon_t) = \frac{\omega_0}{1-\omega_1}$, since $\sigma^2 = \mathrm{E}(\varepsilon_t^2) = \mathrm{E}[\mathrm{E}(\varepsilon_t^2 \mid \varepsilon_{t-1}, \dots)] = \mathrm{E}[\omega_0 + \omega_1 \varepsilon_{t-1}^2] = \omega_0 + \omega_1 \sigma^2$. Therefore, the conditional variance of $\varepsilon_t$ can be written as

$$h_t = \omega_0 + \omega_1 \varepsilon_{t-1}^2 = \sigma^2 + \omega_1(\varepsilon_{t-1}^2 - \sigma^2), \tag{2.41}$$

or in the form of deviation from the unconditional variance as $h_t - \sigma^2 = \omega_1(\varepsilon_{t-1}^2 - \sigma^2)$. So the conditional variance will be above the unconditional variance whenever $\varepsilon_{t-1}^2$ exceeds its unconditional variance $\sigma^2$. If $\varepsilon_t$ were conditionally normally distributed, given the past, the fourth unconditional moment $\mathrm{E}(\varepsilon_t^4)$ will exceed $3\sigma^4$ (the value in the normal distribution), so that the marginal distribution of $\varepsilon_t$ will exhibit fatter tails than the normal distribution. Conditional variances of multi-step ahead forecast errors can also be established to depend on the past squared errors.

These basic results indicate that the properties of the ARCH(1) model also tend to hold for the higher order ARCH($q$) models. A basic impact of the ARCH errors in a process is an assessment of the accuracy of forecast, since the forecast errors will have conditional variances that depend on the past. This allows for formation of correct and more informative (conditional) prediction intervals than that would be obtained under the usual assumption that conditional error variances were constant independent of the past.

Now consider briefly the popular GARCH(1,1) model, $h_t = \mathrm{E}(\varepsilon_t^2 \mid \varepsilon_{t-1}, \ldots) = \omega_0 + \omega_1 \varepsilon_{t-1}^2 + \beta_1 h_{t-1}$. Similar to the result for the ARCH(1) model, assuming that $\omega_1 + \beta_1 < 1$, it is readily shown that the unconditional variance of $\varepsilon_t$ is equal to $\sigma^2 = \mathrm{Var}(\varepsilon_t) = \omega_0/[1 - (\omega_1 + \beta_1)]$. Let $v_t = \varepsilon_t^2 - h_t$, so that the random variables $v_t$ have zero mean, and they are serially uncorrelated since

$$E\big((\varepsilon_t^2 - h_t)(\varepsilon_{t-j}^2 - h_{t-j})\big) = \mathrm{E}[(\varepsilon_{t-j}^2 - h_{t-j})E\{(\varepsilon_t^2 - h_t) \mid \varepsilon_{t-1}, \ldots\}] = 0.$$

Then, since $h_t = \varepsilon_t^2 - v_t$, note that the GARCH(1,1) model can be rearranged as $\varepsilon_t^2 - v_t = \omega_0 + \omega_1 \varepsilon_{t-1}^2 + \beta_1(\varepsilon_{t-1}^2 - v_{t-1})$, or

$$\varepsilon_t^2 = \omega_0 + (\omega_1 + \beta_1)\, \varepsilon_{t-1}^2 + v_t - \beta_1 v_{t-1}. \tag{2.42}$$

This form reveals that the process of squared errors, $\{\varepsilon_t^2\}$, follows an ARMA(1,1) model with uncorrelated innovations $v_t$; however, the $v_t$ are conditionally heteroscedastic. Although this process looks like a linear process, it is not, since $\{v_t\}$ are uncorrelated but dependent. This fact motivates the use of the ACF and PACF of the squares $\varepsilon_t^2$ for model specification and for basic preliminary checking for the presence of ARCH/GARCH effects in the errors $\varepsilon_t$. In practice, a starting point would be examination of the sample ACF and PACF of the squared residuals $\hat{\varepsilon}_t^2$, where $\hat{\varepsilon}_t$ are residuals obtained from fitting of a usual ARMA($p, q$) model. The correspondence between ARCH and ARMA is further noted via the condition, $0 < \sum_{i=1}^{q} \omega_i < 1$, that implies the roots of its characteristic equation lie outside the unit circle (like a causal ARMA).

As there are other excellent sources to learn about ARCH and GARCH models (see Tsay (2010) [315], Chapter 2), we have limited our discussion to some basic details. For maximum likelihood (ML) estimation of models with ARCH or GARCH errors, refer to Engle (1982) [120] and Weiss (1984) [321], among others.

**Preliminary Testing for ARCH Effect:** This is an omnibus test that examines the presence of autocorrelation up to a certain number of lags, which is equivalent to testing, $H_0 : \alpha_1 = \alpha_2 = \cdots = \alpha_q = 0$ in the model

$$\varepsilon_t^2 = \alpha_0 + \alpha_1 \varepsilon_{t-1}^2 + \cdots + \alpha_q \varepsilon_{t-q}^2 + a_t, \tag{2.43}$$

where $a_t$ is a white noise series; note that $\{a_t\}$ does not need to follow a normal distribution. The usual $F$-test used in regression can be applied here; Let $R^2$ be the coefficient of determination that results from (2.43) when $\hat{\varepsilon}_t$'s are used; then

$$F = \frac{R^2/q}{(1 - R^2)/(T - 2q - 1)} \sim \chi^2/q. \tag{2.44}$$

We use $\chi^2$ instead of $F$ due to the fact that $\varepsilon_t$'s are estimated quantities. More informally, one could use the ACF of $\hat{\varepsilon}_t^2$ and if any of the autocorrelations are out of the normal bounds, $\pm 2/\sqrt{T}$, we may examine further for ARCH effects.

**GARCH vs. ARCH:** While ARCH models care for the time dependence in volatilities and have elegant properties such as having excessive kurtosis (matching the empirical finding related to asset returns), they tend to overpredict the volatility and treat both negative and positive, $\varepsilon_t$'s symmetrically. To overcome some of these issues and as well as to come up with a more parsimonious representation (as in AR vs. ARMA considerations) GARCH models stated in (2.42) are proposed.

**IGARCH:** The variance can persist for extended periods and can change at different time spans, thus leading to non-stationary behavior. In the GARCH(1,1) model, it is possible that $w_1 + \beta_1 = 1$, so that IGARCH(1,1) (Integrated GARCH) can be written as

$$\varepsilon_t = \sqrt{h_t} \cdot a_t, \quad h_t = w_0 + w_1 h_{t-1} + (1 - w_1)\, \varepsilon_{t-1}^2. \tag{2.45}$$

If $h_t$ is proxied by $\varepsilon_t^2$, the above model leads to $\varepsilon_t^2 - \varepsilon_{t-1}^2 = w_0 + w_1(\varepsilon_{t-1}^2 - \varepsilon_{t-2}^2)$, which is similar in form to a non-stationary autoregressive model. In practice because of frequent level shifts in volatility, IGARCH appears to be a natural model to fit. Because the IGARCH(1,1) is similar to ARIMA(0,1,1), the model can be taken as an exponential smoothing model for the $\varepsilon_t^2$ series; By iterative substitutions, we can show that

$$h_t = (1 - w_1)\big(\varepsilon_{t-1}^2 + w_1 \varepsilon_{t-2}^2 + w_1^2 \varepsilon_{t-3}^2 \cdots \big), \tag{2.46}$$

where $w_1$ can be taken as a discount factor.

**Prediction Performance of GARCH:** The predictive power of GARCH based on ex-post forecast evaluation criteria tends to suggest that the model provides poor forecasts, although the in-sample parameter estimates turn out to be highly significant. As Andersen and Bollerslev (1998) [18] note, in the model (2.42), the latent volatility, $h_t$, evolves over time. The approach to evaluate the performance of the model for $h_t$ via $\hat{\varepsilon}_t^2$ is not appropriate. While $\hat{\varepsilon}_t^2$ provides an unbiased estimate of $h_t$, it is a noisy measurement due to idiosyncratic error terms associated with it; these error terms may be the result of compounding frictions in the trading process. This will be further clarified when we discuss the analysis of high frequency data.

**Time-Varying ARCH Processes (tvARCH)**

The underlying assumption of ARCH models is stationarity and with the changing pace of economic conditions, the assumption of stationarity is not appropriate for modeling financial returns over long intervals. We may obtain a better fit by relaxing the assumption of stationarity in all the time series models. It is also appropriate for building models with local variations. Dahlhaus and Subba Rao (2006) [99] generalize the ARCH model with time-varying parameters:

$$\varepsilon_t = \sqrt{h_t} \cdot a_t, \quad h_t = w_0(t) + \sum_{j=1}^{\infty} w_j(t)\varepsilon_{t-j}^2, \tag{2.47}$$

which $a_t$'s are i.i.d. with mean zero and variance, one. By rescaling the parameters to unit intervals, the tvARCH process can be approximated by a stationary ARCH process. A broad class of models resulting from (2.47) can be stated as:

$$\varepsilon_{t,N} = \sqrt{h_{t,N}} \cdot a_t, \quad h_{t,N} = w_0\left(\frac{t}{N}\right) + \sum_{j=1}^{p} w_j\left(\frac{t}{N}\right)\varepsilon_{t-j,N}^2 \tag{2.48}$$

for $t = 1, 2, \ldots, N$. This model captures the slow decay of the sample autocorrelations in squared returns that is commonly observed in financial data which is attributed to the long memory of the underlying process. But tvARCH($p$) is a non-stationary process that captures the property of long memory.

Fryzlewicz, Sapatinas and Subba Rao (2008) [152] propose a kernel normalized-least squares estimator which is easy to compute and is shown to have good performance properties. Rewriting (2.48) as

$$\varepsilon_{t,N}^2 = w_0\left(\frac{t}{N}\right) + \sum_{j=1}^{p} w_j\left(\frac{t}{N}\right)\varepsilon_{t-j,N}^2 + (a_t^2 - 1)h_{t,N}^2 \tag{2.49}$$

in the autoregressive form, the least squares criterion with the weight function, $k(u_0, \chi_{k-1,N})$, where $\chi'_{k-1,N} = (1, \varepsilon_{k-1,N}^2, \ldots, \varepsilon_{k-p,N}^2)$ is

$$L_{t_0,N}(\alpha) = \sum_{k=p+1}^{N} \frac{1}{b_N} w\left(\frac{t_0 - k}{b_N}\right) \frac{\left(\varepsilon_{k\cdot N}^2 - \alpha_0 - \sum_{j=1}^{p} \alpha_j \varepsilon_{k-j\cdot N}^2\right)^2}{k(u_0, \chi_{k-1\cdot N})^2}. \tag{2.50}$$

Minimizing (2.50) yields, $\hat{a}_{t_0,N} = R_{t_0,N}^{-1} r_{t_0,N}$, where

$$R_{t_0,N} = \sum_{k=p+1}^{N} \frac{1}{b_N} w\left(\frac{t_0 - k}{b_N}\right) \frac{\chi_{k-1\cdot N}\chi'_{k-1\cdot N}}{k(u_0, \chi_{k-1\cdot N})^2}$$

and (2.51)

$$r_{t_0,N} = \sum_{k=p+1}^{N} \frac{1}{b_N} w\left(\frac{t_0 - k}{b_N}\right) \frac{\varepsilon_{k\cdot N}^2 \chi_{k-1\cdot N}}{k(u_0, \chi_{k-1\cdot N})^2}.$$

The kernel $w$ is a function defined on $[-\frac{1}{2}, \frac{1}{2}]$ which is of bounded variation. The choice of weight function suggested is as follows: It is motivated by the differing variances at various values of $k$. Estimate,

$$\hat{\mu}_{t_0,N} = \sum_{k=1}^{N} \frac{1}{b_N} w \left( \frac{t_0 - k}{b_N} \right) \varepsilon_{k \cdot N}^2$$

and (2.52)

$$s_{k-1 \cdot N} = \sum_{j=1}^{p} \varepsilon_{k-j \cdot N}^2$$

and use $k_{t_0,N}(s_{k-1}, N) = \hat{\mu}_{t_0,N} + s_{k-1 \cdot N}$ as the weight function. Thus the estimation is a two-stage scheme. For the bandwidth selection, the cross-validation method based on a subsample of the observations is suggested.

The methodology presented here is useful because of the changing nature of financial volatility. Real data applications will be discussed in Chapter 5. Here the volatility at the low frequency level is estimated by $r_t^2$, where $r_t$, the returns are based on closing prices of successive days. Alternative approaches to estimating volatility using aggregate prices bars or high frequency data will be discussed in Chapter 4.

## 2.11 Stylized Models for Variance of Asset Returns

Volatility has many important implications in finance. It is one of the most commonly used risk measures and plays a vital role in asset allocation. The estimate of volatility of an asset is obtained using the prices of stock or options or both. Three different measures that are normally studied are stated below:

- Volatility is the conditional standard deviation of daily or low frequency returns.
- Implied volatility is derived from options prices under some assumed relationship between the options and the underlying stock prices.
- Realized volatility is an estimate of daily volatility using high frequency intraday returns.

In this section, we will mainly focus on the first item and the others will be discussed in a later section on high frequency data.

Consider $r_t = \ln(P_t) - \ln(P_{t-1})$, return of an asset. We observed that '$r_t$' exhibits no serial correlation. This does not imply that the series '$r_t$' consists of independent observations. The plot of $r_t^2$ given in Figure 2.4 clearly indicates that volatility tends to cluster over certain time spans. The autocorrelations in Figure 2.4 confirm that there is some time dependence for the volatility. In fact in some cases, there is some long range dependence that spans up to sixty days.

**Stylized Fact 5:** Volatility Clustering: High volatility events tend to cluster in time and this can be measured through positive correlations that exist over several lags.

To formally model the volatility, recall the conditional mean and the conditional variance of the return, $r_t$, are: $\mu_t = \mathrm{E}(r_t \mid F_{t-1})$, $\sigma_t^2 = \mathrm{Var}(r_t \mid F_{t-1}) = \mathrm{E}((r_t - \mu_t)^2 \mid F_{t-1})$. We have observed for the exchange rate series that $\mu_t \approx$ constant and is generally close to zero. But $\sigma_t^2$ clusters around certain time spans and modeling variation in $\sigma_t^2$ is typically done through ARCH/GARCH models and also through stochastic volatility models. The stochastic volatility models have the advantage of modeling jumps in prices that are also often used in derivative pricing. To begin with, we can test for the heteroscedasticity using the test in (2.44) or one of the following which are all equivalent asymptotically:

- **Ljung-Box Omnibus Test:**

$$Q_k = T(T+2) \cdot \sum_{j=1}^{k} \frac{(\hat{e}^j)^2}{T-j} \sim \chi^2_{k-m}, \tag{2.53}$$

  where $\hat{e}^{(j)}$ is the $j$th lag autocorrelation of $r_t^2$; $m$ is the number of independent parameters. Here we test the null hypothesis, $H_0 : \rho_1 = \rho_2 = \cdots \rho_k = 0$.

- **Lagrange Multiplier Test:** Regress $r_t^2$ on $r_{t-1}^2, \ldots, r_{t-q}^2$ and obtain $R^2$ the coefficient of determination; Test the $H_0$: Slope coefficients are all zero, by

$$T \cdot R^2 \sim \chi_q^2. \tag{2.54}$$

These tests were carried out on the exchange rate data; Table 2.3 has the result of the Ljung-Box Test:

Table 2.3: Ljung-Box Chi-Square Statistics for Exchange Rate Data

| h | 12 | 24 | 36 | 48 |
|---|---|---|---|---|
| $Q_h$ | 73.2 | 225.7 | 463.4 | 575.7 |
| df | 6 | 18 | 30 | 42 |
| $p$-value | 0.000 | 0.000 | 0.000 | 0.000 |

The Lagrange Multiplier test with $q = 5$ also gives $R^2 = 0.065$, that $T \cdot R^2 = 243.6$ and compared with $\chi_5^2$ table values has $p$-value $\approx 0$. These results clearly confirm that there is some time dependence in the volatility.

Among the conditional heteroscedasticity models ARCH/GARCH that were discussed in the last section, GARCH is simple to use and also results in more parsimonious modeling. Therefore, we present only the estimated GARCH model here. For the exchange rate data, the estimated GARCH model is $\hat{\sigma}_t^2 = 2.61 \times 10^{-7} + 0.037 r_{t-1}^2 + 0.95 \hat{\sigma}_{t-1}^2$. Observe that the coefficients $r_{t-1}^2$ and $\hat{\sigma}_{t-1}^2$ add up to unity approximately and thus indicating the volatility is best modeled through IGARCH, which seems to hold generally for many asset volatilities.

## 2.12 Exercises

1. Consider the exchange rate daily data from December 4, 2006 to November 5, 2010 (Rupee versus Dollar, Pound and Euro), in file `exchange_rates_1.csv`.

(a) Compute the sample average, standard deviation and the first order autocorrelation of daily returns over the entire sample period. Test if the mean and the first order autocorrelation are significantly different from zero using the tests proposed in Section 2.6.

(b) Plot histograms of the returns over the entire sample period; Does the distribution look normal? Test it through Jarque-Bera test in (2.37).

(c) Aggregate the data at the weekly level; do (a) and (b) on the aggregated data. Compare the result with the result for the daily level.

2. For the returns of all three time series in Problem 1, construct...

(a) ARMA models for the returns; identify the model structure via ACF and PACF.

(b) GARCH models for the squared returns; compare the model coefficients for the three series. Comment on the difference, if any.

(c) Is there a co-movement among the three exchange-rate series? To make the plots on a comparable scale, convert the starting points of the series unity. Does the co-movement vary over different time regimes? (Back up your claim with solid analysis.) Identify the transition states and speculate how you can exploit this for trading decisions.

3. Exchange rates: The file `exchange_rates_2.csv` contains exchange rates between US dollar and twenty-five major currencies. The daily data spans January 3, 2000 to April 10, 2012. Use this data to perform the following tasks: After computing the returns, $r_t = \log(p_t) - \log(p_{t-1})$ for each series.

(a) Plot histograms of the returns and test if the distributions are normal, via Q–Q plots.

(b) Compute the auto-correlation function and identify which lags are significant. What is the tentative ARMA model?

(c) Is there any ARCH effect? Why?

(d) Fit GARCH(1,1) and IGARCH(1,1) models using both normal and $t$-distributions for the innovations. Which volatility model appears to be the best?

4. Exchange rates: Consider the original series, $p_t$, in file `exchange_rates_2.csv` for the period starting June 11, 2003. Answer the following after dividing each series by $p_1$, so that the starting points are the same.

(a) Identify for each series, if there are any regime changes. Test if these changes are statistically significant.

(b) Identify how many clusters are there and if the membership changes under different regimes.

(c) Do the series move together? Provide some intuitive tests (more formal tests will be covered in Chapter 3).

(d) Compare the results in (c) with (b). What is the connection between clustering and co-movement? Briefly discuss.

5. Consider the daily price of Apple stock from January 2, 2015 to January 2, 2018 in file `aapl_daily.csv`. The data have 7 columns (namely, Date, Open, High, Low, Close, Volume, Adj Close). We focus on the adjusted closing price in the last column.

(a) Compute the daily log returns. Is there any serial correlation in the daily log returns? Use the test for white noise as outlined in the text.

(b) Consider the pivot based on the pivot based on the average of high and low price and the pivot based on the average of high, low and close prices. Compute the returns based on the pivot log prices and test for serial correlation. Compare this result with the finding in (a).

(c) Consider the log price series of AAPL stock. Is the log price series unit-root non-stationary? Perform a unit-root (Dickey-Fuller) test to answer the question and present your conclusion.

6. Consider the daily price of Apple stock again in file `aapl_daily.csv`.

(a) Compute various measures of variance computed from the entries of the price bars. Comment on their correlation with log volume.

(b) Use the ARIMA modeling to come up with a parsimonious model for log volume. Comment on the model accuracy by setting aside a validation data set.

7. The *Roll* model for trade prices (in file `aapl_daily.csv`) discussed in the text can be more specifically stated as,

$$\begin{aligned} p_t^* &= p_{t-1} + \epsilon_t, \quad && \epsilon_t \sim N(0, \sigma^2) \text{ and} \\ p_t &= p_t^* + c s_t, \quad && \text{where } s_t \in \{+1, -1\} \end{aligned}$$

Here $p_t^*$ is the "true" value of a security and $p_t$ is the observed trade price, which differ because of the bid-offer spread: $c$ is the half-spread and $s_t$ is the direction of the $t$th trade. $s_t$ and $\epsilon$ are serially independent and independent of each other.

(a) Calculate the serial correlation of the observed prices $p_t$. Construct an MA(1) model with the same autocorrelation structure. Of the two such models, take the invertible one.

(b) For the invertible model, construct the associated AR model.

(c) Use the model to show how to estimate the bid-ask spread and apply the results on the Apple data used in Exercises 5 and 6.

8. Consider the Level III data for the ticker, "BIOF" for a single day in file `biof_levelIII_data.csv`. The data description follows similar details as given in Table 2.1. This exercise consists of two parts; in the first part, we want to construct the limit order book and the second part we want to study the effect of time aggregation.

(a) At any point in time, a limit-order book contains a number of buy and sell orders.

  (i) Develop a program that can visually present a limit order book. Present the order book diagram as of 11:00 a.m. for the first ten positions on both sides.

  (ii) Aggregate demand and supply are represented as step functions of shares accumulated at each price level. The height of step '$i$' on the demand side is the price difference between $(i-1)$th and the $i$th best price. The height of the first step is the mid-price; the length of a step, '$i$', on the demand side is the total number of shares across all orders until $i$th price. Similarly, the supply side can be constructed. Normalize a step height by the cumulative length from the first step to tenth step. Present the aggregate curve as of 11:00 a.m.

  (iii) Using the aggregate graph given in (ii), suggest measures for the order book imbalance—which is a key determinant in some entry/exit algorithms in the high-frequency setting.

(b) Aggregate the data to a 2-minute interval, summarizing the number of transactions, volume weighted average price and average duration between transactions. Also, produce the price bars (open, close, low and high and volume) for each interval.

  (i) Briefly discuss how you would quantify the loss of information due to aggregation.

  (ii) Calculate the variance of the returns based on the aggregated VWAP and compare it with the variance based on all observations.

  (iii) From Exercise 7, it follows that the true price is measured with error, which is due to market friction. The aggregation done above may cancel out the friction and may reflect the true price. Alternatively, one can sample the price every two minutes and compute the variance in the returns. How does this compare to VWAP based descriptives?

# 3

# *Multivariate Time Series Models*

The methods developed in this chapter are natural extensions of the methods presented in Chapter 2 in the multivariate setting. But the underlying data is of large dimension and hence they present some unique challenges. These include unwieldy estimation problems as well as making meaningful interpretation of the estimates. But these methods are essential to address portfolio construction and issues related to trading multiple stocks. Readers will find the main tools described in this chapter are useful to follow the latest research in the areas of commonality and co-movement of multiple stocks. We want to present some basic tools in multivariate regression and tools for dimension-reduction, which provide a direct foundation for modeling multiple time series. Although several multivariate time series models are well-studied in the literature, the focus has been on the multivariate (vector) autoregressive (VAR) models which falls neatly in the framework of multivariate regression with necessary constraints on the regression coefficient matrices that can guarantee stationarity. Thus a good understanding of multivariate regression will lay a solid foundation for studying VAR models.

In the area of multivariate analysis, there are two broad themes that have emerged over time. One that typically involves exploring the variations in a set of interrelated variables and ones that involves investigating the simultaneous relationships between two or more sets of variables. In either case, the themes involve explicit modeling of the relationships or exploring dimension-reduction of the sets of variables for easier interpretation. The multivariate regression introduced in Section 3.1 and its variants are the preferred tools for the parametric modeling and descriptive tools such as principal components or canonical correlations are the tools used for addressing dimension-reduction issues (Section 3.2). Both act as complementary tools and data analysts may want to make use of these tools for an insightful analysis of multivariate data. As mentioned, the methods developed in this chapter would be useful to study some important concepts in finance, such as cross-sectional momentum, pairs trading, commonality and co-movement, topics that will be commented on in subsequent chapters.

## 3.1 Multivariate Regression

We begin with the general multivariate linear regression model,

$$Y_t = CX_t + \epsilon_t, \tag{3.1}$$

where $Y_t = (Y_{1t}, \dots, Y_{mt})'$ is an $m \times 1$ vector of responsive variables and $X_t = (X_{1t}, \dots, X_{nt})'$ is an $n \times 1$ vector of predictors, $C$ is an $m \times n$ coefficient matrix and $\epsilon_t = (\epsilon_{1t}, \dots, \epsilon_{mt})'$ is the $m \times 1$ vector of random errors. In the cross-sectional modeling of asset returns, the popular model introduced by Fama and French (2015) [133], falls into this regression set-up. The response vector has the returns on '$m$' stocks and the predictor set contains the Fama-French factors. If we restrict the first component of $X_t$, $X_{1t} = 1$, the model in (3.1) can include the means of the variables in $Y_t$. We assume $\mathrm{E}(\epsilon_t) = 0$ and $\mathrm{Cov}(\epsilon_t) = \Sigma_{\epsilon\epsilon}$ is a $m \times m$ positive definite covariance matrix. The errors are independent over '$t$', the time index. Thus if we arrange the data and error as matrices,

$$Y = [Y_1, \dots, Y_T], \quad X = [X_1, \dots, X_T] \text{ and } \epsilon = [\epsilon_1, \dots, \epsilon_T], \tag{3.2}$$

and if $e = \mathrm{vec}(\epsilon')$,

$$\mathrm{E}(e) = 0, \quad \mathrm{Cov}(e) = \Sigma_{\epsilon\epsilon} \otimes I_T, \tag{3.3}$$

where the symbol $\otimes$ signifies the Kronecker product.[1]

The unknown parameters in model (3.1) are the $(mn)$ elements of the regression coefficient matrix $C$ and $m(m+1)/2$ elements of the error covariance matrix $\Sigma_{\epsilon\epsilon}$. These can be estimated by the method of maximum likelihood under the assumption of normality of the $\epsilon_t$ which is equivalent to least squares estimation. Note we can rewrite the model (3.1) in terms of the data matrices $\mathbf{Y}$ and $\mathbf{X}$ as

$$\mathbf{Y} = C\mathbf{X} + \epsilon. \tag{3.4}$$

The least squares estimates are obtained by minimizing

$$e'e = \mathrm{tr}(\epsilon\epsilon') = \mathrm{tr}[(\mathbf{Y} - C\mathbf{X})(\mathbf{Y} - C\mathbf{X})'] = \mathrm{tr}\left[\sum_{t=1}^{T} \epsilon_t \epsilon_t'\right], \tag{3.5}$$

where $\mathrm{tr}(A)$ denotes the sum of the diagonal elements of $A$. This yields a unique solution for $C$ as

$$\tilde{C} = \mathbf{YX}'(\mathbf{XX}')^{-1} = \left(\frac{1}{T}\mathbf{YX}'\right)\left(\frac{1}{T}\mathbf{XX}'\right)^{-1}. \tag{3.6}$$

[1]The Kronecker product of two matrices is defined as $A \otimes B = \begin{pmatrix} a_{11}B & \cdots & a_{1m}B \\ \vdots & \ddots & \vdots \\ a_{m1}B & \cdots & a_{mm}B \end{pmatrix}$, where $A = \begin{pmatrix} a_{11} & \cdots & a_{1m} \\ \vdots & \ddots & \vdots \\ a_{m1} & \cdots & a_{mm} \end{pmatrix}$.

Hence, it can be seen from the form of (3.6) the rows of the matrix $C$ are estimated by the least squares regression of each response variable on the predictor variables as $C_j' = (\mathbf{Y}_j'\mathbf{X})(\mathbf{X}\mathbf{X}')^{-1}$ and therefore the covariances among the response variables do not enter into the estimation. Thus, the multivariate regression is essentially a series of multiple regressions and any computing program that can do regression can be used to fit model (3.1) as well.

The ML estimator of the error covariance matrix $\Sigma_{\epsilon\epsilon}$ is obtained from the least squares residuals $\hat{\epsilon} = \mathbf{Y} - \tilde{C}\mathbf{X}$ as $\tilde{\Sigma}_{\epsilon\epsilon} = \frac{1}{T}(\mathbf{Y} - \tilde{C}\mathbf{X})(\mathbf{Y} - \tilde{C}\mathbf{X})' \equiv \frac{1}{T}S$, where $S = \sum_{t=1}^{T} \hat{\epsilon}_t\hat{\epsilon}_t'$ is the error sum of squares matrix, and $\hat{\epsilon}_t = Y_t - \tilde{C}\mathbf{X}_t$, $t = 1, \ldots, T$, are the least squares residual vectors.

### Inference Properties

The distributional properties of $\tilde{C}$ follow from multivariate normality of the error terms $\epsilon_t$. Specifically, we consider the distribution of vec$(\tilde{C}')$, where the "vec" operation transforms an $m \times n$ matrix into an $mn$-dimensional column vector by stacking the columns of the matrix below each other.

**Result 1.** For the model (3.1) under the normality assumption on the $\epsilon_k$, the distribution of the least squares estimator $\tilde{C}$ in (3.6) is that

$$\text{vec}(\tilde{C}') \sim N(\text{vec}(C'), \Sigma_{\epsilon\epsilon} \otimes (\mathbf{X}\mathbf{X}')^{-1}). \tag{3.7}$$

Note, in particular, that this result implies that the $j$th row of $\tilde{C}$, $\tilde{C}_{(j)}'$, which is the vector of least squares estimates of regression coefficients for the $j$th response variable, has the distribution $\tilde{C}_{(j)} \sim N(C_{(j)}, \sigma_{jj}(\mathbf{X}\mathbf{X}')^{-1})$. The inference on the elements of the matrix $C$ can be made using the result in (3.7). In practice, because $\Sigma_{\epsilon\epsilon}$ is unknown, a reasonable estimator such as $\tilde{\Sigma}_{\epsilon\epsilon}$ is substituted for the covariance matrix in (3.7).

In the regression model (3.1) not all predictors are likely to play significant roles. The problem of variable selection currently is an area of active study. Here we present only basic results. We shall indicate the likelihood ratio (LR) procedure that is very useful for testing simple linear hypotheses regarding the regression coefficient matrix. Suppose $\mathbf{X}$ is partitioned as $\mathbf{X} = [\mathbf{X}_1', \mathbf{X}_2']$ and corresponding $C = [C_1, C_2]$, so that the model (3.4) is written as $\mathbf{Y} = C_1\mathbf{X}_1 + C_2\mathbf{X}_2 + \epsilon$, where $C_1$ is $m \times n_1$ and $C_2$ is $m \times n_2$ with $n_1 + n_2 = n$. Suppose we want to test the null hypothesis $H_0 : C_2 = 0$ against the alternative $C_2 \neq 0$. The null hypothesis implies that the predictor variables $\mathbf{X}_2$ do not have any (additional) influence on the response variables $\mathbf{Y}$, given the impact of the variables $\mathbf{X}_1$. Essentially, this involves running two sets of regressions: One with all predictors ($\mathbf{X}$) and one with the subset of the predictors ($\mathbf{X}_1$). The LR test statistic is $\lambda = U^{T/2}$, where $U = |S|/|S_1|$, $S = (\mathbf{Y} - \tilde{C}\mathbf{X})(\mathbf{Y} - \tilde{C}\mathbf{X})'$ and $S_1 = (\mathbf{Y} - \tilde{C}_1\mathbf{X}_1)(\mathbf{Y} - \tilde{C}_1\mathbf{X}_1)'$. The matrix $S$ is the residual sum of squares matrix from fitting of the full model, while $S_1$ is the residual sum of squares matrix obtained from fitting the reduced model with $C_2 = 0$ and $\tilde{C}_1 = \mathbf{Y}\mathbf{X}_1'(\mathbf{X}_1\mathbf{X}_1')^{-1}$. It has been shown (see Anderson (1984) [19, Chapter 8]) that for moderate and large sample size $T$, the test statistic

$$\mathcal{M} = -[T - n + (n_2 - m - 1)/2]\log(U) \tag{3.8}$$

is approximately distributed as Chi-squared with $n_2m$ degrees of freedom ($\chi^2_{n_2m}$); the hypothesis is rejected when $\mathcal{M}$ is greater than a constant determined by the $\chi^2_{n_2m}$ distribution for a given level of significance.

Finally, we briefly consider testing the more general null hypotheses of the form $H_0 : FC = 0$ where $F$ is a known full-rank matrix of dimensions $m_1 \times m$. In the hypothesis $H_0$, the matrix $F$ provides for restrictions in the coefficients of $C$ across the different response variables. These restrictions usually come from a priori economic theory and the statistics outlined below will help to formally test them. The likelihood ratio test is based on $\lambda = U^{T/2}$ where $U = |S|/|S_1|$ and $S_1 = (\mathbf{Y} - \hat{C}\mathbf{X})(\mathbf{Y} - \hat{C}\mathbf{X})'$ with $\hat{C}$ denoting the ML estimator $C$ obtained subject to the constraint that $FC = 0$. In fact, it can be shown that the constrained ML estimator under $FC = 0$ can be expressed as

$$\hat{C} = \tilde{C} - SF'(FSF')^{-1}F. \tag{3.9}$$

With $S_1 = S + H$ and $U = 1/|I + S^{-1}H|$, the test statistic used is $\mathcal{M} = -[T - n + \frac{1}{2}(n - m_1 - 1)]\log(U)$ which has an approximate $\chi^2_{nm_1}$ distribution under $H_0$. Here $|\cdot|$ indicates the determinant of the matrix. A somewhat more accurate approximation of the null distribution of $\log(U)$ is available based on the $F$-distribution, and in some cases (notably $m_1 = 1$ or 2) exact $F$-distribution results for $U$ are available (see Anderson (1984) [19, Chapter 8]). However, for most of the applications the large sample $\chi^2_{nm_1}$ approximation for the distribution of the LR statistic $\mathcal{M}$ will suffice.

### Prediction

Another problem of interest in regard to the multivariate linear regression model (3.1) is confidence regions for the mean responses corresponding to a fixed value $X_0$ of the predictor variables. The mean vector of responses at $X_0$ is $CX_0$ and its estimate is $\tilde{C}X_0$. It can be shown that

$$\mathcal{T}^2 = (\tilde{C}X_0 - CX_0)'\Sigma_{\epsilon\epsilon}^{-1}(\tilde{C}X_0 - CX_0)/\{X_0'(\mathbf{X}\mathbf{X}')^{-1}X_0\} \tag{3.10}$$

has Hotelling's $T^2$-distribution with $T - n$ degrees of freedom, Anderson (1984) [19, Chapter 5]. Thus, $\{(T - n - m + 1)/m(T - n)\}\mathcal{T}^2$ has the $F$-distribution with $m$ and $T - n - m + 1$ degrees of freedom, so that a $100(1 - \alpha)\%$ confidence ellipsoid for the true mean response $CX_0$ at $X_0$ is given by

$$\begin{aligned}(\tilde{C}X_0 - CX_0)'\,\overline{\Sigma}_{\epsilon\epsilon}^{-1}(\tilde{C}X_0 - CX_0)&\\ \leq \{X_0'(\mathbf{X}\mathbf{X}')^{-1}X_0\}\left[\frac{m(T - n)}{T - n - m + 1}\,F_{m,T-n-m+1}(\alpha)\right],&\end{aligned} \tag{3.11}$$

where $F_{m,T-n-m+1}(\alpha)$ is the upper $100\alpha$th percentile of the $F$-distribution with $m$ and $T - n - m + 1$ degrees of freedom. In testing the validity of CAPM model in a large cross-section of stocks, the common test, whether the constant terms are simultaneously zero, is based on (3.10). More details will follow in a later chapter.

A related problem of interest is the prediction of values of a new response vector $Y_0 = CX_0 + \epsilon_0$ corresponding to the predictor variables $X_0$, where it is assumed that $Y_0$ is independent of the values $Y_1, \ldots, Y_T$. The predictor of $Y_0$ is given by $\hat{Y}_0 = \tilde{C}X_0$, and it follows similarly to the above result (3.11) that a $100(1-\alpha)\%$ prediction ellipsoid for $Y_0$ is

$$
\begin{aligned}
&(Y_0 - \tilde{C}X_0)' \overline{\Sigma}_{\epsilon\epsilon}^{-1} (Y_0 - \tilde{C}X_0) \\
&\quad \leq \{1 + X_0'(\mathbf{X}\mathbf{X}')^{-1} X_0\} \left[ \frac{m(T-n)}{T-n-m+1} F_{m,T-n-m+1}(\alpha) \right].
\end{aligned} \tag{3.12}
$$

The application of Hotelling's $T^2$ in the context of testing related to the CAPM model when the number of stocks, '$m$' is larger than the available data, '$T$' has been a subject of active research recently.

## 3.2 Dimension-Reduction Methods

### Reduced-Rank Regression

In many practical situations, there is a need to reduce the dimensions of the variables for lack of data or for parsimonious modeling and for succinct interpretation. A common approach with only one set of variables say, $Y$ or $X$, is to use Principal Components Analysis (PCA), which focuses on finding the linear combination that captures the most variance. Here doing PCA on $Y$ and on $X$ and then relating the PCA via regression is not optimal. We approach the problem through the assumption of lower rank of the matrix $C$ in model (3.1). More formally, in the model $Y_t = CX_t + \epsilon_t$ we assume that with $m < n$,

$$\operatorname{rank}(C) = r \leq \min(m, n) = m. \tag{3.13}$$

The rank condition (3.13) has two related important practical implications. First, with $r < m$ it implies that there are $(m - r)$ linear restrictions on the regression coefficient matrix $C$ of the form

$$l_i'C = 0, \quad i = 1, 2, \ldots, (m - r) \tag{3.14}$$

and these restrictions themselves are often not known a priori in the sense that $l_1, \ldots, l_{m-r}$ are unknown, unlike the known constraints $FC = 0$ discussed in the last section. Premultiplying (3.1) by $l_i'$, we have $l_i'Y_t = l_i'\epsilon_t$. Thus the linear combinations, $l_i'Y_t$, $i = 1, 2, \ldots, (m - r)$, could be modeled without any reference to the predictor variables $X_t$ and depend only on the distribution of the error term $\epsilon_t$. Otherwise, these linear combinations can be isolated and can be investigated separately. These are called structural relationships in macroeconomics and they also play a role in co-integration and concepts related to pairs trading.

The second implication that is somewhat complementary to (3.14) is that with assumption (3.13), $C$ can be written as a product of two lower dimensional matrices that are of full rank. Specifically, $C$ can be expressed as

$$C = AB, \tag{3.15}$$

where $A$ is of dimension $m \times r$ and $B$ is of dimension $r \times n$, but both have rank $r$. The model (3.1) can then be written as

$$Y_t = A(BX_t) + \epsilon_t, \quad t = 1, 2, \dots, T, \tag{3.16}$$

where $BX_t$ is of reduced dimension with only $r$ components. A practical use of (3.16) is that the $r$ linear combinations of the predictor variables $X_t$ are sufficient to model the variation in the response variables $Y_t$ and there may not be a need for all $n$ linear combinations or otherwise for all $n$ variables, as would be the case when no single predictor variable can be discarded from the full-rank analysis. Hence, there is a gain in simplicity and interpretation through the reduced-rank regression modeling, although from a practitioner point of view one would still need to have measurements on all $n$ predictor variables. But some recent developments on 'sparsity' address this issue where the elements of $A$ and $B$ are further shrunk to zeros. This will be discussed briefly later.

For presentation of reduced-rank estimation, we first need an important result known as the Householder-Young or Eckart-Young Theorem (Eckart and Young, 1936 [114]), which presents the tools for approximating a full-rank matrix by a matrix of lower rank. The solution is related to the singular value decomposition of the full-rank matrix. These results form the basis of most of the well-known dimension-reduction techniques such as principal components. The reduced-rank estimates are obtained as approximation of the full-rank estimate of the regression coefficient matrix.

**Result 2.** Let $S$ be a matrix of order $m \times n$ and of rank $m$. The Euclidean norm, $\operatorname{tr}[(S-P)(S-P)']$, is minimum among matrices $P$ of the same order as $S$ but of rank $r$ ($\leq m$), when $P = MM'S$, where $M$ is $m \times r$ and the columns of $M$ are the first $r$ (normalized) eigenvectors of $SS'$, that is, the normalized eigenvectors corresponding to the $r$ largest eigenvalues of $SS'$.

**Remark (Singular Value Decomposition).** The positive square roots of the eigenvalues of $SS'$ are referred to as the singular values of the matrix $S$. In general, an $m \times n$ matrix $S$, of rank $s$, can be expressed in a *singular value decomposition* as $S = V \Lambda U'$, where $\Lambda = \operatorname{diag}(\lambda_1, \dots, \lambda_s)$ with $\lambda_1^2 \geq \cdots \geq \lambda_s^2 > 0$ being the nonzero eigenvalues of $SS'$, $V = [V_1, \dots, V_s]$ is an $m \times s$ matrix such that $V'V = I_s$, and $U = [U_1, \dots, U_s]$ is $n \times s$ such that $U'U = I_s$. The columns $V_i$ are the normalized eigenvectors of $SS'$ corresponding to the $\lambda_i^2$, and $U_i = \frac{1}{\lambda_i} S'V_i$, $i = 1, \dots, s$.

The estimates of model (3.16) are:

$$\hat{A}^{(r)} = \Gamma^{-1/2}[\hat{V}_1, \dots, \hat{V}_r], \quad \hat{B}^{(r)}[\hat{V}_1, \dots, \hat{V}_r]'\Gamma^{1/2}\hat{\Sigma}_{yx}\hat{\Sigma}_{xx}^{-1}, \tag{3.17}$$

where $\hat{\Sigma}_{yx} = (1/T)\mathbf{YX}'$, $\hat{\Sigma}_{xx} = (1/T)\mathbf{XX}'$, and $\hat{V}_j$ is the eigenvector that corresponds to the $j$th largest eigenvalue $\hat{\lambda}_j^2$ of $\Gamma^{1/2}\hat{\Sigma}_{yx}\hat{\Sigma}_{xx}^{-1}\hat{\Sigma}_{xy}\Gamma^{1/2}$, with the choice $\Gamma = \tilde{\Sigma}_{\epsilon\epsilon}^{-1}$. The solution is based on the singular value decomposition of the standardized regression coefficient matrix $\tilde{\Sigma}_{\epsilon\epsilon}^{-1/2}\tilde{C}\tilde{\Sigma}_{xx}^{1/2}$, where $\tilde{C}$ is the least-square estimate of full-rank $C$. The estimates of the constraints, $l_j$'s are obtained from the smallest $(m-r)$ eigenvalues of the same matrix and scaled as, $\hat{l}_j' = \hat{V}_j'\Gamma^{1/2}$.

It should be mentioned, perhaps, that estimation results are sensitive to the choice of scaling of the response variables $Y_t$, and in applications it is often suggested that the component variables $y_{it}$ be standardized to have unit variances before applying the LS procedure.

**Principal Component Analysis (PCA)**

Principal component analysis, originally developed by Hotelling (1933) [200], is concerned with explaining the covariance structure of a large number of variables through a small number of linear combinations of the original variables. The principal components are descriptive tools that do not usually rest on any modeling or distributional assumptions.

Let the $m \times 1$ random vector $Y_t$ have the covariance matrix $\Sigma_{yy}$ with eigenvalues $\lambda_1^2 \geq \lambda_2^2 \geq \cdots \geq \lambda_m^2 \geq 0$. Consider the linear combinations $y_1^* = l_1'Y_t, \ldots, y_m^* = l_m'Y_t$, normalized so that $l_i'l_i = 1$, $i = 1, \ldots, m$. These linear combinations have $\text{Var}(y_i^*) = l_i'\Sigma_{yy}l_i$, $i = 1, 2, \ldots, m$, and $\text{Cov}(y_i^*, y_j^*) = l_i'\Sigma_{yy}l_j$, $i, j = 1, 2, \ldots, m$. In a principal component analysis, the linear combinations $y_i^*$ are chosen so that they are not correlated but their variances are as large as possible. That is, $y_1^* = l_1'Y_t$ is chosen to have the largest variance among linear combinations so that $l_1'l_1 = 1$, $y_2^* = l_2'Y_t$ is chosen to have the largest variance subject to $l_2'l_2 = 1$ and the condition that $\text{Cov}(y_2^*, y_1^*) = l_2'\Sigma_{yy}l_1 = 0$, and so on. It can be easily shown that the $l_i$'s are chosen to be the (normalized) eigenvectors that correspond to the eigenvalues $\lambda_i^2$ of the matrix $\Sigma_{yy}$, so that the normalized eigenvectors satisfy $l_i'\Sigma_{yy}l_i = \lambda_i^2 \equiv \text{Var}(y_i^*)$ and $l_i'l_i = 1$. Often, the first $r$ $(r < m)$ components $y_i^*$, corresponding to the $r$ largest eigenvalues, will contain nearly most of the variation that arises from all $m$ variables. For such cases, in studying variations of $Y_t$, attention can be directed on these $r$ linear combinations only, thus resulting in a reduction in dimension of the variables involved.

The principal component problem can be represented as a reduced-rank problem, in the sense that a reduced-rank model can be written for this situation as $Y_t = ABY_t + \epsilon_t$,with $X_t \equiv Y_t$. From the solutions of reduced-rank regression, by setting $\Gamma = I_m$, we have $A^{(r)} = [V_1, \ldots, V_r]$ and $B^{(r)} = V'$ because $\Sigma_{yx} = \Sigma_{yy}$. The $V_j$'s refer to the eigenvectors of $\Sigma_{yy}$. Thus, a correspondence is easily established. The $l_i$ in PCA are the same as the $V_i$ that result from the solution for the matrices $A$ and $B$. If both approaches provide the same solutions, one might ask, then, what is the advantage of formulating the principal component analysis in terms of reduced-rank regression model. The inclusion of the error term $\epsilon_t$ appears to be superfluous but it provides a useful framework to perform the residual analysis to detect outliers, influential observations, and so on, for the principal components method.

### Canonical Correlation Analysis

Canonical correlation analysis was introduced by Hotelling (1935, 1936) [201, 202] as a method of summarizing relationships between two sets of variables. The objective is to find the linear combination of one set of variables which is most highly correlated with any linear combination of a second set of variables. The technique has not received its due attention because of difficulty in its interpretation. By exploring the connection between the reduced-rank regression and canonical correlations, we provide an additional method of interpreting the canonical variates. Because the role of any descriptive tool such as canonical correlations is in better model building, the connection we display below provides an avenue for interpreting and further using the canonical correlations toward that goal.

More formally, in canonical correlation analysis, for random vectors $X$ and $Y$ of dimensions $n$ and $m$, respectively, the purpose is to obtain an $r{\times}n$ matrix $G$ and an $r{\times}m$ matrix $H$, so that the $r$-vector variates $\xi = GX$ and $\omega = HY$ will capture as much of the covariances between $X$ and $Y$ as possible. The following result summarizes how this is done.

**Result 3.** We assume $\Sigma_{yy}$ is nonsingular. Then the $r \times n$ matrix $G$ and the $r \times m$ matrix $H$, with $H\Sigma_{yy}H' = I_r$, that minimize simultaneously all the eigenvalues of $\mathrm{E}[(HY - GX)(HY - GX)']$ are given by,

$$G = V_*' \, \Sigma_{yy}^{-1/2} \, \Sigma_{yx} \, \Sigma_{xx}^{-1}, \qquad H = V_*'\Sigma_{yy}^{-1/2}, \tag{3.18}$$

where $V_* = [V_1^*, \ldots, V_r^*]$ and $V_j^*$ is the (normalized) eigenvector corresponding to the $j$th largest eigenvalue $\rho_j^2$ of the (squared correlation) matrix

$$R_* = \Sigma_{yy}^{-1/2} \, \Sigma_{yx} \, \Sigma_{xx}^{-1} \, \Sigma_{xy} \, \Sigma_{yy}^{-1/2}. \tag{3.19}$$

Denoting by $g_j'$ and $h_j'$ the $j$th rows of the matrices $G$ and $H$, respectively, the $j$th pair of canonical variates are defined to be $\xi_j = g_j'X$ and $\omega_j = h_j'Y$, $j = 1, 2, \ldots, m$ $(m \le n)$. The correlation between $\xi_j$ and $\omega_j$ $\mathrm{Corr}(\xi_j, \omega_j) = \rho_j$ is the $j$th canonical correlation coefficient between $X$ and $Y$. The canonical variates possess the property that $\xi_1$ and $\omega_1$ have the largest correlation among all possible linear combinations of $X$ and $Y$, $\xi_2$ and $\omega_2$ have the largest possible correlation among all linear combinations of $X$ and $Y$ that are uncorrelated with $\xi_1$ and $\omega_1$ and so on. Because $\Sigma_{\epsilon\epsilon} = \Sigma_{yy} - \Sigma_{yx}\Sigma_{xx}^{-1}\Sigma_{xy}$, the correspondence between canonical correlations and reduced-rank regression for the choice $\Gamma = \Sigma_{yy}^{-1}$ is easily established as both seek to study the linear relationship among the two sets of variables. It can be shown that,

$$\rho_j^2 = \lambda_j^2/(1 + \lambda_j^2) \tag{3.20}$$

and $G$ and $H$ matrices can be obtained from the rescaled versions of $\hat{V}$, used in (3.17). Because of (3.20), testing for the rank of the matrix '$C$' can be made based on the canonical correlations, $\hat{\rho}^2$.

**Partial Least Squares**

The partial least squares (PLS) method was advanced by Wold (1984) [325]. This method is similar to reduced-rank regression. In PLS, the covariance between the response and predictor variables is maximized whereas in reduced-rank regression, the correlation is maximized to arrive at the linear combinations of the predictors. The $m \times 1$ multiple response, $X_t$ $(m > 1)$ version of partial least squares begins with a 'canonical covariance' analysis, a sequential procedure to construct predictor variables as linear combinations of the original $n$-dimensional set of predictor variables $X_t$. At the first stage, the initial predictor variable $X_1^* = b_1' X_t$ is determined as the first 'canonical covariance' variate of the $X_t$, the linear combination of $X_t$ that maximizes $\{(T-1) \times$ sample covariance of $b'X_t$ and $l'Y_t\} \equiv b'\mathbf{X}\mathbf{Y}'l$ over unit vectors $b$ (and over all associated unit vectors $l$). For given vectors $b$, we know that $l = \mathbf{Y}\mathbf{X}'b/(b'\mathbf{X}\mathbf{Y}'\mathbf{Y}\mathbf{X}'b)^{1/2}$ is the maximizing choice of unit vector for $l$, so that we need then to maximize $b'\mathbf{X}\mathbf{Y}'\mathbf{Y}\mathbf{X}'b$ with respect to unit vectors $b$. Thus, the first linear combination vector $b_1$ is the (normalized) eigenvector of $\mathbf{X}\mathbf{Y}'\mathbf{Y}\mathbf{X}'$ corresponding to its largest eigenvalue. This follows easily from Result 2 in Chapter 3. In the univariate case, $m = 1$ and $\mathbf{Y}$ is $1 \times T$, and so this reduces to $b_1 = \mathbf{X}\mathbf{Y}'/(\mathbf{Y}\mathbf{X}'\mathbf{X}\mathbf{Y}')^{1/2}$ with associated "eigenvalue" equal to $\mathbf{Y}\mathbf{X}'\mathbf{X}\mathbf{Y}'$.

At any subsequent stage $j$ of the procedure, $j$ predictor variables $x_{it}^* = b_i' X_t, i = 1, \ldots, j$, are to be determined and the next predictor variable $x_{j+1}^* = b_{j+1}' X_t$; that is, the next unit vector $b_{j+1}$ is to be found. It is chosen as the vector $b$ to maximize the sample covariance of $b'X_t$ and $l'Y_t$ (with unit vector $l$ also chosen to maximize the quantity); that is, we maximize $b'\mathbf{X}\mathbf{Y}'\mathbf{Y}\mathbf{X}'b$, subject to $b_{j+1}$ being a unit vector and $x_{j+1,t}^* = b_{j+1}; X_t$ being orthogonal to the previous set of $j$ predictor variables $x_{it}^*, i = 1, \ldots, j$ (i.e., $b_{j+1}'\mathbf{X}\mathbf{X}'b_i = 0$, for $i = 1, \ldots, j$). The number $r$ of predictors chosen until the sequential process is stopped can be taken as a regularization parameter of the procedure; its value is determined through cross-validation methods in terms of prediction mean square error. The final estimated regression equations are constructed by ordinary LS calculation of the response vectors $Y_k$ on the first $r$ 'canonical covariance' component predictor variables, $X_k^* = (x_{1k}^*, \ldots, x_{rk}^*)' \equiv \hat{B}X_k$, where $\hat{B} = [b_1, \ldots, b_r]'$; that is, the estimated regression equations are $\hat{Y}_k = \hat{A}X_k^* \equiv \hat{A}\hat{B}X_k$ with

$$\hat{A} = \mathbf{Y}\mathbf{X}^{*'}(\mathbf{X}^*\mathbf{X}^{*'})^{-1} \equiv \mathbf{Y}\mathbf{X}'\hat{B}'(\hat{B}\mathbf{X}\mathbf{X}'\hat{B}')^{-1}$$

and $\mathbf{X}^* = \hat{B}\mathbf{X}$ is $r \times T$. In practice, the predictor variables $X_k$ and response variables $Y_k$ will typically already be expressed in terms of deviations from sample mean vectors.

Discussion of the sequential computational algorithms for the construction of the 'canonical covariance' components $x_{jk}^* = b_j' X_k$ required in the partial least squares regression procedure, and of various formulations and interpretations of the partial least squares method are given by Wold (1984) [325], Helland (1988, 1990) [189, 190], and Garthwaite (1994) [159]. In particular, for univariate partial least squares, Garthwaite (1994) [159] gives the interpretation of the linear combinations of predictor variables, $x_{jk}^* = b_j' X_k$, as weighted averages of predictors, where each individual predictor holds the residual information in an explanatory variable that is not contained in earlier linear combination components $(x_{ik}^*, i < j)$, and the quantity to be predicted is the $1 \times T$ vector of residuals obtained from regressing $y_k$ on the earlier components. Also, for the univariate case, Helland (1988, 1990) [189, 190] showed that the linear combination vectors $\hat{B} = [b_1, \ldots, b_r]'$ are obtainable from non-singular $(r \times r)$ linear transformation of $[\mathbf{XY}', (\mathbf{XX}')\mathbf{XY}', \cdots, (\mathbf{XX}')^{r-1}\mathbf{XY}']'$. One can notice the close similarity between reduced-rank regression and partial least squares. The former focuses on maximizing the correlation and the latter on maximizing the covariance.

## 3.3 Multiple Time Series Modeling

Multiple time series analysis is concerned with modeling and estimation of dynamic relationships among '$m$' related time series $y_{1t}, \ldots, y_{mt}$, based on observations over $T$ equally spaced time points $t = 1, \ldots, T$, and also between these series and potential exogenous time series variables $x_{1t}, \ldots, x_{nt}$, observed over the same time period. We shall explore the use of these techniques in leveraging statistical arbitrage in multiple markets in Chapters 5 & 6. We first introduce a general model for multiple time series modeling, but will specialize to vector autoregressive (VAR) models for more detailed investigation. As some concepts are similar to those of the univariate models discussed in Chapter 2, the presentation will be brief.

Let $Y_t = (y_{1t}, \ldots, y_{mt})'$ be an $m \times 1$ multiple time series vector of response variables and let $X_t = (x_{1t}, \ldots, x_{nt})'$ be an $n \times 1$ vector of input or predictor time series variables. Let $\epsilon_t$ denote an $m \times 1$ white noise vector of errors, independently distributed over time with $\mathrm{E}(\epsilon_t) = 0$ and $\mathrm{Cov}(\epsilon_t) = \Sigma_{\epsilon\epsilon}$, a positive-definite matrix. We consider the multivariate time series model

$$Y_t = \sum_{s=0}^{p} C_s X_{t-s} + \epsilon_t, \tag{3.21}$$

where the $C_s$ are $m \times n$ matrices of unknown parameters. In the general setting, the 'input' vectors $X_t$ could include past (lagged) values of the response series $Y_t$. In the context of trading $Y_t$ could denote the prices of related assets, $X_{t-s}$ could represent the past values of $Y_t$ and the values of volume, volatility, market, industry factors of all related assets, etc. Note that the model (3.21) can be written in the form of a multivariate regression model. An important issue that arises in the multiple series modeling is as follows: Even for moderate values of the dimensions $m$ and $n$, the number of parameters that must be estimated in model (3.21), when no constraints are imposed on the matrices $C_s$ can become quite large. Because of the potential complexity of these models, it is often useful to consider dimension reduction procedures such as reduced-rank regression methods described in the last section and more in detail in Reinsel and Velu (1998) [289]. This may also lead to more efficient, interpretable models due to the reduction in the number of unknown parameters. The key quantities that play a role are cross-covariance matrices that relate $X_{t-s}$ to $Y_t$, for appropriate values of '$s$'.

The class of models stated in (3.21) is rather flexible and provides many options for modeling multivariate time series data. Specifically consider the VAR($p$) model,

$$Y_t = \Phi_0 + \sum_{j=1}^{p} \Phi_j Y_{t-j} + \epsilon_t \tag{3.22}$$

with '$p$' lags. As in Section 2.3, we can define the population versions of the mean, variance, and autocovariance functions for stationary series as follows; in addition, we have the cross-correlation matrix that captures the lead-lag relationships among all '$m$' series.

$$\begin{aligned} \text{Mean:} \quad & \mu = \mathrm{E}(Y_t) \\ \text{Variance:} \quad & \Gamma(0) = \mathrm{E}[(Y_t - \mu)(Y_t - \mu)'] \\ \text{Autocovariance:} \quad & \Gamma(l) = \mathrm{E}[(Y_t - \mu)(Y_{t-l} - \mu)'] \\ \text{Cross-correlation:} \quad & \rho(l) = D^{-1}\Gamma(l)D^{-1} \end{aligned} \tag{3.23}$$

Here $D$ is the diagonal matrix consisting of the '$m$' standard deviations, from the square root of diagonal elements of $\Gamma(0)$. It is easy to observe, $\Gamma(l) = \Gamma(-l)'$ and $\rho(l) = \rho(-l)'$. The linear dependence among all the variables in terms of leads-lags is fully captured by examining $\rho_{ij}(l)$. There are excellent sources to study about modeling multiple time series (see Tsay (2010) [315] and Reinsel (2002) [288]). We focus mainly on the use of the VAR($p$) model in studying the relationships among stocks or markets; we elaborate on VAR($p$) below.

Using the back-shift operator $B$, the model (3.22), VAR($p$) can be written as, $\Phi(B)Y_t = \Phi_0 + \epsilon_t$, where $\Phi(B) = I - \Phi_1 B - \cdots - \Phi_p B^p$ is a matrix polynomial. It can be shown for the process $Y_t$ to be stationary, the roots of $|\Phi(\lambda) - I| = 0$ should all lie outside the unit circle. If $Y_t$ is stationary, then $\mu = [\Phi(1)]^{-1}\Phi_0$ where $\Phi(1) = I - \Phi_1 - \cdots - \Phi_p$. An easy consequence of this form is the result of the Yule-Walker equations, which are useful to identify, $\Phi$'s, the autoregressive coefficients from the moment conditions:

$$\begin{aligned} \Gamma(l) &= \Phi_1\,\Gamma(l-1) + \cdots + \Phi_p\,\Gamma(l-p) \\ \rho(l) &= \Phi_1^*\,\rho(l-1) + \cdots + \Phi_p^*\,\rho(l-p). \end{aligned} \tag{3.24}$$

Here $\Phi_i^* = D^{-\frac{1}{2}}\,\Phi_i\,D^{-\frac{1}{2}}$. A general approach to building a VAR model is to sequentially increase the lag value, '$p$' and check if the added lag is significant. The lag coefficients are estimated with their sample counter-parts and they provide good initial estimates.

The VAR($p$) model building at the $i$th stage is fitting the model, $Y_t = \Phi_0 + \Phi_1 Y_{t-1} + \cdots + \Phi_i Y_{t-i} + \epsilon_t$ via least-squares method and estimate the residual covariance matrix via

$$\hat{\Sigma}_i = \left(\frac{1}{T-2i-1}\right) \sum_{t=i+1}^{T} \hat{\epsilon}_t^{(i)} \hat{\epsilon}_t^{(i)'}, \quad i \geq 0, \tag{3.25}$$

where $\hat{\epsilon}_t^{(i)} = Y_t - \hat{\Phi}_0^{(i)} - \hat{\Phi}_1^{(i)} Y_{t-1} - \cdots - \hat{\Phi}_i^{(i)} Y_{t-i}$ are the residuals. The significance of added lag '$i$' is tested with $H_0 : \Phi_i = 0$ versus $H_a : \Phi_i \neq 0$, using the likelihood-ratio criterion:

$$M(i) = -\left(T - m - i - \frac{3}{2}\right) \ln\left(\frac{|\hat{\Sigma}_i|}{|\hat{\Sigma}_{i-1}|}\right). \tag{3.26}$$

This is distributed asymptotically as a Chi-squared distribution with $m^2$ degrees of freedom. Alternatively, the AIC or BIC criteria that penalize for over parameterization can be used;

$$\begin{aligned} \text{AIC}(i) &= \ln\left(|\tilde{\Sigma}_i|\right) + \frac{2m^2 \cdot i}{T} \\ \text{BIC}(i) &= \ln\left(|\tilde{\Sigma}_i|\right) + \frac{m^2 \cdot i \cdot \ln(T)}{T}. \end{aligned} \tag{3.27}$$

Here $\tilde{\Sigma}_i$ is the maximum likelihood estimate of $\Sigma_i$ where the divisor in (3.25) is $T$ instead of $T - 2i - 1$.

An important property of VAR($p$) models is worth noting and will be used later. First, modeling several variables jointly usually results in a lower order vector model as the marginal models for the individual variables or a linear combination of variables can result in a higher order of autoregressive moving average terms that may be hard to interpret. In some practical situations, a certain linear combination of $Y_t$ may be of interest to study, particularly in the dimension reduction aspect whether using the principal components or canonical variables. The result below indicates that a linear combination, constructed without reference to time dependence can still exhibit time series dependence.

**Result 4.** Let $Y_t \sim$ VAR($p$), then $Z_t = l'Y_t$, where $l$ is a $m \times 1$ known vector, follows ARMA with maximum orders $[mp, (m-1)p]$. Thus any single component under the VAR($p$) structure will also follow ARMA with the maximum orders indicated here.

## 3.4 Co-Integration, Co-Movement and Commonality in Multiple Time Series

The financial time series generally exhibit non-stationary behavior. For the stationary series, the dynamic relationships among components of the vector time series, $Y_t$, has been studied in various ways. While the estimation and inference aspects are similar to the univariate series discussed in Chapter 2, because of the sheer dimension of $Y_t$, there are more challenges in the precise estimation of the model coefficients but there are also more opportunities to study the linkages among the components of $Y_t$ that may be insightful. For example, this provides an opportunity to investigate the behavior of a portfolio of stocks. The number of unknown elements in (3.22) is $m^2(p+1)$ which can be quite large even for modest values of '$m$' and '$p$'; therefore considerations of the dimension reduction aspects in modeling (3.22) have been given by Box and Tiao (1977) [54] and Brillinger (1981) [59] among others. The traditional tool used in multivariate analysis for dimension reduction, principal components method is based on the contemporaneous covariance matrix, and thus it ignores the time dependence. While Result 4, stated in the previous section, indicates that the linear combination can exhibit time dependence, we should expect the optimal linear combination to exhibit other desirable features, which are described here.

Identifying the common features among the series have been focused on by several authors; for example, Engle and Kozicki (1993) [127], Vahid and Engle (1993) [318] among others. The series if they are stationary, are studied for their co-movement and if they are non-stationary, they are studied for their co-integration. A method that can capture both aspects can be stated as a reduced-rank autoregressive modeling procedure which we briefly describe below. The basic VAR($p$) model in (3.22) without the constant term can be written as

$$Y_t = CY_{t-1}^* + \epsilon_t, \tag{3.28}$$

where $C = [\Phi_1, \dots, \Phi_p]$ and $Y_{t-1}^* = X_t = (Y_{t-1}, Y_{t-2}, \dots, Y_{t-p})'$. Observe that $'C'$ matrix is of dimension $m \times mp$. We impose a condition on $C$ as in Section 3.4, as $\text{rank}(C) = r < m$, that has implications as observed earlier. Note $B \cdot Y_{t-1}^*$ is the most relevant information from the past $Y_{t-1}^*$ that relates to $Y_t$. The second implication reveals how certain linear combinations, $l_i' \cdot Y_t$ are not related to the past $Y_{t-1}^*$ and can be taken as white noise series. Estimation of $A$, $B$, and $l$'s all follow from the results in Section 3.2. For testing for the rank of the coefficient matrix $C$ in the reduced rank autoregressive model, that is, under the null hypothesis $H_0 : \text{rank}(C) \le r$, the LR test statistic $\mathcal{M} = [T - mp + (mp - m - 1)/2] \sum_{j=r+1}^{m} \log(1 - \hat{\rho}_j^2)$ has an asymptotic $\chi^2_{(m-r)(mp-r)}$ distribution, where $\hat{\rho}_j^2$ are the canonical correlations between $Y_t$ and $Y_{t-1}^*$.

We briefly address an additional issue unique to financial time series analysis. This is the aspect of nonstationarity of the vector time series process $\{Y_t\}$ when a linear combination may result in stationarity. The relation of nonstationarity to the canonical correlation analysis plays a role in testing for the co-integration of multiple series. Thus, in the co-movement (in stationary series) and the co-integration analysis (in non-stationary series), the canonical correlations as descriptive tools play an important role.

An interesting result in Box and Tiao (1977) [54] is the development of the correspondence between nonstationarity in the VAR(1) process, as reflected by the eigenvalues of $\Phi$ and the (squared) canonical correlations between $Y_t$ and $Y_{t-1}$, as reflected in the eigenvalues of the matrix $R_* = \Gamma(0)^{-1/2}\,\Gamma(1)'\,\Gamma(0)^{-1}\,\Gamma(1)\,\Gamma(0)^{-1/2}$. To develop the correspondence more easily, note that we can write the matrix $R_*$ as $R_* = \Gamma(0)^{-1/2}\,\Phi\Gamma(0)\,\Phi'\,\Gamma(0)^{-1/2}$, since $\Phi = \Gamma(1)'\Gamma(0)^{-1}$. The specific result to be established is that the number of eigenvalues of $\Phi$ approaching the unit circle (the nonstationary boundary) is the same as the number of canonical correlations between $Y_t$ and $Y_{t-1}$ that approach unity.

The main implication of the result in Box and Tiao (1977) [54] is that the existence of $d$ canonical correlations $\rho_j$ close to one in the vector AR(1) process implies that the associated canonical components will be (close to) nonstationary series, and these may reflect the nonstationary trending or dynamic growth behavior in the multivariate system, while the remaining $r = m-d$ canonical variates possess stationary behavior. For testing this, we first note that the vector AR(1) model, $Y_t = \Phi Y_{t-1} + \epsilon_t$, can be expressed in an equivalent form as

$$W_t \equiv Y_t - Y_{t-1} = -(I - \Phi)Y_{t-1} + \epsilon_t, \tag{3.29}$$

which is referred to as an error-correction form. Notice that if $\Phi$ has eigenvalues approaching unity then $I - \Phi$ will have eigenvalues approaching zero (and will be of reduced rank). The methodology of interest for exploring the number of unit roots in the AR(1) model is motivated by the model from (3.29), and consists of a canonical correlation analysis between $W_t = Y_t - Y_{t-1}$ and $Y_{t-1}$ ($\equiv X_t$). Under stationarity of the process $Y_t$, note that the covariance matrix between $W_t$ and $Y_{t-1}$ is equal to $\Sigma_{wx} = -(I - \Phi)\,\Gamma(0)$. Thus, the squared canonical correlations between $W_t$ and $Y_{t-1}$ are the eigenvalues of

$$\Sigma_{ww}^{-1}\Sigma_{wx}\Sigma_{xx}^{-1}\Sigma_{xw} = \Sigma_{ww}^{-1}(I - \Phi)\,\Gamma(0)\,(I - \Phi)', \tag{3.30}$$

where $\Sigma_{ww} = \text{Cov}(W_t) = 2\Gamma(0) - \Phi\,\Gamma(0) - \Gamma(0)\,\Phi' = (I - \Phi)\,\Gamma(0) + \Gamma(0)\,(I - \Phi)'$.

Thus for a vector process, $Y_t$, it is possible that each component series $y_{it}$ be nonstationary with its first difference $(1 - B)\,y_{it}$ stationary (in which case $y_{it}$ is said to be integrated of order one), but such that certain linear combinations $z_{it} = b_i' Y_t$ of $Y_t$ will be stationary. The process $Y_t$ then is said to be *co-integrated* with co-integrating vectors $b_i$ (e.g., Engle and Granger (1987) [122]). A possible interpretation, particularly related to economics, is that the individual components $y_{it}$ share some common nonstationary components or "common trends" and, hence, they tend to have certain similar movements in their long-term behavior. A related interpretation is that the component series $y_{it}$, although they may exhibit nonstationary behavior, satisfy (approximately) a long-run equilibrium relation $b_i' Y_t \approx 0$ such that the process $z_{it} = b_i' Y_t$, which represents the deviations from the equilibrium, exhibits stable behavior and so is a stationary process. This fact is exploited in 'pairs trading' or more broadly in trading multiple stocks and can be very useful also in the context of portfolio rebalancing.

Results related to the general VAR($p$) model, $\Phi(B)Y_t = Y_t - \sum_{j=1}^{p} \Phi_j Y_{t-j} = \epsilon_t$ are studied in Engle and Granger (1987) [122] and Johansen (1988, 1991) [221, 222]. As the details are too technical for this book, we suggest that readers may want to refer to these original sources or Reinsel and Velu (1998) [289] for a summary.

Within the context of model (3.29), it is necessary to specify or determine the rank $r$ of co-integration or the number $d$ of unit roots in the model. Thus, it is also of interest to test the hypothesis $H_0$ : $\text{rank}(C) \leq r$, which is equivalent to the hypothesis that the number of unit roots in the AR model is greater than or equal to $d$ ($d = m - r$), against the general alternative that $\text{rank}(C) = m$. The *likelihood ratio test* for this hypothesis is considered by Johansen (1988) [221] and the test statistic is given as

$$-T\log(U) = -T \sum_{i=r+1}^{m} \log(1 - \hat{\rho}_i^2), \tag{3.31}$$

where the $\hat{\rho}_i$ are the $d = m - r$ smallest sample canonical correlations between $W_t = Y_t - Y_{t-1}$ and $Y_{t-1}$. The limiting distribution for the likelihood ratio statistic does not take standard form and has been derived by Johansen (1988) [221], and by Reinsel and Ahn (1992) [287]. The asymptotic distribution of the likelihood ratio test statistic under $H_0$ depends only on the number of unit roots, $d$, and not on any other parameters or on the order $p$ of the VAR model.

Critical values of the asymptotic distribution of (3.31) have been obtained by simulation by Johansen (1988) [221] and Reinsel and Ahn (1992) [287] and can be used in the test of $H_0$. It is known that inclusion of a constant term in the estimation of the nonstationary AR model affects the limiting distribution of the estimators and test statistics. Testing for this and other model variations can be found elsewhere in the literature (see Johansen (1991) [222], Engle and Granger (1987) [122], etc.).

### Co-Movement and Commonality; Further Discussion

Economic time series share some common characteristics such as trends, seasonality and serial correlations. The existence of common elements is identified through the indicators of co-movement. Series that have common stochastic trends can be identified through reduced-rank regression methods as discussed in this section. Centoni and Cubadda (2011) [69] provide a survey of models used in this area. From a modeling point of view, the parsimonious structure that results from the co-movement can be stated as

$$Y_t = C f_t + a_t, \tag{3.32}$$

where $f_t$ is a lower dimensional common feature vector so that a linear combination of $Y_t$, $l'Y_t$ can be independent of this feature. Observe that this implies $l'C = 0$, which is the result of '$C$' being a lower rank matrix. The $f_t$-vector can either be constructed from the past values of $Y_t$, which is the focal view of co-movement, or can be constructed from exogenous variables, say $X_t$, which is the basis of many studies on commonality in finance literature. We briefly summarize the two aspects here.

The co-movement relationships among non-stationary series that result from co-integration are known as long-run equilibrium relationships. Relationships that result from stationary series are known as short-run equilibrium relationships. The co-integrating relationships resulting from (3.29) are given by $\hat{Q}$ that leads to $\hat{Q}'\hat{A} = 0$. A generalization of this concept relates to a serial correlation common feature; to define this, write the VAP($p$) model in its error correction form as

$$W_t = \begin{bmatrix} A & D \end{bmatrix} \begin{bmatrix} BY_{t-1} \\ W^*_{t-1} \end{bmatrix} + \epsilon_t. \tag{3.33}$$

Here $W^*_{t-1} = [W'_{t-1}, \dots, W'_{t-p+1}]'$, with $W_{t-i} = Y_{t-i} - Y_{t-i+1}$. The combination that is free of serial correlation is $\hat{Q}W_t = Q'\epsilon_t$ which implies that $Q'\,[A \quad D] \sim 0$; from the results given in Section 3.2, it easily follows that the linear combination can be constructed from canonical vectors associated with the smallest canonical correlation between $Y_t$ and its past lags.

## 3.5 Applications in Finance

There is a vast literature on commonality in finance and the summary presented here is based on a few select studies. In addition to price and return variables of stocks, the commonality is studied on liquidity and turnover measures as well, which we define below

$$\begin{aligned} \text{Liq}_d &= -\log\left(1 + \frac{|r_d|}{P_d}\right) \\ \text{TO}_d &= \log\left(1 + \frac{v_d}{\text{NSH}}\right). \end{aligned} \tag{3.34}$$

Here $r_d$ is the return on day '$d$', $P_d$ is the price, $v_d$ is the volume traded and NSH is the number of outstanding shares at the beginning of the year. The turnover measure is adjusted by a moving average based on '$K$' days past '$d$'. As outlined in Chordia, Roll and Subrahmanyam (2000) [79] the commonality in liquidity can come from several sources:

- Traders manage their inventory in response to market-wide price swings, thus variation in volume traded can lead to changes in liquidity, which are measured by bid-ask spread, effective spread, quoted depth, etc., besides (3.34).
- Volatility of the stock and presence of informed traders who may have better knowledge about the market-movements.
- Trading cost that includes price impact as a major component.

  The commonality is simply inferred from the coefficients of the regression model

$$\text{DL}_{j,t} = \alpha_j + \beta_j' X_t + \epsilon_{j,t}, \tag{3.35}$$

  where DL is the percentage difference in a liquidity measure and where $X_t$ can include the corresponding market, industry variables. Generally the $\beta_j$'s tend to be positive and a fair number of them are significant which are meant to indicate commonality. Some general observations from related studies are worth noting.

- Commonality in liquidity is high during periods of high market volatility and during high market-wide trading activity; volatility effect is asymmetric; the commonality is more pronounced when market performance is in decline.

The factors in the model (3.32) or $X_t$ in (3.35) are sources of commonality that are broadly grouped into supply or demand side influences. The supply side hypotheses generally evaluate the role of funding constraints of financial intermediates who act as liquidity providers. When markets decline, the reasoning goes, that the intermediaries endure losses and thus reduce providing liquidity. The demand side theory postulates that liquidity commonality arises mainly due to correlated trading behavior of institutional investors. It has been concluded that the demand side explanation is empirically shown to be more plausible than the supply side theory.

A general comment on the type of models used in finance to infer commonality is in order. The liquidity measures at the equity levels are related to market level measures and the commonality is inferred from the strength of their relationships to the market measures. But no explicit modeling attempt is made to see if there is any commonality (similarity) among the coefficients over different equities. This can be done by building a multivariate regression model (3.1) which is essentially a set of multiple regressions—each regression representing a single equity. What binds the regression equations are constraints over coefficients across equities.

Hasbrouck and Seppi (2001) [183] provide a framework that examines the commonality in a comprehensive way. The model relates the order flows to returns via factor models that can be shown to have the reduced-rank regression structure. Let $X_t$ be $n$-vector of the order flows which are influenced by a common vector of order flows:

$$X_t = \theta F_t + \epsilon_t. \tag{3.36}$$

Here $F_t$ is lower (than '$n$') dimensional factor that is assumed to have orthogonal structure. A similar factor model can be written for the returns:

$$r_t = \phi\, G_t + \eta_t. \tag{3.37}$$

The common factors '$F_t$' can be due to dynamic hedging strategies, naïve momentum trading, etc. and the factors, $G_t$, can be due to expectation of future cash flows. To connect the two models (3.36) and (3.37), it is argued that the factor $F_t$ is linearly related to $G_t$, but its relationship to '$r_t$' is subject to further restriction. But in the high frequency context, we expect '$G_t$' and '$F_t$' also share common features. These factors are estimated through the principal components of $r_t$ and $X_t$, respectively. If we assume that the factors '$G_t$' are determined by $X_t$, then

$$G_t = \delta X_t + \eta_t^*. \tag{3.38}$$

Relating $X_t$ to $r_t$ via multivariate regression model, yields

$$r_t = \Lambda X_t + u_t = \phi\, \delta X_t + u_t, \tag{3.39}$$

where the coefficient matrix, $\Lambda$ is of lower rank. Thus the relationships between the returns and order flows can be studied through canonical correlations. The specification in (3.39) can be easily expanded to include other variables related to equity trading such as volume of trade, etc., in the regression framework presented in this chapter. In fact, the direct relationship between returns and order flows in (3.39) is a more efficient way to model than relating the factors from two sets of series as in Hasbrouck and Seppi (2001) [183]. Using the data for 30-Dow stocks for 1994, it is shown that the rank of the matrix '$\Lambda$' in (3.39) can be taken to be unity, as the first canonical correlation between returns and signed trade, an order flow measure tends to dominate.

## 3.6 Multivariate GARCH Models

It is well known that financial volatilities across assets and markets move together over time. It is of interest to study how a shock in one market increases the volatility on another market and in general, how correlations among asset returns change over time. For example, in the computation of hedge-ratio which is the slope of the regression line of spot returns on the futures return, both the covariance between spot and future returns and the variance of future returns can change over time. It must be noted that variances and covariances are generally sensitive to mean shifts. In addition, unlike modeling the mean behavior where a fewer number of quantities are involved, modeling the variance-covariance behavior involves a substantially larger number of quantities and hence various models that account for some redundancy are proposed in the literature. We cover only a select few in this section. Interested readers should refer to the excellent survey paper by Bauwens, Laurent and Rombouts (2006) [34].

The regression model (3.1), to begin with, can be written as

$$Y_t = \mu_t + \epsilon_t, \tag{3.40}$$

where $\mu_t = CX_t$ and $\epsilon_t \sim N(0, \Sigma_{\epsilon\epsilon})$. As $\Sigma_{\epsilon\epsilon}$ is taken to be positive definite, it can be written as $\Sigma_{\epsilon\epsilon} = H^{1/2} \cdot H^{1/2}$, where $H^{1/2}$ is the positive square root of the $\Sigma_{\epsilon\epsilon}$ matrix. Thus $\epsilon_t = H^{1/2} a_t$ where $a_t \sim N(0, I_m)$. When the covariance matrix, $\Sigma_{\epsilon\epsilon}$ changes over time, it can be characterized by

$$\epsilon_t = H_t^{1/2} a_t, \tag{3.41}$$

where $H_t$ is the conditional variance-covariance matrix.

There are two versions of model formulation of the elements of the $H_t$-matrix. Note this $m \times m$ matrix, because of symmetry, has only $m(m+1)/2$ distinct elements. Writing $h_t = \text{Vec}(H_t)$ that stacks these distinct elements as a vector, the Vec(1, 1) model (similar to GARCH(1,1) in the univariate case) is defined as:

$$h_t = c + A\eta_{t-1} + Gh_{t-1}, \tag{3.42}$$

where $\eta_t = \text{Vec}(\epsilon_t \epsilon_t')$, $A$ and $G$ are square matrices of dimension $\frac{m(m+1)}{2}$ and '$c$' is a column vector of the same dimension. Because of the larger dimensionality, various constraints on $A$ and $G$ are suggested. Another formulation that models $H_t$ directly is suggested in Engle and Kroner (1995) [118]. Write

$$H_t = C^{*'} C^* + \sum_{k=1}^{K} A_k^{*'} \epsilon_{t-1} \epsilon_{t-1}' A_k^* + \sum_{k=1}^{K} G_k^* H_{t-1} G_k^*, \tag{3.43}$$

where $C^*$, $A_k^*$ and $G_k^*$ are $m \times m$ matrices with $C^*$ being upper triangular. Here, '$K$' denotes the lag dependence. Further reduction is suggested by assuming $A_k^*$ and $G_k^*$ are rank one matrices. These rank one matrices result from the assumption that the changing variance elements are determined by a single varying factor. Indirectly it implies that the time-varying part of $H_t$ is of reduced rank.

One way to simplify the structure for $H_t$ is to assume that $H_t$ is based on '$K$' underlying factors:

$$H_t = \Omega + \sum_{k=1}^{K} w_k w_k' f_{k,t}, \tag{3.44}$$

where $\Omega$ is a positive definite matrix, $w_k$'s are vectors of factor weights and $f_{k,t}$'s are the factors. The individual factors are then allowed to follow GARCH(1,1) structure:

$$f_{k,t} = a_k + \alpha_k \left(\beta_k' \epsilon_{t-1}\right)^2 + \gamma_k f_{k,t-1}, \tag{3.45}$$

where $a_k, \alpha_k, \gamma_k$ are scalars and $\beta_k$ is a vector. Note this is somewhat a simplified version of (3.43) and is easy to model. See Engle, Ng and Rothschild (1990) [128] for details.

Empirical applications of the models related to covariances face the curse of dimensionality. As the number of variables ($m$) increases, the number of conditional variances and covariances, $\frac{m(m+1)}{2}$ increases exponentially. Also any application should guarantee that the estimated volatility matrices are positive definite. Recall that in a descriptive approach in the case of independent (with no temporal dependence) observations, the principal component analysis (PCA) plays an important role. It is based on the eigenvalue decomposition of the variance-covariance matrix, as discussed in Section 3.2. Hu and Tsay (2014) [204] extend this idea by defining a generalized kurtosis matrix for multivariate volatility and by carrying out the spectral decomposition of this matrix. Note that this matrix is based on the fourth moment, that accounts for both serial and cross-sectional aspects of a vector time series.

To start with, assume that the conditional volatility matrix, $\Sigma_t = \mathrm{E}(y_t y_t' \mid F_{t-1})$, where $F_{t-1}$ denotes the information available at time '$t-1$', can be modeled as

$$\mathrm{Vec}(\Sigma_t) = c_0 + \sum_{i=1}^{\infty} c_i \,\mathrm{Vec}(y_{t-i} y_{t-i}'), \tag{3.46}$$

which is in the form of a ARCH model. The $m^2$-dimensional '$c_0$' vector and $m^2 \times m^2$ matrices '$c_i$' are restricted so that '$\Sigma_t$' is a positive-definite matrix. Define the lag-$l$ generated kurtosis matrix, for $l \geq 0$,

$$\gamma_l = \sum_{i=1}^{m} \sum_{j=1}^{m} \mathrm{Cov}^2(y_t y_t', x_{ij \cdot t-l}), \tag{3.47}$$

where $x_{ij\cdot t-l}$ is a function of $y_{i\cdot t-l} y_{j\cdot t-l}$ for $1 \leq i, j \leq m$. The cumulative kurtosis matrix is then defined as,

$$\Gamma_k = \sum_{l=1}^{K} \gamma_l. \tag{3.48}$$

For the linearly transformed series, $z_t = M y_t$ where $M$ is a $m \times m$ full-rank matrix, note $\mathrm{Cov}(z_t z_t', x_{t-1}) = M' \,\mathrm{Cov}(y_t y_t', x_{t-1}) M$ and if there is no ARCH effect, $\gamma_l$'s and hence $\Gamma_k$ must be singular. The principal volatility components (PVC) matrix, $M$ is obtained from the eigenvalue decomposition of $M$:

$$\Gamma_\infty M = M\Lambda \tag{3.49}$$

with $m_v$ is the $v$th PVC, note $m_v'\gamma_{l\cdot ij}m_v$ is a measure of dependence in volatility. Thus the smaller values of $\Lambda$ would indicate that certain combinations will be independent of volatility dependence which implies that there is a common volatility component. The larger elements of $\Lambda$ would provide the most volatile combinations via the corresponding eigenvectors, $m_v$. This approach has potential to uncover relationships in volatilities that may be helpful for developing good trading algorithms.

## 3.7 Illustrative Examples

As mentioned in the introduction to this chapter, the application of multivariate regression and time series models is in the area of linking multiple markets or multiple assets, and the tools are also useful in understanding the efficiency of the portfolios. There is some commonality among certain equities due to common fundamentals or industry factors that have similar impact on their performance. Whenever there is a deviation from expected commonality or co-movement, it provides trading opportunities. This is the basis of 'pairs trading' that is discussed in detail in Chapter 5. Also the methodologies discussed in this chapter can be useful for portfolio analysis (Chapter 6) as well. The example that we present in this section is mainly for illustrating the methodologies discussed in this chapter as other applications will be taken up elsewhere.

We consider the daily exchange rates of five currencies British Pound (GBP), Euro (EUR), Japanese Yen (JPY), Canadian Dollar (CAD), Singapore Dollar (SGD), and Australian Dollar (AUD) all stated against US Dollar. The data is from January 1, 1999 to May 2, 2016. Thus $Y_t$ is a five dimensional vector of time series. The return series $r_t$ are calculated as $r_t = \ln(Y_t) - \ln(Y_{t-1})$. The plot of $Y_t$ is given in Figure 3.1 and to make the comparison easier, the first observation in all series are set to unity. Clearly the series are non-stationary and exhibit common upward and downward movements in certain durations. The plot of returns and volatilities are given in Figures 3.2 and 3.3, respectively.

Broadly defining the regimes as dollar strengthening—strong versus weak, the following regimes are identified: 1999–2002, 2002–2008, 2008–2009, 2009–2013 and 2013–present. The typical autocorrelation functions (ACFs) for the rate, return and volatility are given in Figures 3.4, 3.5, and 3.6, respectively.

Some general observations can be confirmed as follows:

- Rates are non-stationary.
- Returns are generally white-noise.

- Volatilities exhibit time-dependence.

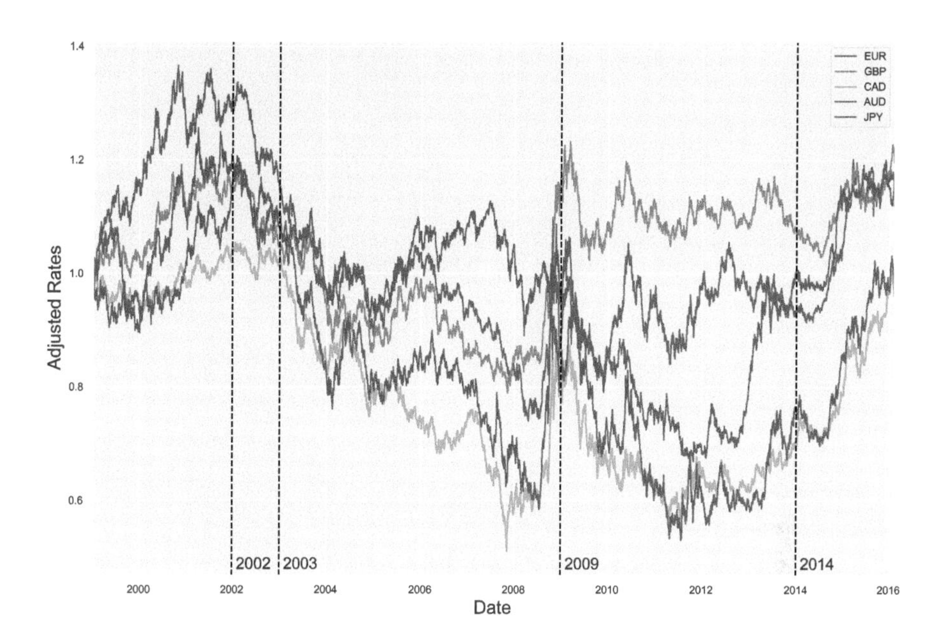


Figure 3.1: Plot of Exchange Rates.

Here, we want to study the co-integrating relationships, if any, in the rates and explore if there are any commonalities in the volatilities. To begin with, we briefly summarize the main ideas of this chapter. Each component series $y_{it}$ of the vector process $Y_t$ is to be nonstationary with its first difference $(1 - B)Y_{it}$ stationary (in which case $Y_{it}$ is said to be integrated of order one), but such that certain linear combinations $z_{it} = b_i' Y_t$ of $Y_t$ will be stationary. Then, the process $Y_t$ is said to be *co-integrated* with co-integrating vectors $b_i$ (e.g., Engle and Granger, 1987 [122]). An elegant interpretation of co-integrated vector series $Y_t$, in economics, is that the individual components $y_{it}$ share some common nonstationary components or "common trends" and, hence, they tend to have certain similar movements in their long-term behavior. A related interpretation is that the component series $y_{it}$, although they may exhibit nonstationary behavior, satisfy (approximately) a long-run equilibrium relation $b_i' Y_t \approx 0$ such that the process $z_{it} = b_i' Y_t$, which represents the deviations from the equilibrium, exhibits stable behavior and so is a stationary process. This is the basis of 'pairs trading' discussed in Chapter 5. For a VAR($p$) model, see the discussion in Section 3.4.

The main steps involved in VAR($p$) modeling are:

- Identification of the order '$p$' of the process either by examining the significance of the elements of the cross-correlation matrix, $\rho(l)$ for various values of '$l$' or by using one of the criteria, AIC or BIC in (3.27).

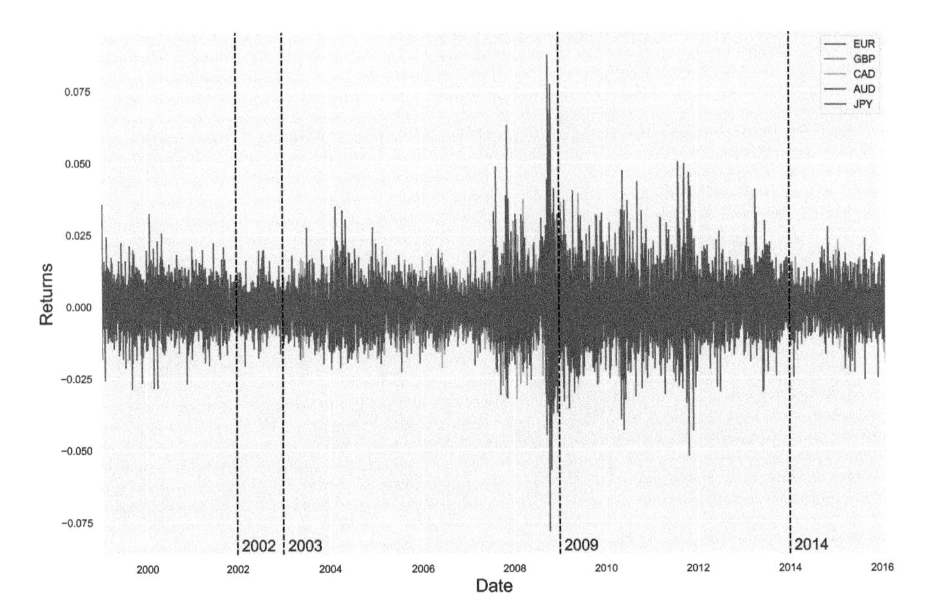


Figure 3.2: Plot of Returns.

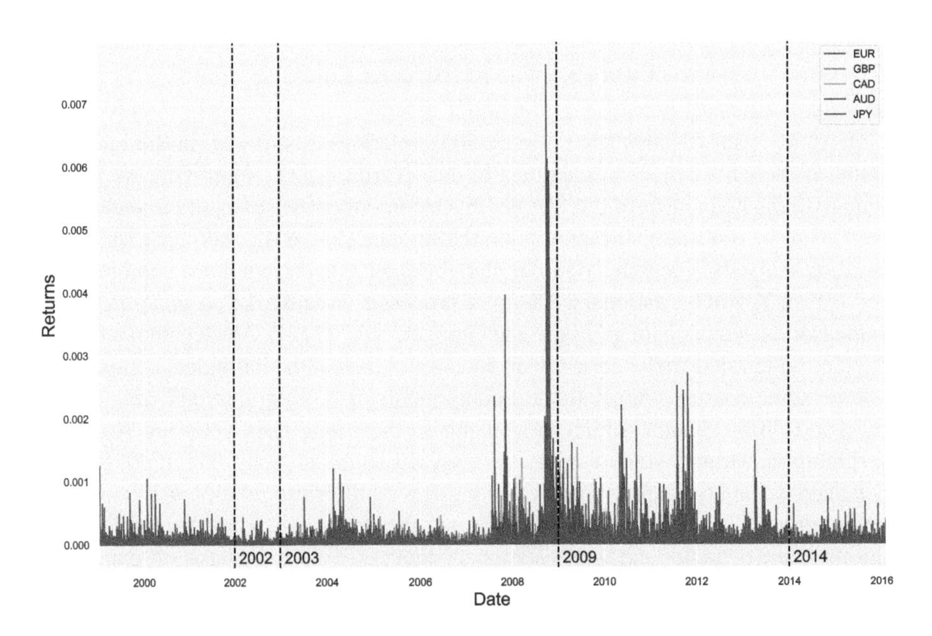


Figure 3.3: Plot of Volatilities.

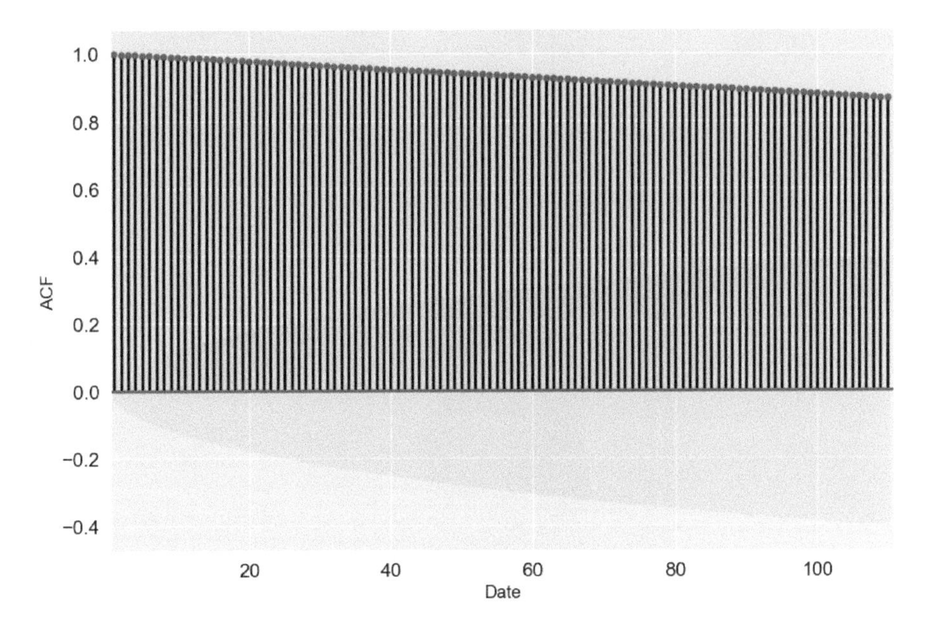


Figure 3.4: Autocorrelation Function for EUR (with 5% significance limits for the autocorrelations).

– For the non-stationary VAR, run the test (2.30) to check if individual series are non-stationary. Then build the model in (3.29), by first testing the co-integrating rank.

The results for testing co-integration for the full duration and the regime (2008–2009) are presented in Table 3.1 and Table 3.2, respectively. All the other regimes have the same pattern as the full duration and therefore the details are omitted. Except for the regime (2008–2009) we did not find any co-integrating relationships. This probably indicates that the exchange rate series may be moving on their own with no co-dependence. For the regime that is isolated here, the co-integrating relationships are plotted in Figure 3.7; there is only *one* stationary transformed series.

The co-integration results clearly indicate that the relationships among the exchange rate series could be regime-dependent. This is an important consideration when developing trading strategies. Now we want to examine if there is a linear combination of the returns that exhibit low volatility using the principal volatility component method (PVCM). Let $r_t = \ln Y_t - \ln Y_{t-1}$ be the five-dimensional returns process. The goal of PVCM, it may be recalled, is to find a matrix, $M$, such that

$$\operatorname{Cov}(Mr_t) = M\Sigma_t M = \begin{bmatrix} \Delta_t & c_2 \\ c_2' & c_1 \end{bmatrix} \tag{3.50}$$

with $\Delta_t$ being time-variant but $c_1$ is time-independent. For a forerunner of this idea in the case of $m = 2$, see Engle and Kozicki (1993) [123] and Engle and Susmel

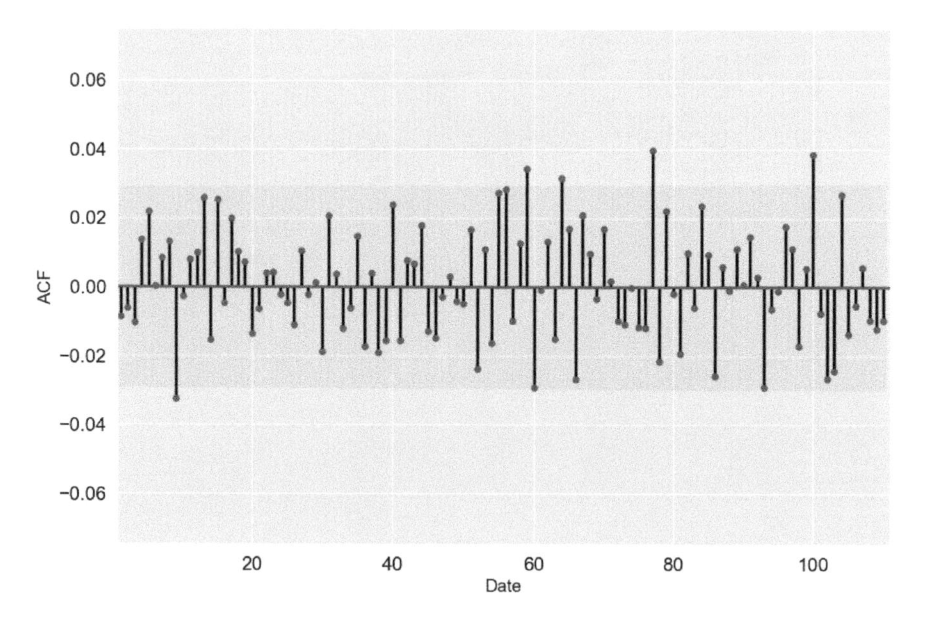


Figure 3.5: Autocorrelation Function for EurR (with 5% significance limits for the autocorrelations).

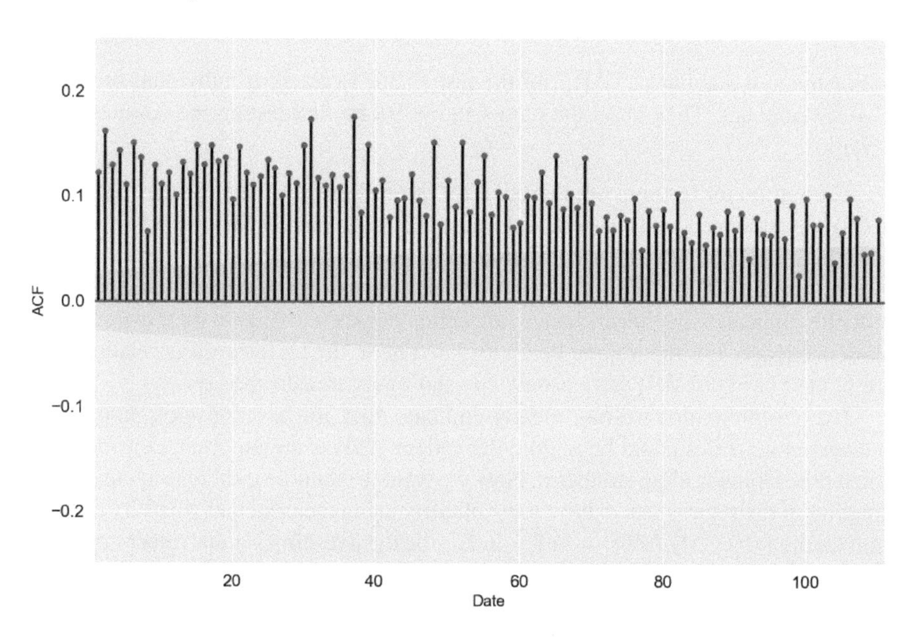


Figure 3.6: Autocorrelation Function for EurRSQ (with 5% significance limits for the autocorrelations).

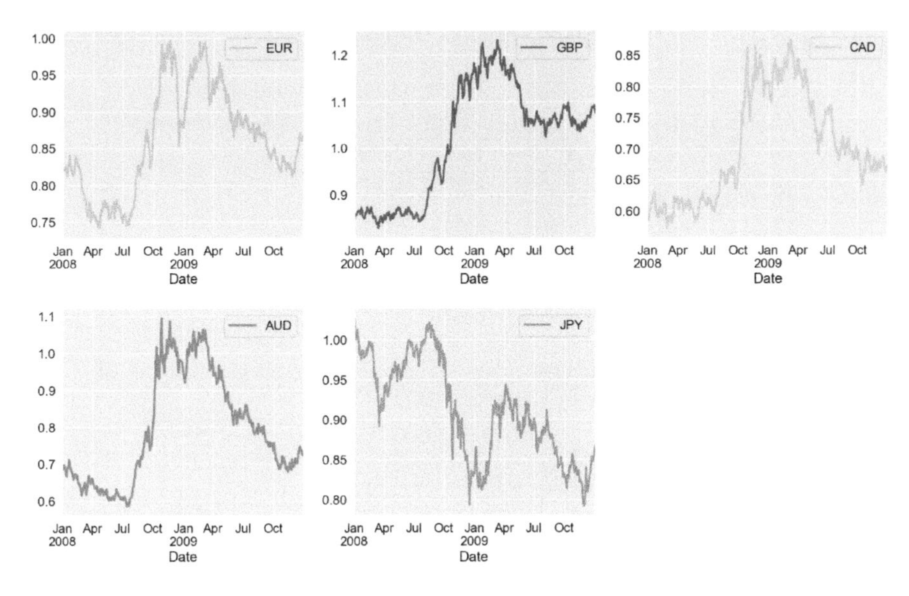


Figure 3.7: Plot of Co-Integrated Series.

Table 3.1: Full Duration

Eigenvalues ($\lambda$): 0.0054, 0.0030, 0.0018, 0.0008, 0.0003

Values of test statistic and critical values of test for various ranks:

| $r$ | test | 5pct |
|---|---|---|
| $r \leq 4$ | 1.26 | 8.18 |
| $r \leq 3$ | 3.71 | 14.90 |
| $r \leq 2$ | 7.87 | 21.07 |
| $r \leq 1$ | 13.47 | 27.14 |
| $r = 0$ | 24.19 | 33.32 |

Eigenvectors, normalised to first column: (These are the co-integrating relations)

| | EUR | GBP | CAD | AUD | JPY |
|---|---|---|---|---|---|
| EUR | 1.000 | 1.000 | 1.000 | 1.000 | 1.000 |
| GBP | −0.172 | −1.105 | −1.721 | 2.163 | −0.487 |
| CAD | −0.890 | 4.537 | −0.810 | 0.707 | 1.666 |
| AUD | −0.343 | −3.838 | 0.829 | −0.431 | 0.214 |
| JPY | 0.305 | −0.318 | −1.300 | −2.242 | 0.889 |

(1993) [126]. To find $M$, write the generalized kurtosis matrix (3.48) as

$$\Gamma_k = \sum_{h=1}^{k} \sum_{i=1}^{m} \sum_{j=1}^{m} \mathrm{Cov}^2(r_t r_t', r_{i,t-h} \cdot r_{j\cdot t-h}). \tag{3.51}$$

Table 3.2: Regime 2007–2009

Eigenvalues ($\lambda$): 0.0592, 0.0367, 0.0298, 0.0039, 0.0038

Values of test statistic and critical values of test of co-integrating rank:

| r | test | 5pct |
|---|---|---|
| r ≤ 4 | 1.97 | 8.18 |
| r ≤ 3 | 2.01 | 14.90 |
| r ≤ 2 | 15.77 | 21.07 |
| r ≤ 1 | 19.50 | 27.14 |
| r = 0 | 31.84 | 33.32 |

Eigenvectors, normalised to first column: (These are the co-integration relations)

| | EUR | GBP | CAD | AUD | JPY |
|---|---|---|---|---|---|
| EUR | 1.000 | 1.000 | 1.000 | 1.000 | 1.000 |
| GBP | −0.628 | 0.392 | −0.606 | 0.379 | −0.241 |
| CAD | 1.733 | −0.615 | −1.009 | −1.189 | −0.456 |
| AUD | −0.919 | −0.452 | 0.519 | 0.103 | −0.588 |
| JPY | −0.289 | 0.522 | −0.500 | −0.567 | −0.856 |

**Result 5.** If $r_t$ is an $m$-dimensional stationary series; the ARCH dimension of $r_t$ is $'r'$ if and only if $\text{rank}(\Gamma_\infty) = r$ and the no-ARCH transformation, satisfies, $M\Gamma_\infty = 0$. Essentially, $M$ is the matrix composed of the eigenvectors of $\Gamma_\infty$.

For the five exchange rate series, the computational details are presented in Table 3.3. It can be seen that the last two eigenvalues are fairly small. The linear combinations, $Mr_t$ are plotted in Figure 3.8. The plot of squared values of the two extremes are plotted in Figure 3.9. The autocorrelation functions of the transformed series are plotted in Figure 3.10 (a) & (b); the combination, 0.6835 EURR + 0.2054 GBPR − 0.3271 CADR + 0.1496 AUDR + 0.6010 JPYR exhibits the lowest volatility. It would obviously be interesting to test this out over different regimes.

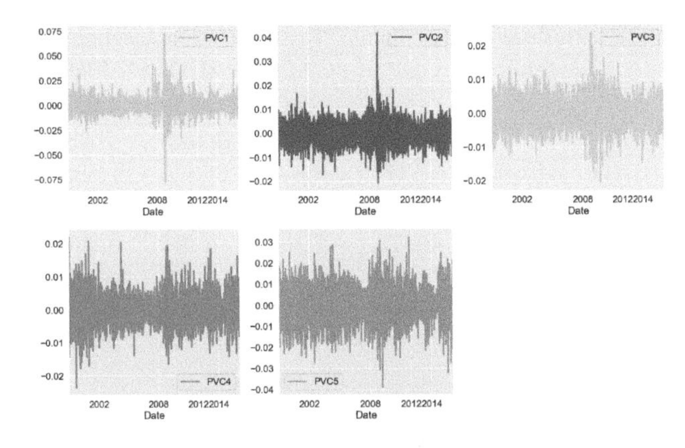


Figure 3.8: Transformed Principal Volatility Component Series.

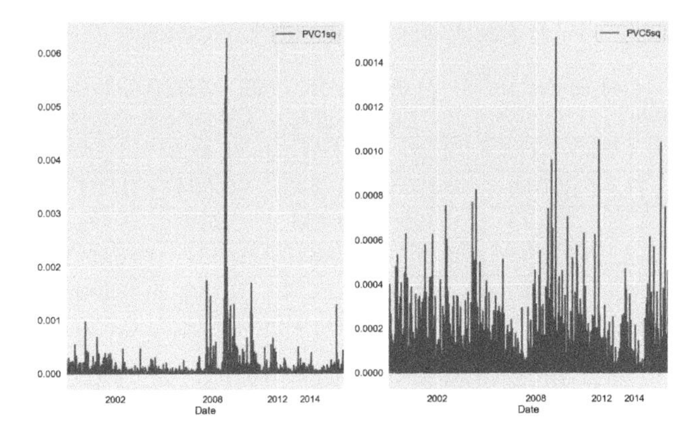


Figure 3.9: Volatilities of Extreme Series.

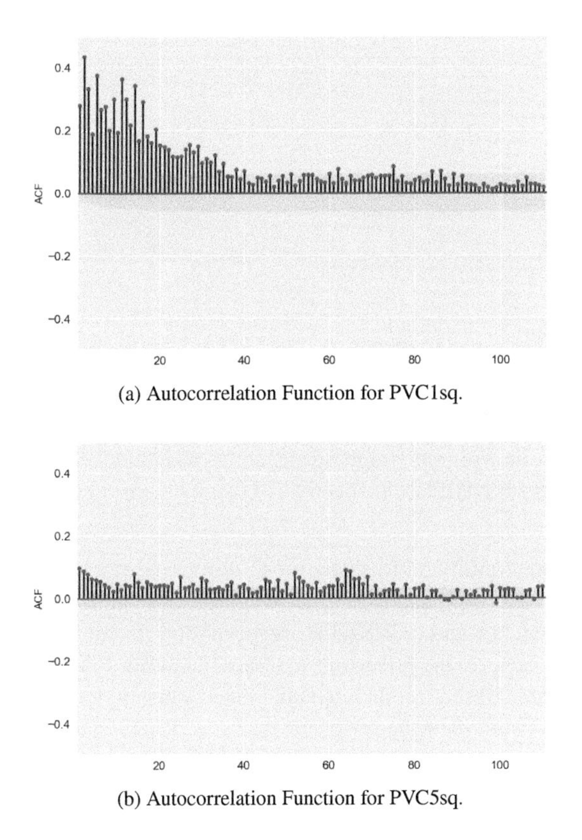


(a) Autocorrelation Function for PVC1sq.

(b) Autocorrelation Function for PVC5sq.

Figure 3.10: Autocorrelation Functions for Principal Volatility Components (with 5% significance limits for the autocorrelations).

Table 3.3: Principal Volatility Components

Eigenvalues ($\lambda$): 21.096, 4.453, 1.623, 0.728, 0.435

Eigenvectors

| | [,1] | [,2] | [,3] | [,4] | [,5] |
|---|---|---|---|---|---|
| [1,] | 0.015 | 0.208 | −0.225 | −0.694 | 0.651 |
| [2,] | 0.173 | 0.495 | 0.753 | 0.219 | 0.331 |
| [3,] | 0.195 | 0.717 | −0.577 | 0.331 | −0.080 |
| [4,] | 0.859 | −0.383 | −0.114 | 0.185 | 0.260 |
| [5,] | −0.440 | −0.228 | −0.191 | 0.571 | 0.626 |

$M$−Matrix

| | [,1] | [,2] | [,3] | [,4] | [,5] |
|---|---|---|---|---|---|
| [1,] | −0.196 | 0.077 | −0.283 | −0.755 | 0.684 |
| [2,] | 0.128 | 0.437 | 0.793 d | 0.319 | 0.205 |
| [3,] | −0.070 | 0.713 | −0.520 | 0.321 | −0.327 |
| [4,] | 0.849 | −0.512 | −0.002 | 0.129 | 0.150 |
| [5,] | −0.468 | −0.180 | −0.142 | 0.455 | 0.601 |

## 3.8 Exercises

1.

- The files `d_15stocks.csv` and `m_15stocks.csv` contain daily and monthly stock returns data for 15 stocks from January 11, 2000 to March 31, 2013.
- The files `d_indexes.csv` and `m_indexes.csv` contain daily and monthly returns data for the volume-weighted and equal-weighted S&P 500 market indices (VWRETD and EWRETD, respectively) from January 11, 2000 to March 31, 2013.

Using daily and monthly returns data for 15 individual stocks from `d_15stocks.csv` and `m_15stocks.csv`, and the equal-weighted and value-weighted CRSP market indexes (EWRETD and VWRETD, respectively) from `d_indexes.csv` and `m_indexes.csv`, perform the following statistical analyses. For the subsample analyses, split the available observations into equal-sized subsamples or at regime changes visually noticeable from the time series plots.

(a) Compute the sample mean $\hat{\mu}$, standard deviation $\hat{\sigma}$, and first-order autocorrelation coefficient $\hat{\rho}(1)$ for daily simple returns over the entire sample period for the 15 stocks and two indexes. Split the sample into 4 equal subperiods and compute the same statistics in each subperiod—are they stable over time?

(b) Plot histograms of daily simple returns for VWRETD and EWRETD over the entire sample period. Plot another histogram of the normal distribution with mean and variance equal to the sample mean and variance of the returns plotted in the first histograms. Do daily simple returns look approximately normal? Which looks closer to normal: VWRETD or EWRETD?

(c) Using daily simple returns for the sample period, construct 99% confidence intervals for $\hat{\mu}$ for VWRETD and EWRETD, and the 15 individual stock return series. Divide the sample into 4 equal subperiods and construct 99% confidence intervals in each of the four subperiods for the 17 series—do they shift a great deal?

(d) Compute the skewness, kurtosis, and studentized range of daily simple returns of VWRETD, EWRETD, and the 15 individual stocks over the entire sample period, and in each of the 4 equal subperiods. Which of the skewness, kurtosis, and studentized range estimates are statistically different from the skewness, kurtosis, and studentized range of a normal random variable at the 5% level? For these 17 series, perform the same calculations using monthly data. What do you conclude about the normality of these return series, and why?

2. For the data in Exercise 1, carry out the following:

(a) Perform a PCA on the 15-stock returns. Using the eigenvalues of the $\hat{\Sigma}$ matrix, identify the number of factors.

(b) Compute the covariance between the returns and the factors and comment.

3. Consider the weekly exchange rate data used in Hu and Tsay (2014) [204] in file `exchange_rates_2.csv`: British Pound ($y_{1t}$), Norwegian Kroner ($y_{2t}$), Swedish Kroner ($y_{3t}$), Swiss Franc ($y_{4t}$), Canadian Dollar ($y_{5t}$), Singapore Dollar ($y_{6t}$) and Australian Dollar ($y_{7t}$) against US Dollar from March 22, 2000 to October 26, 2011.

(a) Test if the exchange rates are stationary individually via Dickey-Fuller test.

(b) Test for co-integration using the likelihood-ratio test given in (3.31). Assume that the order of the VAR model as, $p = 2$.

(c) Compute the log returns: $r_t = \ln(Y_t) - \ln(Y_{t-1})$; Fit VAR(5) model for the returns and extract the residuals.

(d) Using the residuals from (c), compute the $\Gamma_k$ matrix in (3.48) for $k = 5, 10, 20$ and obtain the $M$-matrix in (3.49).

(e) Plot $MY_t$ and comment on the linear combination that has low volatility.

(f) Check (a)–(e) for various regimes; note 2008–2009 is noted for its volatility.

# 4

# *Advanced Topics*

In this chapter, we want to discuss some advanced methods that are applicable in the context of trading. The mechanics of actual trading (entry/exit) decisions call for tracking the market movements in real time and the tools provided in this chapter would be quite useful to better track and understand the market changes. This chapter is broadly divided into two main areas; the first area covers topics which are extensions of models discussed in Chapter 2 in the low frequency context. Although the state-space modeling is discussed in the context of ARIMA, its scope is much wider. We also outline the established methodology for studying regime switching, but indicate the topics of further research where predicting ahead when the regime shift is likely to occur. This is followed by a discussion on using volume information to study price volatility. With the increased exploitation of price behavior data, it has become necessary to look for additional trading signals. We present a theoretical model that shows that there is correlation between volume traded and volatility when there is information flow. The second area of this chapter focuses on the models from point processes to study higher frequency data. Here our approach has been to provide an intuitive discussion of the main tools; as readers will realize, the trade data is quite noisy due to market frictions, so the formal parametric models do not perform well. Thus there is a great deal of scope to do research in non-parametric empirical modeling of arrivals and cancellations in the limit order market.

## 4.1 State-Space Modeling

The state-space models were initially developed by control systems engineers to measure a signal contaminated by noise. The signal at time "$t$" is taken to be a linear combination of variables, called state variables that form the so-called state vector at time, $t$. The key property of the state vector is that it contains information from past and present data but the future behavior of the system is independent of the past values and depends only on the present values. Thus the latent state vector evolves according to the Markov property. The state equation is stated as,

$$Z_t = \Phi_t Z_{t-1} + a_t \tag{4.1}$$

and the observation equation as,

$$Y_t = H_t Z_t + N_t, \tag{4.2}$$

where it is assumed that $a_t$ and $N_t$ are independent white-noise processes; $a_t$ is a vector white noise with covariance matrix $\Sigma_a$ and $N_t$ has variance $\sigma_N^2$. The matrix $\Phi_t$ in (4.1) is an $r \times r$ transition matrix and $H_t$ in (4.2) is an $1 \times r$ vector; both are allowed to vary in time. In engineering applications, the structure of these matrices is governed by some physical phenomena.

The state space form of an ARIMA$(p, d, q)$ process $\Phi(B)Y_t = \theta(B)\epsilon_t$ has been extensively studied (see Reinsel (2002) [288, Section 2.6]). Observe if $\Phi_t \equiv \Phi$ and $H_t \equiv H$ for all $t$ then the system (4.1)–(4.2) is said to be time-invariant. A nice feature of expressing a model in state-space form is that an updating procedure can be readily used when a new observation becomes available to revise the current state vector and to produce forecasts. These forecasts in general tend to fare better.

For the state-space model, define the finite sample estimates of the state vector $Z_{t+1}$ based on observations $Y_t, \ldots, Y_l$, as $\hat{Z}_{t+l|t} = \mathrm{E}[Z_{t+l} \mid Y_t, \cdots, Y_l]$, with $V_{t+l \mid t} = \mathrm{E}[(Z_{t+l} - \hat{Z}_{t+l \mid t})(Z_{t+l} - \hat{Z}_{t+l \mid t})']$. A convenient computational procedure, known as the *Kalman filter equations*, is used to obtain the current estimate $\hat{Z}_{t \mid t}$, in particular. It is known that (see Reinsel (2002) [288]), starting from some reasonable initial values $Z_0 \equiv \hat{Z}_{0 \mid 0}$ and $V_0 \equiv \hat{V}_{0 \mid 0}$, the estimate, $\hat{Z}_{t \mid t}$, can be obtained through the following recursive relations:

$$\hat{Z}_{t \mid t} = \hat{Z}_{t \mid t-1} + K_t(Y_t - H_t \hat{Z}_{t \mid t-1}), \tag{4.3}$$

where $K_t = V_{t \mid t-1} H_t' [H_t V_{t \mid t-1} H_t' + \sigma_N^2]^{-1}$ with $\hat{Z}_{t \mid t-1} = \Phi_t \hat{Z}_{t-1 \mid t-1}, V_{t \mid t-1} = \Phi_t V_{t-1 \mid t-1} \Phi_t' + \Sigma_a$ and $V_{t \mid t} = [I - K_t H_t] V_{t \mid t-1} = V_{t \mid t-1} - V_{t \mid t-1} H_t' [H_t V_{t \mid t-1} H_t' + \sigma_N^2]^{-1} H_t V_{t \mid t-1}$ for $t = 1, 2, \ldots$.

In (4.3), the quantity $\varepsilon_{t \mid t-1} = Y_t - H_t \hat{Z}_{t \mid t-1} \equiv Y_t - \hat{Y}_{t \mid t-1}$ at time $t$ is the new information provided by the measurement $Y_t$, which was not available from the previous observed (finite) history of the system. The factor $K_t$ is termed the "Kalman gain" matrix. The filtering procedure in (4.3) has the recursive "updating" form, and these equations represent the minimum mean square error predictor property. For example, note that $\mathrm{E}[Z_t \mid Y_t, \ldots, Y_1] = \mathrm{E}[Z_t \mid Y_{t-1}, \cdots, Y_1] + \mathrm{E}[Z_t \mid Y_t - \hat{Y}_{t \mid t-1}]$, since $\varepsilon_{t \mid t-1} = Y_t - \hat{Y}_{t \mid t-1}$ is independent of $Y_{t-1}, \ldots, Y_1$. Note from (4.3), the estimate of $Z_t$ equals the prediction of $Z_t$ from observations through time $t-1$ updated by the factor $K_t$ times the innovation $\varepsilon_{t|t-1}$. The quantity $K_t$ can be taken as the regression coefficients of $Z_t$ on the innovation $\varepsilon_{t|t-1}$, with $\mathrm{Var}(\varepsilon_{t \mid t-1}) = H_t V_{t \mid t-1} H_t' + \sigma_N^2$ and $\mathrm{Cov}(Z_t, \varepsilon_{t \mid t-1}) = V_{t \mid t-1} H_t'$. Thus, the general *updating relation* is $\hat{Z}_{t \mid t} = \hat{Z}_{t \mid t-1} + \mathrm{Cov}(Z_t, \varepsilon_{t \mid t-1})\{\mathrm{Var}(\varepsilon_{t \mid t-1})\}^{-1} \varepsilon_{t \mid t-1}$, where $\varepsilon_{t \mid t-1} = Y_t - \hat{Y}_{t \mid t-1}$, and the relation in $V_t$ is the usual updating of the error covariance matrix to account for the new information available from the innovation $\varepsilon_{t \mid t-1}$. Forecasts of future state values are available as $\hat{Z}_{t+l \mid t} = \Phi_{t+l} \hat{Z}_{t+l-1 \mid t}$ for $l = 1, 2, \ldots$.

Note for ARIMA models with $Y_t = HZ_t$ and $H = [1, 0, \ldots, 0]$, this Kalman filtering procedure provides an alternate way to obtain finite sample forecasts, based on data $Y_t, Y_{t-1}, \cdots, Y_1$, for future values in the ARIMA process, subject to specification of appropriate initial conditions to use in the terms in (4.3). For stationary $Y_t$ with $\mathrm{E}(Y_t) = 0$, the appropriate initial values are $\hat{Z}_{0\,|\,0} = 0$, a vector of zeros, and $\hat{V}_{0\,|\,0} = \mathrm{Cov}(Z_0) \equiv V_*$, the covariance matrix of $Z_0$. Specifically, because the state vector $Z_t$ follows the stationary vector AR(1) model $Z_t = \Phi Z_{t-1} + \psi\varepsilon_t$, its covariance matrix $V_* = \mathrm{Cov}(Z_t)$ satisfies $V_* = \Phi V_* \Phi' + \sigma_\varepsilon^2 \Psi\Psi'$, which can be readily solved for $V_*$. For nonstationary ARIMA processes, additional assumptions need to be specified. The "steady-state" values of the Kalman filtering procedure $l$-step ahead forecasts $\hat{Y}_{t+l\,|\,t}$ and their forecast error variances $v_{t+l\,|\,t}$ will be identical to the expressions given in Chapter 2, $\hat{Y}_t(l)$ and $\sigma^2(l) = \sigma^2(1 + \sum_{i=1}^{i-1} \psi_i^2)$.

An alternative set-up is to assume that the deviations in the observation and transition equations are related. With the observation equation stated in (4.2), the transition equation is modified as

$$Z_t = \Phi Z_{t-1} + \alpha N_t. \tag{4.4}$$

This model studied in Ord, Koehler, and Snyder (1997) [278] with a single source of error ($N_t$) is shown to be closely related to various exponential smoothing procedures. There are other formulations of state-space models, such as structural models by Harvey (1989) [179] and dynamic linear models of West and Harrison (1997) [322] that are found to be useful for model building.

We provide two examples to illustrate the application of the models (4.1) and (4.2) in the broader context of this book.

**Example 4.1 (Generalized Local Trend).** The slope in a local linear trend is a random walk. Specifically,

$$\begin{aligned} p_t &= \mu_t + e_t \\ \mu_t &= \mu_{t-1} + \delta_{t-1} + \eta_{0t} \\ \delta_t &= \delta_{t-1} + \eta_{1t} \end{aligned} \tag{4.5}$$

Here the trend is composed of level '$\mu_t$' and slope '$\delta_t$'. The slope '$\delta_t$' when written as

$$\delta_t = D + e(\delta_{t-1} - D) + \eta_{1t} \tag{4.6}$$

in AR(1) form is sometimes known as generalized local linear trend. These models can be written in the form of (4.1) and (4.2); the updating process given in (4.3) can be used to accommodate a wide range of models. ◁

**Example 4.2 (Price Discovery in Multiple Markets).** Menkveld, Koopman and Lucas (2007) [266] study the price discovery process for stocks that are cross-listed in two different markets via state-space models. The question if the trading location matters more than business location is of interest to finance researchers. Consider a stock traded in both Amsterdam and New York. Because of the time difference, there is only a limited overlap in trading. Ignoring this aspect, write the two-dimensional price vector as

$$\begin{aligned} p_t &= \mu_t + e_t \\ &= \mu_t + \theta(\mu_t - \mu_{t-1}) + e_t \end{aligned} \tag{4.7}$$

as the observational equation with delayed price reaction and the state equation is

$$\mu_t = \mu_{t-1} + \epsilon_t. \tag{4.8}$$

Here $\epsilon_t$ can represent a more general covariance structure such as a factor model where a common factor that can influence both prices could be easily accommodated. The above can be written in the form of (4.1) and (4.2) with appropriate choice of '$\Phi$' and '$H$' matrices. For details see Section 2.4 of Menkveld et al. (2007)[266]. The main advantage of state-space models is that they can accommodate the possibility of slowly varying parameters over time, which is more realistic in modeling the real world data. ◁

The estimation of model parameters can be done via likelihood function. For this and how VAR models discussed in Chapter 4 can be written in the state-space form, see Reinsel (2002) [288].

## 4.2 Regime Switching and Change-Point Models

It has long been noted in finance that there are two kinds of recurrences in stock prices: Underreaction and overreaction to a series of good or bad news. The securities that have a good record for an extended period tend to become overpriced and have low returns subsequently. Barberis, Shleifer and Vishny (1998) [30] present a model where the investor believes that the returns can arise from one of two regimes although returns follow random walk: Mean-reverting or trending. The transition probabilities are taken to be fixed and the regime is more likely to stay the same rather than to switch. But the investor's beliefs are updated as and when the returns data are observed. It is presumed that in many instances, we may not discern these shifts directly whether or when they occur but instead can draw probabilistic inference based on the observed behavior posthumously. Hamilton (2016) [174] reviews the area of regime-switching applied in several contexts in macroeconomics, building upon the elegant model introduced in Hamilton (1989) [172]. Here in this section, we will cover only studies that are relevant to finance applications, particularly relevant to trading that exploit anomalies.

Consider the model,

$$y_t = \mu_{s_t} + \epsilon_t, \quad s_t = 1, 2, \tag{4.9}$$

where $y_t$ is the observed variable, such as asset return and $s_t$ represents two distinct regimes and $\epsilon_t$'s are i.i.d. $\sim N(0, \sigma^2)$. Let $\{F_{t-1}\}$ be the information set available as of '$t-1$'. The transition between the regimes is taken to be Markovian,

$$\text{Prob}(s_t = j \mid s_{t-1} = i, F_{t-1}) = p_{ij}, \quad i, j = 1, 2; \tag{4.10}$$

thus, leading to an AR(1) model for $\mu_{s_t}$ as

$$\mu_{s_t} = \phi_0 + \phi_1 \mu_{s_{t-1}} + a_t, \tag{4.11}$$

where $a_t$ by definition can take four possible values depending upon $s_t$ and $s_{t-1}$. Note $\phi_0 = p_{21}\mu_1 + p_{12}\mu_2$ and $\phi_1 = p_{11} - p_{21}$. Also, note the model for $y_t$ in (4.9) is the sum of an AR(1) process and white noise, then from the aggregation Result 2 in Chapter 2, $y_t \sim$ ARMA(1,1); but because of the discrete nature of '$a_t$', (4.9) is a nonlinear process. Observe that the unknown parameters are $(\mu_1, \mu_2, \sigma, p_{11}, p_{22})' = \lambda$ and they can be estimated via maximum likelihood as follows: Note $y_t \mid (s_t = j, F_{t-1}) \sim N(\mu_j, \sigma^2)$. The predictive density of $y_t$ is given as a finite mixture model,

$$f(y_t \mid F_{t-1}) = \sum_{i=1}^{2} \text{Prob}(s_t = i \mid F_{t-1})\, f(y_t \mid s_t = i, F_{t-1}) \tag{4.12}$$

and estimate of $\lambda$ is obtained by maximizing the likelihood function $L(\lambda) = \sum_{t=1}^{T} \log f(y_t \mid F_{t-1}, \lambda)$. Some useful results are:

Predicted regime:

$$\text{Prob}(s_t = j \mid F_{t-1}) = p_{1j}\, \text{Prob}(s_{t-1} = 1 \mid F_{t-1}) + p_{2j}\, \text{Prob}(s_{t-1} = 2 \mid F_{t-1}). \tag{4.13}$$

Optimal forecast:

$$\text{E}(y_t \mid F_{t-1}) = \sum_{i=1}^{2} (\phi_0 + \phi_1 \mu_i)\, \text{Prob}(s_{t-1} = i \mid F_{t-1}). \tag{4.14}$$

The forecast for '$k$' period ahead can be stated as:

$$\text{E}(y_{t+k} \mid F_{t-1}) = \mu + \phi_1^k \sum_{i=1}^{2} (\mu_i - \mu)\, \text{Prob}(s_t = i \mid F_{t-1}), \tag{4.15}$$

where $\mu = \phi_0/(1 - \phi_1)$.

The basic model has been extended in the literature to cover multiple regimes and to vector processes. Interested readers should refer to Hamilton (2016) [174] and references therein. Ang and Timmermann (2012) [21] discuss a model that accounts for changes not only in the mean but also in variances and autocorrelations:

$$y_t = \mu_{s_t} + \phi_{s_t} y_{t-1} + \epsilon_t \tag{4.16}$$

where $\epsilon_t$'s are independent $\sim N(0, \sigma_{s_t}^2)$. Using the data on excess S&P 500 returns, FX returns, etc., it is identified that there are volatility regimes, but no level shifts.

Hamilton (1990) [173] proposes using the popular EM (Expectation-Maximization) algorithm (Dempster, Laird and Rubin (1977) [105]) to obtain the maximum likelihood estimate, instead of the recursive filter approach (see Hamilton (1989) [172]) to optimize the log-likelihood of (4.12), updated with $y_t$. Note by Bayes' Theorem

$$\text{Prob}(s_t = j \mid F_t) = \frac{\text{Prob}(s_t = j \mid F_{t-1})\, f(y_t \mid s_t = j, F_{t-1})}{f(y_t \mid F_{t-1})}. \tag{4.17}$$

To get an estimate of $p_{ij}$ in step '$l+1$', use

$$\hat{p}_{ij}^{(l+1)} = \frac{\sum_{t=1}^{T-1} \text{Prob}(s_t = i, s_{t+1} = j \mid F_t, \hat{\lambda}^{(l)})}{\sum_{t=1}^{T-1} \text{Prob}(s_t = i \mid F_t, \hat{\lambda}^{(l)})} \tag{4.18}$$

and with these estimates obtain the ML estimates of the rest of the parameters ($\mu_1$, $\mu_2$, and $\sigma$) by maximizing the likelihood function. More details can be found in Hamilton (1990) [173].

There is a vast literature on Bayesian approaches to this problem, see for example, Chib (1998) [77] and subsequent related publications. An interesting application is given in Pastor and Stambaugh (2001) [283]. Lai and Xing (2013) [236] study a general model, somewhat similar to (4.16):

$$\text{ARX-GARCH:} \quad y_t = \beta_t' x_t + v_t \sqrt{h_t}\, \epsilon_t, \tag{4.19}$$

where $h_t \sim \text{GARCH}(p, q)$ and $\beta_t$ and $v_t$ are piecewise linear. Using the weekly returns of the S&P 500 index, the AR(1)-GARCH(1,1) (AR for the mean and GARCH for the variance) model is attempted and it is identified that 1998–2003 had higher volatility. This can be seen from Figure 4.1 that provides the plot of closing price, returns, volatility and volume data. In most of the applications of the above regime-switching models, particularly in finance, the plots of relevant quantities clearly reveal the pattern and these models provide a way to confirm the level and size shifts. From the plots, it is easy to see how the volatility in returns increases during the regime changes.

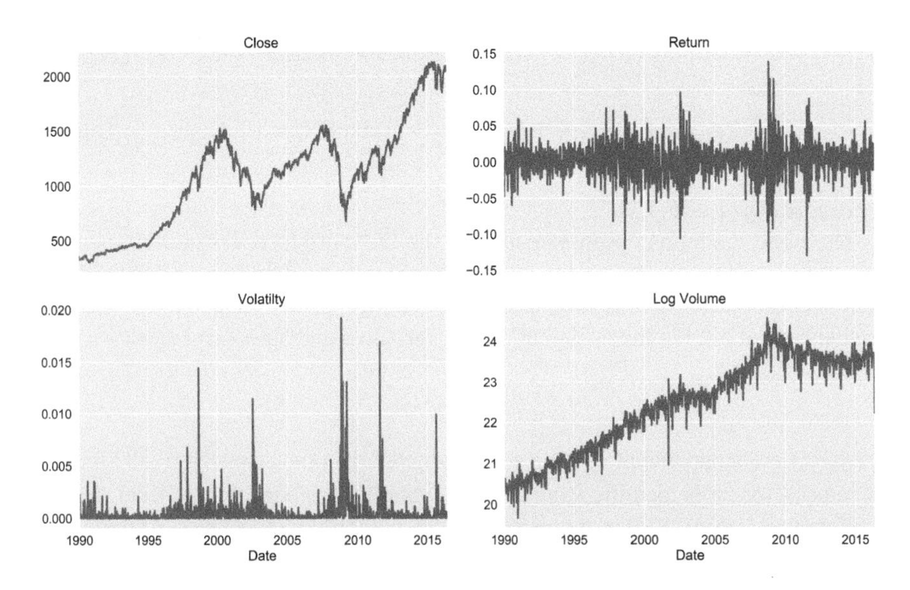


Figure 4.1: Plot of S&P 500 Index.

We want to conclude this section with some simple ideas borrowed from the area of statistical process control. Recall that if the market is efficient, the return $r_t =$

$\ln P_t - \ln P_{t-1} \sim \text{IID}(\mu_t, \sigma^2)$, where $\mu_t = 0$. But assume that $\mu_t = c\delta$ when $t \geq t_0$ and zero elsewhere. The cumulative sum chart (CUSUM) to detect the positive and negative mean shift are based on recursive calculations:

$$\begin{aligned} c_t^+ &= \max\left(0, c_{t-1}^+ + (r_t - k)\right) \\ c_t^- &= \min\left(0, c_{t-1}^- - (r_t - k)\right), \end{aligned} \tag{4.20}$$

where $k = \delta/2$. The chart signals as shown as $c_t^+$ reaches '$h$'—a decision point set by the user. Jiang, Shu and Apley (2008) [219] modify (4.20) with an exponentially weighted moving average (EWMA) type of estimates that is adaptive in nature. It can be stated as,

$$s_t^+ = \max\left(0, s_{t-1}^+ + w(\hat{\delta}_t)\left(r_t - \frac{\hat{\delta}_t}{2}\right)\right), \tag{4.21}$$

where $w(\hat{\delta}_t)$ is a weight function; a simple weight function $w(\hat{\delta}_t) = \hat{\delta}_t$ is suggested. The $\delta_t$ is updated as

$$\hat{\delta}_t = (1 - \lambda)\,\hat{\delta}_{t-1} + \lambda r_t. \tag{4.22}$$

The adaptive CUSUM chart is effective at detecting a range of mean shift sizes. Similar results for the variance need to be worked out.

In practical applications of these regime-shifting models, it is not easy to discern between a single outlier and persistent level shift. Valid measures that can distinguish these two situations on a timely basis could be very valuable to a trader. We also want to remind the reader that the Markov switching models of changing regimes are indeed latent variable time series models. These models are known as hidden Markov models (HMMs) and represent a special class of dependent mixtures of several processes, these are closely related to state-space models in Section 4.1. An unobservable $m$-state Markov chain determines the state or regime and a state-dependent process of observations. The major difference is that HMMs have discrete states while the Kalman filter based approach refers to latent continuous states.

## 4.3 A Model for Volume-Volatility Relationships

While the price information via returns and the volatility can be modeled through methods described earlier, it has become important to incorporate other relevant trading data to augment trading strategies. To that effect, we want to seek some guidance from economic theory on how that information is linked to price movements.

In the market microstructure theory, it is assumed that price movements occur primarily due to the arrival of new information and this information is incorporated efficiently into market prices. Other variables such as trading volume, bid-ask spread and market liquidity are observed to be related to the volatility of the returns. Early empirical studies have documented a positive relationship between daily trading volume and volatility (See Clark (1973) [81]). Epps and Epps (1976) [129] and Tauchen and Pitts (1983) [311] assume that both the volume and price are driven by the underlying 'latent' information ($I$) and provide a theoretical framework to incorporate this information. Assuming that there are fixed number of traders who trade in a day and the number of daily equilibria, $I$ is random because the number of new pieces of information is random, the return and the volume are written as a bivariate normal (continuous) mixture model with the same mixing variable $I$. Conditional on $I$:

$$\begin{aligned} r_t &= \sigma_1 \sqrt{I_t} Z_{1t} \\ V_t &= \mu_V I_t + \sigma_2 \sqrt{I_t} Z_{2t} \end{aligned} \tag{4.23}$$

where $Z_{1t}$ and $Z_{2t}$ are standard normal variables and $Z_{1t}$, $Z_{2t}$, and $I_t$ are mutually independent. Thus the volatility-volume relationship,

$$\begin{aligned} \mathrm{Cov}(r_t^2, V_t) &= \mathrm{E}(r_t^2 V_t) - \mathrm{E}(r_t^2)\ \mathrm{E}(V_t) \\ &= \sigma_1^2 \mu_V\ \mathrm{Var}(I_t) > 0 \end{aligned} \tag{4.24}$$

is positive due to the variance in $I_t$. If there is no new information or if there is no variation in the mixing variable, $\mathrm{Var}(I_t) = 0$ and thus the relationship vanishes. The theory of arrival rates suggest a Poisson distribution and based on empirical evidence, the lognormal distribution is taken to be a candidate for distribution of mixing variable, $I_t$. The parameters of the model in (4.23) are estimated through maximum likelihood. Gallant, Rossi and Tauchen (1992) [155] using a semi-parametric estimate of the joint density of $r_t$ and $V_t$ conduct a comprehensive study of NYSE data from 1925 to 1987. The following summarizes their main findings:

**Stylized Fact 6 (Volume–Volatility Relationship):**

- There is a positive correlation between conditional volatility and volume.
- Large price movements are followed by high volume.
- Conditioning on lagged volume substantially attenuates the leverage (which is an asymmetry in the conditional variance of current price change against past price change) effect.
- After conditioning on lagged value, there is a positive risk-return relation.

These findings can be easily confirmed using the S&P 500 data presented in Figure 4.1. During the period of regime changes (for example, October 1999–November 2000), we should expect higher volatility and therefore higher correlation with volume.

Andersen (1996) [17] modifies the model (4.23) by integrating the microstructure setting of Glosten and Milgrom (1985) [167] with stochastic volatility, that is built on weaker conditions on the information arrival process. While the first equation in (4.23) remains the same, the volume $V_t$ has informed and noise components, $V_t = IV_t + NV_t$. Noise trading component, $NV_t$ is taken as a time homogeneous

Poisson process, $P_0(m_0)$. Therefore, the systematic variation in trading is mainly due to informed volume. Then $IV_t \mid I_t \sim \text{Poisson}(I_t\mu)$ and thus

$$V_t \mid I_t \sim \text{Poisson}(m_0 + I_t m_1). \tag{4.25}$$

It is easy to see $\text{Cov}(r_t^2, V_t) = \sigma^2 m_1 \text{Var}(I_t) > 0$ under this setting as well. These models clearly indicate that the intraday return volatility and volume processes jointly contain some predictable elements.

There are other studies that focus on the trading volume and returns relationship and we mention only a select few here. Campbell, Grossman, and Wang (1993) [65] demonstrate for individual large stocks and stock indices the first-order daily autocorrelation in the returns tend to decline with volume. The authors develop a theoretical model where the economy has two assets, risk-free asset and a stock and there are two types of investors, one with constant risk aversion and the other with risk aversion changing over time. Under this set-up, it is implied that a decline in a stock price on a high-volume day is more likely than a decline on a low-volume day to be associated with an increase in the expected stock return.

Gervais, Kaniel, and Mingelgrin (2001) [163] investigate the role of trading activity in providing information about future prices. It is shown that periods of extremely high (low) volume tend to be followed by positive (negative) excess returns. The formation period for identifying extreme trading volume is a day or a week, but the effect lasts at least twenty days and holds across all stock sizes. To test if this information can be used profitably, the authors construct a trading strategy, by sending buy (sell) limit orders at the existing bid (ask) price, at the end of the formation period. If the orders are not taken, they are converted into market orders at the closing. The strategy is shown to result in profits, in particular with the small-medium firm stocks, after adjusting for transaction costs. The model used for the empirical framework is the vector autoregressive model discussed in Chapter 3, with an addition of a set of exogenous control variables:

$$Y_t = \Phi_0 + \sum_{j=1}^{p} \Phi_j Y_{t-j} + \sum_{l=1}^{L} B_l X_{t-l} + \epsilon_t. \tag{4.26}$$

The components of $Y_t$ include stock or market related variables such as detrended log of stock turnover, the stock return and the value weighted market return. The control variables are market volatility based on the daily return standard deviation and the dispersion which is the cross-sectional standard deviation of the security returns. The impulse response functions are used to aggregate the overall relationship among the endogenous variables. Statman, Thorley, and Vorkink (2006) [306] show that the trading volume is dependent on past returns over many months and it is argued that this may be due to the overconfidence of investors.

These and other studies generally confirm the information content of the volume and turnover of stocks traded in the low frequency context; strategies to exploit these, especially in a higher frequency context, will be mentioned in Chapter 5. We will discuss how intra-day flow of volume can be related to price changes.

In algorithmic trading a key ingredient of many strategies is the forecast of intra-day volume. Typically a parent order is split into several child orders and the timing of the submission of child orders depend on the volume forecast for an interval of time that is considered for trading. Brownlees, Cipollini and Gallo (2011) [62] provide a prediction model for intra-day volume which will be described in detail in the next chapter.

The model in (4.23) is extended to cover multiple stocks in He and Velu (2014) [188]. The role of common cross-equity variation in trade related variables is of interest in financial economics. The factors that influence the prices, order flows and liquidity are most likely to be common among equities that are exposed to the same risk factors. Exploring commonality is useful for institutional trading such as portfolio rebalancing. The mixture distribution in (4.23) is developed by assuming that factor structures for returns and trading volume stem from the same valuation fundamentals and depend on a common latent information flow.

## 4.4 Models for Point Processes

In many fields of study, observations occur in a continuum, space or time in the form of point events. The continuum can be multi-dimensional, but our focus is on the one-dimensional time scale with points distributed irregularly along the time scale. The main interest lies in estimating the mean rate of occurrence of events or more broadly on the patterns of occurrences. There are excellent monographs on this topic: Cox and Lewis (1966) [93], Daley and Vere-Jones (2003) [100], etc. But the interest in financial applications was revived by the seminal paper by Engle and Russell (1998) [125]. Financial market microstructure theories as discussed in Kyle (1985) [234], Admati and Pfleiderer (1988) [3] and Easley and O' Hara (1992) [113] suggest that the frequency and the timing of trade-related activities, that include posting, canceling and executing an order, carry information about the state of the market. The transactions generally tend to cluster during certain times of the day and the change in the mean rate of occurrences may suggest a new information flow about a stock.

To illustrate, recall Figure 2.1, on trading activities in Section 2.1. If we denote the observed intervals between successive activities as durations by $d_1, d_2, \ldots, d_r$, the time of occurrences are obtained by forming the cumulative sums of the $d$'s, $t_1 = d_1, t_2 = t_1 + d_2, \ldots, t_r = t_{r-1} + d_r$. Cox and Lewis (1966) [93] suggest two ways to present this type of data graphically. One method is based on cumulative number of events that have occurred at or before '$t$' against '$t$'. The slope of the line between any two points is the average number of events per unit for that period. One way to standardize the plot would be to approximate the graph by a line, '$a\,t$' when '$a$' is the slope of the graph indicating the average rate of occurrence for the entire duration. The second method calls for dividing the time scale into equally spaced time intervals and count the number of events in each interval; this also can be alternatively studied by fixing a certain number of events and count on the time it takes for this number to occur. In the context of stock data, this could mean simply recording not when the trade occurs but when the price changes. The advantage of the second plot is that the local fluctuations are readily observed and the advantage of the first plot is that it enables us to see systematic changes in the rate of occurrence.

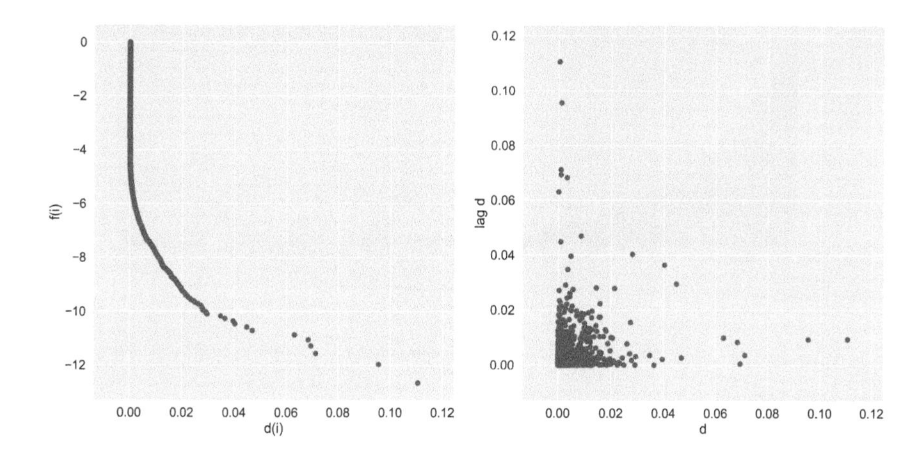


Figure 4.2: Empirical Survivor Function and Serial Correlation plots.

The baseline distribution for the durations is exponential resulting from Poisson arrivals for a specified interval that assume a constant rate. It is suggested to plot $\log(1 - \frac{i}{n_0+1})$ against $d_{(i)}$, where $d_{(i)}$ results from the ordered durations, $d_{(1)} \leq d_{(2)} \leq \cdots \leq d_{(n_0)}$ calculated between successive $n_0 + 1$ occurrences. Departures from linearity would indicate that the exponential distribution may not hold. In addition it is also suggested to plot $d_{i+1}$ against $d_i$ to check on the dependance of durations. Cox and Lewis (1966) [93, p.14] state that, "It is not clear how correlation in such data should be measured and tested; in particular it is not obvious that the ordinary correlation coefficient will be the most suitable measure with such data." Engle and Russell (1998) [125] show how the dependence in the duration data can be explicitly modeled in the context of high frequency transaction data. We illustrate this with the trading data for Microsoft (MSFT) for the month of January 2013. The graph (Figure 4.2) clearly indicates that the durations do not follow an exponential distribution and do exhibit some dependence. This is confirmed by the autocorrelation function for durations (Figure 4.3).

Cox and Lewis (1966) [93] also observe a few issues that can arise with the point-process type data. It is possible that two or more events recorded are happening at the same time. This may be due to latency issues but if they occur due to genuine coincidences it is better to analyze the number of events attached to each occurrence time as a variate separately. Further complications arise from the fact there may be events, such as price and volume that may be interdependent and events that occur

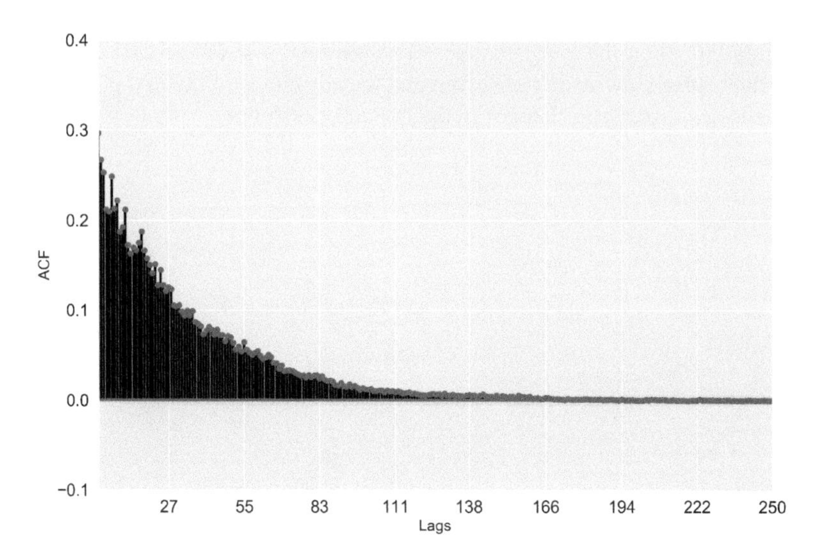


Figure 4.3: Autocorrelation Function for Duration (with 5% significance limits for the autocorrelations).

with related series that may be useful for modeling and predicting the future occurrence times.

### 4.4.1 Stylized Models for High Frequency Financial Data

There has been a great deal of interest in studying market microstructure to better understand the trading mechanisms and the process of price formation. With the availability of Trades and Quotes (TAQ) data that contains all equity transactions on major exchanges, it is possible to better model the key elements of market microstructure. The extensive details from the order books over multiple exchanges provide a massive amount of data, the analysis of which will be taken up in later chapters as well. Here we want to present some unique characteristics of high frequency data and present the most commonly used models in practice. To reiterate, the transactions (trades, quotes, bids, etc.) may occur at any point in time (Figure 4.4) during the exchange hours as given below.

High frequency data refer to the 'tick' '$t_i$' data which contains, in addition to the exact time of even occurrence, '$y_i$' called marks that may refer to all other elements of the limit order book such as traded price, quote, book imbalance, etc., that may be associated with '$t_i$'. The traditional time series methods that we discussed in the context of low frequency data analysis are not applicable here as the ticks can occur at any point in time when the exchange is open. Standard econometric techniques are for data in discrete time intervals, but aggregating high frequency data to some fixed

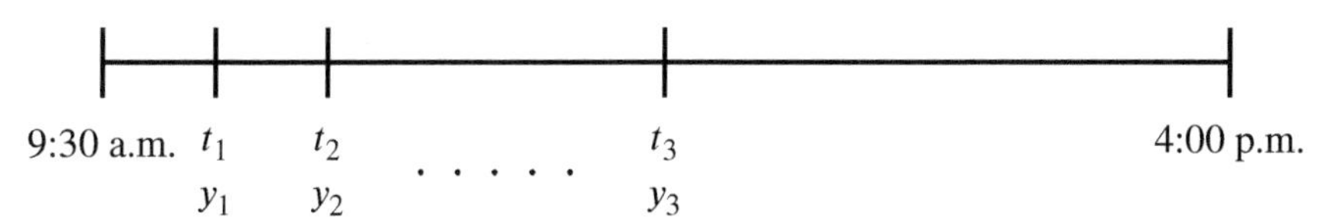


Figure 4.4: Exchange Hours.

time interval would not capture the advantages of having access to detailed time-stamped transaction data. Even for some heavily traded stocks, if the aggregation intervals are chosen to be short, there may be many intervals with no data and if the intervals are long, the microstructure features will be lost. Also certain key features such as the imbalance between bid side and ask side of the limit order book and when and how the market orders cross the spread are not easy to aggregate in a fixed time interval, however short or long it may be. Moreover, the timings of transactions provide valuable information for trading. Some noted features of high frequency data are:

- **Nonsynchronicity:** Different stocks have different trade intensities and even for a single stock, the intensity can vary during the course of a day. For the aggregated low frequency (daily) data, thus, we cannot assume that daily returns occur in equally-spaced time series.

- **Multiple Transactions with the Same Time Stamp:** It is possible that in periods of heavy trading especially around the opening and closing times of the exchange, each tick may contain multiple transactions. For the analysis of this type of occurrence, simple aggregate summary measures to more elaborate measures such as the variation in prices in addition to average prices are suggested.

- **Multiple Exchanges:** In the US market there are currently over twelve lit exchanges; due to latency issues there could be time delays in recording the submission times of orders that get dispatched at the same time. Also with Reg NMS mandating getting the best price anywhere in the market, an aggregated view of the market given a fixed time interval can miss capturing the other dependencies across exchanges.

- **Intra-day Periodicity:** Generally, it is observed for stocks, transaction activities are higher near the open and the close than in the middle of the day. Thus volatility is higher, immediately after the opening and before the closing of the market resulting in U-shape pattern (See Figure 4.5) of activity and volume.

- **Temporal Dependence:** High frequency data generally exhibit some dependence. The dependence is due to:

  - Price discovery
  - Bid-ask bounce
  - Execution clustering of orders

- **Volatility:** Volumes are higher near the open and close, spreads are larger near the open and narrower near the close.
- **Time between Trades:** Shorter near the market open and close.

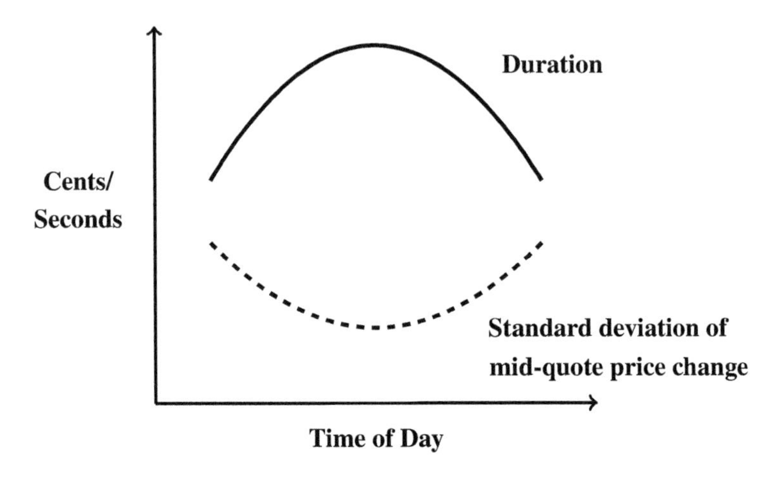


Figure 4.5: Typical Diurnal Patterns.

Under aggregation with low frequency data generally the dependence tends to decrease.

**Stylized Fact 7 (Continued from Chapter 2): Negative Autocorrelation in Returns:** Non-synchronous trading and bid-ask bounce both can introduce negative lag-1 autocorrelation in the returns. See Roll (1984) [291] and Lo and MacKinlay (1990) [249].

We want to briefly again discuss the Roll model that was mentioned in Chapter 1; on a typical day of trading, stocks exhibit price changes numerous times. These changes merely represent market frictions between the demand and supply sides and are not necessarily due to arrival of new information about the stock. Assume that the true price (unobservable) follows a random walk model, Example 2.3 Chapter 2, as

$$p_t^* = p_{t-1}^* + \epsilon_t \tag{4.27}$$

and the observed price,

$$p_t = p_t^* + u_t \tag{4.28}$$

with '$\epsilon_t$' denoting the true information about the stock and '$u_t$' denoting the market frictions. But the observed return,

$$r_t = p_t - p_{t-1} = \epsilon_t + u_t - u_{t-1} \tag{4.29}$$

inducing the negative autocorrelation of lag 1 in the return due to the carry over term, '$u_{t-1}$'. The term on the right-hand side because of zero autocorrelation beyond lag-1 can be written as MA(1) model:

$$r_t = a_t - \theta a_{t-1}. \tag{4.30}$$

The value of '$\theta$' generally tends to be small, but nevertheless significant.

**Commonly used Duration Models:** The timing aspect of transactions was discussed in Easley and O'Hara (1992) [113] and a rigorous set of tools was developed by Engle and Russell (1998) [125], as an alternative to fixed interval analysis. Treating the ticks as random variables that follow a point process, the intensity of transactions in an interval of length, $\Delta$, is defined as

$$\lambda(t) = \lim_{\Delta \to 0} \frac{\mathrm{E}[N(t+\Delta) - N(t) \mid F_t]}{\Delta}, \tag{4.31}$$

where $N(t)$ denotes the number of ticks until '$t$' and $F_t$ is the information available until that time. The commonly used model for such arrival rates is Poisson that implies that the time between two successive ticks is independent of pairwise ticks and follows an exponential distribution. But the duration data, $d_i = t_i - t_{i-1}$ exhibit some dependence, thus implying that time moves faster sometimes, faster than the clock time. The intensity rate, $\lambda(t)$ is not a constant, which is a typical characteristic of the exponential distribution. We briefly outline how the durations are modeled. For detailed discussion, see Tsay (2010) [315] or Lai and Xing (2008) [235, Section 11.2]. The autoregressive conditional duration (ACD) model is defined as follows:

$$d_i = t_i - t_{i-1} = \psi_i \varepsilon_i,$$

and (4.32)

$$\psi_i = \alpha + \sum_{j=1}^{p} \alpha_j d_{t-j} + \sum_{v=1}^{q} \beta_v \psi_{i-v},$$

where $\varepsilon_i$ are i.i.d. with $\mathrm{E}(\varepsilon_i) = 1$; If $\varepsilon_t' s$ are assumed to follow an exponential distribution the model (4.32) is called E(xpotential)ACD. Two alternative models WACD and GGACD are based on the following specification for the errors:

$$\begin{aligned} &\text{Weibull: } h(x) = \frac{\alpha}{\beta^\alpha} x^{\alpha-1} \exp\left\{-(\frac{x}{\beta})^\alpha\right\} \\ &\text{Generalized Gamma: } y = \lambda \frac{x}{\beta}, \end{aligned} \tag{4.33}$$

where $\lambda = \frac{\sqrt{K}}{\sqrt{K+\frac{1}{\alpha}}}$. The hazard function of Generalized Gamma is quite flexible and may fit better various patterns. The stationary conditions would require that the roots of $\alpha(B) = 1 - \sum_{j=1}^{q} (\alpha_j + \beta_j) B^j$ are outside the unit circle when $g = \max(p, q)$ and $B$ is the back-shift operator. It is clear that the model (4.32) is similar to the GARCH model given in (2.39). The estimation is typically done through conditional maximum likelihood and all inferences are asymptotic.

An alternative more direct ARMA type model for log durations is suggested by Ghysels, Gourieroux and Jasiak (2004) [165]:

$$\ln(d_i) = \omega + \sum_{j=1}^{p} \alpha_j \ln(d_{t-j}) + \sum_{j=1}^{v} \beta_j \varepsilon_{i-j} + \varepsilon_i \tag{4.34}$$

and $\varepsilon_i = \sqrt{h_i^v} u_i, h_i^v = \omega^v + \sum_{j=1}^{p^v} \alpha_j^v \varepsilon_{i-j}^2 + \sum_{j=1}^{q^v} \beta_j^v h_{i-j}^v$, where $u_i \sim N(0,1)$ i.i.d.; here the conditional mean and variance of durations are assumed to be separable. The duration volatility that is modeled using the second equation in (4.32) is interpreted as 'liquidity' risk.

**Other Duration Related Models:** So far the discussion has been around modeling the durations, but the information set, $F_t$ has information on 'marks', the information associated with the past ticks, such as the number of units transacted, price, etc. McCulloch and Tsay (2001) [262] suggest the following model:

$$\ln d_i = \beta_0 + \beta_1 \ln(d_{t-1}) + \beta_2 s_{i-1} + \sigma \varepsilon_i, \tag{4.35}$$

where $s_i$ is the size of the $i$th price change measured in ticks; other relevant variables can be easily added to this model.

### 4.4.2 Models for Multiple Assets: High Frequency Context

There is a considerable interest in extending the univariate duration models to multiple stocks or to multiple types of arrival processes. Because investment managers generally consider a portfolio of stocks rather than a single one, it is important for them to follow the transaction processes of several stocks simultaneously. Because of non-synchronous trading (see Figure 4.6 for the arrival of information for a single stock) relating two stocks with different trading intensities on a common time scale is somewhat difficult. Engle and Lunde (2003) [119] establish the following model for a bivariate point process for trades and quotes: Define $x_i = t_i - t_{i-1}$ as trade duration and $y_i = t_i^{fw} - t_{i-1}^{fw}$ as the forward quote duration; our goal is to model $(x_i, y_i)$

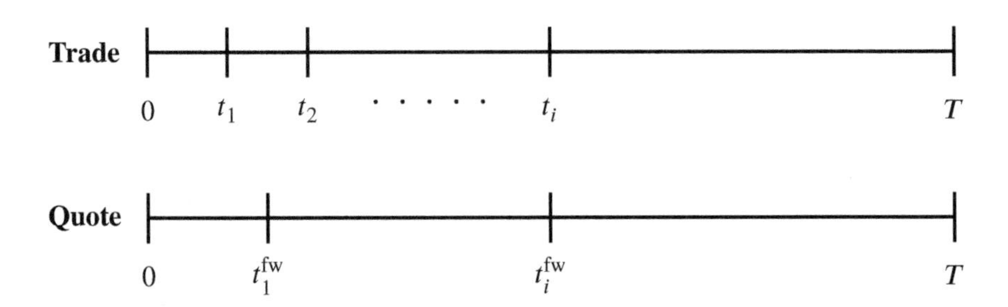


Figure 4.6: Trades and Quotes.

as a bivariate duration process but the underlying timings, as can be noted, are not synchronized. Thus, if $y_i > x_i$, define

$$\widetilde{y}_i = (1 - d_i)\, y_i + d_i x_i, \tag{4.36}$$

where $d_i$ is an indicator variable, $d_i = I_{\{y_i > x_i\}}$. Model $(x_i, \widetilde{y}_i)$ as a censored process. The Equation (4.36) provides a mechanism in a way to align the two processes. The joint distribution is given by

$$p(x_i, \widetilde{y}_i \mid F_{i-1}, \omega) = g(x_i \mid F_{i-1}, \omega_1)\, f(\widetilde{y}_i \mid x_i, F_{i-1}, \omega_2). \tag{4.37}$$

The first term is the trade density and the second term is the quote density. The $\psi_i$ terms as in (4.32) are modeled as EACD, $\psi_i = f(\psi_{i-1}, x_{i-1}, z_{i-1})$ and quote density is also modeled as EACD after adjusting for censoring:

$$f(y_i \mid \cdot) = [h(y_i \mid \cdot)]^{1-d_i} \times S(x_i \mid \cdot)^{d_i}. \tag{4.38}$$

The estimation of the parameters is done through quasi maximum likelihood. Using data for eight stocks, Engle and Lund (2003) [119] observe that high trade arrival rates, large volume per trade, wide bid-ask spreads, all predict more rapid price revisions. The model given in Engle and Lund (2003) [119] can be conceptually applied to any two point processes. Extending this concept to more than two processes could be of interest to traders who may want to monitor multiple trading exchanges simultaneously, but modeling this type of data is quite tedious.

## 4.5 Analysis of Time Aggregated Data

The commonly available data from public sources provide select price information, High, Low, Open and Close prices along with the total transaction volume in a trading day. The usual estimate of volatility is based on closing prices, but having this additional price (bar) data can help to get a better estimate of volatility. As shown in the last section, using high frequency data leads to estimators that need to be corrected for market frictions. But using daily price bar data may not be helpful to make entry/exit decisions as they are to be made in real time; so as a compromise, many traders construct price bar data from high frequency point process data for a shorter time intervals such as 5–30 minutes. This has the effect of smoothing the noise arising from market frictions but also results in data in discrete time units allowing the traditional discrete time series methods discussed in Chapter 2 and Chapter 3 to be effectively employed.

### 4.5.1 Realized Volatility and Econometric Models

Estimating volatility using high frequency data has received a great deal of attention as intra-day strategies became more prevalent. In the low frequency (daily) level some estimators based on select prices were presented in Chapter 2. Intra-day dynamics of volatility is of interest to traders and is usually not fully captured by the stock price indicators sampled during the day. If higher frequency data are used to estimate volatility of lower frequency data, it is important to know the model for the return at the lower frequency. To illustrate this, we consider data observed at two time scales; although they are somewhat at low frequency level, the concept can be easily extended to the high frequency context. If $r_{t,i}$ is the $i$th day return in $t$th month, the $t$th month return assuming there 'n' trading days, $r_t^m = \sum_{i=1}^n r_{t,i}$. Note that $\sigma_m^2 = \text{Var}(r_t^m \mid F_{t-1}) = \sum_{i=1}^n \text{Var}(r_{t,i} \mid F_{t-1}) + 2\sum_{i<j} \text{Cov}(r_{t,i}, r_{t,j} \mid F_{t-1})$. If $r_{t,i}$ is a white noise sequence, then

$$\hat{\sigma}_m^2 = \frac{n}{n-1} \sum_{i=1}^{n} (r_{t,i} - \bar{r}_t)^2. \tag{4.39}$$

But we had observed that in the high frequency data, returns do exhibit some serial correlation and so adjusting (4.39) for serial correlation is important to get a more accurate estimate of volatility. A simpler estimate of $\sigma_m^2$ is the so-called realized volatility ($\text{RV}_t$)

$$\text{RV}_t = \sum_{i=1}^{n} r_{t,i}^2. \tag{4.40}$$

But in the estimation of intra-day volatility, the number of sampling intervals, '$n$' and hence, '$\Delta$', the interval size can affect the estimate. If $\Delta \to 0$, the value $\text{RV}_t$ in the $\Delta$-interval goes to infinity. Thus the optimal choice of '$\Delta$' is crucial and is reviewed below, for a judicious choice.

Bandi and Russell (2006) [29] argue that the observed price is the sum of the efficient price and a friction price (see (4.28)) that may be induced by bid-ask bounce, price discreteness, etc. Thus the observed variance such as (4.39) is the sum of variance of the efficient returns and the variance of microstructure noise. As they indicate, both are useful: "The variance of the efficient return process is a crucial ingredient in the practice and theory of asset valuation and risk management. The variance of microstructure noise components reflects the market structure and the price setting behavior of market participants and thereby contains information about the market fine-grain dynamics." Both components can be estimated using high frequency data but sampled at different frequencies. Data sampled at low frequency are used to estimate the efficient return variable and higher frequency data are used to estimate the microstructure noise variance. Additional references on this topic include, Aït-Sahalia, Mykland and Zhang (2005) [4] and Barndoff-Nielsen and Shepard (2002) [31].

On any given day, assume that the observed price, at time '$i$' is as in (4.28)

$$p_i = p_i^* + u_i, \quad i = 1, 2, \dots, n, \tag{4.41}$$

where $p_i^*$ is the efficient price and $u_i$ is the microstructure noise. Dividing a trading day into $M$ sub-periods, where $\delta = 1/M$ is the size of the sub-period, write the observed returns as

$$r_{j,i} = r_{j,i}^* + \varepsilon_{j,i}, \quad j = 1, 2, \dots, M. \tag{4.42}$$

Assume that the frictions, $u$'s are i.i.d. mean zero and variance $\sigma_u^2$. Because $\varepsilon_{j,i} = u_{(i-1)+j\delta} - u_{(i-1)+(j-1)\delta}$, $\mathrm{Var}(\varepsilon_{j,i}) = 2\sigma_\eta^2$. It has been noted already that the return series exhibit negative first order autocovariance and $\varepsilon$'s can be taken to represent a variable, $\eta$, with MA(1) structure. The returns are computed at higher frequency at which the new information arrives. Thus $\sum_{j=1}^{M} r_{j,i}^2/M$ can be used to estimate noise variance.

To motivate the optimal sampling frequency choice, we consider two possible situations. If there is no microstructure noise, the usual estimate of variance, $\hat{\sigma}^2 = \frac{1}{T}\sum_{i=1}^{n} r_i^2$ has the asymptotic variance '$2\sigma^4\delta$' which suggests selecting '$\delta$' as small as possible. The implication of the presence of microstructure noise is that returns are likely to follow a MA(1) process with a negative autocorrelation and the proportion of total variance due to market microstructure noise is, $\Pi = \frac{2\sigma_\eta^2}{\sigma^2\delta + 2\sigma_\eta^2}$ and as '$\delta$' gets smaller, this proportion gets closer to unity. This indicates that the optimal sampling frequency should be finite when noise is present.

For the estimation of efficient price variance, we need to consider an optimal division of a day into '$M$' number of subperiods. The optimal sampling frequency over '$n$' days is chosen as $\delta^* = 1/M^*$ with $M^* = (\hat{Q}_i/\alpha)^{1/3}$, where $\hat{\alpha} = \sum_{i=1}^{n}\sum_{j=1}^{M} \bar{r}_{j,i}^2/(nM)$ and $\hat{Q}_i = \frac{M}{3}\sum_{j=1}^{M} \hat{r}_{j,i}^4$. Based on the analysis of a sample of S&P 100 stocks' mid quotes for the month of February 2002, Bandi and Russell (2006) [29] conclude that 5-min frequency is optimal and the 15-min interval may provide variance estimates 'excessively volatile'. The economic benefit of optimal sampling is that high frequency traders can employ realistic variance forecasts in formulating entry and exit strategies, by separating the variance due to market frictions.

## 4.5.2 Volatility and Price Bar Data

The price as assumed earlier is to follow a random walk model. Therefore, price changes (thus returns) over a time interval are distributed with mean zero and variance that is proportional to the length of the interval. Assuming that the prices follow continuous sample paths although the trading is closed for a certain duration and when trading is open, the actual transactions occur at discrete points in time. Treating the trading day as represented in a unit interval [0, 1] with $[0, f]$ representing the 'market close' time and $[f, 1]$ as 'open' time, it is shown that an estimator of volatility

$$\hat{\sigma}_1^2 = \frac{(O_1 - C_0)^2}{2f} + \frac{(C_1 - O_1)^2}{2(1-f)}, \tag{4.43}$$

has efficiency two compared to the usual estimator, $\hat{\sigma}_0^2 = (C_1 - C_0)^2$ based on the closing prices of two successive trading days. This suggests clearly the inclusion of additional data points, such as opening price, which can be quite informative. Garman and Klass (1980) [158] suggest these and other estimators which are superior to the classical estimator of volatility $\hat{\sigma}_0^2$. It is argued that the high and low prices contain information regarding the volatility during the trading period and a composite estimator that is proposed,

$$\hat{\sigma}_2^2 = a\,\frac{(O_1 - C_0)^2}{f} + (1-a)\,\frac{(H_1 - L_1)^2}{(1-f)\ln 2} \tag{4.44}$$

with optimal choice of '$a$'$= 0.17$ yields even higher efficiency.

The estimators like the above assume that either the price process has no drift or no price jumps between the previous day's closing price and current day's opening price. The former assumption leads to overestimating the volatility and the latter to underestimating volatility. Yang and Zhang (2000) [327] provide an estimator based on several periods of low, high, open and close prices that is independent of the drifts and the jumps. Although several estimators are discussed in this paper, we mention only two of them that have high efficiency. Denoting

$$\begin{aligned}
o &= \ln O_1 - \ln C_0, \text{ the normalized open;}\\
u &= \ln H_1 - \ln O_1, \text{ the normalized high;}\\
d &= \ln L_1 - \ln O_1, \text{ the normalized low;}\\
c &= \ln C_1 - \ln O_1, \text{ the normalized close,}
\end{aligned}$$

the classical estimator based on $n$-period historical data is

$$V_{cc} = \frac{1}{n-1} \cdot \sum_{i=1}^{n} [(o_i + c_i) - (\overline{o} + \overline{c})]^2 \tag{4.45}$$

and the estimator given in Rogers and Satchell (1991) [290]

$$V_{\text{RS}} = \frac{1}{n} \cdot \sum_{i=1}^{n} [u_i(u_i - c_i) + d_i(d_i - c_i)]. \tag{4.46}$$

The estimator suggested in Yang and Zhang (2000) [327] takes the form,

$$V_{\text{YZ}} = V_0 + kV_c + (1-k)V_{\text{RS}} \tag{4.47}$$

where $V_0 = \frac{1}{n-1}\sum_{i=1}^{n}(o_i - \overline{o})^2$ and $V_c = \frac{1}{n-1}\sum_{i=1}^{n}(c_i - \overline{c})^2$, with the optimal choice of $k = 0.34/(1.34 + (n+1)(n-1))$.

**Applications:** It is well-known that the standard stochastic volatility models used in finance are difficult to estimate due to their non-Gaussian nature. The proxies, the absolute or squared returns, because of the measurement errors tend to be somewhat inefficient. Alizadeh, Brandt and Diebold (2002) [8] argue that range as a proxy for volatility is superior as it is relatively free of the two problems associated with other measures. In the estimator (4.44), the second term which is essentially range-based is bound to be always positive. It can also be shown that the standard deviation of the log range is nearly one-fourth the standard deviation of the log absolute return. The range is also more robust to microstructure effects mostly due to bid-ask bounce compared to popular realized volatility measure (4.40). Martens and van Dijk (2007) [261] define realized range over an interval of time as,

$$\mathrm{RR}_t^{\Delta} = \frac{1}{4\ln 2} \cdot \sum_{i=1}^{I} (\ln H_{t,i} - \ln L_{t,i})^2, \tag{4.48}$$

where '$\Delta$' is the intra-day interval. This estimator is biased and the bias is corrected by scaling $\mathrm{RR}_t^{\Delta}$ with the ratio of average levels of the daily range and the realized range over the previous '$q$' days. The scaled realized range,

$$\mathrm{RR}_{s,t}^{\Delta} = \left( \frac{\sum_{l=1}^{q} \mathrm{RR}_{t-l}}{\sum_{l=1}^{q} \mathrm{RR}_{t-l}^{\Delta}} \right) \mathrm{RR}_t^{\Delta} \tag{4.49}$$

is known to estimate the volatility more accurately.

The range based statistics can also be used to estimate bid-ask spreads. Recall from (4.28), the Roll model assumes that the observed price is the sum of the true price and the error, that reflects the market frictions. The frictions can be measured by the bid-ask spread which is a proxy for market liquidity. Trading decisions are made by closely monitoring this measure that varies during a trading day. Corwin and Schultz (2012) [90] show how this can be estimated using low-frequency range-based statistics. The reasoning goes as follows: As noted in (4.29), the return is the sum of information and the frictions—sources of volatility and the bid-ask spread. The volatility increases with the length of the trading period but not the bid-ask spread. Thus the price range over a two-day period reflects two days' volatility and one spread but the sum of two consecutive days' price ranges reflect two days' volatility and twice the spread. This reasoning is used to separate the two effects.

Assume that the true or actual price follows a diffusion process. Actual price differs from the true price by $s/2$, where $s$ is the spread that remains constant over two consecutive days. Further, it is assumed that daily high price is a buyer-initiated trade while the daily low price is a seller initiated trade. Write,

$$\begin{aligned} \left[ \ln\left( \frac{H_t^0}{L_t^0} \right) \right]^2 &= \left[ \ln\left( \frac{H_t^A(1+s/2)}{L_t^A(1-s/2)} \right) \right]^2 \\ &= \left[ \ln\left( \frac{H_t^A}{L_t^A} \right) \right]^2 + 2\alpha \ln\left( \frac{H_t^A}{L_t^A} \right) + \alpha^2, \end{aligned} \tag{4.50}$$

where $\alpha = \ln\left(\frac{1+s/2}{1-s/2}\right)$ is taken to be fixed. Taking expectations on both sides, we have

$$E\left[ \ln^2\left( \frac{H_t^0}{L_t^0} \right) \right] = K_1\sigma^2 + 2\alpha K_2 \sigma + \alpha^2, \tag{4.51}$$

where $K_1 = 4\ln 2 = E\left( \ln^2\left( \frac{H_t^A}{L_t^A} \right) \right)$ and $K_2 = \sqrt{\frac{8}{\pi}} = E\left( \ln\left( \frac{H_t^A}{L_t^A} \right) \right)$. Taking expectations over two days, if we let $\beta = E\left[\sum_{j=0}^{1} \ln^2\left(\frac{H_{t+j}^0}{L_{t+j}^0}\right)\right]$ results in

$$\beta = 2K_1\sigma^2 + 4K_2\alpha\sigma + \alpha^2, \tag{4.52}$$

which has two unknowns '$\alpha$' and '$\sigma$' to solve. If we let $\gamma = \ln^2(H_{t,t+1}^0/L_{t,t+1}^0)$, where the subscripts $(t, t+1)$ indicate that the values are computed over a two day period '$t$' and '$t+1$'. This leads to

$$\gamma = 2K_1\sigma^2 + 2\sqrt{2}K_2\alpha\sigma + \alpha^2. \tag{4.53}$$

Both (4.52) and (4.53) are two quadratic equations in '$\alpha$' and '$\sigma$' that can be iteratively solved. The explicit solutions are given in Corwin and Schultz (2012) [90].

Abdi and Ranaldo (2017) [1] provide an estimator that includes the closing price that is easier to calculate and performs well in terms of the criteria considered earlier. Define the mid-range estimate as, $\eta_t = \frac{\ln H_t + \ln L_t}{2}$ and $c_t = \ln C_t$. Then the effective spread can be obtained from,

$$s^2 = 4E\left[(c_t - \eta_t)(c_t - \eta_{t+1})\right]. \tag{4.54}$$

An estimate of '$s$' assuming '$N$' trading days in a month is:

$$\hat{s} = \frac{1}{N}\sum_{t=1}^{N} \hat{s}_t, \tag{4.55}$$

where $\hat{s}_t = \sqrt{\max\left\{4(c_t - \eta_t)(c_t - \eta_{t+1}), 0\right\}}$. This estimate tends to perform well in terms of both cross-sectional correlates and time series correlations. It also seems to have some predictive power.

There are two main reasons for the exposition of price bar data to capture the essential aspects of market movement. The price bar data in discrete time intervals is easier to analyze; it can also smooth out the market frictions. By the choice of short time intervals, it can closely approximate the high frequency data. But how the entry/exit decisions are made and their accuracy based on price bar data needs further investigation.

## 4.6 Analytics from Machine Learning Literature

The technical trading rules that are to be discussed in Chapter 6 involve a relatively small information set to carry out predictions. It treats each of the '$n$' time series of asset returns as autonomous. As pointed out by Malkiel (2012) [258], these rules can be easily implemented by most market participants if such opportunities should emerge, and market efficiency would rule them out as winning strategies. More powerful prediction methods using a very large set of potential predictors and yet capable of avoiding overfitting are needed to take advantage of transient opportunities. We gave an overview of recent advances in high-dimensional regression in Chapter 4. The classification techniques can also be formulated with a high-dimensional feature vector. In machine learning and in computer science the focus is to handle rapidly different types of data (including text). The literature in this field is vast and so richly detailed that a single chapter wouldn't do justice to the topic. Consequently, here we will only briefly mention tools that are deemed to be relevant to trading and invite interested readers to further their knowledge with dedicated references such as Goodfellow, Bengio, and Courville (2016) [168] and Lopez de Prado (2018) [252].

Machine learning is a still-growing area of computer science that encompasses other well-established areas such as statistics, computational algorithms, control theory, etc. The focus in machine learning is on developing efficient algorithms for prediction or for classification using large data sets. The efficiency is gauged by predictive validation accuracy. The inferential aspects of statistical theory, such as standard error of the estimates, confidence intervals, etc., are generally not of much concern. Some areas where machine learning methods have led to significant contribution are classification, clustering and multi-dimensional regression. The attractiveness of these methods lies in the fact that they do not need any a priori theory to suggest which relevant variables to consider. Therefore with no prescription of variables, the variable or feature selection becomes an important process in machine learning. The statistical foundations of machine learning methods are also alternatively referred to as statistical learning methods. We provide a brief description of a select few methods in this section.

### 4.6.1 Neural Networks

This is one of the main methods of machine learning where the performance of a task is learned by analyzing training examples that have been hand-labeled in advance. The neural network in its simplest form can be represented as follows:

(4.56)

Here, the square boxes contain '$n$' dimensional input or feature vector, $X$ and '$m$' dimensional output vector, $Y$ and the circular box indicates hidden or unknown layer in-between that connects the input to the output. Structurally, neural networks are a two-stage regression or classification model with intermediary layers, and conceptually the model is similar to the reduced-rank regression model 3.16. But, $Z$ is generally a non-linear function of $X$. In the ordinary least squares regression model, the goal is to get '$m$' linear combinations of '$X$' that best predicts '$Y$'. Here '$r$' dimensional intermediaries can vary based on the assumption of 'hidden' layers, but the 'key' is that the output vector, '$Y$', can be a non-linear function of '$X$' via the hidden layers and '$r$' can be larger than '$n$'. In its simplest form the neural network model can be written as follows:

$$
\begin{aligned}
Z_i &= \sigma(\beta_i' X), \quad i = 1, 2, \ldots, r \\
Y_j &= g_j(Z), \quad j = 1, 2, \ldots, m,
\end{aligned}
\tag{4.57}
$$

where the function $\sigma(u) = \frac{1}{1+e^{-u}}$ is the sigmoid function. Note that this is the function used in logistic regression. In the regression set-up, $g_j(Z) = \alpha_j' Z$, but in the $m$-class classification,

$$
g_j(Z) = \frac{e^{\alpha_j' Z}}{\sum_{l=1}^{m} e^{\alpha_j' Z}}, \tag{4.58}
$$

called the softmax function, is used. In the set-up given in (4.56), we assume only one hidden layer, but in practical applications many layers are assumed which leads to non-uniqueness problems. As shown in Hastie, Tibshirani and Friedman (2009) [184] the neural network problem is closely related to a non-parametric method called projection pursuit regression. In its simplest form where the information flows in only one direction, it is called a feed-forward neural net, which is commonly used in many applications.

To estimate the '$r \times n$' parameters $\beta$'s and '$m \times r$' parameters of $\alpha$'s, the criterion in the regression setting is the usual sum of squares,

$$
W(\theta) = \sum_{l=1}^{n} \sum_{k=1}^{m} \left(y_{lk} - f_k(x_l)\right)^2 = \sum_{i=1}^{m} W_i(\theta) \tag{4.59}
$$

and in the classification setting the criterion is,

$$
W(\theta) = -\sum_{i=1}^{n} \sum_{k=1}^{m} y_{ik} \ln f_k(x_i), \tag{4.60}
$$

which is also termed a measure of deviance. The method to estimate the unknown parameters, '$\theta$' is a gradient descent, also-called as back-propagation method. This is essentially based on the first differential of $W_i(\theta)$; the '$r \times 1$' iteration is

$$
\hat{\theta}^{(r+1)} = \hat{\theta}^{(r)} - \gamma_r \sum_{i=1}^{n} \frac{\delta W_i(\theta)}{\delta \theta}, \tag{4.61}
$$

where $\gamma_r$ is called the learning rate. Details on the choice of the learning rate are discussed in Hastie et al. (2009) [184].

A more general form of neural net is given in Figure 4.7 below. The number of layers is to be determined conceptually or empirically, and because these models are over-parametrized, the optimization in (4.61) can be a nonconvex function. As a result, some practical considerations are worth keeping in mind when fitting these models, such as the non-exhaustive examples given below.

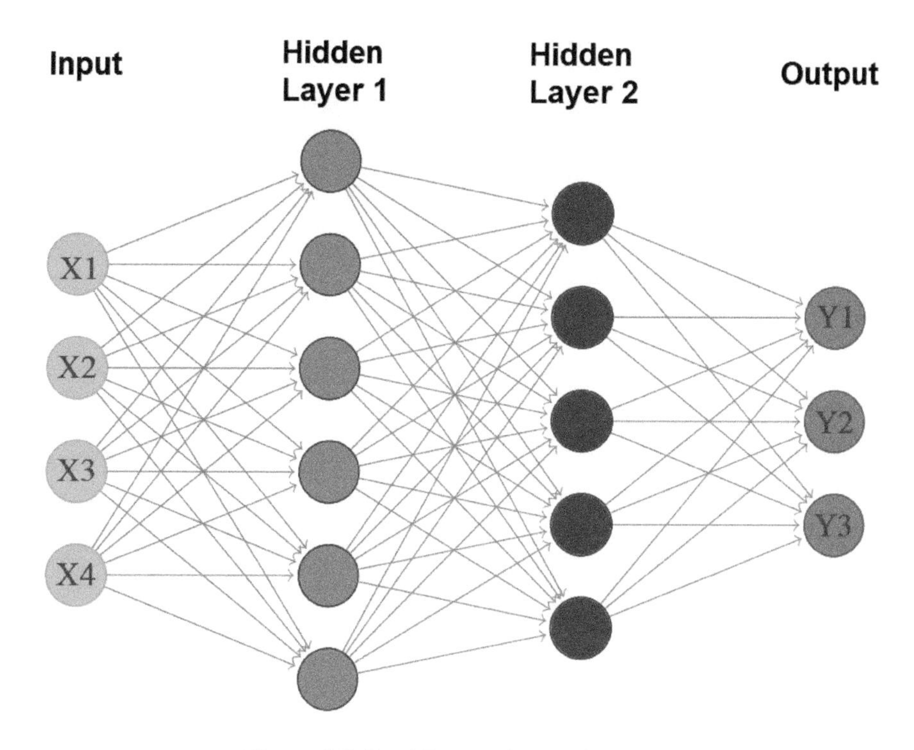


Figure 4.7: Feed-Forward Neural Nets.

- All inputs are to be standardized, as the scaling of these inputs determines the scaling of the weights in the bottom layers.
- Starting values of '$\theta$' are suggested to be random values near zero, as large values tend to result in inferior solutions.
- Neural Network models tend to overfit the data, so early stopping rates in the iterative process of (4.61) are suggested. To avoid overfitting, consider minimizing

$$W^*(\theta) = W(\theta) + \lambda\theta'\theta, \tag{4.62}$$

where '$\lambda$' is a positive tuning parameter.

- Many hidden units ($Z$'s) are preferred as they can capture various non-linearities.
- Multiple minima in (4.61) are possible. A suggested approach to avoid this is a method known as 'bagging' which calls for computing averages of predictions obtained from random perturbation of training data.

### 4.6.2 Reinforcement Learning

The set of techniques covered in this area is quite broad and generally refers to learning from interaction to achieve a goal. The learner is called an 'agent' who learns from interacting with its 'environment' through trial and error. The learning is assumed to progress through 'rewards'. More formally, the environment is in a state, $s_t \in S$, the set of all possible states; the agent's decision or action, $a_t \in A(s_t)$, the set of actions available in state '$s_t$'. As a consequence, the agent gets a reward $r_{t+1} \in R$ and a new state of the environment, '$s_{t+1}$' emerges. Reinforcement learning methods study how the agent changes the policy, $\Pi_t(s_t, a_t)$ with a goal towards maximizing the reward, a result of the interaction in the long run. For an excellent survey on this topic, refer to Kaelbling, Littman and Moore (1996) [224]. The common assumption is that the state-space is stationary which means that the transition probabilities from one state to another remain constant and so are reward signals. The long-run reward function discounted by learning rate that is optimized by the agent is denoted as, $E\left(\sum_{t=0}^{\infty} \gamma^t r_t\right)$. Other reward functions such as regret, a measure of the expected decrease in reward gained due to executing an algorithm have been considered in the literature. For a comprehensive treatment of the subject, see Sulton and Barto (2018) [309]. Much of the notes in this section follow from the references cited here.

**Markov Assumption:** We assume the sets $S$, $A$ and $R$ are all finite. We assume that any response at '$t+1$' depends on the state and action at '$t$'. More specifically,

$$P(s_{t+1} = S', r_{t+1} = r' \mid F_t) = P(s_{t+1} = S', r_{t+1} = r' \mid s_t, a_t), \tag{4.63}$$

where $F_t$ is the set of values of all the prior events until '$t$'. It is useful to note that the key quantities that define the dynamics of the decision process under the assumptions are the transition probabilities, $P_{ss'} = P(s_{t+1} = s' \mid s_t = s, a_t = a)$ and the anticipated value of the next reward, $R_{ss'} = \mathrm{E}(r_{t+1} \mid s_t = s, a_t = a, s_{t+1} = s')$. The reinforcement learning algorithms are based on estimating future value functions of state-action pairs which depend on the agent's policy, $\Pi(s, a)$. Formally, the state-value function is,

$$V^{\Pi}(s) = \mathrm{E}_{\Pi}\left[\sum_{k=0}^{\infty} \gamma^k r_{t+k+1} \mid s_t = s\right], \tag{4.64}$$

and the action-value function is,

$$q^{\Pi}(s, a) = \mathrm{E}_{\Pi}\left[\sum_{k=0}^{\infty} \gamma^k r_{t+k+1} \mid s_t = s, a_t = a\right]. \tag{4.65}$$

While the above formulation appears to be quite general, consistency conditions imposed to satisfy some recursive relationships lead to a form for $V^{\Pi}(s)$ called the Bellman equation:

$$V^{\Pi}(s) = \sum_{a} \Pi(s, a) \sum_{s'} P_{ss'}[R_{ss'} + \gamma V^{\Pi}(s')]. \tag{4.66}$$

The second part of the above equation indicates how the values that result from future states are discounted and weighted by their probabilities. The optimal state-value function is denoted as

$$V^*(s) = \max_{\Pi} V^{\Pi}(s) = \max_{a} \sum_{s'} P^a_{ss'}(R^a_{ss'} + \gamma V^*(s')). \tag{4.67}$$

The uniqueness and existence of $V^{\Pi}$ are possible when $\gamma < 1$; it is guaranteed that eventual termination will result from all states under the policy, Π.

Similarly, the optimal action-value function can be written as

$$Q^*(s,a) = \max_{\Pi} Q^{\Pi}(s,a) = \sum_{s'} P^a_{ss'}(R^a_{ss'} + \gamma \max_{a'} Q^*(s',a')). \tag{4.68}$$

Exact solutions are possible only under idealized conditions. A possible approach to solve the above is to use dynamic programming, but in practice the methods to solve the problem tend to be heuristic and computational such as using a neural network (Deep Queue Network—DQN) to approximate the action-value function.

More generally speaking, Reinforcement Learning algorithms can be trained either using Value-based methods ($Q$-Learning) or Policy-based methods (Policy Gradient), so it's worth mentioning a few words about the differences.

In $Q$-learning, as we just described, for a given state we calculate the $Q$ value for every action in the action space and we pick the action corresponding to the max value. This means that choosing actions depends on the $Q$ value at each decision point: $Q$-Learning attempts to learn the value of being in a given state (s), and taking a specific action (a) there with regards to the stated reward structure (per the Bellman equation, the expected long-term reward for a given action is equal to the immediate reward plus the discounted expected reward from the best future action taken at the following state).

In Policy-based methods, we directly learn the policy function Π without explicitly modeling the value function, resulting in actions being taken without directly calculating $Q(s, a)$. This tends to be closer to the actual objective of learning the optimal policy (e.g., learning to trade optimally) directly from experiences, rather than through intermediary steps which are subject to higher noise and uncertainty. An interesting specificity of financial applications—like trading—in favor of a Policy-based approach is that the true reward for the agent tends to be terminal (e.g., P&L, Execution Slippage, etc.), but the rewards of each individual action are less obvious, nor do they necessarily matter. For instance, when learning to optimally slice a large trade into a multitude of child orders, what is the value of having posted a top-of-book passive order two hours into the execution compared to having posted one level below the top-of-book? While it's hard to get to these values, one would want the agent to learn to give a value to the intermediate actions in order to guide it toward the ultimate reward.

While still a nascent field in finance, Reinforcement Learning is evolving fast and holds promise for trading applications that naturally lend themselves to be modeled as Markov Decision Processes. We encourage the curious reader to explore further the vast literature available on this topic.

### 4.6.3 Multiple Indicators and Boosting Methods

In developing models that capture the main features of the data, it must be noted that there are no unique models. Thus they yield different forecasts. It is well-known in the forecasting literature that a combined forecast generally does perform overall better than the individual forecasts. If there are '$T$' models and thus '$T$' forecasts, $\hat{f}_1, \ldots, \hat{f}_T$, how do we combine them to get a single forecast? Because these are all based on the same data, the estimates are likely to be correlated, resulting in a covariance matrix, $\hat{\Sigma}_f$. Three estimates that are proposed in the literature are described in Table 4.1.

Table 4.1: Three Proposed Estimates.

| | |
|---|---|
| **Weighted Estimator 1:** | $\hat{f}_{w_1} = \hat{f}' \hat{\Sigma}_f^{-1} \hat{f}$ |
| **Weighted Estimator 2:** | $\hat{f}_{w_2} = \hat{f}'(\text{diag}\hat{\Sigma}_f)^{-1} \hat{f}$ |
| **Simple Average Estimator:** | $\sum_{i=1}^{T} \frac{\hat{f}_i}{T}$ |

Due to the uncertainty in the estimation of $\Sigma_f$, especially in regime changes that may result in large estimation error, the simple average estimator is sometimes advocated. In the context of modeling the asset prices the elements of $\Sigma_f$ represent how well the methods do in forecasting. In order to use the above weighted estimators we need to keep track of the performance of the individual forecasting methods and in principle, the methods that yield inferior forecasts would get less weight. But a question about the span of their performance, that is, whether to use more recent versus past data, remains open.

#### Ensemble Learning

As mentioned, the idea of combining multiple forecasts (classifiers) into a single one is a longstanding goal of many decision-makers. This is especially important when methods that are successful in different regimes are complementary. There are two types of ensemble methods, bagging and boosting that we discuss here. Suppose the goal is to determine if we want to enter the market or not based on the information available at that time. It is somewhat easier to come up with simple rules but is harder to find a rule that has high accuracy. Under boosting, we study how to combine the simple rules uniformly into a 'stronger' rule. It is suggested to focus on examples that are hardest to forecast by simple rules and take a weighted majority of these rules to come up with a stronger rule.

To motivate, we will assume that we want to determine whether the price of an asset will go up or down. Under the random walk model, the error in guessing randomly is 0.5. But any rule should result in an error less than 0.5. We begin with a training set, $S = \{(x_i, y_i), i = 1, 2, \ldots n\}$ where $x_i$ is the information and $y_i$ is a binary variable that indicates whether to enter the market or not. Assume that there are '$T$' rules, $h_t(x)$ are the classifiers and $D_t$ is the distribution over the training set with the error,

$$\text{err}_D(h) = Pr_{(x,y)\sim D}(h(x) \neq y). \tag{4.69}$$

The rule $h_t$ is supposed to have error $\text{err}_{D_t}(h_t)$ with respect to '$D_t$'.

The boosting algorithm works as follows:

**Initialize:** $D_t(i) = \frac{1}{n}$ for all $i$

**Iterate:** $D_{t+1}(i) = \frac{D_t(i)}{z_t} \cdot \exp[-\alpha_t y_i h_t(x_i)]$,

where $\alpha_t = \frac{1}{2}\ln\left(\frac{1-\text{err}_{D_t}(h_t)}{\text{err}_{D_t}(h_t)}\right)$, $z_t$ = normalizing constant

**Stopping Time:** Stop when the error on a validation data set does not improve

**Final Rule:** $\text{Sign}\left(\sum_{t=1}^{T} \alpha_t h_t(x)\right)$

It is possible to overfit with boosting; its performance clearly depends upon how well the two decisions (enter or not enter) are clearly separated based on the past data. The use of kernels can improve local separability. It is also important to optimally extract features, $x$, that can be discriminating between the two options for $y$. The main advantage of ensembles of different rules is that it is not likely that all rules will make the same error. The algorithm described above known as, "adaptive boosting algorithm (AdaBoost)" is by Freund and Schapire (1995, 1996,1997) [145, 146, 147] and it automatically "adapts" to the data at hand. If the $t$th classifier $h_t(x)$ is parameterized and the overall predictor takes the form

$$h(x) = \text{Sign}\left(\sum_{t=1}^{T} \alpha_t \text{Sign}(w^{(t)}x + b^{(t)})\right) \tag{4.70}$$

is a two-layer neural network with '$T$' hidden units. One can see the identical underlying nature of the structure between neural network and boosting algorithm.

The idea of bagging comes from training the rules on multiple data sets that supposedly have similar structure. The multiple data sets are obtained by bootstrap resampling, meaning the data sets are not likely to be too similar. For data on asset prices, which are represented by an AR(1) model with the coefficient close to one, the asymptotic distribution of the sample estimate of the coefficient is non-normal. So the bootstrap samples are also useful to construct the appropriate sampling variance. The bootstrap samples are created by resampling the innovations or empirical residuals. From the single data set, $S$ we create '$M$' bootstrapped samples, $S_1, \ldots, S_M$ with each set containing '$n$' samples. Then apply the boosting method on all these samples and average them. As noted in the literature, bagging tends to reduce variance and thus it provides an alternative to regularization, see Breiman (1996) [57].

The gradient boosting method introduced by Breiman (1999) [58] and generalized procedure called Logitboost, by Friedman et al. (2000) [149] and Friedman (2001) [148] further improves on the ADA boost. The method starts by fitting an additive model (ensemble), $\sum_t \alpha_t h_t(x)$ in a forward stage-wise manner. In each stage, a weak learner is introduced to compensate for the shortcomings of existing weak learners. In gradient boosting, gradients are used to identify the shortcomings and in the regression context, these are residuals. Now the training set, $S^* = \{(x_i, y_i - h(x_i)), i = 1, \dots, n\}$ is treated like, $S$, with residuals playing the role of the data. The optimization occurs by moving in the opposite direction of the gradient of the function.[1]

## 4.7 Exercises

The data set in file `csco_levelIII_data.csv` on the entire trading session, including early and late hours from 6:00 a.m. to 8:00 p.m. EST, contains intraday activity for up to 5 major national exchanges: Nasdaq, Direct Edge, NYSE, ARCA, and BATS (exact number of exchanges depends on the historical period). All messages are consolidated into one file ordered by timestamp. This data set allows to build a full national depth book (super book) at any moment intraday.

The details of Level III data are given in Table 1.10.

1. You need to aggregate the data into 5-min intervals before answering the following.

(a) Let $x_t$ denote the number of trades in the $t$th 5-minute interval. Ignore the time gaps between trading days. Plot the time series and its ACF. Determine if there are intraday period patterns in the series.

(b) Using the last transaction in the $t$th 5-minute interval as the stock price in that interval, plot the time series $y_t$ of 5-minute returns during the period and the corresponding ACF.

(c) Consider the bivariate time series $(x_t, y_t)$. How does $y_t$ vary with $x_t$? Are there any intraday periodic patterns in $(x_t, y_t)$?

(d) Plot the durations of the transaction times for Mondays and Fridays. Is there any difference in the patterns?

(e) Fit GARCH models for the durations and interpret the coefficients.

(f) Is there any particular exchange that offers better price?

2. Use the transaction data in Exercise 1. There are seventy-eight five-minute intervals in a trading day. Let $d_i$ be the average of all log durations for the $i$th 5-minute interval across all trading days. Let $t_j$ be tick time of a trade during the $i$th 5-minute interval, define an adjusted duration as $\Delta t_j^* = \Delta t_j / \exp(d_i)$ where $\Delta t_j = t_j - t_{j-1}$.

(a) Is there a diurnal pattern in the adjusted duration series?

[1] Gradient boosting is readily implemented in the open-source XGBoost library.

(b) Build EACD and WACD models for the adjusted duration and compare them. Now aggregate the data over the 5-minute intervals; Let $r_t$ be the return and $V_t$ is the volume of transactions in the $t$th interval.

(c) Is there any relationship between the volume $V_t$ and volatility $r_t^2$. If there is develop a model that will capture the dependence of volatility on volume. Let $\overline{p}_t$ be the average price in the $t$th 5-minute interval.

3. Use the CISCO execution data only; aggregate at various time intervals: 1 min, 2 min, $\ldots$, 15 min. For these levels record:

(a) Average price, VWAP, realized volatility and aggregate volume.

(b) By relating the returns and volatility estimates based on average price to realized volatility, identify the optimal duration for aggregation that will cancel out the market frictions but pick up the true information. (You may want to refer to Bandi and Russell (2006) [29].)

# Part III

# Trading Algorithms

In this part of the book, we present trading algorithms based on statistical analyses of market data. These analyses are also guided by established principles in financial economics. The market is assumed to consist of informed and noise traders. It is also postulated that the market is efficient, and the informed traders generally gain at the expense of noise traders. Any information about a stock is quickly impounded in price by the Efficient Market Hypothesis (EMH). If there are anomalies, statistical arbitrage techniques can be employed to successfully detect and exploit them for profitable opportunities. These methodologies that were developed in the low frequency setting need to be carefully reconsidered to be successfully extended to the high frequency setting. In that space, the focus of traders has also been on Execution Strategies which are covered in Part IV .

In Chapter 5, we present some commonly used strategies based on the statistical methods described in Part II . Although the literature is replete with numerous strategies, most of them are some variations or combinations of basic strategies described here. As price-based strategies have been fully exploited, it is important to consider strategies that use other information such as volume, market volatility, related stock performance, etc. Also presented are improved strategies based on machine learning methods. In Chapter 6, we discuss portfolio theory—fundamental to diversification of risk and how the investment/trading strategies are handled. As portfolio rebalancing is quite common and because rebalancing requires trading, this chapter provides a comprehensive view of investment strategies with transaction costs. We conclude this part with Chapter 7 that deals with the analysis based on news and sentiment. This is essential as short-term price movements are not easily explained by the economic indicators that are slow to vary and the stock related events such as earnings announcements occur only on periodic basis. The sentiment analysis captures an understanding of how noise traders trade and can be considered as part of the emerging area of Behavioral Finance.

# 5

# Statistical Trading Strategies and Back-Testing

## 5.1 Introduction to Trading Strategies: Origin and History

**Statistical Arbitrage Opportunity and Strategies:** Trading strategies are central to the notion of statistical arbitrage. Strategies that claim to have better performance than the market date back to the inception of trading in finance. This area under the name of 'technical analysis' is based on using past prices and other stock and market-related information to make investment decisions. The ideas date back to a Japanese speculator, Munehisa Homma, who made a fortune in the rice market using techniques called candlestick patterns. The famous Dow Theory is based on the assertion that the market in certain phases can be predictable. Thus, there is a history of studies that empirically conclude that the behavior of stock prices appear to contradict the efficient market hypothesis, at least in the short term. Broadly, there are two types of strategies, one the momentum strategy that advocates buying well-performing stocks and selling poor-performing stocks (see Jegadeesh and Titman (1993) [215]) and the other, contrarian or reversal strategy that advocates buying past losers and selling past winners. Other strategies, known as value strategies suggest buying value stocks and selling glamour stocks (see Lakonishok et al. (1994) [238]). They use the company fundamental data such as price earnings ratios, dividends, book-to-market values, cash flows and sales growth as indicators of value.

Hogan, Jarrow, Teo and Warachka (2004) [198] define "a statistical arbitrage as a long horizon trading opportunity that generates a riskless profit." Four conditions to be met are: It is a self-financing strategy with a zero initial cost, in the limit has positive expected discounted profits, a probability of a loss converging to zero and a time-averaged variance converging to zero; the last condition implies that an arbitrage opportunity eventually produces riskless profit. Citing anomalies to reject market efficiency has been criticized on the ground that such tests depend upon a specified model for equilibrium returns. Hogan et al. (2004) [198] argue that statistical arbitrage can be defined without any reference to an equilibrium model and reject market efficiency. They provide a statistical test to determine if a trading strategy constitutes statistical arbitrage based on short horizon incremental profits. Some of the common issues that arise in developing and testing strategies are: Handling the transaction costs, both direct costs, such as brokerage fee, and indirect costs due to the market entry/exit; cross-validation to avoid over-fitting; no guaranteed performance in different market settings. In this chapter, we will cover some of these trading strategies, propose evaluation criteria and provide the details of these algorithms. In practice, no single strategy works well all the time. So it is important to develop a suite of strategies; we will cover some of them here.

## 5.2 Evaluation of Strategies: Various Measures

In this section, we want to lay out how trading strategies are generally evaluated. These measures are different from measures used for evaluating execution strategies, which will be taken up in Part IV. Here the focus is on benchmarks that are considered in the industry and which provide a range of dimensions that need to be looked at when a strategy is failing. The standard measures stem from the capital asset pricing model:

$$r_t - r_f = \alpha + \beta(r_{mt} - r_f) + \varepsilon_t, \tag{5.1}$$

where $r_t$ is the return on the asset and $r_f$ is the return from a risk-free asset; the slope coefficient, '$\beta$' measures the riskiness of the asset and '$\alpha$' denotes the excess return. Also $\bar{r}$ and $s$ be the sample average and standard deviation of the returns over the evaluation period and these are substituted for $\mathrm{E}(r_t)$ and '$\sigma$', respectively, in the measures given below.

- **Excess Return:** $\alpha$: Positive value indicating better performance, as it reflects risk adjusted return.

- **Sharpe Ratio:** $\dfrac{\mathrm{E}(r_t) - r_f}{\sigma}$: This measure is standardized by the volatility of the asset. This is a widely used benchmark measure.

- **Treynor Ratio:** $\dfrac{\mathrm{E}(r_t) - r_f}{|\beta|}$: This measure is an alternate to the Sharpe ratio where the excess return is standardized by market risk.

- **Information Ratio:** $\dfrac{\alpha}{\sigma(\varepsilon_t)}$: This is a modified version of excess return.

While the above ratios take a long-term view of the investments, two others that are used in the industry focus on the execution aspects of the strategies:

- **Sortino Ratio:** $\dfrac{\mathrm{E}(r_t) - r_{\text{target}}}{\mathrm{DR}(r_t)}$, where $r_{\text{target}}$ is the target rate chosen by the investor and $\mathrm{DR}(r_t)$ denotes the downside volatility or the draw down. The target can be $r_f$ or the average market return or return of the competitors. Another measure is,

– **Calmar Ratio:** $\dfrac{\text{Annualized Compound Rate of Return}}{|\text{ Maximum Drawdown }|}$

The attention to drawdown is important because there is no guarantee that a strategy chosen through simple back-testing will work in the future. The last two ratios combine both the idea of maximizing the return (numerator part) and the idea of minimizing the loss (denominator part). The drawdown numbers are essentially determined by the traders based on their own experience and their risk tolerance. There are some studies that focus on the statistical properties of their criteria (see Bertrand and Protopopescu (2010) [36] on the information ratio and Lo (2002) [248] on the Sharpe ratio). Another measure is to simply compare the strategy returns to the returns from a buy and hold investment.

The standard reporting of backtesting results beyond the above measures includes: Number of winning and losing trades, largest winning and losing trades, net slippage, total commissions along with annualized versions of the above ratios. The so-called equity curve which is the cumulative return graph along with the buy and hold graph can clearly indicate the effectiveness of the strategies to be discussed in the next sections.

## 5.3 Trading Rules for Time Aggregated Data

Trading Rules that are broadly labeled under technical analysis, involve the prediction of future short-term price movements from an inductive analysis of past movements using either quantitative techniques, such as moving averages or qualitative methods such as pattern recognition from visual inspection of a time-series plot of the asset price data or a combination of both via semi-parametric procedures such as kernel smoothing. In the foreign exchange market, Menkhoff and Taylor (2007) [265] observe that the widespread use of these techniques is puzzling, "since technical analysis eschews scrutiny of economic fundamentals and relies on information on past exchange rate movements that, according to the weakest notion of market efficiency, should already be embedded in the current exchange rate, making its use either unprofitable or implying that any positive returns that are generated are accompanied by an unacceptable risk exposure." In general there are economic fundamentals that can explain long term movements, there is still no fundamental-based exchange rate model that is capable of forecasting short-term movements.

Technical analysis (also known as Chartist analysis) is a set of techniques for obtaining forecasts based on past history of price and volume patterns. Clearly for the technical analysis to be useful, price movements should follow some regular, recurring patterns and these patterns must hold for long enough periods to sustain the transaction costs and false signals so that the investment, results in a profit. What constitutes a useful forecast from an investor point of view could be, if the past history of prices can be used to increase expected gains. But as Fama and Blume (1966) [132] point out, "In a random walk market, with either zero or positive drift, no mechanical trading rule applied to an individual security would consistently outperform a policy of simply buying and holding the security." The past movements could be useful only when there is some degree of dependence in successive price changes. The serial correlation measures may not capture the complex dependence that may exist. In what follows we briefly describe some commonly used trading rules and comment on their noted performance.

### 5.3.1 Filter Rules

Alexander (1961, 1964) [6, 7] studied the filter technique that is intuitive and is easy to implement; the basic idea is summarized below:

**x% Filter:** If a daily closing price moves up at least x% from its past low, buy and hold, until its price moves down at least x% from a subsequent high, sell and go short. Maintain the short position until price rises at least x% above a subsequent low, buy. Moves less than x% in either direction are ignored.

The general logic behind this rule is that if the price moved up $x\%$, it is likely to move up more than $x\%$ further before it moves down by $x\%$; Thus the random walk model is not likely to be upheld by the data. As observed by Mandelbrot (1963) [259], because of the price jumps that occur at the time of transaction, the purchase price and sale price will not exactly match as required by the rule.

The initial application of this rule on daily data for S&P industrials showed that the trading rules with 5%, 6% and 8% filters generated larger gross profits than the buy and hold strategy. This study did not account for transaction costs. But a later study that incorporated a commission of 2% for each round-trip, showed only the larger filter $\sim$ 46% rule beat the buy and hold strategy and all the other filters adjusting for transaction costs did not perform any better than simple buy and hold strategy.

Fama and Blume (1966) [132] after applying Alexander's filter technique for each of the individual securities in the Dow Jones Industrial Average for the daily data during a five year period, 1957–1962, concluded, "When commissions are taken into account, the larger profits under the filter technique are those of the brokers. When commissions are omitted, the returns from the filter techniques are, of course, greatly improved but are still not as large as the returns from simply buying and holding." It is argued that under the random walk (RW) model, the average return on long positions should be approximately equal to the buy and hold average returns. This may not be true if the drift occurs at the time of entry or at the time of exit; the filter rule can be stated under the RW model as follows; let the return, $r_t = \ln P_t - \ln P_{t-1} = \mu_t + a_t$, when $a_t \sim N(0, \sigma_a^2)$ and '$\mu_t$' is the random drift term and $\sigma_a^2$ is the volatility; as per the filter rule, enter if $r_t > r$ and suppose this occurs at time $t_0$. If the drift term is approximately picked up at time '$t_0$', so that $\mu_t = 0$, for $t < t_0$ but $\mu_t = \mu > 0$ for $t \geq t_0$, the returns from the filter should fare better. The lasting of the drift is to be carefully monitored and using the exit strategy when $r_t < r$ will guarantee any undue losses.

When the time intervals are small it has been observed that there is evidence of positive dependence in returns and that may be useful to predict short term dependence. This may signal opportunities for more frequent transactions but the cost of transactions can erode the profits.

**Risk-Adjusted Filter:** Sweeney (1988) [310] develops a test for profits when risk-adjusted filters are used. The stocks considered in Fama and Blume (1966) [132] that are "winners" in one period tend to persist as winners in later periods as well. Thus focusing only on winners rather than entire set of stocks gives better results. Assume that the investor uses the filter rule each day to decide to enter or exit based solely as past information. Consider the span of $N$ days and with a filter strategy, that the asset is held on $N_{\text{in}}$ days and is not held on $N_{\text{out}}$ days. At the end of $N$ days, the filter's performance is evaluated. The average excess rate of buy and hold is $\overline{R}_{\text{BH}} = \frac{1}{N} \sum_{i=1}^{N} (r_t - r_f)$. Assume that '$f$' is the proportion of days, the asset is not held; then the average excess rate of the filter is, $\overline{R}_F = \frac{1-f}{N_{\text{in}}} \sum_{t \in I} (r_t - r_f)$. Assume that under CAPM, expected excess rate of return equals a constant risk premium, $\text{PR}_j$. Further, that $\text{E}(\overline{R}_{\text{BH}}) = \text{PR}$ and $\text{E}(\overline{R}_F) = (1-f)\,\text{PR}$; Use the adjusted statistic

$$x = \overline{R}_F - (1-f)\,\overline{R}_{\text{BH}} \tag{5.2}$$

to judge the filter rule; observe $\text{E}(x) = 0$ and $\sigma_x = \sigma[\frac{f(1-f)}{N}]^{1/2}$, where '$\sigma$' is the standard deviation of excess return $r_t - r_f$. Fama and Blume (1966) [132] use

$$X = \overline{R}_{FI} - \overline{R}_{BH} \tag{5.3}$$

where $\overline{R}_{fI} = \frac{1}{N_{\text{in}}} \sum_{t \in I} (r_t - r_F)$. Further, note that $\text{E}(X) = 0$ and $\sigma_x = \sigma_X$.

The main advantage of $X$ is that it measures profits for each day in the duration of '$N$' days and $x$ measures profits per day for those days when the stock is held. The above results hold for the case of time varying risk premia as well; Suppose $R_f - r_f = \text{PR} + a_t$, where $a_t$ is stochastic; thus

$$x = \frac{1}{N_{in}} \sum_{t \in I} a_t - \frac{1}{N} \sum_{t=1}^{N} a_t + (\overline{\text{PR}}_{N_{in}} + \overline{\text{PR}}_N). \tag{5.4}$$

If the investor has no insight into changes in risk premia, then $\text{E}(\overline{\text{PR}}_{N_{\text{in}}} - \overline{\text{PR}}_N) = 0$; If the investor has no insight about the stochastic, $a_t$, the filter rule may be forecasting the risk premia.

It has been pointed out that the Fama and Blume measure is biased against finding successful filter rules. The statistics $X$ and $x$ are meant for long strategies where the investor has the funds in the risk free asset when not holding positions. If we define the difference in cumulative excess returns between days investor holds and does not hold positions,

$$\overline{R}_{F'} = \frac{1}{N}\left(\sum_{t\in I}(r_t - r_f) - \sum_{t\in O}(r_t - r_f)\right) \tag{5.5}$$

and it is shown that $E(\overline{R}_{F'} - \overline{R}_{BH}) = (1-2f)PR - PR = -2fPR$, a proportion of constant risk premium, thus resulting in a bias.

### 5.3.2 Moving Average Variants and Oscillators

The moving average trading rules are generally based on comparing the current price with a form of moving average of past prices. Consider the prices, $P_t, P_{t-1}, \dots, P_{t-m+1}$ with a moving window of width '$m$'. Various versions of moving average are available in the literature and their relative merits depend on the price patterns:

$$\text{Simple Moving Average (SMA): } \overline{P}_t^{(m)} = \frac{1}{m}\sum_{i=0}^{m-1} P_{t-i}$$

$$\text{Exponentially Weighted Average (EWA): } \overline{P}_t^{(m)} = \left(\frac{1}{\sum_{i=0}^{m-1}\lambda^{i+1}}\right)\sum_{i=0}^{m-1}\lambda^{i+1}P_{t-i} \tag{5.6}$$

$$\text{Bollinger Weighted Moving Average (BWMA): } \hat{P}_t^{(m)} = \left(\frac{1}{\sum_{i=1}^{m} i}\right)\sum_{i=0}^{m-1}(m-i)P_{t-i}$$

Note that the smoothing constant, '$\lambda$', must be between zero and one. The rules based on indicators of the above kind tend to be profitable in markets that exhibit definite trends and so they are generally known as "trend-following" or sometimes, called "momentum" indicators. Some simple rules that are used in practice are:

**SMA Rule: If** $\dfrac{\overline{P}_t^{(m)}}{P_t} > U$ **or** $< L$, **trade;**

**EWA Rule: If** $\dfrac{\overline{P}_t^{(m)}(\lambda)}{P_t} > U$ **or** $< L$, **trade**

The two types of potential strategies are broadly known as "Reversal" vs. "Momentum" and they differ mainly on when the entry is carried out. The contrarian or reversal, in a way, looks for local minimum to enter the market and the momentum follower looks for a local maximum to enter.

While the above rules are easy to implement, one may observe that they do not take into account the variance or the volatility of the price (return) process. The widely used Bollinger band, conceptually similar to monitoring the quality of a manufacturing process, is based on a weighted moving average and the variance. Define the standard deviation as, $\hat{\sigma}_t^{(m)} = \left\{ \frac{1}{m-1} \sum_{i=0}^{m-1} (P_{t-i} - \overline{P}_t^{(m)})^2 \right\}^{1/2}$ and the rule can be stated as follows:

> **BWMA Rule: If$\mathbf{P_t > P_t^+ = \hat{P}_t^{(m)} + 2\hat{\sigma}_t^{(m)}}$, the upper band, then sell; if $\mathbf{P_t < P^- = \hat{P}_t^{(m)} - 2\hat{\sigma}_t^{(m)}}$ , the lower band, then buy; No action is taken in-between.**

Broadly the key quantities such as moving average, weighted (exponentially or otherwise) average, etc., correspond to the running-mean smoother discussed in Lai and Xing (2008) [235, Section 7.2.1]. Instead of the running-mean, some technical rules use the running maximum (i.e., $\max_{0 \le i \le m} P_{t-i}$) or the running minimum (i.e., $\min_{0 \le i \le m} P_{t-i}$) in the above rules. Properties of these rules can be easily investigated.

**Moving Average Oscillator:** The so-called moving average oscillator rule requires computing two moving averages of short and long time spans. Let $\overline{P}_t^{(m)}$ and $\overline{P}_t^{(n)}$ be the short and long averages, where $m < n$. The rule would signal a break in trend when the longer moving average is crossed by a shorter moving average (or for example spot rate). The upward trend is signaled when short moving average intersects from below the long average. Conversely, a downward break in trend is signaled if the crossing occurs from above. Precisely the rule is:

> **Oscillator Rule:**
>
> **Enter if $\mathbf{\overline{P}_{t-1}^{(m)} < \overline{P}_{t-1}^{(n)}}$ and $\mathbf{\overline{P}_t^{(m)} > \overline{P}_t^{(n)}}$**
>
> **Exit if $\mathbf{\overline{P}_{t-1}^{(m)} > \overline{P}_{t-1}^{(n)}}$ and $\mathbf{\overline{P}_t^{(m)} < \overline{P}_t^{(n)}}$**

**RSI Oscillator:** Another oscillator that is designed to measure the strength of price movement in a particular direction is the relative strength indicator (RSI). The logic being that if the changes in price are too rapid in one direction, a correction in the opposite direction is likely to follow and may occur soon. Thus, these types of rules are called 'reversal' or 'overbought/oversold' indicators. To define RSI, let $U_t = \sum_{i=1}^{m} I(P_{t-i} - P_{t-i-1} > 0)\,|P_{t-i} - P_{t-i-1}|$ and $D_t = \sum_{i=1}^{m} I_t(P_{t-i} - P_{t-i-1} < 0)\,|S_{t-i} - S_{t-i-1}|$ where $I(\cdot)$ is an indicator function. Thus $U_t$ and $D_t$ aggregate the positive and negative side of the price changes. Now define $\text{RS}_t = \frac{U_t}{D_t}$, which can be taken as a measure of odds ratio, and define

$$\text{RSI}_t = 100 \cdot \frac{U_t}{U_t + D_t} = 100 - 100 \left( \frac{1}{1 + \text{RS}_t} \right). \tag{5.7}$$

By construction, RSI is normalized and it takes values between 0 and 100.

**Reversal Rule:**

**If $\text{RSI}_t > 70$, it is overbought signaling a downward correction and hence exit or short sell.**

**If $\text{RSI}_t < 30$, it is oversold signaling an upward correction and hence enter.**

These rules that appear to be somewhat ad hoc can all be related back to the random walk model (see Chapter 2) possibly with a drift:

$$p_t = \mu + p_{t-1} + a_t, \tag{5.8}$$

where $a_t \sim \text{N}(0, \sigma^2)$, $\sigma^2$ is the volatility of the stock and note the return, $r_t = p_t - p_{t-1}$; thus, $r_t \sim \text{N}(\mu, \sigma^2)$. If there is any autocorrelation in '$r_t$', and if $\overline{r}_n$ is the average based on the last '$n$' observations, then $\overline{r}_n \sim \text{N}\left(\mu, \sum_{|h|<n}\left(1 - \frac{|h|}{n}\right)\frac{\gamma(h)}{n}\right)$, where $\gamma(h)$ is the autocovariance function of lag '$h$', with $\gamma(0) = \sigma^2$. When $\gamma(h) = 0$ for $h > 0$, that $\text{P}(\overline{r}_n > 0) = \text{P}(\overline{a}_n > -\mu) = \text{P}\left(\frac{\overline{a}_n}{\sigma} > \frac{-\mu}{\sigma}\right) = 1 - \Phi\left(-\frac{\mu}{\sigma}\right)$, where $\Phi$ is the c.d.f. of the standard normal distribution. Thus we can evaluate $\text{RS}_t$ by the odds ratio, $\left(1 - \Phi\left(-\frac{\mu}{\sigma}\right)\right)/\Phi(-\frac{\mu}{\sigma})$. In fact the reversal rule stated above implies, for $\text{RSI}_t$ to be greater than 70, the odds ratio, $\text{RS}_t > 2.33$. All the rules stated in this section can be written in the form of the c.d.f. of the underlying distribution of the return. This model-based interpretation of the rules can help us in accommodating other factors such as volume more formally in the model and evaluate their value added to the strategy performance. While there are many areas where technical rules are shown to perform well, the use of the technical rules in the foreign exchange market is well documented (see Menkhoff and Taylor (2007) [265]).

## 5.4 Patterns Discovery via Non-Parametric Smoothing Methods

Charting prices has been a part of financial practice for many decades; but because of its highly subjective nature of interpretation of geometric shapes, reading charts is "often in the eyes of the beholder." Lo, Mamaysky and Wang (LMW) (2000) [247] propose a systematic approach to pattern recognition using nonparametric kernel regression. The general goal of this analysis is to identify regularities in the price series by extracting nonlinear patterns. The smoothing estimators that can extract the signal by averaging out the noise are ideally suited for this task.

Assume that price follows,

$$P_t = m(x_t) + a_t, \tag{5.9}$$

where $m(x_t)$ is an arbitrary function of the state variable, $x_t$. For most practical purposes, assume $x_t = t$ as the data is typically chronological. Kernel regression, orthogonal series expansion, projection pursuit, neural networks are all examples of smoothing methods that are applied to estimate $\hat{m}(x_t)$;

$$\hat{m}(x) = \frac{1}{T}\sum_{t=1}^{T} w_t(x)P_t, \tag{5.10}$$

where the weights $\{w_t(x)\}$ are large for those $P_t$'s paired with $x_t$'s near $x$ and are small for those away from $x$. The weight function in (5.10) defined for a neighborhood of width '$h$' around '$x$' can be stated as

$$\hat{m}_h(x) = \frac{1}{T}\sum_{t=1}^{T} w_{t,h}(x)P_t = \frac{\sum_{t=1}^{T} K_h(x - x_t)P_t}{\sum_{t=1}^{T} K_h(x - x_t)}, \tag{5.11}$$

where $K(x)$ is a probability density function and is also-called 'kernel', and $\hat{m}_h(x)$ is called the Nadaraya-Watson kernel estimator. A popular density is that of a standard normal distribution, $K(t) = \frac{e^{-t^2/2}}{\sqrt{2\pi}}$.

Selecting an appropriate bandwidth, '$h$' is important as a small '$h$' may not smooth much and a large '$h$' may result in too smooth a graph. The choice of '$h$' is made through cross-validation, by minimizing

$$\mathrm{CV}(h) = \frac{1}{T}\sum_{t=1}^{T}(P_t - \hat{m}_h(t))^2, \tag{5.12}$$

where

$$\hat{m}_{h,t} = \frac{1}{T}\sum_{\tau \neq t}^{T} w_{\tau,h}P_\tau. \tag{5.13}$$

Thus, the bandwidth choice depends on empirically how well the estimator fits each observation, $P_t$, when it is not used in the construction of a Kernel estimator. LMW (2000) [247] observe that generally the bandwidth choice made through (5.12) when applied to price series, tend to be larger and they suggest using $0.3 \times h$.

To automate the detection of charting patterns, first the pattern is defined in terms of its geometric properties, then $\hat{m}(x)$ is estimated via (5.11), to match with it. LMW (2000) [247] define ten different patterns based on local extremes. The window size is $(t, t+l+d-1)$, where $l$ and $d$ are fixed parameters, so that a pattern is completed within $l + d$ trading days. Within each window, estimate a kernel regression using prices in that window:

$$\hat{m}_h(\tau) = \frac{\sum_{s=t}^{t+l+d-1} K_h(\tau - s)P_s}{\sum_{s=t}^{t+l+d-1} K_h(\tau - s)}, \quad t = 1, \ldots, T - l - d - 1 \tag{5.14}$$

and the local extremes are identified through sign changes in $\hat{m}_h(\tau)$.

To test if the kernel regression yields better prediction, LMW (2000) [247] compare the unconditioned distribution of returns to the conditioned distribution, conditioned on the existence of a pattern. These comparisons are made through a goodness-of-fit test such as Kolmogorov-Smirnov test. It is generally found that the Kernel regression method yields somewhat better predictions. Jegadeesh (2000) [214] in his discussion of LMW (2000) [247] questions the chartists' belief, that selected patterns of the past tend to repeat, as there is a great deal of noted subjectivity in the interpretation of charts. It is further noted that the practitioners tend to use the visual patterns in conjunction with other information such as macro level events that exert large influence compared to stock specific moves, to time their trades.

## 5.5 A Decomposition Algorithm

The rules that were discussed in previous sections can be termed mostly trend following because they use past behavior of price or returns data to study future behavior. The predictability of returns generally tends to be weak, consistent with the efficient market theory. But the direction of change and volatility of returns tend to exhibit somewhat stronger dependence over time. Anatolyev and Gospodinov (2010) [15] exploit this and decompose the return into a product of sign and absolute value components. Their joint distribution is obtained through a continuous variable model for absolute value of the return and a binary choice model, for direction of the return and a copula for their interaction. Precisely the returns, '$r_t$', is written as

$$r_t = c + |r_t - c| \operatorname{sgn}(r_t - c) = c + |r_t - c|(2 \cdot I(r_t > c) - 1), \tag{5.15}$$

where $I(r_t > c)$ is an indicator function. The values of $c$ are determined by the user; it could be a multiple of the standard deviation of '$r_t$' or it could represent the transaction cost or the risk free rate or any other benchmark of interest. This formulation is conceptually similar to Rydberg and Shephard (2003) [295] and potential usefulness of the decomposition is also mentioned in Anatolyev and Gerko (2005) [14].

An advantage of the decomposition can be illustrated as follows. Assume for convenience, $c = 0$. From the model for $|r_t|$ and $I(r_t > 0)$, write $E_{t-1}(|r_t|) = \alpha_{|r|} + \beta_{|r|}\text{RV}_{t-1}$ and $\Pr_{t-1}(r_t > 0) = \alpha_{|I|} + \beta_{|I|}\text{RV}_{t-1}$, where RV denotes the realized volatility defined in (4.40). This results in

$$E_{t-1}(r_t) = -E_{t-1}(|r_t|) + 2E_{t-1}(|r_t| \cdot I(r_t > 0)) = \alpha_r + \beta_r \text{RV}_{t-1} + \sigma_r \text{RV}_{t-1}^2. \tag{5.16}$$

Thus, the decomposition in (5.15) leads to accommodating non-linearity and the model in (5.16), can be expanded to include other asset related variables as well. Hence any dynamic modeling that captures the hidden non-linearities is likely to improve the prediction of $r_t$. Even when the random walk model for price, return $r_t$ is white noise, the above formulation helps to capture some time dependence. We describe below other useful formulations.

The marginal distributions of the absolute value and sign terms are modeled as follows.

$$|r_t - c| = \psi_t \eta_t, \tag{5.17}$$

where $\eta_t$ is a Weibull distribution (4.33), and

$$\ln \psi_t = w_v + \beta_v \ln \psi_{t-1} + r_v \ln|r_{t-1} - c| + \rho_v I(r_{t-1} > c) + x'_{t-1}\delta_v. \tag{5.18}$$

Note the above specification is the same as the autoregressive conditional duration (ACD) model discussed in Chapter 4. The $x_t$'s are macroeconomic predictors that may have an effect on volatility dynamics. The specification of the indicator variable is done as follows: $I(r_t > c)$ is assumed to follow Bernoulli($p_t$), where $p_t = \text{logit}(\theta_t)$, where

$$\theta_t = w_d + \phi_d I(r_{t-1} > c) + y'_{t-1}\delta_d, \tag{5.19}$$

where the $y_t$'s can, besides the macro economic variables, include the realized variance, $\text{RV}_t = \sum_{s=1}^{m} r_{t,s}^2$, bipower variation, $\text{BPV}_t = \frac{\pi}{2} \cdot \frac{m}{m-1} \cdot \sum_{s=1}^{m-1} |r_{t,s}|\,|r_{t,s+1}|$ and the realized third and fourth moments of returns, representing skewness and kurtosis measures, $\text{RS}_t = \sum_{s=1}^{m} r_{t,s}^3$ and $\text{RK}_t = \sum_{s=1}^{m} r_{t,s}^4$.

With the marginals specified as above with the corresponding, c.d.f.'s $F(\cdot)$, the joint density is given as

$$F_{r_t}(u, v) = c\left[F_{|r_t - c|}^{(u)}, F_{I(r_t>0)}^{(v)}\right] \tag{5.20}$$

where $c(w_1, w_2)$ is a Copula, a bivariate c.d.f. For a general discussion on Copulas, refer to Lai and Xing (2008) [235, Section 12.3.3].

In the trading context, the value of a strategy to buy or sell depends on how good the short term prediction of $r_t$ is. Note that the conditional mean can be expressed as, $E_{t-1}(r_t) = c - E_{t-1}(|r_t - c|) + 2E_{t-1}(|r_t - c|\, I(r_t - c))$. Hence, $\hat{r}_t = c - \hat{\psi}_t + 2\hat{\varepsilon}_t$, where $\varepsilon_t$ is the expected value of the cross product (second) term in $E_{t-1}(r_t)$. If the cross-product terms are conditionally independent, $\varepsilon_t = \psi_t p_t$ and thus $\hat{r}_t = c + (2\hat{p}_t - 1)\hat{\psi}_t$. Note, the prediction of $r_t$ includes both the directional and volatility predictions.

A possible strategy is to invest in stocks if the predicted excess return is positive or invest in bonds if it is negative. In terms of established measures of performance, the decomposition method is shown to yield somewhat superior results.

**A Related Neural-Network Model:** There is more and more interest in using Neural Networks for developing technical rules that have better predictive power than the baseline models such as random walk or the following AR($p$) model:

$$r_t = \alpha + \sum_{i=1}^{p} \phi_i r_{t-i} + a_t, \tag{5.21}$$

where $a_t \sim N(0, \sigma_a^2)$. This model is augmented with a term used in the oscillator rule that was discussed earlier in this chapter. Define $s_t^{m,n} = P_t^m - P_t^n$, the oscillator signal resulting from the short and long term moving averages of the prices and the model,

$$r_t = \alpha + \sum_{i=1}^{p} \phi_i r_{t-i} + \sum_{i=1}^{p} \eta_i s_{t-i}^{m,n} + a_t \tag{5.22}$$

is an extended linear model and is shown to increase the predictive power. Extending the above to the feed forward neural network model with '$d$' hidden units is written as

$$r_t = \alpha_0 + \sum_{i=1}^{p} \phi_{ij} r_{t-i} + \sum_{j=1}^{d} \eta_j G \left[ \alpha_j + \sum_{i=1}^{p} r_{ij} s_{t-i}^{m,n} \right] + a_t, \tag{5.23}$$

where the activation function $G(x) = \dfrac{1}{1+e^{-\alpha x}}$ is of the logistic functional form. Gencay (1998) [162] examining the daily Dow Jones Industrial Average Index from 1877 to 1998 finds the evidence of nonlinear predictability in returns by using the oscillator signal.

## 5.6 Fair Value Models

The trading strategies are typically evaluated by comparing their performance against some benchmarks discussed in Section 5.2. Traders often face challenges to find trading opportunities that will outperform these benchmarks. In the model considered in Section 5.5, the quantity, '$c$' is taken to be transaction costs or simply it can be set to risk-free rate. The fair value models are based on the assumption that the fair price of a stock is influenced by some common factors. These factors are general market factors; examples of the linear factor models of returns are CAPM, Fama-French three- and five-factor models and Carhart factor models (see below and also Chapter 6 for a detailed discussion of these factors). Consistent with algorithmic trading principle, large deviations that are not explained by the dynamics of the established factors are taken to be abnormal and can be exploited as trading opportunities. These models are termed 'fair value' models because they refer to reasonable benchmarks for the comparison of the strategy performance.

For the CAPM model (5.1), it is expected that $\alpha = 0$ if the model holds. Any deviation from the null hypothesis $H_0 : \alpha = 0$ indicates the stock (alpha) performance after adjusting for the risk-return relationship. Note that $\alpha$ can be written as,

$$\alpha = \mathrm{E}(r_t) - \beta\,\mathrm{E}(r_{mt}) + (\beta - 1) r_f. \tag{5.24}$$

When $\beta > 1$, the stock is taken to be riskier than the market. If the stock is closely aligned with the market, that is $\beta = 1$, $\alpha = \mathrm{E}(r_t) - \mathrm{E}(r_{mt})$ and thus a positive '$\alpha$' would indicate performance superior to market performance.

In addition, it is expected in a high frequency setting, $\epsilon_t$'s are likely to exhibit some serial dependence:

$$\epsilon_t = a_t - \theta a_{t-1}. \tag{5.25}$$

Estimating the model parameters can be accomplished by the methodology given in Chapter 2; the procedure is generalized least squares and is iterative. The CAPM model can be extended to include other factors to provide a general model:

$$r_t = \alpha + \beta' f_t + \epsilon_t, \tag{5.26}$$

where $f_t$ is an '$n$'-dimensional vector of factors that may include in addition to the market index: SMB is the return on a diversified portfolio of small stocks minus the return on a diversified portfolio of big stocks, HML is the difference between the returns on diversified portfolios of high and low Book to Market (B/M) stocks, RMW is the difference between the returns on diversified portfolios of stocks with robust and weak profitability and CMA is the difference between the returns on diversified portfolios of the stocks of low (conservative) and high (aggressive) investment firms (see Fama and French (2015) [133]). Carhart (1997) [68] in studying the persistence of mutual fund performance uses one-year momentum in stock returns as an additional factor to Fama and French's original three-factor model.

Another variant of this type is considered in Avellaneda and Lee (2010) [24]. They describe a small portfolio trading algorithm. The returns of stocks in the portfolio are regressed on the returns of exchange traded funds (ETFs) or portfolios that are predictive. The resulting residuals are then examined for signals. The main practical difficulty here is that some stocks can be a part of several ETFs and so the set of predictors can be large.

A general comment is in order: The so-called CAPM factors mostly meant for explaining cross-sectional variation are slow-moving and hence the only factor that is likely to change in the high frequency context is the market factor. Also the error term may contain mostly the information about market frictions and hence the CAPM model itself may not hold in practice.

## 5.7 Back-Testing and Data Snooping: In-Sample and Out-of-Sample Performance Evaluation

With the publication of an influential paper by Brock, Lakonishok and LeBaron (1992) [60] on technical rules, the mainstream finance academics started paying attention to studying the properties of these rules. The rules considered are mostly the variants of rules related to moving average and trading range break. Applying these rules on the Dow Jones Index from 1897 to 1986, it is shown that technical strategies perform well. Results obtained are not consistent with the usual benchmark models: The random walk, AR(1), GARCH-M and Exponential GARCH. Buy signals generally tend to fare better than sell signals. The $p$-values for the rules are estimated through bootstrap method. But Ready (2002) [286] argues that the apparent success after transaction costs of the Brock et al. (1992) [60] moving average rules is a spurious result of data snooping. It is shown that the rules did poorly post 1986 and the average returns for 1987–2000 were higher on days when strategies suggest to be out of the market. Thus the performance of the technical rules must be carefully evaluated.

Data snooping occurs if the same data is used for both model selection and model validation. The significant results obtained can be due to pure chance. As Sullivan, Timmermann and White (1999) [308] suggest data snooping may be due to survivorship bias as only successful trading rules tend to be replicated. Thus it is important to consider almost the entire universe of trading rules to study the snooping bias. It is suggested using a bootstrap procedure studied in White (1999) [323] to check on data snooping bias.

The bias comes mainly from the application of multiple trading rules on the same set of data and thus by pure chance can yield some significant results. If there are '$l$' trading rules under consideration and they are validated over next '$n$' periods, define

$$\overline{f} = \frac{1}{n} \sum_{j=1}^{n} \hat{f}_{t+j}, \tag{5.27}$$

where $\hat{f}_{t+j}$ is the observed performance measure for period '$t+j$', based on the data up to '$t$'. For the $k$th trading rule, let

$$f_{k,t+j} = \ln(1 + r_{t+j} s_k) - \ln(1 + r_{t+j} s_0), \tag{5.28}$$

where $r_{t+j}$ is the return, $s_k$ and $s_0$ are signal functions for the $k$th trading rule. The baseline or benchmark rule can take three positions: Long, short and neutral, taking values 1, $-1$, and zero, respectively. The signal functions are of course based on data and information available as of '$t$'. The proposed test assumes the null hypothesis that the best technical rule does no better than the benchmark rule:

$$H_0 : \max_{k=1,\ldots,l} \mathrm{E}(f_k) \leq 0. \tag{5.29}$$

Rejection of $H_0$ would establish that the best technical rule is superior to the benchmark rule. Hypothesis similar to (5.29) can be formulated, for the Sharpe ratio as well, which will be a measure of risk-adjusted return. The $p$-values for $H_0$ are obtained through bootstrap resampling procedures, to construct the following statistics: If there are '$B$' bootstrapped values of $\overline{f}_k$ denoted by $\overline{f}^*_{k,i}$, $i = 1, 2, \ldots, B$ construct the following statistics:

$$\overline{v}_l = \max_{k=1,\ldots,l} \{\overline{f}_k \sqrt{n}\} \tag{5.30}$$

$$\overline{v}_{l,i} = \max_{k=1,\ldots,l} \{\sqrt{n}(\overline{f}^*_{k,i} - \overline{f}_k)\}. \tag{5.31}$$

The values of $\overline{v}_l$ are compared to the quantities of $\overline{v}^*_{l,i}$ to obtain the $p$-values.

The number of trading rules, $l = 7846$, considered by Sullivan et al. (1999) [308] is fairly exhaustive. The benchmark rule is being 'always out of the market'. The difference between the two $p$-values, i.e., between the bootstrap resampling method and the nominal method is fairly substantial, for the out-of-sample periods for the universe of trading rules considered. However, there are certain trading rules, that are found to outperform the benchmarks both in terms of the average return and the Sharpe ratio.

Two main drawbacks of the above test are: One the test is conservative and may lose power when poor performing models are included in the universe of the tests; second the test does not identify significant models other than the best performing model.

Hsu, Hsu and Kuan (2010) [203] propose a stepwise modification of the 'superior predictive ability (SPA)' test suggested in Hansen (2005) [176] and adapt the modification suggested in Romano and Wolf (2005) [292]. The modified method works as follows; the test statistic is

$$\mathrm{SPA}_l = \max\left( \max_{k=1,\ldots,l} (\overline{f}_k \sqrt{n}, 0) \right). \tag{5.32}$$

By resampling, the critical values of $\mathrm{SPA}_l$ are determined. Following the approach:

(a) Rearrange $\overline{f}_k$ in a descending order.

(b) Reject the top model '$k$' if $\sqrt{n}\overline{f}_k$ is greater than the critical value; if no model is rejected stop; else go to the next step.

(c) Remove $\overline{f}_k$ of the rejected models; regenerate the critical values. Now do the test and if no model is rejected, end the procedure; else go to the next step.

(d) Repeat (c.) until no model can be rejected.

The hypotheses stated in (5.29) are joint hypotheses and hence the tests are judged by Family Wise Error Rate (FWE) which is the probability of rejecting at least one correct null hypothesis. Hsu et al. (2010) [203] consider two types of technical rules, filter and moving-average based, for several market indices and ETFs. There are also several rules that result in significant increase in mean return and also fare favorably in terms of Sharpe Ratio.

Bajgrowicz and Scaillet (2012) [26] use the concept of false discovery rate (FDR) to reach a negative conclusion as in Ready (2002) [286]. The FDR is studied in detail in the context of mutual funds performance by Barras, Scaillet and Wermers (2010) [32]. The methodology that we discussed last indicates if a rule outperforms a benchmark after accounting for data snooping and it does not provide any information on the other strategies. But in reality investors gather signals from multiple rules and these rules are not conceptually or empirically independent. The FDR approach by allowing for a certain number of false discoveries, accounts for the dependency and also provides an array of high performing rules. Bajgrowicz and Scaillet (2012) [26] address also the issue of persistence of these rules associating a realistic transaction cost with each trading signal. Their main finding, based on the analysis of Dow Jones Industrial Average daily data (1897–1996) is that 'we are not able to select rules whose performance is persistent and not canceled by transaction costs'. They conclude: "...investors should be wary of the common technical indicators present in any investment website...," but acknowledge, "...our results say little about the existence of profitable trading strategies in other markets, using different frequencies or more sophisticated rules." These warnings do not seem to have deterred use of technical rules in practice.

## 5.8 Pairs Trading

Pairs trading is a market neutral trading strategy that enables traders to make profits from theoretically any market conditions. It was pioneered by a group of quants at Morgan Stanley led by Gerry Bamberger and Nunzio Tartaglia. Statistical Arbitrage (StatArb) pairs trading, first identifies pairs of similar securities whose prices tend to move together and thus uncovers arbitrage opportunities if there is a divergence in prices. If their prices differ markedly, it may be taken that one may be overvalued, the other is undervalued or both situations may occur together. Taking a short position on the relatively overvalued security and a long position on the undervalued security will result in profit when the mispricing corrects itself in the future. The magnitude of the spread between the two securities indicates the extent of mispricing and the range of potential profit. This type of trading is sometimes called noise trading as it is not correlated to any changes in the fundamental values of the assets. A pairs trading strategy which assumes the fundamental values of the two stocks are similar can provide a measure of the magnitude of noise traders risk.

To illustrate, we consider the so-called "Siamese twins" stocks: Royal Dutch/Shell and Unilever NV/PLC. Prior to 2005, Royal Dutch and Shell were jointly held by two holding companies. Royal Dutch which was domiciled in the Netherlands had an interest of 60% of the group and Shell based in the United Kingdom had a 40% interest. When the unification occurred it was agreed to divide the aggregate dividends in the ratio, 60:40. If the law of one price holds, the theoretical parity ratio expressed in a common currency is,

$$\text{Theoretical Parity} = \frac{0.6\,(\text{ outstanding Shell Shares })}{0.4\,(\text{ outstanding Royal Dutch Shares })}.$$

Scruggs (2007) [300] defines mispricing as simply being the deviation of the actual price ratio from the theoretical parity. Figure 5.1 plots the deviation from the theoretical parity which is taken to be 1.67.

Scruggs (2007) [300] also discusses another pair. Unilever NV and Unilever PLC are the twin parent companies of the Unilever group and to operate as a single company, they are governed by an equalization agreement, "one NV ordinary share equals to 1.67 ordinary shares of PLC." Thus, the theoretical parity is fixed at 1.67:1. Figure 5.2 plots the deviation from this parity. The deviation could be as high as 27% in this case, which happens in February of 1999. Both pairs clearly indicate the opportunities to exploit the divergence in the stock prices. Froot and Dabora (1999) [151] also studied the pairs of stocks, but their focus was on relating the changes in stock prices to the location of trade; the movements are more aligned with the movements of the markets where they are most actively traded.

The first step in developing a pairs trading strategy is selecting two or more stocks whose price movements are closer to each other. Two examples discussed here are natural-pairs, as the parent company is the same and they can be called twins. Finding such natural twins is hard in practice. As substitutes generally, mature companies

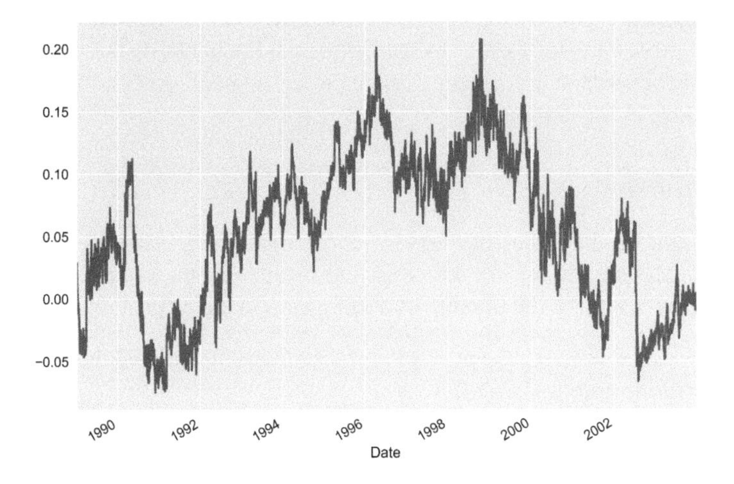


Figure 5.1: Royal Dutch/Shell.

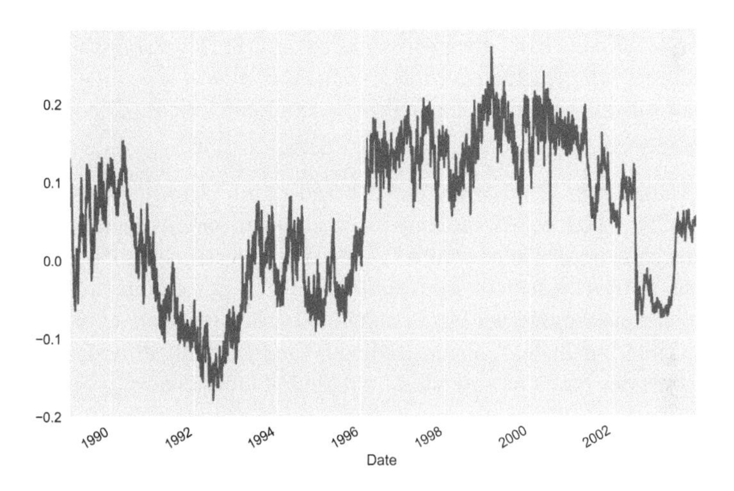


Figure 5.2: Unilever NV/PLC.

with long trading histories and similar exposure to macro factors are considered as candidates for forming pairs; usually they tend to be from the same industry. We will discuss in detail how pairs can be formed later. First we present some formal procedures for implementing the pairs trading strategy.

### 5.8.1 Distance-Based Algorithms

Gatev, Goetzmann and Rouwenhorst (2006) [160] suggest using the distance based algorithm on normalized prices to identify the periods when the prices diverge.

The normalization factor is a total cumulative returns index of the stock over the pairs formation period. Pairs are formed by computing the Euclidean distance between any two stocks and by selecting the pairs that are closer. A simple pairs trading algorithm is stated below, as illustration:

Pairs Trading Algorithm:

- Normalize the prices of every stock to equal unity, at the end of each calendar month, m; use the preceding twelve months as the estimation period; let $T_m$ be the number of trading days in the estimation period.
- For each $t = 1, \dots, T_m, p_t^i = \prod_{T=1}^{t} [1 \times (1 + r_T^i)]$, where $T$ is the index for all the trading days till $t$;
- $d_{i,j,m} = \sum_{i=1}^{T_m} \dfrac{(p_t^i - p_t^j)}{T_m}$

  St.Dev $d_{i,j,m} = \sqrt{\dfrac{1}{T_m - 1} \left( \sum_{i=1}^{T_m} [(p_t^i - p_t^j)^2 - d_{i,j,m}]^2 \right)}$
- If price difference ($d$) diverges by more than 2 St.Dev $d$, buy the cheaper stock and sell the expensive one; if pairs converge at later time, unwind; if pairs diverge but do not converge in six months, close the position.

Gatev et al. (2006) [160] study the risk and return characteristics, using daily data from 1962 to 2002, of pairs trading for stocks from various sectors. The general finding is that pairs trading is profitable in every broad sector category. In relating the excess returns to the risk factors, market, SMB, HML, momentum and reversal, it is found that the exposures are not large enough to explain the average returns of pairs trading. Thus they conclude, "...pairs trading is fundamentally different from simple contrarian strategies based on reversion," although it looks conceptually the same.

## 5.8.2 Co-Integration

There is some evidence that stock prices are co-integrated (for a discussion on co-integration, see Chapter 3) for the US Stock market (see Bossaerts (1988) [48]). If the price vector, $p_t$ is co-integrated with co-integrating rank, $r$ which imply that there are '$r$' linear combinations of $p_t$ that are weakly dependent; thus these assets can be taken as somewhat redundant and any deviation of their price, from the linear combination of the other assets' prices is temporary and can be expected to revert. Writing the pairs trading concept in a model form (co-integrated VAR) has several advantages. It allows us to extend pairs to triplets, etc., and also the model can accommodate incorporating other relevant factors such as related indices.

A model based co-integration approach to pairs trading is elegantly presented in Tsay (2010) [315, Section 8-8]. It is based on the assumption that prices follow a random walk and on the additional expectation that if two stocks have similar risk factors, they should have similar returns; writing $p_{it} = \ln(P_{it})$ and with $p_{it}$ following a random walk, $p_{it} = p_{it-1} + r_{it}$ where $r_{it}$ is the return. Thus, bivariate price series are said to be co-integrated, if there exists a linear combination, $w_t = p_{1t} - \gamma p_{2t}$ which is stationary and therefore it is mean reverting.

With the error-correlation form of the bivariate series (follows from (3.29) in Chapter 3),

$$\begin{bmatrix} p_{1t} - p_{1t-1} \\ p_{2t} - p_{2t-1} \end{bmatrix} = \begin{bmatrix} \alpha_1 \\ \alpha_2 \end{bmatrix} \begin{bmatrix} w_{t-1} - \mu_w \end{bmatrix} + \begin{bmatrix} a_{1t} \\ a_{2t} \end{bmatrix}, \tag{5.33}$$

where $\mu_w = \text{E}(w_t)$ is the mean of $w_t$, which is the spread between the two log prices. The quantity $w_{t-1} - \mu_w$ denotes the deviation from equilibrium, $\alpha$'s show the effect of this past deviation on the current returns and should exhibit opposite signs. Going long with one share of stock 1 and shorting $\gamma$ shares of stock 2, will result in the return of the portfolio, $r_{p,t+i} = w_{t+i} - w_t$, at time 't'. The increment does not depend on $\mu_w$.

Based on the above discussion the following strategy can be followed.

Let $\eta$ be the trading cost and $\Delta$ be the target deviation of $w_t$ from $\mu_w$ and assume $\eta < 2\Delta$.

- Buy a share of stock 1 and sell $\gamma$ shares of stock 2 at time t if $w_t = \mu_w - \Delta$
- Unwind at time '$t + i$' ($i > 0$) if $w_{t+i} = \mu_w + \Delta$

Thus the return of the portfolio would be $2\Delta$ which is greater than the trading cost.

For other modified versions of the strategy refer to Tsay (2010) [315, Section 8.8]. An alternative formulation through vector moving average is given in Farago and Hjalmarsson (2019) [138].

From a practitioner's standpoint, for testing if there is a co-integrating relationship between pairs, it is easier to use the two-step test proposed in Engle and Granger (1987) [122],

(a) Run a regression model: $p_{1t} = \alpha + \beta p_{2t} + u_t$. The residual term in the co-integrating relation.

(b) Test if the errors, $u_t$ are stationary, via the model $u_t = \gamma + \phi u_{t-1} + a_t$, with the null hypothesis $\text{H}_0 : \phi = 1$ vs. the alternative hypothesis $\text{H}_a : \phi < 1$.

This test is easy to use particularly in the case of bivariate series.

To illustrate the pairs trading strategy, we consider the Royal Dutch/Shell data. The daily data is from January 3, 1989 to December 31, 2003. The rescaled (with starting point to be the same for better visual comparison) plot is given in Figure 5.3; the series are broadly aligned as expected but the gap between them at different time points is also quite evident. Let $p_{1t}$ denote the log price of Royal Dutch and let $p_{2t}$ denote the log price of Shell. Results can be summarized as follows:

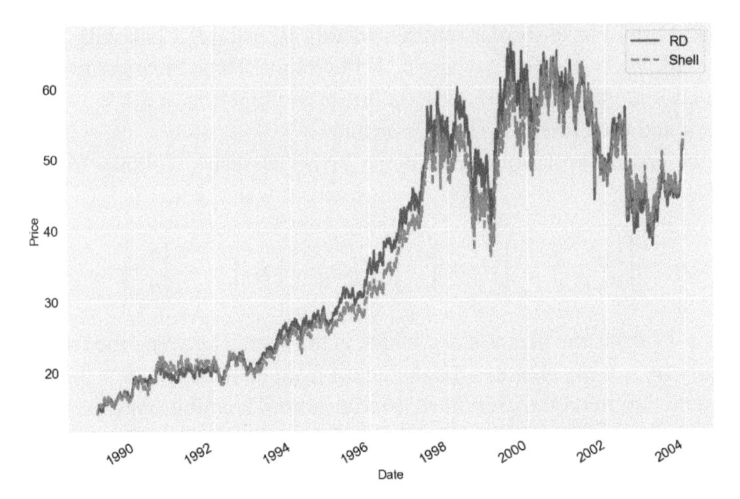


Figure 5.3: Rescaled Plot of Royal Dutch, Shell.

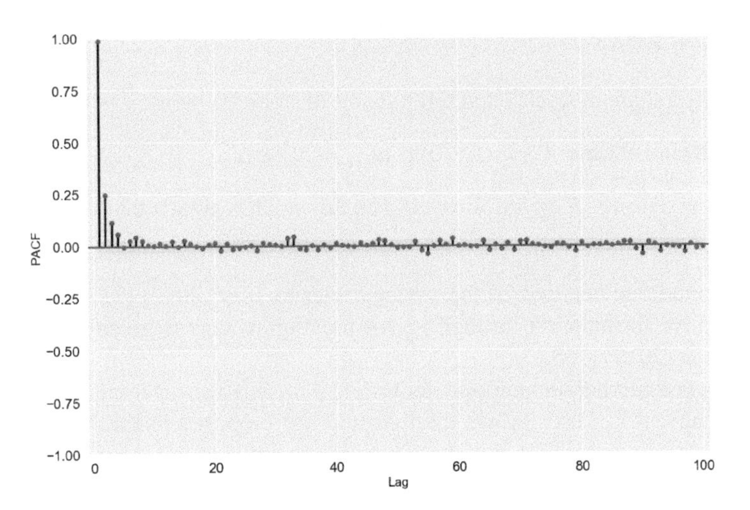


Figure 5.4: Partial Autocorrelation Function for RESI.

- $\hat{p}_{1t} = -0.061^* + 1.025^* p_{2t}$, $R^2 = 0.988$, Durbin-Watson (DW) statistic is 0.024 which indicates that the errors are not independent. (Recall from Chapter 2, DW∼2 for independent errors.)

- Partial autocorrelation function of the estimated residual $\hat{u}_t = p_{1t} - \hat{p}_{1t}$ is given in Figure 5.4, indicate that the residuals are stationary, probably with an AR(2) or AR(3) structure. Recall the augmented Dickey-Fuller test using the error correction form, where $v_t = u_t - u_{t-1}$ is regressed on $u_{t-1}$ and $v_{t-1}$ and $v_{t-2}$ and if the coefficient of $u_{t-1}$ is significant, it would confirm the co-integrating relationship.

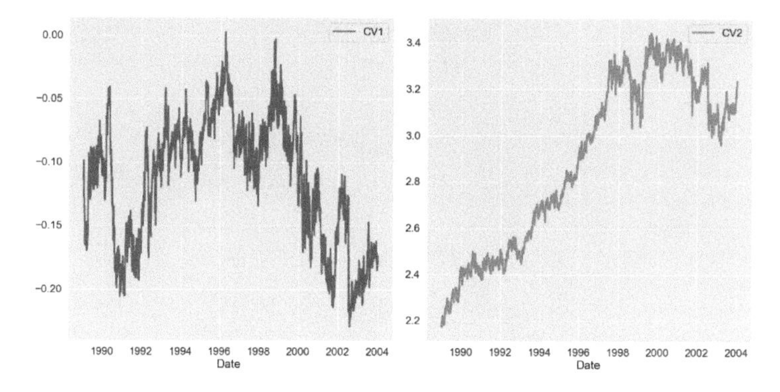


Figure 5.5: Plot of Canonical Series.

The estimated model is, $\hat{v}_t = -0.008^* u_{t-1} - 0.273^* v_{t-1} - 0.1154^* v_{t-2}$ and the $t$-ratio that corresponds to the coefficient of $u_{t-1}$ is $-3.21$ which is significant.

– The error correction model (5.33), which is $p_t - p_{t-1} = \Phi^* p_{t-1} + a_t$, where $p_t = (p_{1t}, p_{2t})'$ and $\Phi^* = \Phi - I$, can be shown to have rank$(\Phi^*) = 1$ using the test statistic in (3.31); the two linear (canonical variate) combinations are $\text{CV1} = p_{1t} - 1.04p_{2t}$ and $\text{CV2} = p_{1t} - 0.18p_{2t}$ which are plotted in Figure 5.5. The first series appears to be stationary which is $w_t = \text{CV1}$ in (5.33). This can be used for trading purposes using the rules discussed in the earlier part of this chapter.

To illustrate opportunities in pairs trading in real time, we consider the daily exchange rates between Indian rupee and US dollar, GBP and Euro from December 4, 2006 to November 5, 2010. We normalize the first day's observation as unity and also note that the initial ratio is 1:1.98:1.33 in rupees for US dollar, GBP and Euro respectively. From their plots in Figure 5.6, it is obvious that all three series move together until the middle of 2007, then the Euro takes off and then the US dollar which is closely aligned with GBP takes off to join the Euro. For testing co-integration, one has to look at the three regimes individually. Thus tracking the co-integrating relationships over time, traders can generate profits by properly timing the swapping of currencies.

The study by Farago and Hjalmarsson (2019) [138] also focuses on using the co-integration method to identify pairs for trading and evaluates the strategy using Sharpe ratio. Writing the return process as $r_t = \mu + u_t$ and $u_t$-process capturing the autocorrelation in the returns, the price process $p_t$ is decomposed into a deterministic trending component (equivalent to a drift), a non-stationary martingale component and a transitory noise component. The co-integration process generally eliminates both the trending and martingale components and thus leaving the convergence of the pairs to mainly transitory component. The co-integration of any two stocks in the market which implies long-run relationships is not likely to occur. To contrast, stocks in the Swedish markets, listed as $A$ and $B$ shares, which share the same features, yield high Sharpe ratios compared to other stock pairs.

Figure 5.6: Rupee Exchange Rate Data.

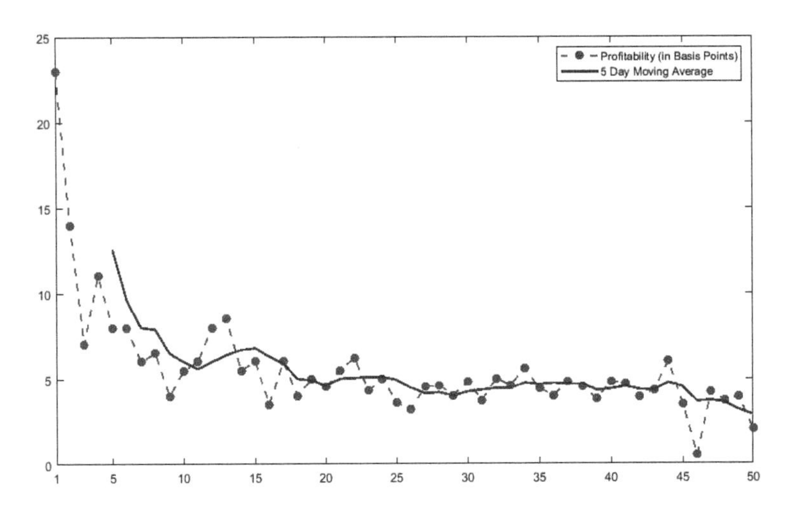


Figure 5.7: Pairs Trading Profitability in Event Time.

### 5.8.3 Some General Comments

**Finding Pairs:** An alternative approach is to use fundamentals of the firms to select two stocks that have similar risk factor exposures. Based on the Asset Pricing Theory (APT) model,

$$r_{it} - r_{ft} = \alpha_i + \beta_i' f_t + a_{it}, \tag{5.34}$$

where $r_{ft}$ is the risk-free rate and $f_t$ are the factor scores. Now collect the coefficient vector $\hat{\theta}_i' = (\hat{\alpha}_i, \hat{\beta}_i)'$ for $i = 1, 2, \ldots, N$ assets under consideration and form a standardized distance measure,

$$d_{ij}^2 = \hat{\theta}_i' \,\mathrm{Cov}^{-1}(\hat{\theta}_i)\hat{\theta}_i \tag{5.35}$$

and select pairs with the smaller distance. This can be expected to result in similar assets after adjusting for the risk factors. Generally the factors tend to vary much more slowly than the excess returns, $r_{it} - r_{ft}$, the distance based on $\hat{\theta}_i$ tends to move slower than short-term price movements. Recent studies confirm that the simple Euclidean distance-based pairing may be inadequate. Do and Faff (2010) [106] suggest adding two additional metrics: Industry homogeneity and the frequency of the historical reversal in the price spread; the measure in (5.35) can accommodate that easily.

**High Frequency Equity Pairs Trading:** While there is evidence of the profitability of pairs trading strategies at lower frequencies, not much is known in academia (but quite known in practice) about high frequency intra-day pairs trading. Bowen, Hutchinson and O'Sullivan (2010) [52] consider data from the London Stock Exchange for the most liquid stocks, the constituents of the index FTSE100. The duration is divided into a formation period, that is used to identify pairs of stocks that move together, and a trading period when the actual transactions occur. The distance based criteria using normalized prices is used to identify the top pairs during the formation period. As in any other strategy, pairs trading profits are affected by transaction costs, and this becomes more acute as trading frequency increases. Results appear to be also quite sensitive to the speed of execution, as excess returns disappear quickly if delay in execution is greater than one period. The majority of profits seem to occur in the first and last hour of trading, when relatively larger volatility and liquidity occur.

Although pairs trading has been used extensively by traders, it also has its limits. In a recent study, Engelberg, Gao, Jagannathan (2009) [116] demonstrate that the opportunities for medium frequency pairs trading are not many and opportunities for profit-making may vanish within a few days (see Figure 5.7). Scruggs (2007) [300] attributes the demise of Long-Term Capital Management to the hedge fund overconfidence in the strategy and their undue bet sizes compared to other market participants.

### 5.8.4 Practical Considerations

After identifying a potential pairs strategy via co-integration testing for instance, traders need to pay particular attention to the expected profitability of the strategy once real-life trading frictions have been taken into consideration. In practice, a significant portion of potential pairs trading strategies experience negative P&L in production due to large execution slippage, whether originating from aggressive hedging at trade initiation, or position closing out.

When back testing a pairs strategy, one of the immediate costs that might render the strategy unprofitable is the spread cost. In other words, how much will it cost to extract liquidity when the signal triggers, in particular if the signal is too short-lived to allow for sourcing liquidity passively, via limit order book posting. For intraday pairs trading, it is also valuable to analyze if there is any asymmetry of cost between the initiation trade and the unwind trade. Oftentimes, it is not uncommon to witness significantly higher exit costs once the market has normalized following a short-lived entry trade opportunity.

The individual instrument's microstructure also plays an important role in the strategy profitability. For instance long queue names (instruments for which the top of book liquidity is large in comparison to daily volume) are harder to trade, because the time it takes for an order to reach the top of the queue and trade passively is generally larger than the time that is acceptable for the strategy to mitigate its legging risk.[1] As a result, most of the executions on the leg that displays a long queue will end up having to pay the full spread to be executed.

As an example, when computing the spread ratio between two instruments using the market mid-point, one might be led to believe that the ratio is currently above $2\sigma$ of its distribution, warranting to initiate a trade. However, when considering the size of the spread and amount of the quoted volume on the passive side at the top of the book, it is not impossible that the realized ratio when crossing the spread is only a $0.5\sigma$ event, while getting a passive execution when quoting at the end of a long queue would represent a $5\sigma$ event that one shouldn't expect to realistically happen. These anecdotal examples give an idea of what practitioners should pay particular attention to when designing their strategies. Below, we mention a non-exhaustive list of implementation challenges also worth considering:

- **Difference in Liquidity Between the Two Legs:** Usually the selected pair has a slower moving leg and a faster moving one. This means that, in practice, one of the instruments might be ticking faster, thereby having a greater influence on the moves of the measured ratio. Additionally, faster ticking instruments also tend to have narrower spreads, meaning that in absence of control, the instrument the most likely to trigger a trade is also the one that is the cheapest to transact, forcing the strategy to then rapidly hedge out its exposure by paying the larger spread on the slowest ticking leg. One way of handling this situation is, for instance, to only let the strategy quote on the slow moving leg as long as it is "in ratio," and then only trigger the aggressive spread-crossing hedge trade if and when the passive leg receives a fill.

- **Rounding/Tick Size Granularity:** While in theory it is possible to come up with a signal threshold that is as granular as desired, in practice the step increment of the signal might be constrained by the microstructure of the instruments traded. Most exchanges globally impose a minimum tick size increment, that can vary based on the liquidity of the instrument. Here too, all things being equal, a practical approach would be to quote on the instrument, which constrained tick size contributes the most to the ratio moves, and only trigger the aggressive spread-crossing hedge trade if and when the most constrained tick size leg receives a fill.

---

[1]Legging risk can be defined as the risk of not being able to timely fill a particular leg of a pair strategy at the required price.

- **Signal Stability:** As a result of both the difference in liquidity between long and short legs, and the potential frictions induced by the instrument microstructure described above, it is not uncommon for the signal to jump around the trigger ratio frequently. Trading in and out on each signal crossing of the pre-determined ratio, would incur large transaction costs that would render many strategies unprofitable. To address this issue, a common approach is to define two different thresholds sufficiently apart for entering and exiting the trade, so that signal noise does not trigger excess in and out trading. It is also possible to enforce a minimum holding duration to prevent the stop loss signal to trigger just due to local noise around the ratio threshold.

- **Time of Day:** In the first fifteen to thirty minutes after a market opens, many instruments have significantly wider bid-ask spreads while price discovery sets in after a night of information accumulation and no possible trading. If this significant difference in transaction costs is not properly accounted for, the trades initiated during that period can have an impact on the overall profitability of the strategy.

- **Market Asynchronicity:** While pairs are usually constructed between instruments trading concomitantly, it is also possible to construct pairs strategies with instruments from different markets. For instance, a pair between a US and a Japanese stock. This raises an additional challenge, which is: How to assess the theoretical value on an instrument that is not currently traded? Practitioners commonly address this issue by first modeling the relationship between the instrument's daily time series, and then inferring its stability at more granular frequencies. The common risk can be hedged with a market wide instrument such as an index future, and be unwound overnight when the second leg is entered, thereby limiting unwanted exposure to market risk.

- **Traditional Index Arb / ETF Arb:** When trading a pairs strategy between a composite instrument (Index, ETF, ...) and its constituents, the fast and efficient nature of electronic markets might render trading opportunities too small to be exploited profitably across all instruments. In particular if the number of constituents is quite large, obtaining passive fills on most of them within a reasonable time frame is not a realistic assumption. As a result, traders can implement partial replication strategies whereby they only trade an optimized tracking basket of the instruments, that allows for the best replication under some liquidity and cost assumptions (spread size, top of book size, ...).

- **Local Market Idiosyncrasies:** For instance, in Japan, sector classification (which is often used as a starting point for pairs selection or as a way to avoid picking names that would have spurious correlation) does not match international sector classification one-to-one. Local market participants rely on the 33 TOPIX sectors that have a closer fit to the actual structure of the Japanese economy than the commonly used 11 GICS®sectors.

- **Coupling Market Making and Pairs Trading Strategies:** One possible approach to long queue and wide spread instruments pairs trading, is to use a market making strategy to first gain queue priority in the longest queue name and achieve a certain child order density in the order book at regular volume-time intervals. For example, if only one of the instruments has a long queue, as long as the pairs strategy ratio is below the trigger threshold, the market making strategy can earn the spread on the long queue instrument while staying out of the market on the other instrument. When the ratio reaches the threshold, one can cancel the child orders created on the opposite side of the book by the market making strategy and only keep the ones on the side relevant to the pairs strategy. These orders, placed over time while market making, should already have accrued some queue priority, which will increase the probability of receiving a passive fill once the ratio is favorable. Then, the pairs strategy is able to trade aggressively the liquid leg as soon as the passive fill is received. Based on the liquidity and microstructure of the instruments in the pairs strategy, different equivalent strategies can be combined to achieve higher profitability.

## 5.9 Cross-Sectional Momentum Strategies

The trading strategies discussed thus far depend exclusively on the existence of time-series patterns in returns or on the relationships between returns and the risk factors. Time series strategies generally exploit the fact that stock prices may not at times follow a random walk, at least in the short run. Even if price follows a random walk, it has been shown that strategies based on past performance contain a cross-sectional component that can be favorably exploited. For instance, momentum strategies demonstrate that the repeated purchase of past winners, from the proceeds of the sale of losers can result in profits. This is equivalent to buying securities that have high average returns at the expense of securities that yield low average returns. Therefore if there is some cross-sectional variance in the mean returns in the universe of the securities, a momentum strategy is profitable. On the other hand, a contrarian strategy will not be profitable even when the price follows a random walk. For the demonstration of the momentum strategy in the low frequency setting, refer to Jegadeesh and Titman (1993) [215], 2002 [217]) and Conrad and Kaul (1998) [82].

To better understand the sources of profit from a momentum strategy Lo and MacKinlay (1990) [249] decompose the profit function into three components. We will follow the set-up as outlined in Lewellen (2002) [242]. Let $r^k_{i,t-1}$ be the asset $i$'s, '$k$' month return ending in '$t-1$' and let $r^k_{m,t-1}$ be the return from the equal-weighted (market) index's '$k$' month return ending in '$t-1$'. The allocation scheme over '$N$' assets under the momentum strategy is,

$$w^k_{i,t} = \frac{1}{N}(r^k_{i,t-1} - r^k_{m,t-1}), \quad i = 1, \ldots, N. \tag{5.36}$$

For ease of interpretation, $w_{i,t}^k = w_{i,t}$ and focus on only one previous month's return for allocation; Note that for each '$t$', the weights $w_{i,t}$ sum to zero. Let $r_t$ be a $N \times 1$ vector of returns, with the mean return $\mu = \mathrm{E}(r_t)$ and $\Omega = \mathrm{E}[(r_{t-1} - \mu)(r_t - \mu)']$ the autocovariance matrix of returns. Let $w_t$ be $N \times 1$ vector of weights, $w_{it}$; the portfolio return for month 't' is $\pi_t = w_t' r_t$. It is easy to show

$$\begin{aligned} \mathrm{E}(\pi_t) &= \frac{1}{N}\,\mathrm{E}\left[\sum_i r_{i,t-1} \cdot r_{it}\right] - \frac{1}{N}\,\mathrm{E}\left[r_{m,t-1} \cdot \sum_i r_{it}\right] \\ &= \frac{1}{N}\sum_i (\rho_i + \mu_i^2) - (\rho_m + \mu_m^2), \end{aligned} \tag{5.37}$$

where $\rho_i$ and $\rho_m$ are lag 1 autocovariances of asset '$i$' and the equal weighted index. Note that the average autocovariance is $\mathrm{tr}(\Omega)/N$ and the autocovariance for the market as a whole is $(1'\Omega 1)/N^2$. The $\mathrm{tr}(\Omega)$ denotes the sum of the diagonal elements of $N \times N$ matrix $\Omega$; recall that the diagonal elements are the autocovariance, $\rho_i$, of individual stocks. Hence (5.37) can be written as,

$$\mathrm{E}(\pi_t) = \left(\frac{N-1}{N^2}\right) \cdot \mathrm{tr}(\Omega) + \frac{1}{N^2}[\mathrm{tr}(\Omega) - 1'\Omega 1] + \sigma_\mu^2. \tag{5.38}$$

The three terms in (5.38) precisely identify three sources of profits (see Lo and MacKinlay (1990) [249]):

- Stock returns are positively autocorrelated, so each stock's past return predicts high future return.
- Stock returns are negatively correlated with the lagged returns of other stocks, so poor performance of some stocks predicts high returns for others.
- Some stocks simply have high unconditional mean returns relative to other stocks.

The finance literature is replete with empirical verifications of the momentum strategy. However, the analytical decomposition in (5.38) is not quite informative about the economic sources of momentum. The success of the momentum strategy has been attributed to various theories. The literature argues that firm-specific returns can drive momentum. Firm-specific returns can be persistent because investors under-react to firm-specific news or stock returns covary strongly due to a common influence of a macro economic factor or a macro policy. Lewellan (2002) [242] shows that the Size and B/M portfolios also exhibit strong momentum; as these portfolios are well-diversified, their returns reflect systematic risk and therefore it is argued macroeconomic factors must be responsible for the momentum. Hvidkjaer (2006) [210] suggests that momentum could be partly driven by the behavior of small traders who under-react initially, followed by delayed reaction to buying and selling pressures. The trading pressure tends to translate into price pressures that may affect the trading behavior of the small traders.

Although cross-sectional momentum is widely documented, among the three components in (5.38), time series momentum as represented by the first term tends to dominate for every liquid instrument in a study of fifty-eight instruments spread over a variety of asset classes, equity index, currency, commodity and bond futures. Moskowitz, Ooi and Pedersen (2012) [273] show that the autocovariance term, $\text{tr}(\Omega)/N$ is the dominant factor, particularly in foreign exchange and equity markets. To make meaningful comparison across asset classes, the excess returns are scaled by their ex-ante volatilities and the following models are estimated for predicting price continuation and reversal:

$$\frac{r_t^s}{\sigma_{t-1}^s} = \alpha + \beta_h \frac{r_{t-h}^s}{\sigma_{t-h-1}^s} + \varepsilon_t^s \tag{5.39}$$

and

$$\frac{r_t^s}{\sigma_{t-1}^s} = \alpha + \beta_h \, \text{sgn}(r_{t-h}^s) + \varepsilon_t^s, \tag{5.40}$$

where $h = 1, 2, \ldots, 60$ months and the ex-ante annualized variance $\sigma_t^2$ is calculated as follows:

$$\sigma_t^2 = 261 \sum_{i=0}^{\infty} (1-\delta)\, \delta^i \, (r_{t-i} - \overline{r}_t)^2. \tag{5.41}$$

Here the weights $\delta^i(1-\delta)$ add up to one, $\overline{r}_t$ is the exponentially weighted average return computed in a similar fashion. The exponential smoothing parameter is chosen to be $\sum(1-\delta)\,\delta^i = \frac{\delta}{1-\delta} = 60$ days. Observe that using $\sigma_{t-1}$ in (5.39) and (5.40) avoids the look-ahead bias.[2]

We can also investigate the profitability of strategies based on time series momentum with varying look-back period ($k$) and holding period ($h$). The trading strategy is that if the excess return over the past $k$ months is positive, go long or otherwise go short. The position size is taken to be, $1/\sigma_{t-1}^s$, inversely proportional to the ex-ante volatility of the instruments, '$s$'. With this choice it is argued that it will be easier to aggregate the outcome of the strategies across instruments with different volatilities and it will also lead to somewhat stable time series. For each trading strategy $(k, h)$, a single time series of returns, $r_t^{\text{TSMOM}(k,h)}$, which is the average return across all portfolios active at time, '$t$' is computed. To evaluate the overall value of the time series momentum strategies, the returns are adjusted for possible risk factors:

$$\begin{aligned} r_t^{\text{TSMOM}(k,h)} &= \alpha + \beta_1 \text{MKT}_t + \beta_2 \text{BOND}_t + \beta_3 \text{GSCI}_t + \beta_4 \text{SMB}_t \\ &= +\beta_5 \text{HML}_t + \beta_6 \text{UMD}_t + a_t, \end{aligned} \tag{5.42}$$

where $\text{MKT}_t$ is the excess return on the MSCI World index, $\text{BOND}_t$ is Barclay's Aggregate Bond Index, $\text{GSCI}_t$ is S&P GSCI commodity index and the others are Fama-French factors for the size, value and cross-sectional momentum premiums.

Using monthly data spanning January 1985 to December 2009, Moskowitz et al. (2012) [273] find that the time series momentum strategies provide additional returns

[2]This occurs if data otherwise not available at the time of study is used in simulation or developing strategies.

over and above a passive buy and hold strategy. It is also shown that the model (5.42) results in a large and significant alpha. The analysis also illustrates that speculators are likely to benefit from the trend following at the expense of hedgers. The momentum appears to last about a year and then begins to reverse. It appears that the momentum strategies remain profitable even after adjusting for price impact induced by trading. The price impact (see Chapter 9) is the impact of the current trade on the subsequent trades. Although the trading decisions are done at a macro level, it should be kept in mind that the actual trading takes place at a micro level.

The role of trading volume in explaining the results of momentum strategies is studied in Lee and Swaminathan (2000) [240]. They find that, in equilibrium, both price and volume are simultaneously determined and thus both can exhibit momentum. In addition to the decile classification based on the past returns, the universe of stocks was further divided into three volume size categories, yielding thirty groups. The study provides evidence that low volume stocks tend to be under-valued by the market. Thus investor expectations affect not only a stock return but also its trading volume. High volume firms generally earn lower future returns. The general notion that volume fuels momentum is not supported but is true for losers and is further noted that "it helps information 'diffusion' only for winners."

It is worth mentioning that while most documented applications of the momentum strategy are in equity markets, there is a growing academic interest in their use in foreign exchange (FX) markets. The FX markets are traditionally cited as markets where technical rules such as trend following have been successful. But studies such as Menkhoff, Sarno, Schemeling and Schrimpf (2012) [264] have explored cross-sectional momentum strategies and have demonstrated their profitability (up to 10%) in FX markets. The momentum portfolios in FX markets are observed to be skewed toward minor currencies but trading these currencies have somewhat higher transaction costs. The base currency returns, $r_t = \ln(S_t) - \ln(S_{t-1})$ where $S_t$ is the spot exchange rate, and the interest-adjusted return, where $F_{t-1}$ is the future price at $t$-1,

$$\begin{aligned} r_t^* &= \ln(S_t) - \ln(F_{t-1}) \\ &= \ln(S_t) - \ln(S_{t-1}) + \ln(S_{t-1}) - \ln(F_{t-1}) \\ &\sim r_t + \frac{r_f - r_d}{12}, \end{aligned} \tag{5.43}$$

which is the sum of pure currency appreciation and the return due to interest rate appreciation. Here $r_f$ is the foreign interest rate and $r_d$ is the domestic interest rate; See Okunev and White (2003) [277] for the details of this decomposition. It is found that the profitability is clearly originating from spot rate changes themselves and not necessarily driven by the interest rate differential.

In the equity markets, it is argued that momentum strategies are successful because of slow information processing and investor overreaction. However FX markets are dominated by professionals and thus any information should be impounded in the FX rate almost instantaneously. But if the portfolios are mostly made up with minor currencies, they are likely to exhibit higher volatility, higher country risk and higher interest-rate changes, thus imposing effective limits to arbitrage. It is further observed that the cross-section based currency momentum is not related to benchmark technical trading strategies and even after adjusting for them, the cross-sectional momentum profits are found to be higher.

### Other Sorting Criteria

The classical momentum studies described above tend to focus on stratifying the stocks in ten deciles based on the past average returns. Han, Yang and Zhou (2013) [175] suggest sorting based on volatility which reflects the information uncertainty. Other sorting criteria may include credit rating, distance to default, etc. Generally, after sorting by a criterion, the decile portfolios are formed. The strategy then is to use the moving average based rule (see Section 5.3) as follows: If the price is above the moving average, buy or continue to hold or else invest in a risk-free asset. The performance of this strategy is then compared with buy and hold method. It is shown that the volatility based strategy results in positive returns and generally these returns increase with the volatility. If the returns are further adjusted for market risk (CAPM) and Fama-French factors, the average returns are greater and increase with the volatility levels. Two items of note here: As the returns are adjusted for the risk factors, they are likely to result in fewer trading opportunities as these models become more restrictive; if results are not adjusted for transaction costs, buying and selling a portfolio of stocks can be expensive and so the results may be too optimistic.

### Practical Considerations

Academic research indicates that the momentum factor is quite robust and works across markets with average return that is ten percent higher than the market return per year, before taxes and fees. But it is a short-term, high-turnover strategy with deep drawdowns. Israel, Moskowitz, Ross and Serban (2017) [211] discuss the implementation cost for a live momentum portfolio. Their results indicate that with judicious choice of execution algorithms, the profitability of momentum can still be maintained, after accounting for trading costs. Some suggestions that come from practitioners are:

- Rebalance more frequently, say monthly.
- Give more weight to stocks with higher momentum rather than a pure market capitalization.
- Give some importance to stocks with cheaper trading costs.
- Consider intermediate not just immediate past performance of the stocks for rebalancing.

Other suggestions include considering dual momentum, absolute and relative momentum, that use different benchmarks and work well both in upward and downward market trends.

## 5.10 Extraneous Signals: Trading Volume, Volatility, etc.

The trading algorithms that are presented in this chapter thus far focus on the price and return behavior of an equity. In pairs trading, we considered the behavior of price-return of a related equity to decide when to buy one and sell the other. But there is also information, as noted in Section 4.3, present in the trading volume. The model discussed there relates the volume to volatility. If there is no variation in the information flow about an equity, there should not be any correlation between volume traded and the return volatility. If the correlation exists, we could indirectly infer that there is information and we may want to exploit that for trading. There is a vast literature in finance studying the relationship between trading volume and serial correlation in returns. We briefly review some select studies and provide an illustrative example with some possible algorithms.

Blume, Easley and O'Hara (1994) [45] investigate the informational role of volume. If the impounding of information into the stock price is not immediate, volume may provide information about the evolving process of a security's return. The model discussed in Section 4.3 by Tauchen and Pitts (1983) [311] and the associated empirical tests clearly document the strong relationship between volume and the absolute return. "But why such a pattern exists or even how volume evolves in markets is not clear." Recall that the model assumes that the information content affects both the return and the volume specifications; yet the correlation between the return and volume can be shown to be zero. If the information content can be studied through the sequence of security prices and the associated volume, it may provide insights into the inefficiency in the market and how statistical arbitrage rules can exploit the inefficiency especially if there is an empirical lead-lag relationship. The quality of traders information can be captured best by combining price change with volume change. It is shown that the relationship between volume and price is in the form of V-shape thus indicating a non-linear relationship. Both 'bad' and 'good' information about the stock is likely to result in higher volume of trade.

Campbell, Grossman and Wang (1993) [65] also explore the relationship between volume and returns; volume information is used to distinguish between price movements that occur due to publicly available information and those that reflect changes in expected returns. It is predicted that price changes with high volume tend to be reversed and this relationship may not hold on days with low volume. Llorente, Michaely, Saar and Wang (2002) [246] extend this work by postulating that the market generally consists of liquidity and speculative traders. In periods of high volume but speculative trading, return autocorrelation tends to be positive and if it is liquidity trading, return autocorrelation tends to be negative. In the former case, returns are less likely to exhibit a reversal. Defining the volume turnover ($V$) as the ratio of number of shares traded to the total number of outstanding shares, the following regression model is fit for each stock:

$$r_t = \beta_0 + \beta_1 r_{t-1} + \beta_2 V_{t-1} \cdot r_{t-1} + a_t. \tag{5.44}$$

For stocks associated with more speculative trading (larger bid-ask spreads), $\beta_2$ is expected to be positive. The model can be extended to add other variables such as a market proxy ($r_{mt}$). Note the first order autocorrelations that we expect to hold different signs under different volume-return scenarios are captured by '$\beta_1$' coefficient. The model's predictions are tested for NYSE and AMEX stocks and the results are supportive of the predictions.

Gervais, Kaniel and Mingelgrin (2001) [164] show that stocks with unusually large trading activity over periods tend to experience large returns over the subsequent periods. It is argued that higher trading activity creates higher visibility which in turn creates subsequent demand for that stock. This approach where contrarian or momentum is characterized by trend in volume is also studied by Conrad, Hameed and Niden (1994) [84] and Lee and Swaminathan (2000) [240]. Assume that the change in trading activity is measured by

$$u_{it} = (V_{it} - V_{it-1})/V_{it-1} \tag{5.45}$$

and a positive $u_{it}$ implies a higher transaction security and a negative $u_{it}$ is taken to imply a low transaction security. Now form four portfolios with stocks that exhibit positive versus negative returns in conjunction with positive versus negative transaction activity. The weights for each of the formed portfolios are given as:

$$w^*_{i,t} = \frac{r_{it-1}(1 + u_{it-1})}{\sum_{i=1}^{N_p} r_{it-1}(1 + u_{it-1})}, \tag{5.46}$$

where $N_p$ is the number of securities in each combination ($p = 1, \dots, 4$). The empirical results confirm that autocorrelation of low and high transaction securities are different; high transaction stocks experience price reversals and low transaction stocks experience positive autocorrelations in returns. These relationships are noted to be stronger for smaller firms.

The discussion thus far refers to the use of low frequency data. Most of the studies present results for daily or weekly aggregated data. Simply using the end of the period closing price and the volume for the period is not appropriate as transactions occur at specified prices and at specified times over the trading period. The availability of high frequency data has enabled to study the evolving influence of volume after controlling for order book imbalances. Because the parent order now gets sliced into small child orders, both the number of trades and size of trades need to be considered together in place of volume. As argued in Kyle (1985) [234], the price volatility is induced by net order flow. Market makers tend to infer information from the net order flow and revise their prices. For a study of return and volume using a random sample of approximately three hundred stocks in NYSE, Chan and Fong (2000) [72] consider a two-step model for daily data:

$$\begin{aligned} r_{it} &= \sum_{k=1}^{4} d_{ik} D_{kt} + \sum_{j=1}^{12} \beta_{ij} r_{it-j} + \epsilon_{it} \\ \epsilon_{it} &= \phi_{i_0} + \phi_{im} M_t + \sum_{j=1}^{12} \phi_{ij} |\epsilon_{it-j}| + \gamma_i (V_i \text{ or } T_i) + \eta_{it}. \end{aligned} \tag{5.47}$$

Here, $D$ is the day-of-the week indicator, $M_t$ is an indicator for Monday, $V$ and $T$ are share volume and number of trades. The second equation that measures the volume-volatility relationship indicates that the number of trades ($T$) has a somewhat better explanatory power than shares volume ($V$). The relationship between order book imbalance and volatility is captured by modifying the first equation in (5.47) as,

$$r_{it} = \sum_{k=1}^{5} \alpha_{ik} D_{kt} + \sum_{j=1}^{12} \beta_{ij} r_{it-j} + \sum_{h=1}^{5} \lambda_{h,i} NT_{h \cdot it} + \varepsilon_{it}, \tag{5.48}$$

where $NT_{h,it}$ is the net number of trades in size category '$h$'. The net number is calculated by classifying the trade as buyer-initiated or seller-initiated using Lee and Ready's (1991) [241] rule. It is shown that the coefficient '$\lambda$' is positive and significant for most stocks. But the second-stage regression results in weaker '$\gamma$' values.

Chordia, Roll and Subrahmanyam (2002) [80] extend this and provide a more comprehensive empirical analysis; resulting stylized facts which can act as benchmark expectations. These are listed below:

- Order imbalances are strongly related to market returns; imbalances are high after market declines and low after market advances.
- Liquidity is not predictable from past imbalances but is predictable from market returns.
- Order imbalances are strongly related to volatility measures after controlling for volume and liquidity.

These stylized facts would be useful for spotting anomalies and hence for devising appropriate trading strategies.

### 5.10.1 Filter Rules Based on Return and Volume

The rules that were discussed in Section 5.3 can all be modified to include volume as a weighing factor for the returns. That would involve tracking both the price and volume changes over time; it gets particularly cumbersome to track the volume deviations. They should be appropriately normalized for the time of the day. Therefore the rules focus on defining and tracking the changes in both quantities individually and combine them with necessary thresholds. Cooper (1999) [89] suggests four strategies based on weekly returns and another four strategies that incorporate the return (thus price) and the volume information. First the return rules are for weeks '$t-1$' and '$t-2$':

First-Order Price Filters:

$$\text{Return states} = \begin{cases} \text{For } k = 0, 1, \ldots, 4: & \begin{cases} \text{loser}_{k^\star A} & \text{if } -k^\star A > r_{i,t-1} \geq -(k+1)^\star A \\ \text{winner}_{k^\star A} & \text{if } k^\star A \leq r_{i,t-1} < (k+1)^\star A \end{cases} \\ \text{For } k = 5: & \begin{cases} \text{loser}_{k^\star A} & \text{if } r_{i,t-1} < -k^\star A \\ \text{winner}_{k^\star A} & \text{if } r_{i,t-1} \geq k^\star A \end{cases} \end{cases}$$

Second-Order Price Filters: (5.49)

$$\text{Return states} = \begin{cases} \text{For } k = 0, 1, \ldots, 4: & \begin{cases} \text{loser}_{k^\star A} & \text{if } -k^\star A > r_{i,t-1} \geq -(k+1)^\star A \\ \text{and} & -k^\star A > r_{i,t-2} \geq -(k+1)^\star A \\ \text{winner}_{k^\star A} & \text{if } k^\star A \leq r_{i,t-1} < (k+1)^\star A \\ \text{and} & k^\star A \leq r_{i,t-2} < (k+1)^\star A \end{cases} \\ \text{For } k = 5: & \begin{cases} \text{loser}_{k^\star A} & \text{if } r_{i,t-1} < -k^\star A \\ \text{and} & r_{i,t-2} < -k^\star A \\ \text{winner}_{k^\star A} & \text{if } r_{i,t-1} \geq k^\star A \\ \text{and} & r_{i,t-2} \geq k^\star A \end{cases} \end{cases}$$

Here $r_{i,t}$ is the non-market-adjusted return for security '$i$' in week '$t$', $k$ is the filter counter, $k = 0, 1, 2, \ldots, 5$ and $A$ is the lagged return grid width, set equal to 2%.

To relate if return reversals are related to trading volume, define the growth volume as the percentage change in volume adjusted for the outstanding shares:

$$\%\Delta v_{i,t} = \left[\frac{V_{i,t}}{S_{i,t}} - \frac{V_{i,t-1}}{S_{i,t-1}}\right] / \left[\frac{V_{i,t-1}}{S_{i,t-1}}\right] \tag{5.50}$$

Here $S_{i,t}$ is the number of outstanding shares. Next, we define filters based on this:

Volume Filters:

$$\text{Growth in volume states} = \begin{cases} \text{For } k = 0, 1, \ldots, 4: & \begin{cases} \text{low}_{k^\star B} \text{ if } -k^\star B > \%\Delta v_{i,t-1} \\ \qquad\qquad \geq -(k+1)^\star B \\ \text{high}_{k^\star C} \text{ if } k^\star C \leq \%\Delta v_{i,t-1} \\ \qquad\qquad < (k+1)^\star C \end{cases} \\ \text{For } k = 5: & \begin{cases} \text{low}_{k^\star B} & \text{if } \%\Delta v_{i,t-1} < -k^\star B \\ \text{high}_{k^\star C} & \text{if } \%\Delta v_{i,t-1} \geq k^\star C \end{cases} \end{cases} \tag{5.51}$$

Here $B$ is the grid width for low growth in volume and is set to fifteen percent and $C$ is the grid width for high growth in volume and is set to a value of fifty percent. The asymmetry in the limits is due to skewness in the volume distribution. The value of '$k^*$' can be calibrated through backtesting.

Using the data for large cap stocks, Cooper (1999) [89] examines the profitability of the various filters based on the portfolios that meet the conditions of the filters. Some general conclusions are listed below:

– Winner price-high volume strategy performs the best.

- Loser portfolios that condition on two consecutive weeks losses tend to do better than the strategies based on one-week losses.
- Loser price-low volume strategy results in higher profit than the strategy based on loser price only.

### 5.10.2 An Illustrative Example

The methodology discussed in the last section on price-based and volume-based filters on low frequency data can be extended to high frequency data as well. The key issue with the volume data is to come up with proper standardization to identify abnormal level of activity in a high frequency setting. The intra-day seasonality and the overall increase or decrease in inter-day volumes need to be tracked carefully to study and fully exploit the deviations. We highlight the practical issues involved using 30-min price-based data for Treasury yields from June 8, 2006 to August 29, 2013. This can be taken as medium-frequency data. The price bars consist of high, low, open and close prices and the volume traded during the interval are considered. Because of the aggregated nature of the data, the traditional time series methods can be readily used to study the trend in price and volume series. Daily data is just a snapshot at 4 p.m. and the total volume for the day is the sum of all 30-min volumes since the previous day.

Let $p_{t.m}$ be the log price in the $m$th interval of the day '$t$'. Note that '$m$' indexes the 30-min interval in a day; the value of $m$ can range from 1 to 13. Let $r_{t.m} = p_{t.m} - p_{t.m-1}$ be the return and let $v_{t.m}$ be the log volume in that interval. We define volatility within the $m$th time unit based on the price bar data as,

$$\hat{\sigma}^2_{t.m} = 0.5[\ln(H_{t.m}) - \ln(L_{t.m})]^2 - 0.386[\ln(C_{t.m}) - \ln(\mathrm{O}_{t.m})]^2 \tag{5.52}$$

similar to a measure defined for the daily data in Section 4.5.

The plot of price and volume data is given in Figure 5.8. The price data generally shows an upward trend, but it is difficult to see the pattern in volume, especially during the course of the day. A simple average volume plot clearly indicates most of the trading occurs between 8 a.m. and 4 p.m. (Figure 5.9). The autocorrelation function of the log volume ($v_{tm}$) clearly demonstrates the intra-day variations in the data. There is a great deal of interest among practitioners to model the volume accurately as volume is commonly used in entry/exit decisions in a trading strategy. VWAP execution strategy, for example, assumes that volume is known a priori. Other benefits cited in Satish, Saxena and Palmer (2014) [297] are if a trader receives a large order ten minutes before the close with an instruction to complete the order, having a reliable estimate of expected volume that is possible before the market close can help evaluate the feasibility of that order execution. The other benefit that is mentioned is how improved volume forecast can aid alpha capture. Brownlees, Cipollini and Gallo (2011) [62] propose a dynamic model,

$$V_{tm} = \eta_t \cdot \phi_m \cdot \mu_{t.m} \cdot \epsilon_{t.m}, \tag{5.53}$$

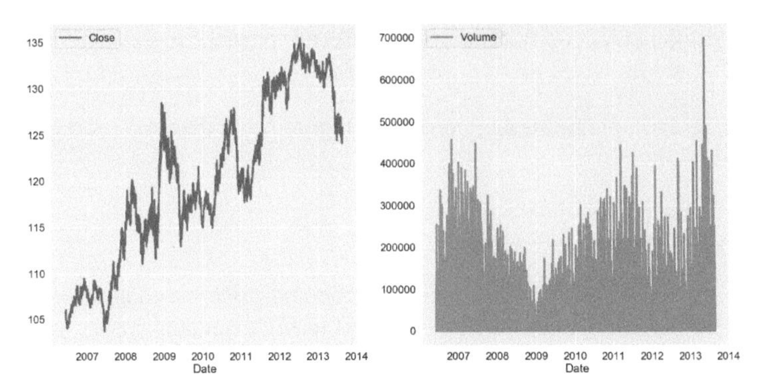


Figure 5.8: Treasury Yields: Price and Volume Plot.

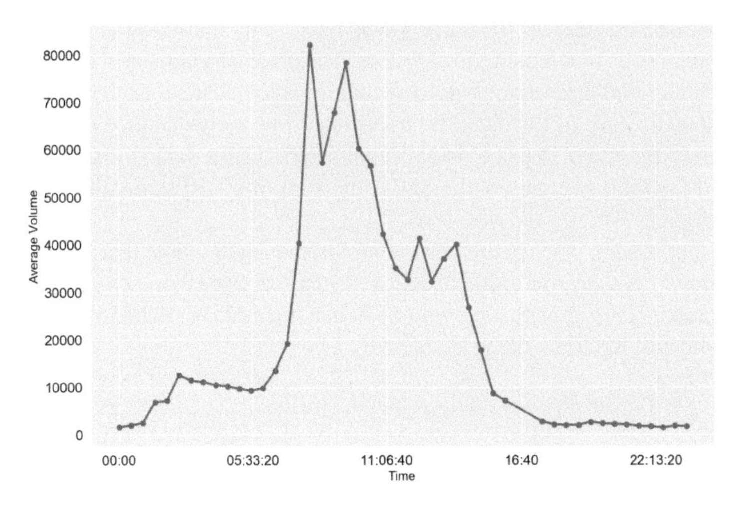


Figure 5.9: Plot of Average Volume.

where $\eta_t$ is a daily component, $\phi_m$ is the periodic component and $\mu_{t.m}$, an intra-daily dynamic, non-periodic component. The periodic component $\phi_m$ modeled as a Fourier (sine/cosine) function; the daily component is modeled as a function of past values of daily volumes and the same is true with $\mu_{t.m}$; they are modeled as a function of $\mu_{t.m-1}$ and $V_{t.m-1}$. From the structure of the model (5.53), it is obvious that the multiplication components can be modeled additively through log transformations.

We suggest a model somewhat similar to (5.53) but more directly related to models discussed in Chapter 2. Before we state the results, we want to recall the main findings from past empirical studies:

- Volatility, measured by squared returns is positively correlated with volume.
- Large price movements are generally followed by high volume.

- Conditioning on lagged volume attenuates the leverage effect.
- After conditioning on lagged volume, there is a positive risk-return relationship.

Our approach to modeling is stated as follows:

- Model the volume at different frequency levels. (Here $m = 30$ min, $t =$ day).
- Use the low frequency data to develop macro strategies.
- Use the high frequency data to make micro decisions.

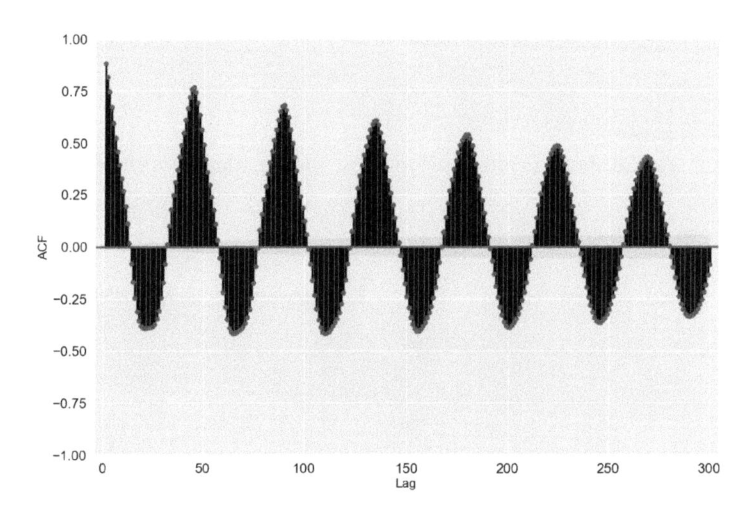


Figure 5.10: Autocorrelation Function for Log Volume.

Given the strong intra-day periodicity that is evident from Figure 5.10, we can simply create indicator variables for the 30-min periods to represent the periodicity levels and regress the volume on the indicators to capture the time of the day variation. This explains almost sixty-five percent of the variation in log volume. The residuals capture non-periodic patterns and the partial autocorrelation function is given in Figure 5.11. This suggests that the residuals can be modeled as higher order autoregressive process. As these components are additive, the forecasts from indicator regression and this autoregression of residuals can be summed up, to arrive at a reasonable volume forecast. The plot of logged daily volume is given in Figure 5.12 and it is clearly stationary. The partial autocorrelation function in Figure 5.13 has somewhat a simpler structure for the daily aggregated data.

How do we use these results optimally and for practical purposes?

- Forecast next day volume using the macro model or by aggregating the forecasts from the micro model.

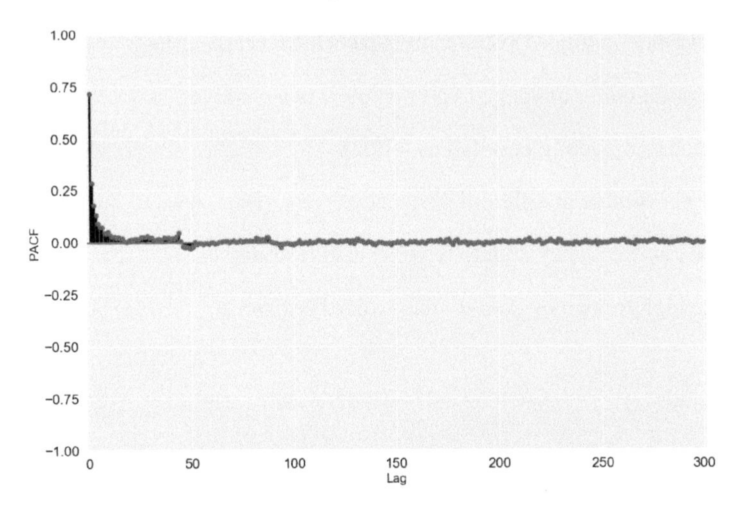


Figure 5.11: Partial Autocorrelation Function for Residuals (with 5% significance limits).

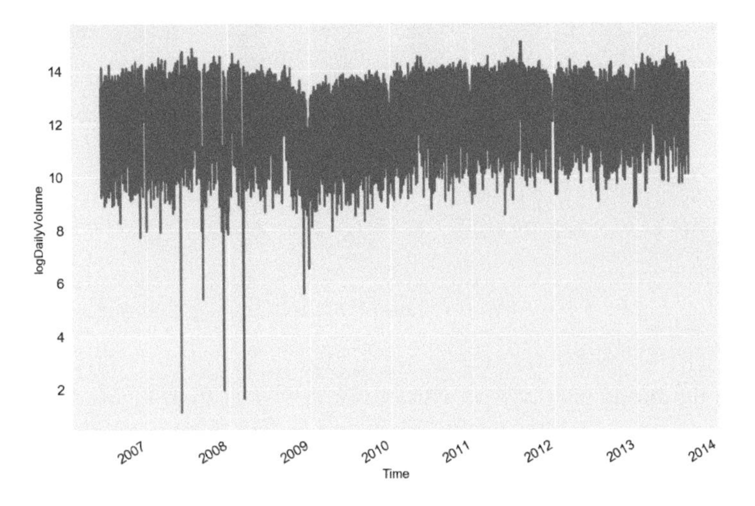


Figure 5.12: Plot of Daily Volume.

- Using the 30-min volume forecast using the micro model, construct cumulative volume forecasts for the $m$th time interval.

- If the actual cumulative volume is not in line with the forecast of cumulative value, search if any new information has arrived! This can help to decide if the stock is oversold or underbought.

These methods obviously require the use of time series programs. As an alternative easy to use method, as an approximation, we suggest using an exponential smoothing approach. The details are as follows:

- **Volume Forecast 1 (Direct):** Use daily data to forecast $\hat{v}_{t+1}$, next day's volume; use exponential smoothing constant in the range of 0.25–0.35.
- **Volume Forecast 2 (Aggregated):** We will use a two-step procedure here:
  - (a) Compute the average volume $\overline{v}$ over certain number of days for each '$m$'. Usually the industry uses '30' days.
  - (b) Compute the residuals: $e_{t \cdot m} = v_{t.m} - \overline{v}$; use exponential smoothing on $e_{t \cdot m}$ with a smoothing constant in the range of 0.25–0.35; estimate $\hat{e}_{t+1.m}$;
  - (c) Forecast for next period '$m$': $\hat{v}_{t+1.m} = \overline{v} + \hat{e}_{t+1.m}$.
  - (d) Aggregate $\hat{v}_{t+1 \cdot m}$ over $m$ to get $\hat{v}_{t+1}$.

Now we have two forecasts of daily volume $\hat{v}_{t+1}$ and $\hat{v}_{t+1}$.

We want to relate the volume to volatility using the measure in (5.52). Figure 5.14, based on the plot of Figure 5.12 indicates that there are clustering effects. From the partial autocorrelation function given in Figure 5.15, there is some memory in the variance and can be modeled via GARCH. Here our goal is to show how volume and volatility are likely to be related. The simple cross-correlation function, Figure 5.16, shows how the volume and volatility are closely related; here the intra-day periodicity in the residuals in volume and volatility is shown. Again in the spirit of an intuitive, simpler approach to model the volatility, we suggest obtaining volatility forecasts as:

$$\begin{cases} \hat{\sigma}^2_{t+1} = \text{weighted average of } \sigma^2_t, \sigma^2_{t-1}, \dots, \sigma^2_{t-30} \\ \hat{\sigma}^2_{t+1 \cdot m} = \text{weighted average of } \sigma^2_{t \cdot m}, \sigma^2_{t-1 \cdot m}, \dots, \sigma^2_{t-30 \cdot m} \end{cases}$$

The following algorithm that combines the price, volume and volatility is suggested; we assume that price-based rules given in Section 5.3 are being followed.

An Omnibus Rule:

- Enter if
  (a) price-based rules favor and if
  (b) $\sigma^2_{t+1 \cdot m} < \hat{\sigma}^2_{t+1 \cdot m}$, (realized volatility < predicted volatility) and if
  (c) $\sum_{i=1}^{m} v_{t+1} < \sum_{i=1}^{m} \hat{v}_{t+1}$, cumulative actual volume < cumulative predicted volume.
- Exit if either (a) or (b) or (c) is not true.

Although the above rule may appear to be stringent the practical applications yield significant results.

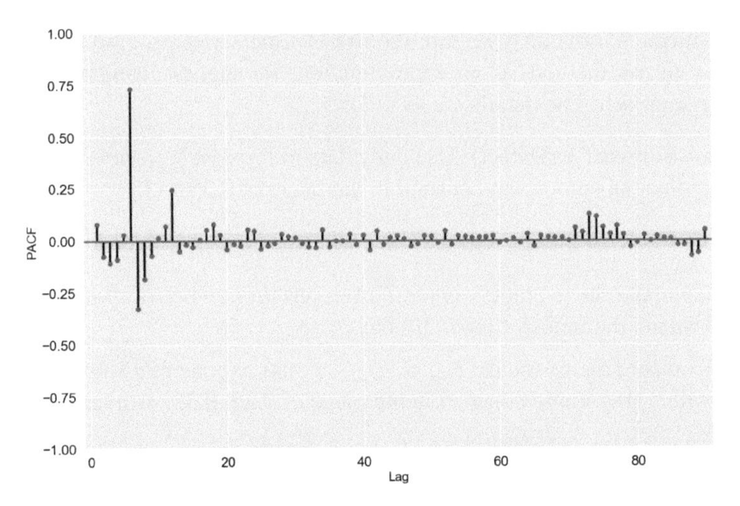


Figure 5.13: Partial Autocorrelation Function for Log Daily Volume (with 5% significance limits).

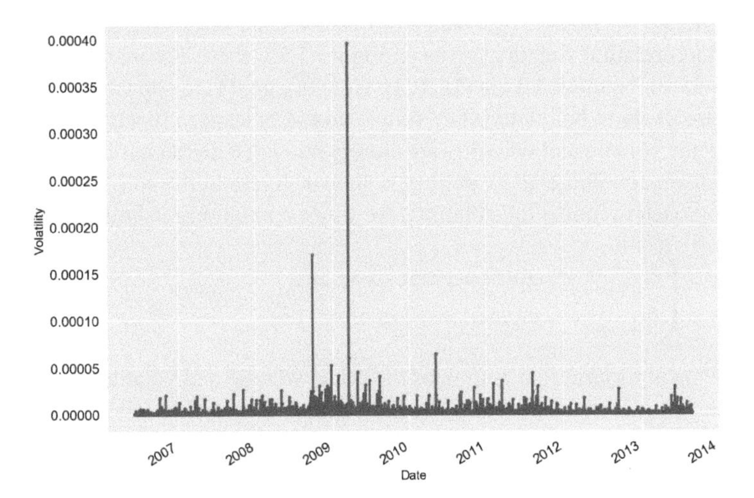


Figure 5.14: Plot of 30-min Volatility.

## 5.11 Trading in Multiple Markets

There has been a great deal of interest in studying the performance of cross-listed stocks in several markets, that operate in overlapping or in non-overlapping hours of trading. Both in NYSE and in NASDAQ the listing of non-US firms has increased over time. The general motivation behind cross-listings is to manage the cost of raising

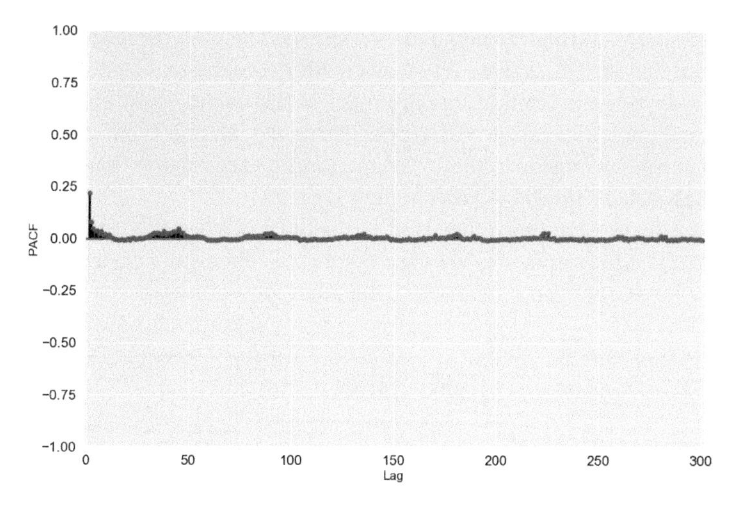


Figure 5.15: Partial Autocorrelation Function for Volatility (with 5% significance limits).

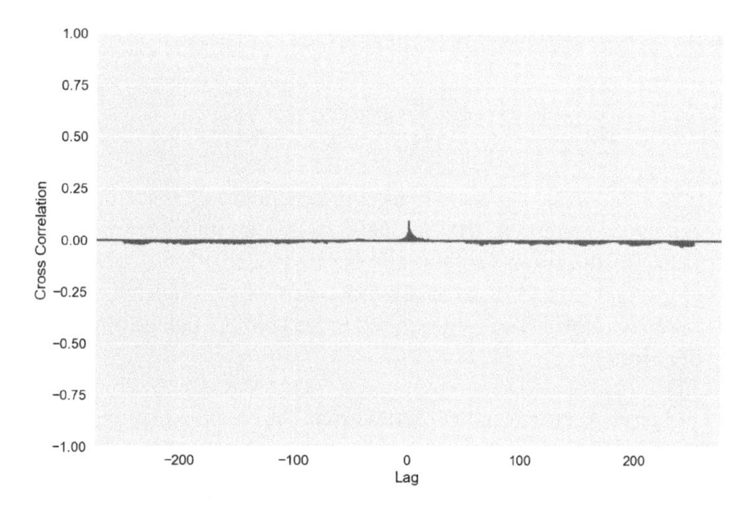


Figure 5.16: Cross Correlation Function between Residuals and Volatility.

capital. Some unique features of trading stocks in multiple markets include the non-overlapping trading hours and potential legal restrictions on moving capital from one market to another. But our focus here is to study the imperfect pricing that may arise from cross-listed markets and see how trading opportunities in these markets can be exploited. More broadly it is of interest to study if and how the information flow occurs from one market to another.

Foreign companies list shares for trading in US exchanges in the form of ADR (American Depositary Receipts) that ensures the convertibility of home-market shares in US shares and back to home-market shares. There is generally no conversion fee and there may be restrictions on cross-country ownership. Gagnon and Karolyi (2010) [154] investigate arbitrage opportunities in these markets by comparing intra-day prices and quotes and observe that price parity deviation is quite small (4.9 basis points) but can reach large extremes when the markets are volatile. Trading strategies in the markets can be similar to pairs trading strategies.

To model the price (co)movements in a calendar day when exchanges operate, the hours can be stated in terms of GMT (See Figure 5.17 for a stock traded in exchanges spread over three continents): Denoting the ADR log returns as $r_{At}$ and the home

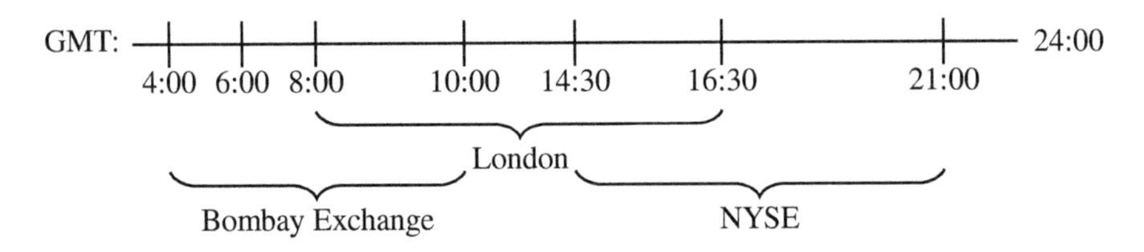


Figure 5.17: Opening Hours of Some Stock Exchanges.

market returns as $r_{Ht}$, these returns are modeled as:

$$r_{A-H.t} = \alpha + \sum_{i=-1}^{1} \beta_i^{\text{US}} r_{M,t+i}^{\text{US}} + \sum_{i=-1}^{1} \beta_i^{\text{H}} r_{M,t+i}^{\text{US}} + \sum_{i=-1}^{1} \beta_i^{Fx} r_{fx.t+i} + a_t. \tag{5.54}$$

How $r_{A-H.t} = r_{A.t} - r_{H.t}$, $r_M$'s are market indices and $r_{fx}$ is the currency return. Using the standard deviation of the residuals, '$a_t$', as an idiosyncratic risk proxy in the model (5.54) results from Gagnon and Karolyi (2010) [154] who indicate 'the presence of important sources of excess co-movement of arbitrage returns with market index and currency returns'. Rules described in Section 5.3 can be applied to $r_{A-H.t}$ for trading decisions.

**Dynamic Linkages Across Vector Time Series:** As observed, some markets operate at times that may or may not overlap with others. On any given calendar day, for example, Asian markets open and close before the US markets open. Efficient market theory would suggest that all available information about a stock would be impounded in the stock price and there should be no systematic lagged inter-market adjustments that can be effectively exploited. But empirical studies reveal that there could be significant correlations across markets, during the same calendar day and possibly in lagged relationships as well. This phenomenon is also known as a *data synchronization problem.* Some studies focus on the overlapping durations between two markets if they exist (for example between London and NYSE), but the information flow during overlapping and non-overlapping periods may not be processed similarly.

To motivate, consider two markets that do not overlap and let us denote the closing log prices as, $p_t = (p_{1t}, p_{2t})'$; if the markets are efficient and operate independently, each component of $p_t$ will follow a random walk model; but if there is any dependency, it is usually captured by the co-integrating relationship among the components as in pairs trading set-up. We will assume that $p_{2t}$ occurs after $p_{1t}$ in a calendar day, '$t$'. Koch and Koch (1991) [231] discuss a dynamic simultaneous equations model that takes into account both the contemporaneous and the lead-lag relationships. In its simplest form, the model can be stated as,

$$\begin{pmatrix} 1 & 0 \\ -\phi_{21} & 1 \end{pmatrix} \begin{pmatrix} p_{1t} \\ p_{2t} \end{pmatrix} = \begin{pmatrix} \phi_{11} & \phi_{12} \\ 0 & \phi_{22} \end{pmatrix} \begin{pmatrix} p_{1t-1} \\ p_{2t-1} \end{pmatrix} + \begin{pmatrix} a_{1t} \\ a_{2t} \end{pmatrix},$$

or (5.55)

$$\Phi_{01} p_t = \Phi_{11} p_{t-1} + a_t.$$

The above model can also be written as VAR(1) model, $p_t = \Phi_1^* p_{t-1} + \epsilon_t$. Note there are four elements in $\Phi_1^*$. But the same number of elements in (5.55) as well. It can be shown that one in $\Phi_{01}$ and three in $\Phi_{11}$ are exactly identified from VAR(1) model. Note if we denote $\Phi_{01}^{-1} = (\phi_{01}, \phi_{02})$ and $\Phi_{11} = \begin{pmatrix} \phi'_{11} \\ \phi'_{21} \end{pmatrix}$ when $\phi$'s are $2 \times 1$ column vectors, $\Phi_1^* = \phi_{01}\phi'_{11} + \phi_{02}\phi'_{21}$ is the sum of two rank one matrices. One could use rank factorization to identify the element $\phi_{21}$ and the elements in $\Phi_{11}$ as well.

Some research issues in this context are:

- How do we identify the triangular coefficient matrices, both contemporaneous and the lagged, from the general VAR(p) coefficient matrices?
- What is the effect of ignoring the contemporaneous dependence on the estimation of co-integrating relationships?
- For stocks traded in multiple markets does it make any difference in forecasting the price direction based on information flows by accounting for the data synchronization approximately?

The last aspect is studied here for a select stock.

To illustrate, we consider the data on WIPRO—an Indian stock that is traded in both US and Bombay exchanges. The plot of the closing prices in both markets (CU—closing price in US market, VU—volume in US market, etc.) is given in Figure 5.18. The exchange rate has been generally in favor of the dollar during this period (2005–2014), but the volatility is higher in the US market. To illustrate the flow of information in one market to another we fit the following models; let $p_{1t}$ be the log closing price in India and $p_{2t}$ be the log closing price in US; Table 5.1 below summarizes the results. These results would indicate that the markets are not fully efficient and there is some information transfer from one market to other. The dependence of the US market price on the closing price of the Indian market is clearly evident. This can be favorably used for better US price forecasts.

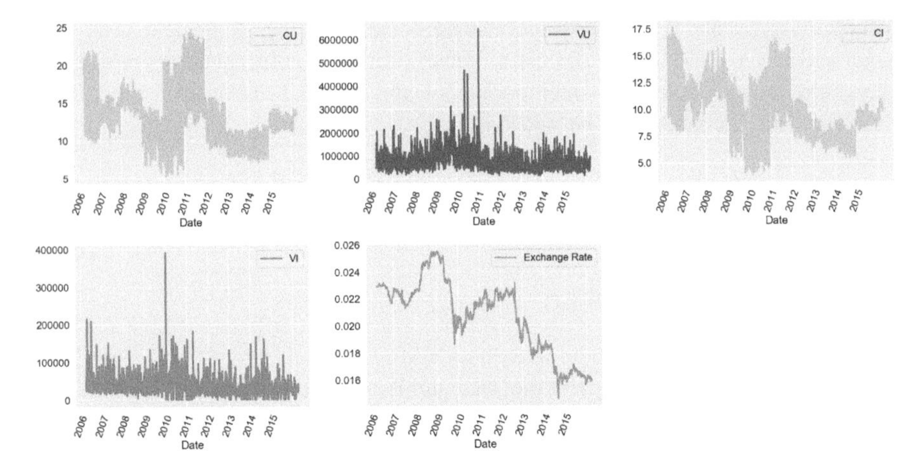


Figure 5.18: Plot of Closing Prices, Volumes and Exchange Rates.

Table 5.1: WIPRO Regression Results.

| Predictors | Response Variable $p_{1t}$ | $p_{2t}$ |
|---|---|---|
| Constant | −0.056* | −0.019* |
| $p_{1t}$ | — | 0.6328* |
| $p_{1t-1}$ | 0.91* | — |
| $p_{2t-1}$ | −0.026 | 0.3626* |
| $R^2$ | 0.79 | 0.91 |

*Significant at 5% levels

## 5.12 Other Topics: Trade Size, etc.

### Trade Size

In the trading strategies described above, once entry-exit decisions are made, the size—the number of units to buy and sell—is usually assumed to be the total value of the investment. Many traders tend to do staggered entry or exit, so that all funds are not invested or taken out at once. The question then is how to determine the size of the bet on each transaction. We want to explore some practical ideas in this section.

A criterion that is attributed to Kelly (1956) [229] that was introduced in the context of gambling determines the fraction of investment capital to bet, '$f$', with the initial investment, $X_0$ and '$n$' trials of investment opportunities, is based on the wealth accumulation as

$$X_n = X_0(1 + fW)^{pn}(1 - fL)^{qn}, \tag{5.56}$$

where '$W$' and '$L$' are the number of successes and failures and '$p$' and '$q$' are the respective probabilities. Thus the growth rate per bet is given as

$$\frac{1}{n}\ln\left(\frac{X_n}{X_0}\right) = p\ln(1 + fW) + q\ln(1 - fL) \tag{5.57}$$

and when maximized, this results in the optimal bet size as,

$$f^* = p - \frac{q}{W/L}. \tag{5.58}$$

In order to implement this in practice, good estimates of '$p$' and '$q$' are needed; the bet size can be revised based on the (recent) empirical observation of '$W$' and '$L$' during the course of investment. A general version of the problem could be to determine the optimal betting vector, $f^* = (f_1^*, \dots, f_n^*)$, with some dependency structure on the sequential games, but that appears to be quite complex.

The usefulness of statistical arbitrage rules can be extended beyond the entry-exit decisions, but the size of entry-exit decisions must be made with alternatives such as investing in risk-free assets. It is shown in Zhu and Zhou (2009) [332] that when there is uncertainty about the future returns, fixed allocation rules can be enhanced by technical analysis. The fixed allocation rule follows from the portfolio theory presented in Chapter 6. With a single risky asset, whose behavior is monitored through technical analysis and a risk free asset, the optimal allocation rule follows from Chapter 6 in the form of Sharpe ratio structure, $w = (\mu - r_f)/\sigma^2$, where $\mu$ and $\sigma^2$ are estimated from the past data and are expected to remain stable in the near future. Assuming that the investor with the initial wealth, $X_0$, wants to maximize the expected value,

$$\mathrm{E}(U(X_n)) = \mathrm{E}\left[\frac{X_n^{1-\gamma}}{1-\gamma}\right], \tag{5.59}$$

where '$\gamma$' is the risk aversion parameter, with the budget constraint,

$$\frac{dX_t}{x_t} = r_f\, dt + w_t(\mu_0 + \mu_1 S_t - r_f)\, dt + w_t\sigma\, dB_t. \tag{5.60}$$

Here it must be observed that $B_t$ is a Brownian motion, $\mu_0$ and $\sigma$ are the mean and standard deviation of the returns and $S_t$ is the state variable that can provide information on the future stock returns. The fixed allocation rate under the assumption that returns are random is

$$w(\gamma) = \frac{\mu - r_f}{\gamma \cdot \sigma^2} \tag{5.61}$$

which is the Sharpe ratio adjusted for the risk-aversion. If the information from the state variable, although uncertain, is considered then the optimal investment size is given as

$$w(\gamma) = \frac{\mu - r_f}{\gamma \cdot \sigma^2 - (1-\gamma)(d)}. \tag{5.62}$$

The term '$d$' depends on '$\mu_1$' in (5.60) that indicates the influence of state variable on the return process and volatility in the state variable; for details see Zhu and Zhou (2009) [332]. As this quantity '$d$' can take a range of values, the adjustment factor $(1-\gamma)d$ can take positive or negative values. Note when $\gamma = 1$, $w(\gamma) = w$, (5.62) leads to the standard measure of Sharpe ratio.

To evaluate the additional value of tracking the price over time to decide on the entry strategy, we follow the simple rule based on the moving average of fixed length. If the current price, $p_t$ is greater than the moving average, enter the market:

$$\eta_t = \eta(p_t, A_t) = \begin{cases} 1, & \text{if } p_t > A_t \\ 0, & \text{otherwise} \end{cases} \tag{5.63}$$

Under this set-up the optimal bet-size is shown to be

$$w^* = \frac{\mu - r_f}{\sigma^2} + \frac{\mu_1}{\sigma^2} \cdot \frac{b_1}{b_2}, \tag{5.64}$$

where $b_2 = \mathrm{E}(\eta_t)$ and $b_1 = \mathrm{E}((S_t - \overline{S})\eta_t)$. Observe '$\mu_1$' is the effect of information-based state variable, $S_t$, on the price formation, '$b_2$' is the average number of entries and '$b_1$' provides the average excess of the state variable. While other quantities in (5.64) have been historically well-studied, the key ratio that is new here is over $(\mu_1 \cdot b_1)/b_2$ which can be taken as the total size $(\mu_1 b_1)$ that can be distributed over '$b_2$' entries. If $\mu_1 = 0$, observe that the moving average rule may not add any flexibility to the fixed size entry.

## 5.13 Machine Learning Methods in Trading

In Section 4.6, we covered some machine learning techniques. Here we briefly discuss the boosting methods that are applied to trading. Other methods require more detailed space for a thorough discussion and these methods are better described elsewhere in the literature. In order to follow the details here, refer to Section 4.6.2. We outline an automated layered trading method with machine learning and on-line layers with an overall risk management view. Details can be found in Creamer and Freund (2010) [94]. Recall that the main idea behind boosting is to combine several weak rules of classification or prediction to arrive at a stronger rule. The authors suggest the Logitboost method. For the sake of consistency in describing both the Adaboost and Logitboost, we repeat the former method again here.

Associated with past features, $x_i$ (prices: close, open, high and volume, etc.), there is a current response variable '$y_i$' that could be binary, such as the price increases or decreases, directional indicator. The training set, $S = \{(x_i, y_i) : i = 1, 2, \ldots, n\}$. We will further assume that there are '$T$' rules, and $h_i(x_i)$ are the classifiers. These rules are variants of the technical rules described in this chapter. We begin with weights $w_i^t = \frac{1}{n}$ for all $t$. With $h_t(x_i)$ as the prediction rule, define the prediction score $F_t(x) = \alpha_t h_t$. Two methods, Adaboost and Logitboost are given in Creamer and Freund (2010) [94] stated below in Figure 5.19:

$$
\begin{aligned}
&F_0(x) \equiv 0 \\
&\text{for } t = 1, \dots, T \\
&\quad w_i^t = e^{-y_i F_{t-1}(x_i)} \text{ (Adaboost, Freund and Shapire 1997)} \\
&\quad w_i^t = \frac{1}{1 + e^{y_i F_{t-1}(x_i)}} \text{ (Logitboost, Friedman et al. 2000)} \\
&\quad \text{Get } h_t \text{ from } \textit{weak learner} \\
&\quad \alpha_t = \frac{1}{2} \ln \left( \frac{\sum_{i : h_t(x_i)=1, y_i=1} w_i^t}{\sum_{i : h_t(x_i)=1, y_i=-1} w_i^t} \right) \\
&\quad F_{t+1} = F_t + \alpha_t h_t
\end{aligned}
$$

Figure 5.19: The Adaboost and Logitboost algorithms.

$y_i$ is the binary label to be predicted, $x_i$ corresponds to the features of an instance $i$, $w_i^t$ is the weight of instance $i$ at time $t$, $h_t$ and $F_t(x)$ are the prediction rule and the prediction score at time $t$, respectively.

Thus the machine learning layer generates several 'experts' using historical stock related data. Each 'expert' has a rule that combines the technical indicators and generates an investment signal. Train an initial expert ($\psi_i$) with logitboost. Every '$d$' days train a new expert, and assume we have over time several trained experts. The second layer provides a way to combine the signals from the experts.

Let $s_t^i$ be the prediction score of expert '$\psi_i$' at time '$t$'. Then the cumulative abnormal return (CAR) for the $i$th expert at time '$t$' is

$$\text{CAR}_t^i = \sum_{s=t_i+1}^{t} \text{Sign}(s_t^i) r_s^i. \tag{5.65}$$

Here $t_i$ is the time when the $i$th expert is added to the pool, and $r_s^i$ is the abnormal return for expert '$i$' at time '$s$'. Because the returns are white noise, $\text{CAR}_i^t \sim N(\epsilon, 1)$, where '$\epsilon$' is the advantage that the expert exhibits over pure random guessing. Give a bigger weight to an expert with larger, '$\epsilon$' value. The weight of the first expert is,

$$w_t^1 = \exp\left[\frac{c \cdot \text{CAR}_t^1}{\sqrt{t}}\right], \tag{5.66}$$

where '$c$' is an exogenous parameter.

When a new expert gets added at $t > t_i$, the weight is

$$w_t^i = I_i \times \text{ramp}(t - t_i) \times \exp\left[\frac{c \cdot \text{CAR}_t^i}{\sqrt{t - t_i}}\right], \tag{5.67}$$

where $I_i = \sum_{j=1}^{i-1} w_{t_i}^j / (i-1)$ is the weight, which is the average value of the previous experts before time $t_i$ and $\text{ramp}(t - t_i) = \min\left[\frac{t - t_i}{t_{i+1} - t_i}, 1\right]$ that brings the new experts gradually to the pool. The overall experts' weight at time '$t$' is,

$$w_t = L_t - S_t, \tag{5.68}$$

where $L_t$ is the fraction of experts that suggest a long position. Thus,

$$L_t = \sum_{i\,:\,s_t^i>0} w_t^i / \sum_i w_t^i. \tag{5.69}$$

Essentially, the second layer algorithms entertains new experts. The CAR for the expert system is

$$\text{CAR} = \sum_t (w_t \cdot r_t - (w_t - w_{t-1}) \cdot \text{tc}), \tag{5.70}$$

where 'tc' is the transaction cost.

While the above two layers provide information about entry/exit decisions, the performance of these decisions are evaluated using the benchmark ratios given in Section 5.2. Creamer and Freund (2010) [94] demonstrate that these methods yield profitable results. They do warn that, "...boosting requires powerful control mechanisms in order to reduce unnecessary and unprofitable trades that increase transaction costs. Hence, the contribution of new predictive algorithms by the computer science or machine learning literature to finance still needs to be incorporated under a formal framework of risk management." This area remains to be explored.

## 5.14 Exercises

1. **Technical Rules:** Consider the data on exchange rates (in file, `exchange_rates_1.csv`) between the Indian Rupee and the US Dollar, British Pound and Euro; it is daily data from December 4, 2006 to November 5, 2010. In order to compare the performance of a strategy, appropriately normalize the entry point exchange rate: Invest equal amounts of capital (counted in INR, which we will call the "base currency") in each trade. Total portfolio return on each day should be computed as a weighted sum of returns from each active trade (weighted by the current value of each trade). Evaluate the signals discussed below using the Sharpe ratios and PNL with and without transaction costs of 15 basis points. Assume a risk free rate of 3%. (1 basis point = $10^{-4}$; so each dollar traded costs 0.0015 in transaction costs).

(a) **Moving Average Trading Rules:** Trading signals are generated from the relative levels of the price series and a moving average of past prices.

- **Rule 1:** Let $\overline{p}_t^{(m)}$ as defined in (5.6); use the strategy to sell if $p_t > \overline{p}_t^{(m)}$ and to buy if $p_t < \overline{p}_t^{(m)}$. Evaluate this strategy. What is the optimal '$m$'?
- **Rule 2 (Bollinger Rule):** Use the Bollinger rule as outlined in the text; what is the optimal '$m$'?
- **Rule 3 (Resistance-Support):** Buy if $p_t > \max_{0 \le i < m} p_{t-i}$ and sell if $p_t < \min_{0 \le i < m} p_{t-i}$. Evaluate this rule. What is the optimal '$m$'?
- **Rule 4 (Momentum):** Compute a short-term moving average over the last $m$ prices, $\overline{p}_t^{(m)}$ and a long-term average over the last $n$ prices, $\overline{p}_t^{(n)}$ with $m < n$. If $\overline{p}_t^{(m)}$ crosses $\overline{p}_t^{(n)}$ from below, sell and if it crosses $\overline{p}_t^{(n)}$ from above, buy. Evaluate this strategy. What are the optimal values of $m$ and $n$?

(b) **RSI Oscillator Rule:** Use $\text{RSI}_t$ as defined in Section 5.3.
If $\text{RSI}_t < 30$, buy and if $\text{RSI}_t > 70$, sell. Evaluate this trading strategy. What is the optimal '$m$'?

(c) **Pairs Trading: Distance Method**

We want to develop a pairs trading strategy and check if it does any better than the strategies in (i) and (ii). At the last trading day of the month, look back 3 months and identify if any pairs are worth trading. Use the distance measures, thresholds, as given in the text. (See Section 5.8.1).

(d) **Pairs Trading: Co-Integration Method (Section 5.8.2)**

In the co-integration method, we can consider the possibility of trading more than 2 assets. Check using a moving window of three months which series are co-integrated. Develop appropriate trading strategies to trade portfolios formed from the co-integrating vectors and compare with the pairs trading strategies based on other criteria. Concretely define the trading indicators and thresholds for entry and exit. Summarize the performance results.

(e) **Pairs Trading: APT Method**

Consider the data with the S&P 500 market index returns and assume that the risk free rate is 3%. Regress each currency's excess return (over the risk free rate) on the excess market return over a moving window of 3 months. Find pairs based on the similarity of the regression coefficients.

(a) Compare the pairs that result from the APT method with those obtained in Problem 2.

(b) After finding the pairs, use the same rules as in Exercise 2 to decide when to enter and exit specific trades.

Summarize the performance results.

(f) **Summing Up**

Provide a summary table of portfolio returns of all methods with the main findings and drawbacks of all the methods that are explored in this exercise. In particular, comment on the performance of single-asset strategies versus multi-asset strategies. Also comment on whether the market factor adds any value.

2. The file `exchange_rates_2.csv` contains exchange rates between US$ and twenty-five major currencies; the daily data spans January 3, 2000 to April 10, 2012. Consider the data from January 3, 2002 for the following currencies: EUR, GBP, DKK, JPY and CNY. Compute $r_t = \ln p_t - \ln p_{t-1}$, returns for each series.

**A. Evaluation of Trading Rule:** Consider the data starting June 11, 2003.

(a) Compute the average return for each day. We will treat this as a 'market' return.

(b) Consider the two currencies: EUR and JPY and assume that your investment is restricted to only these two currencies. We will follow the simple algorithm: If

$$\frac{r_i - \bar{r}}{sd(r_i)} = \begin{cases} > 2, & \text{sell} \\ < -2, & \text{buy} \\ \text{otherwise}, & \text{hold} \end{cases}$$

The average and standard deviation of the returns are based on last 30-day observations. Evaluate this algorithm by comparing it with buy and hold.

(c) What is the value of the 'market' return? How could this information be used?

**B. Trading Rule with Other Factors:** In addition to exchange rates from June 11, 2003, consider the data on commodity prices: Gold, Oil, Copper and a commodity index. Among the following currencies, CAD, BRL, ZAR, AUD, NZD, CAF and RUB, are likely to be influenced by the commodity prices. Discuss how you would incorporate this into a trading strategy. You may want to consider fair-value type of models.

3. The file `industry_pairs.csv` contains daily stock prices, returns, trading volume and shares outstanding from January 2000 to June 2014 for the following companies:

Oil stocks: HAL, SLB, EQT, OXY, CVX

Tech stocks: AAPL, CSCO, GOOG, INTL, MSFT

Bank stocks: BAC, JPM, WFC, C, STI

Retail stocks: ANF, GES, SKS, SHLD, URBN

Begin forming pairs in January 2001 based on the estimation period, January 2000 to December 2000 and evaluate the performance of the pairs after holding them for six months in the year 2001. Do this on a moving window basis over time. The last estimation period is January 2014 to December 2014. Eligible pairs corresponding to the last estimation remain eligible till December 2014. Pairs that opened in December 2013 and did not converge with the stipulated maximum of six months would have been closed in June 2014.

As you recall, the pairs trading strategy works as follows (refer to the text for more details):

- Within each sector, match pairs based on normalized price differences over the past year.
- When a tracked pair diverges by more than two standard deviations of their normalized price difference, buy the "cheap" stock and sell the "expensive" one, after a one-day wait period (to mitigate the effects of any market irregularities).
- If prices for a traded pair re-converge, exit the paired positions.
- If prices for a traded pair do not re-converge within six months, close the positions. An alternative strategy closes all traded pairs in ten trading days.

Questions of Interest:

(a) Quantify the profitability of the pairs trading strategy and on average, how long it takes for pairs to re-converge?

(b) Can you provide some insight into why the strategy works based on trading volume, shares outstanding and the market trend?

Another Question of Interest:

(c) Is there a pair across sectors that can outperform the pairs formed within each sector? Speculate how this is possible.

4. The file `pairs_trading.csv` contains daily stock prices, returns, trading volume and shares outstanding from January 1992 to June 2010 for the following companies:

DELL, HP, ADBE, MSFT, AAPL
JPM, WFC, BAC, C, BRK.A
JNJ, WMT, COST, KO, PEP
GM, TM, F
XOM, COP

Begin forming pairs in January 1993 based on the estimation period, January 1992 to December 1992. The last estimation period is January 2009 to December 2009. Eligible pairs corresponding to the last estimation remain eligible till December 2009. Pairs that opened in December 2009 and did not converge with the stipulated maximum of six months would have been closed in June 2010.

Questions of Interest:

(a) Quantify the profitability of the pairs trading strategy and on average, how long it takes for two pairs to reconverge?

(b) Can you provide some insights into why the strategy works based on trading volume, shares outstanding and the market trend?

5. One of the most intriguing asset pricing anomalies in finance is the so-called "price momentum effect" that was covered in this chapter. Using data in file `momentum_data.csv`, the goal is to replicate the price momentum strategy described in Jegadeesh and Titman (2001) [216] using more recent data extending to December 2014. [This problem is based on the notes from Ravi Jagannathan, Northwestern University.]

**A. Source of Data:** Using the monthly stock returns from January 1965 to December 2014 you should construct the decile ranking by yourself. For future reference, the decile ranking file is available from Ken French's web site. The NYSE market capitalization decile contains all common shares regardless of the end of the month price.[3] You should use each stock's permanent identification code (PERMNO) as a unique identifier for that stock.

Note that not all stocks might have valid monthly returns for various reasons. One reason is that the stock gets delisted somewhere during the month. For the construction of your momentum portfolios, you can simply ignore the delisting returns in your portfolio construction and return accumulations for the sake of simplicity. More specifically, for this exercise, if the end of the month return of any stock is missing, you may simply assume the return is zero in your compounding. You should keep track of such cases (identify the reason for delisting/missing observation) and prepare a table summarizing how many such cases occurred.

---

[3] The Fama-French factors can be downloaded from Ken French's website. Note that you should only use the data from the "US Research Returns Data" section.

**B. Stocks Selection Criterion:** Consider all the common stocks traded on AMEX, NYSE and NASDAQ but subject to the following filtering rules:

- Exclude all stocks priced lower than $5 at the end of month ($t$).
- Exclude all stocks belonging to the smallest NYSE market decile at the end of month ($t$). You should first find the cutoff for the NYSE smallest market decile, then delete all the stocks (including stocks on AMEX, NYSE, and NASDAQ) with a market cap that's below the cutoff.
- Another thing to take into consideration is that when you form your portfolios, you should only apply the filters to the stocks when you form the portfolio (i.e., at the end of month $t$). The future returns (stock returns at month $t+2$ to $t+7$) should not be filtered. Otherwise it creates a look-ahead bias. For example, for portfolio returns at month $t+2$ to $t+7$, drop stocks with a price <5 at any month between $t+2$ to $t+7$. First, when forming the portfolio at time t, you use information you don't have at month t. Second, you may eliminate stocks with likely poor performance (low price) in the future. The portfolio returns (especially the loser portfolio returns) are likely to be biased upward from this action.

Again, you should keep track of such cases and prepare a table summarizing how many such cases have occurred.

**C. Portfolio Formation:** At the end of each month ($t$), sort the stocks into 10 decile portfolios based on the six-month cumulative returns between month ($t-5$) to ($t$). The first decile contains the stocks with the lowest past six-month returns, and the tenth decile contains the stocks with the highest past six-month returns.

- Start by accumulating the return of each decile portfolio between ($t+2$) and ($t+7$), i.e., skipping the first month following the portfolio formation.
- At the beginning of each month, construct overlapping decile portfolios as equally weighted averages of the past six months' decile portfolios. That is, at each month $\tau$, the $i$th overlapping decile portfolio consists of an equal weighted average of the $i$th decile portfolios of the previous six month ($\tau-5$) to ($\tau$).[4]
- For each month and of the 10 overlapping decile portfolios, record the portfolio returns between January 1995 and December 2014.

**D. Questions of Interest:** After constructing the relevant portfolios, answer the following questions. When you answer these questions, note any additional assumptions/choices you make.

---

[4]These 10 overlapping decile portfolios are the testing assets in our asset pricing test exercises.

**Momentum Strategy Profits**

(a) Report the monthly average returns of your overlapping decile portfolios, as well as the winner ($P1$, highest past six-month return portfolios) minus loser ($P10$, lowest past six-month return portfolios).

(b) Report the monthly average returns of your overlapping decile portfolios for January and non-January months respectively, as well as the winner ($P1$, highest past six-month return portfolios) minus loser ($P10$, lowest past six-month return portfolios).

(c) Use CAPM and the Fama-French version of multifactor asset pricing model to test the spreads between winner and loser portfolios ($P10$ - $P1$). The Fama-French factors can be obtained from Ken French's web site.

(d) Tabulate these results in a table similar to table $I$ in Jegadeesh and Titman (2001) [216].

(e) Compare your results to the results in Jegadeesh and Titman (2001) [216] *and comment on your findings.*

**Momentum Strategy and Market Capitalization**

(f) Repeat the above exercise but split the sample into small-cap and large-cap portfolios, where the small-cap portfolio contains the stocks in the lower five NYSE market capitalization deciles.

(g) Tabulate these results in a table similar to table $I$ in Jegadeesh and Titman (2001) [216].

(h) Compare the simple momentum strategy's profits, and momentum strategy cut by market capitalization profits to the value-growth strategy return. The value-growth strategy return can be proxied by the HML factor obtained from Ken French's website.

(i) *Comment on your findings*. In particular, briefly suggest at least three reasons why momentum profits may be related to market capitalization, and how you would go ahead and test it further.

6. This is another exercise using data for sixteen stocks to replicate the price momentum strategy albeit in a small scale. The daily returns from January 11, 2000 to November 20, 2005 for sixteen stocks are in the file `d_logret_16stocks.csv`; to this add two more columns: SP500 returns and risk free rate.

**Portfolio Formation**

On the last day of each month, sort the stocks into four quartile portfolios based on the past six-month cumulative returns. The first quartile contains the stocks with the lowest past six-month returns, and the fourth quartile contains the stocks with the highest past six-month returns.

(a) Start by accumulating the return of each quartile portfolio for the next six months.

(b) For each month and of the four quartile portfolios, record the average portfolio returns between January 2002 and November 2005.

Questions of Interest: Same as in Exercise 5.

7. Consider the 30 min price bar data for Treasury Yields from June 8, 2006 to August 29, 2013 in file `treasuries.csv`. Daily data for Treasury Yields is just a snapshot at 4 p.m. and the total volume for the day is the sum of 30 min volumes since the previous day. This is the same data used in Section 5.10.2 in the text.

(a) Develop ARMA time series models for both daily and the 30 min volumes. We can use daily data for developing macro strategies (whether to enter next day) and 30 min data for micro strategies (when to enter next day).

(b) Compare the exponential smoothing approach suggested in the text with ARMA modeled in (a).

(c) Compute volatility measure, $\sigma_t^2 = 0.5[\ln \frac{H_t}{L_t}]^2 - 0.386[\ln \frac{C_t}{O_t}]^2$ for both daily and 30 min levels and compare them.

(d) Compute volatility forecasts, $\hat{\sigma}_{t+1}^2$ = weighted average of $\sigma_t^2, \sigma_{t-1}^2, \ldots, \sigma_{t-22}^2$, for daily data and $\hat{\sigma}_{t+1 \cdot m}^2$ = weighted average of $\sigma_{t \cdot m}^2, \ldots, \sigma_{t-22 \cdot m}^2$ for $m^{th}$ 30-min interval. [In (a), you should notice strong periodicity in the 30-min data.]

(e) Develop trading strategies that incorporate (see the Omnibus Rule in the text):

   (a) price information only

   (b) price and volume information and

   (c) price, volume and volatility information

(f) Demonstrate that volume and volatility add value to the enter/exit strategies.

# 6

# *Dynamic Portfolio Management and Trading Strategies*

## 6.1 Introduction to Modern Portfolio Theory

Of fundamental interest to financial economists is to examine the relationship between the risk of a financial security and its return. While it is obvious that risky assets can generally yield higher returns than risk-free assets, a quantification of the trade-off between risk and expected return was made through the development of the Capital Asset Pricing Model (CAPM) for which the groundwork was laid by Markowitz (1959) [260]. A central feature of the CAPM is that the expected return is a linear function of the risk. The risk of an asset typically is measured by the covariability between its return and that of an appropriately defined 'market' portfolio. Models of expected return-risk relationships include the Sharpe (1964) [301] and Lintner (1965) [245] CAPM, the zero-beta CAPM of Black (1972) [43], the arbitrage pricing theory (APT) due to Ross (1976) [293] and the intertemporal asset pricing model by Merton (1973) [268]. The economy-wide models developed by Sharpe, Lintner and Black are based on the work of Markowitz which assumes that investors would hold a mean-variance efficient portfolio. The main difference between the work of Sharpe and Lintner and the work of Black is that the former assumes the existence of a risk-free lending and borrowing rate whereas the latter derived a more general version of the CAPM in the absence of a risk-free rate.

In this section, we briefly review the CAPM model and the implications for empirical research in the area of portfolio construction and testing.

### 6.1.1 Mean-Variance Portfolio Theory

A portfolio which is composed of individual assets has its risk and return characteristics based on its composition and how the individual asset characteristics correlate with each other. The optimal combination is designed to produce the best balance between risk and return. For a given level of return, it will provide the lowest risk and for an acceptable level of risk, it will provide the maximum return. The locus of the combination of risk and reward that characterizes the optimal portfolios is called the "Efficient Frontier."

We follow the same conventions as before to denote the return as $r_t = \ln(P_t) - \ln(P_{t-1})$; assume we have '$m$' assets in the portfolio with '$w_i$' denoting the share value invested in asset, '$i$', if $R_t = (P_t - P_{t-1})/P_{t-1}$, the portfolio return is $R_{pt} = \sum_{i=1}^{m} w_i R_{it}$ and thus $r_t = \ln\left(1 + \sum_{i=1}^{m} w_i R_{it}\right) \simeq \sum_{i=1}^{m} w_i r_{it}$. If $r_t$ denotes the vector of '$m$' asset returns and with weights stacked up as a vector, $w$, then the portfolio return, $r_{pt} = w' r_t$ resulting in $\mu_p = E(r_{pt}) = w'\mu$ and $\sigma_p^2 = w'\Sigma w$, where $\Sigma$ is the $m \times m$ variance-covariance matrix of $r_t$. If the returns are uncorrelated or negatively correlated, then observe that

$$\sigma_p^2 = w'\Sigma w \leq \sum_{i=1}^{m} w_i \operatorname{Var}(r_{it}) \leq \frac{v}{m}, \tag{6.1}$$

where '$v$' is the maximum of $\operatorname{Var}(r_{it})$, clearly indicating that diversification tends to reduce risk. The power of the diversification can be seen clearly if we assume the covariance matrix, $\Sigma$, has all variances equal and if all off-diagonal covariance elements are the same. Then, $\sigma_p^2 = \frac{1}{n} \cdot \sigma^2 + \frac{n-1}{n} \cdot \rho \cdot \sigma^2$. Observe that if $\rho = 0$, $\sigma_p^2 \to 0$ as $n \to \infty$ and if $\rho = 1$, $\sigma_p^2 = \sigma^2$ that results in no benefit. When $\rho < 0$ as shown in (6.1), the portfolio variance is less due to diversification. But it should be noted that we cannot completely eliminate portfolio risk when the correlations among the assets are positive. Observe that $\sigma_p^2 = \frac{1}{m^2} \sum_{i=1}^{m} \sigma_i^2 + \frac{1}{m^2} \sum_{i \neq j} \sigma_{ij} \leq \frac{\sigma_{\max}^2}{m} + \frac{m-1}{m} \cdot A \to A$ as $m \to \infty$. Some amount of risk will remain if the portfolio consists of assets that move with the market.

**Minimum Variance Portfolio:** The efficient portfolio is obtained through the constrained optimization:

$$\min_{w} w'\Sigma w \ni w'\mu = \mu^*, \quad w'1 = 1. \tag{6.2}$$

Here 1 is the $m \times 1$ unit vector. The solution to this can be obtained via Lagrangian function and can be found in Campbell, Lo and MacKinley (1996) [66, Section 5.2]. With $A = 1'\Sigma^{-1}\mu$ (weighted mean), $B = \mu'\Sigma^{-1}\mu$ ($F$-ratios), $C = 1'\Sigma^{-1}1$ and $D = BC - A^2$, the 'weight' vector, $w$, is given as,

$$w_{\text{eff}} = \{B\Sigma^{-1}1 - A\Sigma^{-1}\mu + \mu^*(C\Sigma^{-1}\mu - A\Sigma^{-1}1)\}/D$$

and (6.3)

$$\sigma_{\text{eff}}^2 = (B - 2\mu^* A + \mu^{*2} C)/D.$$

The graph of the efficient frontier $(\mu^*, \sigma_{\text{eff}})$ in the mean-standard deviation space is the right side of a hyperbola (see Figure 6.1). It must be noted that in the formulation (6.2), the weights can be negative implying that short selling is allowed. With no short selling, $w_i \geq 0$ for all '$i$', there is no explicit solution but it can be numerically solved via quadratic programming.

Suppose an investor is interested in a portfolio with maximum Sharpe ratio (= average return/standard deviation), which represents the return per unit risk. Graphically, the portfolio is the point where a line through the origin is tangent to the efficient frontier, because this point represents the highest possible Sharpe ratio. So it is called the tangency portfolio. The inverse of the slope is obtained by setting $\frac{\delta \sigma_{\text{eff}}^2}{\delta \mu^*} = 0$

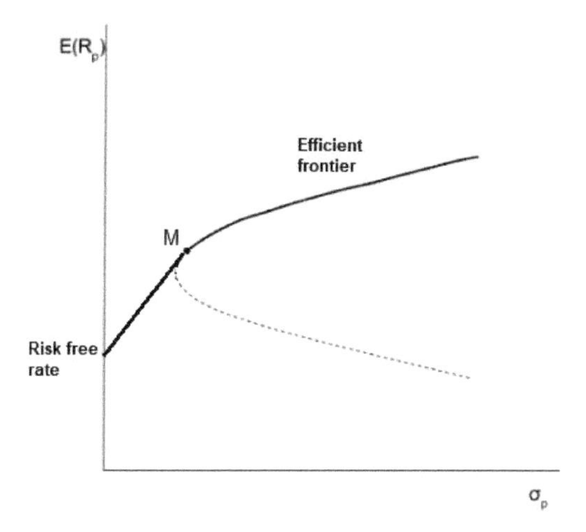


Figure 6.1: Efficient Frontier with No Short Selling.

which yields $\mu^* = A/C$ and $\sigma^2_{\text{MV}} = \frac{1}{C}$ and the efficient combination is simply, $w_{\text{MV}} = \Sigma^{-1}\mu/C$. Thus we have described two portfolios that an investor can prefer. If minimum amount of risk is desired, the minimum variance portfolio is taken and if the objective is to maximize the Sharpe ratio, the tangency portfolio is taken. These can be formulated from the point of view of unified theory of utility maximization.

Assume that the utility function is given as,

$$u = w'\mu - \frac{\lambda}{2}\, w'\,\Sigma\, w, \tag{6.4}$$

where '$\lambda$' is called the parameter of absolute risk-aversion. The greater the '$\lambda$', the more risk averse the investor is. Thus assuming all investors are risk-averse, the objective function is,

$$\max_{w} u \ \ni \ w'1 = 1. \tag{6.5}$$

It can be easily shown that the resulting weight vector is,

$$w_{\text{eff}} = \frac{1}{\lambda}\left(\Sigma^{-1}\mu + \Sigma^{-1}1 \cdot \gamma^*\right), \tag{6.6}$$

where $\gamma^* = (\lambda - A)/C$. The results can be obtained again using Lagrangian multipliers and the matrix differentiation results $\left(\frac{\delta \Sigma w}{\delta w} = \Sigma;\ \frac{\delta(w'\Sigma w)}{\delta w}) = 2 \cdot w'\Sigma\right)$. From (6.6), observe that it is a combination of minimum variance portfolio and the tangency portfolio. At the optimum point, note

$$\mu_{\text{opt}} = w'\mu = \frac{D}{\lambda C} + \mu_{\text{MV}}$$

and (6.7)

$$\sigma^2_{\text{opt}} = w'\Sigma w = \sigma^2_{\text{MV}} + \frac{1}{\lambda^2} \cdot \frac{D}{C}.$$

If an investor is fully risk-averse, $\lambda \to \infty$ and the optimal portfolio is the minimum variance portfolio and when $\lambda = A$, the optimum portfolio is the tangency portfolio. Thus, both are special cases of Markowitz strategy.

For realistic situations related to trading we need to consider borrowing and lending at the risk-free rate, '$r_f$'. We assume that in addition to '$m$' risky assets, where investment is made, there is a possibility of putting the funds in risk-free assets such as treasury bonds or certificates of deposits, etc. The optimization problem can be stated as,

$$\min_{w} w'\Sigma w \ni w'\mu + (1 - w'1)r_f = \mu^*. \tag{6.8}$$

This results in efficient weights:

$$w_{\text{eff}} = \frac{\mu^* - r_f}{(\mu - r_f 1)'\Sigma^{-1}(\mu - r_f 1)} \cdot \Sigma^{-1}(\mu - r_f 1). \tag{6.9}$$

If $w_i$'s are set $\geq 0$, there is no explicit analytical solution and Figure 6.1 depicts the optimal allocation. When $w_i$'s are unrestricted, the efficient frontier based on all risky assets is given as,

$$\mu = r_f + \frac{\mu_m - r_f}{\sigma_m} \cdot \sigma \tag{6.10}$$

which is known as the capital market line or the security market line; here $\mu_m$, the mean market return can be the return from the S&P 500 index. The form of (6.10) lends itself to empirical testing via the linear regression model,

$$r_{it} - r_f = \beta_i(r_{mt} - r_f) + \epsilon_{it}, \tag{6.11}$$

where $\beta_i = \text{Cov}(r_{it}, r_{mt})/\sigma^2_m$; note the volatility in '$r_i$' can be decomposed as

$$\sigma_i^2 = \beta_i^2\sigma_m^2 + \sigma_\epsilon^2. \tag{6.12}$$

If $\beta_i > 1$, the asset is taken to be aggressive. The commonly used performance ratios (Section 5.2) follow from the population version of the model (6.11) which is $\mu - r_f = \alpha + \beta(\mu_m - r_f)$. Sharpe ratio is given as $(\mu - r_f)/\sigma$, Treynor's ratio is $(\mu - r_f)/\beta$ and Jensen's ratio is simply the '$\alpha$'. If $\alpha > 0$, it is taken that the portfolio performs better than the market, after adjusting for the risk.

The CAPM model is empirically tested via (6.11) with added intercept terms:

$$r_{it} - r_f = \alpha_i + \beta_i(r_{mt} - r_f) + \epsilon_{it}. \tag{6.13}$$

If CAPM holds then $H_0 : \alpha_i = 0, (i = 1, 2, \dots, m)$ should be supported. But empirical evidence does not seem to support $H_0$; for which various explanations are possible. It is argued that the proxies such as the S&P 500 index are not theoretically true market portfolios. A true market portfolio should be based on all assets besides equities. Also, a proxy for the risk-free rate that is uncorrelated with the returns of the assets and the market is equally subject to some debate. It has been shown that other asset related information provides better explanatory power for the cross-sectional variation of the asset returns. This has led to multi-factor models that are discussed next.

### 6.1.2 Multifactor Models

Ross (1976) [293] introduced the Arbitrage Pricing Theory (APT) which is more general than CAPM; the cross-sectional variation in the asset returns is postulated to be due to multiple risk factors. There is no need to identify a market portfolio nor the risk-free return. The model can be stated as:

$$r_{it} = \alpha_i + \beta_i' f_t + \epsilon_t \tag{6.14}$$

and given $f_t$ is a $n \times 1$ vector of factors, $E(\epsilon_{it}) = 0$, $E(\epsilon_{it}\epsilon_{jt}) = \sigma_{ij}$, $i, j = 1, 2, \dots, m$ and $t = 1, 2, \dots, T$. In the absence of arbitrage, Ross (1976) [293] shows that the following relationship should hold:

$$\mu_i \sim r_f + \beta_i' \lambda_k, \tag{6.15}$$

where $\lambda_n$ is a $n \times 1$ vector of factor risk premia.

Generally, there are two approaches to empirically model (6.14) and test if (6.15) holds. In the first approach, it is assumed that the factors $(f_t)$ are unknown and are estimated from the $(m \times m)$ covariance matrix of the returns via principal components. The number of factors, $n$, is generally determined via subjective considerations. In the second approach, it is taken that the factors are determined by macroeconomic variables that reflect systematic risk and by asset characteristics. Chen, Roll and Ross (1986) [76] consider expected inflation, spread between high and low grade bond yields, spread between long and short interest rates for US government bonds, etc., as known factors. The asset characteristics that are noted to be useful are: Ratio of book to market value, price-earnings ratio, momentum, Fama-French factors. These factor models with known or unknown factors tend to fare better than CAPM. As we will demonstrate, they are also useful for the construction of portfolios as these factors are likely to represent different risk dimensions.

### 6.1.3 Tests Related to CAPM and APT

To test if any particular portfolio is ex ante mean-variance efficient, Gibbons, Ross and Shanken (1989) [166] provide a multivariate test statistic and study its small sample properties under both null and alternative hypotheses. This follows from the multivariate regression model and the associated test procedures discussed in Chapter 3. Recall the null hypothesis of interest in CAPM is:

$$H_0 : \alpha_i = 0 \quad \forall i = 1, 2, \dots, m. \tag{6.16}$$

The test statistic is a multiple of Hotelling's $T^2$ stated as,

$$F = \frac{T - m - 1)}{m} \cdot \frac{\hat{\alpha}'\hat{\Sigma}_{\epsilon\epsilon}^{-1}\hat{\alpha}}{1 + \hat{\theta}_p^2} \sim F(m, T - m - 1), \tag{6.17}$$

where $\hat{\theta}_p = \bar{r}_p / s_p$, the ratio of sample average and standard deviation of $r_{pt}$. The noncentrality parameter depends on $\alpha'\Sigma_{\epsilon\epsilon}^{-1}\alpha$ which is zero under $H_0$. The statistical power of the $F$-test thus can be studied. The test statistic in (6.17) can be easily derived from the results on multivariate regression in Section 3.1. Thus, we can compute the residual covariance matrices both under the null and alternative hypotheses to arrive at the LR statistic.

A practical question that needs to be resolved is on the choice of '$m$' and '$T$'. With over 5000 stocks listed in NYSE and with daily data available since 1926, the choice is restricted only by the condition $T > m$ for the variance-covariance matrix in the estimation of the model (6.13) to be non-singular. Gibbons et al. (1989) [166] suggest '$m$' should be roughly a third to one half of '$T$'. Then it is also important to decide which '$m$' assets are to be chosen. One suggested approach is to use a subset of beta-sorted assets but it may not necessarily give better power to the $F$-test.

Note that $H_0 : \alpha = 0$, where $\alpha' = (\alpha_1, \dots, \alpha_m)$ is violated if and only if some portfolio of the assets considered in the regression model (6.13) has non-zero intercept. If we write $\hat{\alpha}$ as the $m \times 1$ vector of estimates of the intercepts, then for a linear combination of $a'r_t$, where $r_t$ is the $m \times 1$ vector of asset returns, [observe that if we let $Y_t = r_t - r_f 1$ and $X_t = [1, r_{mt} - r_f]' = [1, x_t]'$, $a'Y_t = a'\alpha + (a'\beta)x_t + \epsilon_t$, and thus] $\text{Var}(a'\hat{\alpha}) = (1 + \hat{\theta}_p^2)\frac{a'\hat{\Sigma}a}{T}$ and hence

$$t_a^2 = \frac{T(a'\hat{\alpha})^2}{(1 + \hat{\theta}_p^2)(a'\hat{\Sigma}a)}. \tag{6.18}$$

Maximizing $t_a^2$ is therefore equivalent to minimizing $a'\hat{\Sigma}a$ subject to $a'\hat{\alpha} = c$, which is equivalent to the portfolio minimization problem. The solution is

$$a = \frac{c}{\hat{\alpha}'\hat{\Sigma}^{-1}\hat{\alpha}} \cdot \hat{\Sigma}^{-1}\hat{\alpha} \tag{6.19}$$

and thus $t_a^2 = T \cdot \frac{\hat{\alpha}'\hat{\Sigma}^{-1}\hat{\alpha}}{1+\hat{\theta}_p^2} = \frac{T}{T-m-1} \cdot m \cdot F$. Thus, the portfolio based on '$a$' can provide information about creating a better model. The portfolio based on '$a$' is termed an active portfolio and Gibbons et al. show that when this is combined with the 'market' portfolio it results in ex-post efficient portfolio. To make the comparison easier between the three portfolios, average returns for all three are set equal which results in the choice of $c = \bar{r}_m \cdot \frac{\hat{\alpha}\hat{\Sigma}^{-1}\hat{\alpha}}{\hat{\alpha}'\hat{\Sigma}^{-1}\hat{\gamma}}$ and thus the active portfolio results in

$$a = \bar{r}_m \cdot \frac{\hat{\Sigma}^{-1}\hat{\alpha}}{\hat{\alpha}'\hat{\Sigma}^{-1}\bar{r}} \tag{6.20}$$

which can be compared with $w_{\text{eff}}$ in (6.3). It must be noted that the weights obtained in (6.3) are not in reference to any other portfolio and not adjusted for any comparison with other portfolios. Thus one could expect that the two portfolios would be the same only when the assets selected are not correlated to market index or to any benchmark portfolio.

The test for APT model for known factors in (6.14) also follows easily from the multivariate regression results discussed in Chapter 3. The model can be restated as

$$r_t = \alpha + Bf_t + \epsilon_t. \tag{6.21}$$

If $r = [r_1, \ldots, r_T]'$ and $f = [f_1, \ldots, f_T]'$ are $m \times T$ and $n \times T$ data matrices, then

$$\hat{\alpha} = (r' M_f 1_T)(1_T' M_f 1_T), \tag{6.22}$$

where $M_f = I_T - f'(ff')^{-1} f$ is a $T \times T$ idempotent matrix ($M_f^2 = M_f$). The null hypothesis, $H_0 : \alpha = 0$ is tested through Jensen's test:

$$F = \left(\frac{T - m - n}{m}\right)(1 + \overline{f}'\hat{\Omega}^{-1}\overline{f})^{-1}\hat{\alpha}'\hat{\Sigma}^{-1}\hat{\alpha} \sim F(m, T - m - n), \tag{6.23}$$

where $\overline{f} = \frac{1}{T}\sum_{t=1}^{T} f_t$, $\hat{\Omega} = \frac{1}{T}\sum_{t=1}^{T}(f_t - \overline{f})(f_t - \overline{f})'$ and $\hat{\Sigma}$ is the residual covariance matrix. Recall that the known factors could be macroeconomic based or the financial variables such as Fama and French factors discussed in Section 5.6 or industry related factors. Commercial factor risk models will be briefly discussed in the last section.

The key tool for estimation of unknown factors is Principal Component Analysis (PCA). For a brief discussion on PCA, refer back to Chapter 3. The main difference between Factor Analysis (FA) and PCA is that the former is meant to capture the joint covariances among the returns, $r_t$ whereas PCA's focus is on capturing the sum of variances or total volatility of $r_t$. Assume that $\text{Var}(r_t) = \Sigma_{rr}$, $m \times m$ covariance matrix. Now transform the returns by $Pr_t$ where $P$ is orthogonal, $P'P = I_m$ and $P\Sigma_{rr}P' = \Lambda$, where $\Lambda$ is the diagonal matrix. Note the transformed returns, $Z_t = Pr_t$ has $\text{Var}(Z_t) = \Lambda$ and thus the principal components are uncorrelated. The goal is to recover much of the total variance $\sum_{i=1}^{m}\sigma_i^2$ with a few linear combinations of $r_t$.

**Result.** The solution to $\max_w w'\Sigma w \ni w'w = 1$ is obtained from solving $\Sigma w = \lambda w$. Thus '$w$' is the eigenvector that corresponds to the largest eigenvalue of $\Sigma$ obtained from $|\Sigma - \lambda I| = 0$ where $|\cdot|$ denotes the determinant of the matrix; note $\text{tr}(\Sigma) = \sum_{i=1}^{m}\sigma_i^2 = \sum_{i=1}^{m}\lambda_i$ and the values of '$\lambda_i$' are disproportional such that we can approximate $\text{tr}(\Sigma) \sim \sum_{i=1}^{r}\lambda_i$, where '$r$' is much less than '$m$'.

In the factor model setup given in (6.20), it is assumed that the factors $f_t$ are random with $E(f_t) = 0$, $\text{Var}(f_t) = \Omega$ and $E(f_t'\epsilon_t) = 0$. This results in,

$$\text{Cov}(r_t) = B\Omega B' + V = \Sigma_{rr}, \quad \text{Cov}(r_t, f_t) = B\Omega, \tag{6.24}$$

where $\Omega$ and $V$ are taken to be diagonal; note because $r_t \sim N(\alpha, \Sigma_{rr})$ the generalized least-squares estimates are given as:

$$\hat{f}_t = (\hat{B}\hat{V}^{-1}\hat{B})^{-1}\hat{B}'\hat{V}^{-1}r_t, \quad \hat{\alpha} = \overline{r} - \hat{B}\overline{f}. \tag{6.25}$$

From (6.25), it follows that $\text{Var}(\hat{f}_t) = (B'V^{-1}B)^{-1}/T$ and $\text{Var}(\hat{\alpha}) = \frac{1}{T}(V - B(B'V^{-1}B)^{-1}B')$. To decide on the number of factors, the following test statistics is suggested:

$$\chi^2 = -\left(T - \frac{4r+5}{6}\right)\left(\log|\hat{\Sigma}_{rr}| - \log|\hat{B}\hat{B}^{-1} + \hat{V}|\right) \sim \chi^2_{\{[(m-r)^2-m-r]/2\}}. \tag{6.26}$$

The number of adequate factors can also be verified through the significant '$\lambda$'s in the result stated above. We will discuss an application in the next section that will clarify many of these concepts.

### 6.1.4 An Illustrative Example

We consider monthly data from July 1926 to October 2012 on returns from six portfolios that are formed based on Size and Book-to-Market (BM) ratio. This data is taken from French's data library. The titles follow the naming convention, for example, SML contains small size plus low BM to BIGH contains big size with high BM ratio. The data also contains $r_f$, $r_{mt} - r_f$ and the two Fama-French factors SMB (small-minus-big size factors) and HML (high-minus-low BM factor).

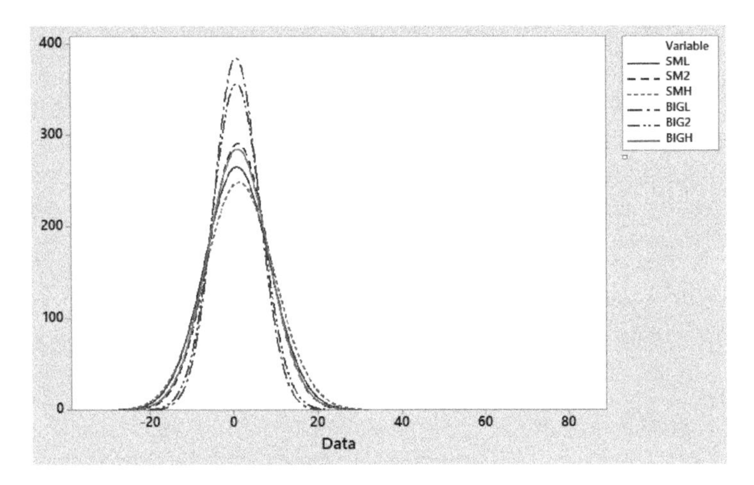


Figure 6.2: Histogram of Monthly Returns.

The histogram of the portfolio returns is given in Figure 6.2. The descriptive statistics are stated in Table 6.1. The returns are generally positively skewed with more sharply peaked than the normal distribution. The Jarque-Bera test (Equation 2.37) rejects the null hypothesis that returns are normally distributed. The efficient frontier with the risk-free rate using the weights in (6.9) along with the portfolios is plotted in Figure 6.3, by setting $\mu^* = \hat{\mu}_m$, the average of the market portfolio. The weights for the assets are $(-1.06, 1.08, 0.40, 0.58, -0.36, -0.32)$ and the share of the risk-free asset is 0.68. The scatter plot of the actual mean value versus standard deviation of the six portfolios (Figure 6.4) yields the slope of 0.1471 which has a larger empirical

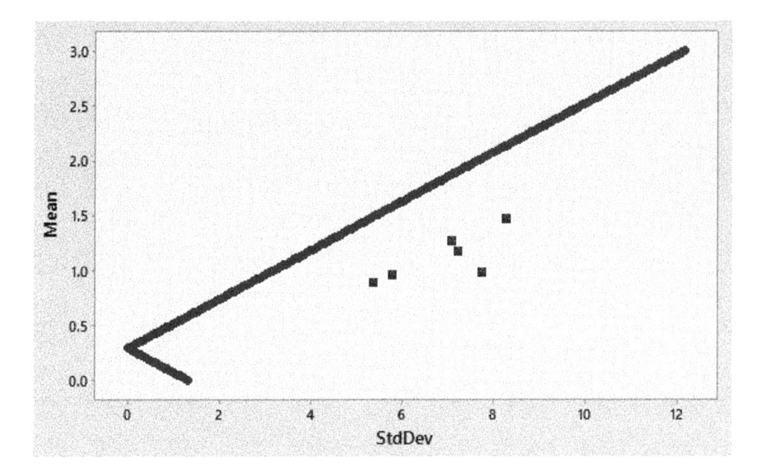


Figure 6.3: Efficient Frontier-Empirical.

Table 6.1: Descriptive Statistics ($T = 1036$)

| Variable | Mean | St. Dev. | Median | Skewness | Kurtosis |
|---|---|---|---|---|---|
| SML | 0.983 | 7.771 | 1.100 | 0.98 | 10.23 |
| SM2 | 1.271 | 7.092 | 1.530 | 1.27 | 14.51 |
| SMH | 1.472 | 8.303 | 1.660 | 2.13 | 21.64 |
| BIGL | 0.892 | 5.365 | 1.195 | −0.13 | 5.16 |
| BIG2 | 0.960 | 5.799 | 1.225 | 1.25 | 16.82 |
| BIGH | 1.173 | 7.243 | 1.395 | 1.53 | 18.05 |

Sharpe ratio $\left(\frac{\bar{r}_m - r_f}{\hat{\sigma}_m}\right)$, 0.1147. This clearly indicates that the market portfolio is not mean-variance efficient. Note that while three portfolios (BIGL, BIG2, BIGH) lie on the line, two others (SM2 and SMH) lie above the line. This generally implies that it is possible to outperform the empirical Markowitz portfolio.

The three known-factor model (6.14) results are presented in Table 6.2. There are several conclusions that can be drawn from this table. With the three-factors, the variations in the cross-sectional returns are well-captured by these factors. The coefficients of the market factors are closer to unity indicating that the six portfolios are closely aligned with the market index. But the estimates of $\alpha$ do indicate some mixed results. The value of the test statistics in (6.23) is 5.54 which is larger than $\chi^2_6(0.05) = 2.1$ and hence indicates that there should be some consideration to other factors besides Fama-French factors.

To identify the number of factors required to capture the total variance, we perform PCA of the covariance matrix. The Scree plot is given in Figure 6.5. The first component is responsible for 91% of the total variance and if three components are included we can capture up to 98% of the overall variance. The first factor loading is almost proportional to the unit vector indicating that it can be interpreted as a market

Table 6.2: Fama-French Model (Estimates)

| Portfolio | Const | Market | SMB | HML | $R^2$ |
|---|---|---|---|---|---|
| SML | −0.168* | 1.09* | 1.05* | −0.167* | 0.974 |
| SM2 | 0.0625 | 0.977* | 0.818* | 0.702* | 0.978 |
| SMH | 0.0202 | 1.03* | 0.933* | 0.783* | 0.991 |
| BIGL | 0.077* | 1.02* | −0.0965* | −0.231* | 0.978 |
| BIG2 | −0.0507 | 0.966* | −0.122* | 0.729* | 0.954 |
| BIGH | −0.112* | 1.079* | 0.0205 | 0.819* | 0.968 |

*Significant at 5% level

factor. It is easy to show with the three-factor model, the difference between $\hat{\Sigma}$ and $\hat{B}\hat{B}' + V$ is very small. This sort of parameter reduction will become useful when we deal with a large number of assets, a topic that is taken up in Section 6.2.

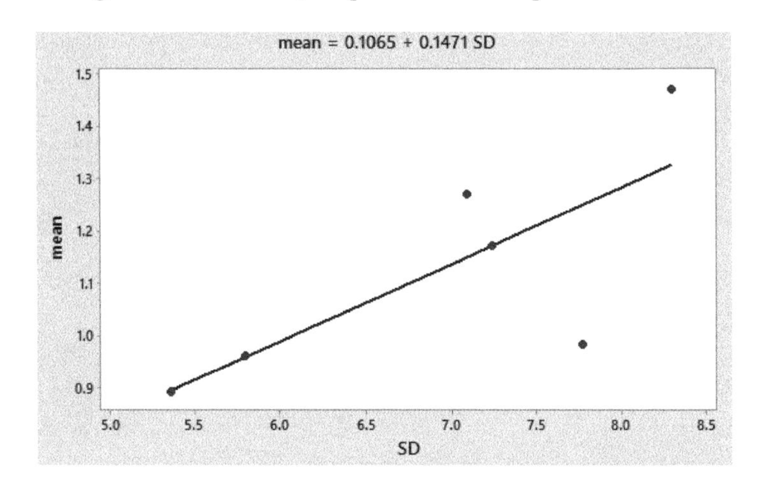


Figure 6.4: Capital Market Line.

### 6.1.5 Implications for Investing

As the focus of this book is on trading and investing we want to summarize some main takeaways from the widely used theoretical models. Some implications of these models are listed below:

**One-Fund Theorem:** There is a single investment of risky assets such that any efficient portfolio can be constructed as a combination of this fund and the risk-free asset. The optimal one-fund is the market fund which implies that the investor should purchase every stock which is not practical; this is a central foundation behind the rise of passive index funds.

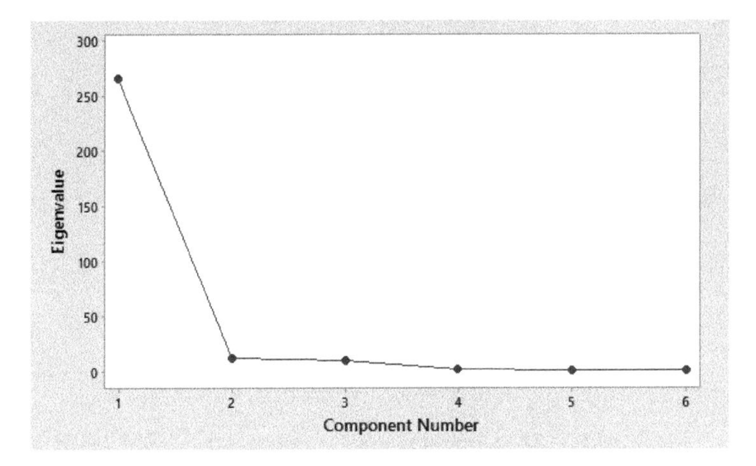


Figure 6.5: Scree Plot of Eigenvalues.

**Two-Fund Theorem:** Two efficient portfolios (funds) can be established so that any other efficient portfolios can be duplicated through these funds; all investors seeking efficient portfolios need only to invest in combinations of these two funds.

**No Short-Selling:** When the weights, $w_i$, are restricted to be positive, typically many weights tend to be zeros, by contrast with short-selling, most of the weights tend to be non-zeros. The resulting concentrated portfolios are more likely to have higher turnover rates. This has implications for the increase in trading costs.

**CAPM:** Recall the model is $r_i - r_f = \beta_i(r_m - r_f) + \epsilon_i$ with $\sigma_i^2 = \beta_i^2\sigma_m^2 + \sigma_\epsilon^2$; $\beta_i = \frac{\sigma_{im}}{\sigma_m^2}$ is the normalized version of covariance. If $\beta_i = 0$, $r_i \sim r_f$ even if $\sigma_i^2$ is large. There is no risk premium; the risk that is uncorrelated with the market can be diversified away. If $\beta_i < 0$, $r_i < r_f$ even if $\sigma_i^2$ is large; such an asset reduces the overall portfolio risk when it is combined with the market portfolio; they provide a form of insurance. Observe that the portfolio 'Beta' is $\beta_p = \sum_{i=1}^m w_i\beta_i$. The CAPM can be used also as a pricing formula. Note if the purchase price of an asset is '$P$' (known) which may be sold at '$Q$' (unknown), from CAPM, $r = \frac{Q-P}{P} = r_f + \beta(r_m - r_f)$ which results in the pricing formula

$$P = \frac{Q}{1 + r_f + \beta(r_m - r_f)}. \tag{6.27}$$

This procedure for evaluating a single asset can be extended to jointly evaluating multiple assets as well.

**Investment Pattern:** The minimum-variance portfolio method is likely to invest into low residual risk and low beta stocks. The CAPM model, (6.11) written in the vector form leads to

$$r_t - r_f 1 = \alpha + \beta(r_{mt} - r_f) + \epsilon_t \tag{6.28}$$

and with the assumption that the error-covariance matrix is diagonal, the covariance matrix of the excess returns is

$$\Sigma_{rr} = \beta\beta'\sigma_m^2 + D, \tag{6.29}$$

where $D$ is diagonal with the element, $\sigma_{\epsilon i}^2$. The inverse of the covariance matrix has a simpler structure:

$$\Sigma_{rr}^{-1} = D^{-1} - \frac{\sigma_m^2}{1+a} \cdot bb', \tag{6.30}$$

where $a = \sigma_m^2 \cdot \sum_{i=1}^m b_i\beta_i$ and $b_i = \beta_i/\sigma_i^2$. We have shown earlier that the portfolio weights under minimum-variance is

$$w = \frac{\Sigma_{rr}^{-1}1}{1'\Sigma_{rr}^{-1}1}, \quad \sigma_{\text{MV}}^2 = \frac{1}{1'\Sigma_{rr}^{-1}1}. \tag{6.31}$$

Substituting (6.30) in (6.31), notice that the vector weights,

$$w = \sigma_{\text{MV}}^2 \left( D^{-1}1 - \frac{\sigma_m^2}{1+a} \cdot bb'1 \right), \tag{6.32}$$

with a typical element, $w_j = \frac{\sigma_{\text{MV}}^2}{\sigma_j^2} \left( 1 - \beta_j \left( \frac{\sigma_m^2}{1+a} \cdot \sum_{i=1}^m b_i \right) \right)$. It is easy to show that the second term in parenthesis above, $\frac{\sigma_m^2}{1+a} \cdot \sum_{i=1}^m b_i \sim 1$ and so

$$w_j = \frac{\sigma_{\text{MV}}^2}{\sigma_j^2}(1 - \beta_j). \tag{6.33}$$

Thus when $\sigma_j^2$ is small and low $\beta_j$, $w_j$ will be large.

**Diversifying the Diversifiers:** In practice, finding proxies for the true tangency portfolio or a completely risk-free asset has been difficult. Practitioners traditionally have used cap-weighted equity indices as proxies for tangency portfolios but that is shown to result in inefficient risk-return trade-off. Recent studies (see Amenc, Goltz, Lodh and Martellini (2012) [13]) focus on diversifying across various forms of efficient portfolios, notably those formed using the maximum Sharpe ratio and the minimum variance criteria. It is empirically shown that the minimum variance portfolios provide conservative exposure to stocks that perform well in adverse conditions and Sharpe ratio portfolios provide access to higher performance stocks. Combining the two with controls on tracking error is shown to result in a smoother performance.

## 6.2 Statistical Underpinnings

The implementation of the portfolio theory in practice when the means and covariances of returns are unknown has raised some challenging issues. It is a theory that applies to a single period and the choice of portfolio weights that would determine the portfolio returns for a future period. Because of uncertainty involved in the future returns the use of the estimates of the means and covariances of the returns based on past data must be carefully evaluated. When the sample average and the sample covariance matrix (which are also maximum likelihood estimates under the assumption that returns are normally distributed) are plugged in, the main thesis that it would result in an efficient portfolio does not seem to hold empirically.

The following numerical illustration taken from Bruder, Gaussel, Richard and Roncalli (2013) [63] is quite telling. Consider the universe of four assets with the average return vector and the covariance matrix as given below.

$$\begin{aligned} \bar{r}' &= (0.05, 0.06, 0.07, 0.08) \\ \hat{\Sigma} &= \begin{pmatrix} 0.01 & 0.0084 & 0.0098 & 0.0105 \\ & 0.0144 & 0.01176 & 0.0126 \\ & & 0.0156 & 0.0147 \\ & & & 0.0225 \end{pmatrix} \end{aligned} \tag{6.34}$$

If the risk tolerance parameter '$\lambda$' is set at 0.1, then the resulting optimal portfolio has the following weights: $w^* = (0.2349, 0.1957, 0.1878, 0.2844)$. If the standard deviation (volatility) of the second asset increases by 3%, the weight attached to this asset changes to $-0.1404$ and if the return of the first asset changes from 5% to 6%, the optimal weight goes to 0.6319 from 0.2349. The sensitivity of the efficient portfolio weights even to minor perturbations is quite obvious.

This may be due to a number of reasons. We have to keep in mind that the performance of the portfolio depends on the future means and covariances of the returns. As seen from the plot of the six portfolio returns, Figure 6.2, the distributions may not be normal distributions. These are portfolios formed as linear combinations of several individual assets and therefore individual assets can even more easily deviate from normality. Secondly, this could be the result of the so-called curse of dimensionality. When '$m$', the number of assets, is fairly large the number of unknown parameters, $m(m+1)/2$ is also large and to estimate them accurately, it requires to have a fairly large data set. An extreme example would be that tracking the 3000 stocks Russell index would require estimating 4.5 million covariances. Another aspect that needs to be modeled is the time dependence of the returns. Under efficient market hypothesis, returns are expected to be random; but as seen in previous chapters there is some time dependence, an anomaly exploited by trading algorithms. This needs to be taken into account in the portfolio construction as well. We address some of these issues in this section.

It has been shown that because of the large sampling error of the estimates, an equally weighted portfolio can perform closer to the efficient portfolio. Studies have incorporated the estimation risk via Bayesian approach and evaluated the impact of this risk on the final choice of the portfolios. The exact distribution of the maximum likelihood estimators of the means and variance is well-known, but the exact distribution of the returns and the volatility of the efficient portfolio is not easy to obtain. Jobson and Korkie (1980) [220] derive the asymptotic distribution and examine its properties via extensive simulations. The convergence of the sample estimates is shown to depend upon the magnitude of $T$ to $1/A^2$, where $A = 1'\Sigma^{-1}\mu$, the sum of weighted return averages. Generally if the covariances among the assets are relatively small but the mean returns are large, the estimators tend to perform better. In any case, good estimates of '$\mu$' and '$\Sigma$' are essential for the construction of the efficient portfolio. Michaud (1989) [269] suggests bootstrap methods to repeatedly resample with replacement from the observed sample of returns $(r_1, \dots, r_m)$ and use the resulting average and the variance of the bootstrap replications in the portfolio construction.

**Shrinkage and Bayesian Methods:** The shrinkage methods follow the same optimization criterion but impose a penalty on the size of the coefficients. For example, we can add the following additional constraints to (6.2):

$$\text{Ridge: } \sum_{i=1}^{m} w_i^2 \leq w, \qquad \text{Lasso: } \sum_{i=1}^{m} |w_i| \leq w. \tag{6.35}$$

Both procedures can be unified as, $\sum_{i=1}^{m} |w_i|^q \leq w$ for $1 \leq q \leq 2$. As it has been shown that these methods yield estimates that are Bayes estimates with different priors. We will now discuss a Bayesian approach.

The Bayesian approach combines the prior distribution of the unknown parameters and the data likelihood to obtain posterior distribution of the parameters. Usually the mode of this distribution is taken as a reasonable estimate. In an influential paper, Black and Litterman (1992) [44] describe the need for Bayesian approach. The straight solution to (6.2) almost always results in large short positions in many assets which would increase trading costs. If the weights are restricted to be positive the solution results in zero weights for many assets and large weights in the assets with small capitalizations. The results generally are not reasonable, mainly due to poor estimate of expected future returns. In practice, investors augment their views using their prior experience and their intuition to forecast the likely returns based on other asset related data. The Bayes theorem is a natural way combines these two aspects. The details that follow are from Lai, Xin and Chen (2011) [237] and we will comment on the practical implementation issues after presenting the mathematical results.

The model begins with the following distributions:

$$r_t \sim N(\mu, \Sigma), \quad \mu \big| \Sigma \sim N(\nu, \Sigma/\kappa) \text{ and } \Sigma \sim IW_m(\psi, n_0), \tag{6.36}$$

where $IW_m(\psi, n_0)$ is the inverted Wishart with '$n_0$' degrees of freedom with $E(\Sigma) = \psi/(n_0 - m - 1)$ (for details on inverted Wishart, see Anderson (1984) [19]). These priors are known as conjugate priors as their functional forms closely match the likelihood functional form. When the degree of freedom '$\kappa$' goes to zero, the prior is non-informative. The posterior estimates are:

$$\begin{aligned} \hat{\mu} &= \frac{\kappa}{T+\kappa} \cdot \nu + \frac{T}{T+\kappa} \bar{r} \\ \hat{\Sigma} &= \frac{\psi}{T + n_0 - m - 1} + \frac{T}{T + n_0 - m - 1} \cdot \left\{ S + \frac{\kappa}{T+\kappa} (\bar{r} - \nu)(\bar{r} - \nu)' \right\}, \end{aligned} \tag{6.37}$$

where $S = \frac{1}{T}\sum_{t=1}^{T}(r_t - \bar{r})(r_t - \bar{r})'$, the residual covariance matrix. If '$\kappa$' and '$n_0$' are taken to be large, it implies that the investor puts more emphasis on the priors. To use (6.37), the investor has to specify these values.

Black and Litterman (1992) [44] propose shrinking the prior estimate of $\mu$ to a vector '$\pi$' representing excess equilibrium returns, such as the one implied by CAPM which is termed the return on the global market. Thus

$$\pi = \frac{\mu_m - r_f}{\sigma_m^2} \cdot \Sigma \cdot w_m, \tag{6.38}$$

where $w_m$ are the weights on the global market and $\sigma_m^2$ is the variance of the return on the global market. To get the posterior distribution of $\mu \big| \pi$ via Bayes Theorem, as mentioned in Satchell and Snowcraft (2000) [296], the approach by Black and Litterman (1992) [44] was to place this in a tractable form that investors can operationalize. Assume that $k \times m$ matrix $P$ represents investor's beliefs, then $P\mu \sim N(q, \Omega)$, where '$\Omega$' is diagonal. The parameters '$q$' and '$\Omega$' are known as hyperparameters. Also it is assumed that $\pi \big| \mu \sim N(\mu, \tau\Sigma)$, where $\tau$ is a scaling factor. Thus the p.d.f. of '$\mu$' is given as $\mu \big| \pi \sim N(\hat{\mu}^{\text{BL}}, \hat{\Sigma}^{\text{BL}})$, where

$$\hat{\mu}^{\text{BL}} = (\hat{\Sigma}^{\text{BL}})^{-1}[(\tau\Sigma)^{-1}\pi + P'\Omega^{-1}q], \quad \hat{\Sigma}^{\text{BL}} = (\tau\Sigma)^{-1} + P'\Omega^{-1}P. \tag{6.39}$$

For various interesting practical applications of this result, see Satchell and Snowcraft (2000) [296]. Although the requirement for subjective judgments may appear to make it difficult to use the Black-Litterman model, in asset management practice, it is widely used. It is shown that there is improvement in portfolio performance of an informed trader who also learns from market prices over a dogmatic informed trader who solely relies on private information. A misspecified model can result in better performance even if private signals are not accurate. The Black and Litterman model provides a unified framework.

It has been noted in the literature that estimating the expected returns using the past returns fails to account for likely changes in the level of market risk. Merton (1980) [267] based on a detailed exploratory study concludes, that the investors should recognize that the expected market return in excess of risk free return should be positive and that there is heteroscedasticity in the realized returns. It must be noted that the price series is a random walk and therefore, with shifting means, the successive differences (essentially the returns) can exhibit different levels of variances. Predicting future returns under the random walk model for price is still an open issue.

Recent focus has been on the improved estimation of the covariance matrix, $\Sigma$. As the sample covariance matrix with $m(m+1)/2$ elements requires substantially large '$T$' for estimation, alternatives for a structural covariance matrix with fewer parameters are proposed. They are all in the same spirit of shrinkage estimators:

$$\hat{\Sigma} = \alpha S_0 + (1 - \alpha)S, \tag{6.40}$$

where '$\alpha$' is called a shrinkage constant and '$S_0$' matrix is restricted to have only a few 'free' elements. With the CAPM model that leads to covariance structure, (6.29), the resulting covariance matrix which is the sum of a diagonal matrix and a matrix of rank one, will have only '$2m$' independent elements that need to be estimated. Ledoit and Wolf (2003) [239] describe a procedure to arrive at the estimate of $\alpha$ in (6.40), which we briefly describe below.

Minimizing the following risk function,

$$R(\alpha) = \sum_{i=1}^{m} \sum_{j=1}^{m} E(\alpha s_{ij}^0 + (1-\alpha)s_{ij} - \sigma_{ij})^2, \tag{6.41}$$

we arrive at '$\alpha$' as

$$\alpha^* = \frac{\sum_i \sum_j \left[\text{Var}(s_{ij}) - \text{Cov}(s_{ij}^0, s_{ij})\right]}{\sum_i \sum_j \left[\text{Var}(s_{ij}^0 - s_{ij}) + (\phi_{ij} - \sigma_{ij})^2\right]}, \tag{6.42}$$

where $\phi_{ij} = E[\alpha s_{ij}^0 + (1-\alpha)s_{ij}]$. It is further shown that

$$\alpha^* \sim \frac{1}{T} \cdot \frac{\pi - \rho}{\gamma}, \tag{6.43}$$

where the asymptotic quantities, $\pi = \sum_i \sum_j \text{Asy Var}(\sqrt{T} s_{ij})$, $\rho = \sum_i \sum_j \text{Asy Cov}(\sqrt{T}\, s_{ij}^0, s_{ij})$ and $\gamma = \sum_i \sum_j (\phi_{ij} - \sigma_{ij})^2$. Note that with CAPM specification which can be taken as a single index model, the '$\gamma$' quantity represents the misspecification of the single index.

Fan, Fan and Lv (2008) [135] consider high dimensional covariance matrix estimation via factor model (6.14). The multifactor model generally captures well the cross-sectional variations and thus the number of parameters that need to be estimated can be significantly reduced. It is possible that the number of factors '$n$' can grow with the number of assets, '$m$', and the sample period, '$T$'. Comparing $\hat{\Sigma}_{rr} = \hat{B}\hat{\Omega}\hat{B}' + \hat{V}$ in (6.24) with the sample covariance matrix of the returns, '$S$', it is shown that the former performs better if the underlying factor model is true. In calculations that involve the inverse of the covariance matrix, such as the calculation of portfolio weights, factor model based estimates tend to perform better but it does not make a difference in the actual risk calculation ($\hat{w}'\hat{\Sigma}\hat{w}$), which involves the estimate of the covariance matrix. Recent studies have focused on providing tighter bounds on the risk; the calculations are somewhat complicated as they involve studying estimates of the weight vector as well as the estimates of the covariance matrix. See Fan, Han, Liu and Vickers (2016) [136] and the references therein for additional details.

### 6.2.1 Portfolio Allocation Using Regularization

It is noted that the poor performance of Markowitz's conceptual framework is due to the structure of the optimization problem. It is an ill-conditioned inverse problem as the solution requires the inverse of the covariance matrix that may involve highly correlated securities. A number of regularization procedures have been proposed in the literature to address the instability in the solution and we present a few of them here.

**The $L_2$ Constraint:** This is also known as ridge constraint that imposes a penalty on the size of the estimates as given in (6.35). Formally, the problem is defined as follows:

$$\min_{w} w'\Sigma w \ni w'\mu = \mu^*, w'\mu = 1 \text{ and } \sum_{i=1}^{m} w_i^2 \leq w_0. \tag{6.44}$$

When there are many correlated assets, the weights are poorly determined and can exhibit high variance. A large positive weight for an asset can be canceled by a large negative weight for the correlated asset. The size constraint added in (6.44) can address this issue. If the added constraint has the associated Lagrangian multiplier, $\gamma^*$, the solution in (6.6) is modified by replacing $\Sigma$ by $(\Sigma + \gamma^* I_m)$. The solution adds a positive constant to smaller eigenvalues of possibly a singular matrix, $\Sigma$. Observe that the eigenvalue decomposition of $\Sigma = V\Lambda V'$ essentially gives us the principal components of the asset universe considered for the formation of the portfolio. This optimization approach can also be extended to accommodate mimicking a target portfolio with weights $w^*$. The set-up would be the same except for the last size constraint which is replaced by

$$(w - w^*)'A(w - w^*) \leq w_0, \tag{6.45}$$

where $A$ is a general specified matrix. If '$\lambda$' and '$\gamma$' are the Lagrangian multipliers associated with the mean specification constraint in (6.45) and the constraint above, the solution is

$$\hat{w}(\lambda, \gamma) = (\hat{\Sigma} + \gamma A)^{-1}(\lambda\hat{\mu} + \gamma A w_0), \tag{6.46}$$

which results from shrinking both the mean vector and the covariance matrix.

**The $L_1$ Constraint:** The shrinkage method LASSO (least absolute shrinkage and selection operator) is like the ridge method (see (6.35) above) but there are some important differences. The problem can be stated as follows:

$$\min_{w} w'\Sigma w \ni w'\mu = \mu^*, w'\mu = 1 \text{ and } \sum_{i=1}^{m} |w_i| \leq w_0. \tag{6.47}$$

The added constraint in (6.47) leads to nonlinear solution and no explicit solution as stated in (6.46) in the case of ridge penalty is possible. The numerical solution is obtained via quadratic programming.

Both $L_1$ and $L_2$ constraint versions yield solutions that can be justified as Bayes with different priors. The LASSO method produces generally a sparse solution that is easy to interpret. Sparser solution in the portfolio context means lower transaction costs when the portfolios are to be rebalanced. A convenient way to solve these optimization problems is to formulate the original setup in (6.2) in the regression framework; this will help to borrow regularization ideas used in the general area of multivariate regression. Observe that $\Sigma = E(r_t r_t') - \mu\mu'$ and thus the minimization problem is equivalent to (in the sample version)

$$\hat{w} = \arg\min_{w} \frac{1}{T}\|\mu^* 1_T - R'w\|_2 \ni w'\hat{\mu} = \mu^*, w'1_m = 1. \tag{6.48}$$

Here, $R$ is the $m \times T$ matrix of the returns where rows represent assets and $\hat{\mu} = \frac{1}{T}\sum_{t=1}^{T} r_t$. The regularization constraint in (6.47) when added to (6.48) results in the following optimization problem:

$$\hat{w}^{(\tau)} = \arg\min_{w} [\|\mu^* 1_T - R'w\|_2 + \tau\|w\|_1] \ni w'\hat{\mu} = \mu^*, w'1_m = 1. \tag{6.49}$$

As stated in Brodie, Daubechies, De Mol, Giannone and Loris (2009) [61] adding the $L_1$-constraint results in several useful advantages. In addition to introducing sparsity that is helpful for managing large portfolios; the procedure provides a framework for regulating the amount of shorting which can be a proxy for transaction costs. More importantly it provides a solution that is more stable which was a main concern in the original optimization setup as small perturbations in the estimates of $\mu$ and $\Sigma$ can lead to very different solutions.

The algorithm that is most convenient to solve (6.49) is Least Angle Regression (LAR) developed in Efron, Hastie, Johnstone and Tibshirani (2004) [115]. It starts with a large value of '$\tau$' in (6.49) and as '$\tau$' decreases the optimal solution, $\hat{w}^{(\tau)}$, changes in a piecewise linear fashion and thus it is essential to find only the critical points where the slope changes. More relevant details can be found in Brodie et al. (2009) [61]. The LASSO procedure was evaluated using Fama-French 100 portfolios sorted by size and book-to-market as the universe. The efficient portfolio was constructed on a rolling basis using monthly data for five years and its performance is evaluated in the following year. It is concluded, see Brodie et al. (2009) [61, Figure 2], that optimal sparse portfolios that allow short portfolios tend to outperform both the optimal no-short-positions portfolio and the naïve evenly weighted portfolio.

### 6.2.2 Portfolio Strategies: Some General Findings

The following wisdom advocated by Rabbi Isaac bar Aha that "one should always divide his wealth into three parts: A third in land, a third in merchandise and a third ready to hand," has been tested out in the case of equity allocation as well in the form of equal division among assets. Although this diversification strategy appears to be naïve because it is based on neither any theory nor any data, it has done relatively well and has stood out as a folk wisdom. DeMiguel, Garlappi and Uppal (2009) [104] compare this naïve strategy with other theory-based strategies that were discussed in earlier sections in terms of its performance using various data sets. The summary of their findings is worth noting: "...of the strategies from the optimizing models, there is no single strategy that dominates the $1/N$ strategy in terms of Sharpe ratio. In general, $1/N$ strategy has Sharpe ratios that are higher ... relative to the constrained policies which in turn have Sharpe ratios that are higher than those for the unconstrained policies."

The naïve portfolio has also very low turnover. In addition to keeping naïve portfolio as a benchmark, a practical implication is that the estimation of the moments of asset returns, such as the mean and the variance, needs improvement, using other asset characteristics.

**Combination of Strategies:** Tu and Zhou (2011) [317] combine rules based on Markowitz Theory with the naïve strategy to obtain new portfolio rules that can uniformly perform better. The combination can be interpreted as a shrinkage estimator, shrinking toward the naïve estimator. The linear combination

$$\hat{w}_c = \delta\hat{w} + (1-\delta)w_N, \tag{6.50}$$

where $\hat{w}$ is the weights that come from variations of efficient portfolios and $w_N$ has all its elements equal. With the criterion of minimizing the loss function,

$$L(w^*, \hat{w}_c) = U(w^*) - E[U(\hat{w}_c)], \tag{6.51}$$

where $U(\cdot)$ is the expected utility, the combination coefficient $\delta$, is obtained. This strategy was tested out using the same data as in DeMiguel et al. (2009) [104] and generally the combination rule performs well. This idea is in the same realm as diversification of different strategies.

**Familiarity:** It has been advocated by many economists and practitioners such as Warren Buffet that investors should focus on a few stocks that they are familiar with rather than a wide scale diversification. Boyle, Garlappi, Uppal and Wang (2012) [55] develop a theoretical framework to answer some relevant questions related to familiarity and diversification. Writing the return as sum of two components, systematic and idiosyncratic,

$$r_i = r_s + u_i \qquad i = 1, 2, \ldots, m \tag{6.52}$$

the utility function in (6.4) is modified to take into account the investors familiarity with some assets. This is done via the assumption that the investor is ambiguous about the mean, $\mu$, but has perfect knowledge about $\sigma$. This information alters the criterion in (6.5) as,

$$\max_{w} \min_{\mu} \left( w'\mu - \frac{\lambda}{2} \cdot w'\Sigma w \right) \ni \frac{(\mu - \hat{\mu})^2}{\sigma_{\hat{\mu}}^2} \leq \alpha_i^2, \quad i = 1, 2, \ldots, m, \tag{6.53}$$

where $\sigma_{\hat{\mu}}^2 = (\sigma_s^2 + \sigma_u^2)/T$. Two key quantities that play important roles are the level of ambiguity '$\alpha_i$' and inverse of the coefficient of variation, $\hat{\mu}/\sigma_{\hat{\mu}}$. Some conclusions from this study are, when ambiguity is present, the investor tends to hold disproportionate (to Markowitz allocation) amount in the familiar assets and the number of assets held tends to be smaller.

Huberman (2001) [208] provides several instances of investment options that investors consider based on their familiarity, rather than choices that lead to efficient outcome. Workers in firms generally tend to choose their employer's stock as they probably think that they know about the company and may believe they have information about its performance that the market does not have. Thus various examples are provided that indicate that, although investment in the familiar stocks contradicts the advice to diversify, familiar stocks are commonly held. This adds a dimension to the traditional risk-return formulation that one has to recognize.

**Large Portfolio Selection with Exposure Constraints:** Several techniques have been discussed in this section to reduce the sensitivity of the optimal portfolio weights to uncertainty in the inputs, such as the mean returns and the covariance matrix of returns. Generally these techniques are not adequate to address the effect of estimation errors that accumulate due to large portfolio size. Jagannathan and Ma (2003) [212] impose no short-sales constraints as well as upper bounds on portfolio weights as commonly imposed by practitioners. The empirical evidence indicates that simply imposing upper bounds does not lead to significant improvement in the out-of-sample performance. For an estimated covariance matrix, $S$, the formulation of the model is as follows:

$$\min_{w} w'Sw \ni w'1 = 1, \quad 0 \leq w_i \leq \tilde{w} \quad \text{for } i = 1, 2, \ldots, m. \tag{6.54}$$

Note the variation in its formulation; no expectation on the average return of the portfolio as in the Markowitz setup is specified. Fan, Zhang and Yu (2012) [137] argue that the resulting optimal portfolio, although it performs well, is not diversified enough.

Defining the total proportions of long and short positions by $w^+ = (\|w\|_1 + 1)/2$ and $w^- = (\|w\|_1 - 1)/2$, we have $w^+ + w^- = \|w\|_1$ and $w^+ - w^- = 1$. The optimization problem is stated as,

$$\max_{w} E[U(w'r)] \ni w'1 = 1, \|w\|_1 \leq C \text{ and } Aw = a. \tag{6.55}$$

Here $U(\cdot)$ is a utility function and the extra-constraints $Aw = a$ can capture allocations to various sectors. But the key added constraint, $\|w\|_1 \leq C$ can accommodate from no short sales ($C = 1$) to unrestricted sales ($C = \infty$) constraints. The solution to (6.55) can be obtained via quadratic programming. Fan et al. (2012) [137] advocate data-driven choice of '$C$' via minimizing the risk for a sample learning period. Using select 600 stocks from the pool of stocks that constitute the Russell 3000, the constructed portfolios tend to perform well for $C \sim 2$.

## 6.3 Dynamic Portfolio Selection

The static Markowitz framework, as discussed in the last section, is not realistic for practical application. Dynamic approaches that have been discussed in the literature will be presented here briefly. The single period optimization problem can be formulated as,

$$\max_{w_t} E_t\left[w_t' r_{t+1} - \frac{\gamma}{2}(w_t' r_{t+1})^2\right] \tag{6.56}$$

which results in the optimal solution,

$$w_t = w = \frac{1}{\gamma} \cdot E\left[r_{t+1} r_{t+1}'\right]^{-1} E[r_{t+1}] \tag{6.57}$$

that can be estimated using the sample counterpart. Here $r_{t+1}$ is taken to be the excess returns $R_{t+1} - R_t^f$. It is assumed here that the returns are independent and identically distributed. Brandt and Santa-Clara (2006) [56] extend the above setup to multiperiods. Simply stated the two-period mean-variance objective function is,

$$\max E_t \left[ r^p_{t\to t+2} - \frac{\gamma}{2}(r^p_{t\to t+2})^2 \right], \tag{6.58}$$

where the excess return for the two period investment strategy is,

$$r^p_{t\to t+2} = (R_t^f + w_t' r_{t+1})(R_{t+1}^f + w_{t+1}' r_{t+2}) - R_t^f R_{t+1}^f. \tag{6.59}$$

It can be seen that the above expression can be written as

$$r^p_{t\to t+2} = w_t'(r_{t+1} R_{t+1}^f) + w_{t+1}'(r_{t+2} R_t^f) + (w_t' r_{t+1})(w_{t+1}' r_{t+2}) \tag{6.60}$$

with the first two terms denoting excess returns from investing in the risk free asset in the first period and in the risky asset in the second period and vice versa. The third term that captures the effect of compounding is of smaller magnitude. Ignoring that term results in the optimal portfolio weights,

$$w = \frac{1}{\gamma} E \left[ r^p_{t\to t+2} r'^p_{t\to t+2} \right] E[r^p_{t\to t+2}]. \tag{6.61}$$

The solution reflects the choice between holding the risky assets in the first period only and holding them in the second period only and thus is termed "timing portfolios." The above result can be extended to any number of periods.

**Conditioning Information in Portfolio Construction:** It has been debated that portfolio managers may use conditioning information they have about assets in forming their portfolios, not simply using the mean and variance of the past returns. Thus the managers' conditionally efficient portfolios may not appear efficient to the investor with no knowledge of that information. Ferson and Siegel (2001) [141] study the properties of unconditional minimum-variance portfolios when the conditioning information is present and suggest that, "For extreme signals about the returns of risky assets, the efficient solution requires a conservative response." Managers can reduce risk by taking a small position in the risky asset while keeping the same level of portfolio performance. In its basic formulation, assuming that the investment alternatives are a risky asset and risk-free return, with $w(\tilde{S})$—the proportion invested in the risky-asset, the portfolio return and the variance are:

$$\mu_p = E[r_f + w(\tilde{S})(r_t - r_f)], \quad \sigma_p^2 = E[(w(\tilde{S})(r_t - r_f))^2] - (\mu_p - r_f)^2. \tag{6.62}$$

Here $\tilde{S}$ is the observed signal and $w(\tilde{S})$ is a function of that signal that the manager considers at the time of allocation. The minimization criterion leads to the following weight for the risky asset:

$$w(\tilde{S}) = \frac{\mu_p - r_f}{C} \left( \frac{\mu(\tilde{S}) - r_f}{(\mu(\tilde{S} - r_f)^2 + \sigma_\epsilon^2(\tilde{S})} \right), \tag{6.63}$$

where $C = E\left[\frac{(\mu(\tilde{S})-r_f)^2}{(\mu(\tilde{S})-r_f)^2+\sigma_\epsilon^2(\tilde{S})}\right]$ and the minimized variance is $\sigma_p^2 = (\mu_p - r_f)^2 \left(\frac{1}{C} - 1\right)$. Here $\sigma_\epsilon^2(\tilde{S})$ is the conditional variance given the signal. Note when this is zero, which signifies that there is no conditional information, $w(\tilde{S}) = w = (\mu_p - r_f)/(\mu - r_f)$, that is the result of Markowitz theorem. The above results can be easily extended to multiple risky asset scenarios.

To implement this idea, we assume that the weight vector is a linear function of '$k$' state variables:

$$w_t = \theta z_t \tag{6.64}$$

and the maximization criterion in (6.56) can be modified as,

$$\max_{\theta} E_t \left[(\theta z_t)' r_{t+1} - \frac{\gamma}{2}(\theta z_t)' r_{t+1} r_{t+1}' (\theta z_t)\right]. \tag{6.65}$$

Denoting $\tilde{r}_{t+1} = z_t \otimes r_{t+1}$ and $\tilde{w} = \text{vec}(\theta)$, we can obtain the optimal solution as,

$$\tilde{w} = \frac{1}{\gamma} E[\tilde{r}_{t+1}\tilde{r}_{t+1}']^{-1} E(\tilde{r}_{t+1}) = \frac{1}{\gamma} E[(z_t z_t') \otimes (r_{t+1} r_{t+1}')]^{-1} E[z_t \otimes r_{t-1}]. \tag{6.66}$$

Here $\otimes$ is meant for a Kronecker product. Brandt and Santa-Clara (2006) [56] also show how both single period and multi-period problems can be formulated as the constrained VAR.

Using the established evidence from the literature that market returns can be predicted by conditioning on dividend-price ratio, short-term interest rate, term spread and credit spread, to study the market returns for both stocks and bonds, the authors evaluate the conditional and unconditional policies at monthly, quarterly and annual holding periods, with the value of $\gamma = 5$. Some general conclusions are:

- Conditional policies are quite sensitive to state variables. Because of the predictability of returns, conditional policies allow for the investor to be more aggressive on average. Sharpe ratios are generally higher for conditional policies.

- Results are less pronounced for the longer holding periods. Conditional policies can be very different at different time frequencies.

The above stated conclusions hold for multi-period portfolio policies as well with monthly rebalancing.

## 6.4 Portfolio Tracking and Rebalancing

The proportions of capital allocated to various assets can change over time due to asset performances and due to change in investors' risk aversion. Thus, investors often engage in portfolio rebalancing to keep the portfolio on target risk exposure and exploit market opportunities to enhance returns in the long run. First, we formulate the problem with constraints that account for these and then present the results of some empirical studies.

The portfolio returns ($r_t$) are generally compared with some index returns ($I_t$); the excess returns (ER) and the tracking error (TE) are defined as follows:

$$\text{ER} = \frac{1}{T}\sum_{t=1}^{T}(r_t - I_t), \quad \text{TE} = \frac{1}{T}\sum_{t=1}^{T}(r_t - I_t)^2. \tag{6.67}$$

The optimization function is to minimize $\lambda\sqrt{\text{TE}} - (1-\lambda)\text{ER}$, via the choice of the portfolio weights. Another optimization setup is to balance between tracking error and transaction costs. Let $x = w - w_0$, where '$w_0$' is the current portfolio weights and '$w$' is the target weights; the problem with constraints can be stated as:

$$\min_{w}(w - w_0)\,\Omega(w - w_0) \ \ni w'1 = 1 \tag{6.68}$$

subject to the following constraints:

$$\begin{aligned} &\text{long only:} \quad w \geq 0 \\ &\text{turn-over:} \quad \sum_{i=1}^{m} |x_i| \leq u \\ &\text{maximum exposure to a factor:} \quad \sum_{i=1}^{m} \beta_{ik} w_i \leq u_k \\ &\text{holding:} \quad l_i \leq w_i \leq u_i \end{aligned} \tag{6.69}$$

Note the criterion in (6.68) can also be taken as a version of tracking error.

Some practical implementation ideas are discussed in Pachmanova and Fabozzi (2014) [279]. In addition to the naïve allocation strategy discussed earlier in this chapter, other ad-hoc methods simply imitate the composition of some market indices such as S&P 500 that weighs stocks by their market capitalizations or Dow-Jones that weighs stocks by their prices. To imitate, select the stocks with the largest weights in the benchmark. Other methods include stratification of benchmarks into groups and select stocks from each group that ensures diversification. The group can be industry-based or if it is based on past performance, the selected assets are called basis assets. The difference between the original allocation model and the rebalancing models can be due to change in the objectives and also due to change in transaction costs and taxes. If a manager wants to move toward higher alpha, the restriction on the tracking error should be added to (6.69). Like other calculations related to portfolio management, the rebalancing also involves backward-looking estimates. To reflect the future more accurately, managers should arrive at forward-looking estimates of tracking error using conditioning arguments given in the earlier section.

**Transaction Costs:** Bid-ask spreads, brokerage fees, taxes and the price impact of trading are covered in Chapter 9. Here we present some observations related to trading costs as noted in the literature. Defining the turnover as,

$$\mathcal{T} = \sum_{i=1}^{m} |w_i - w_{0i}| = \sum_{i=1}^{m} \tau_i \tag{6.70}$$

and the costs of rebalancing as

$$C = \sum_{i=1}^{m} k_i \tau_i. \tag{6.71}$$

Note $w_{0i} = \frac{w_i r_i}{\sum w_i r_i}$ and $\tau_i = \left|\frac{w_i(r_i - r_{0i})}{\sum w_i r_i}\right|$ and thus a key term here is $r_i - r_{0i}$, the difference between the actual and target return for asset '$i$'. Kourtis (2014) [232] studies the distribution of $\tau$ and $C$ under the naïve allocation strategy. Denoting $r_p = \sum w_i r_i$ the portfolio return, the distribution of $\tau_i$ depends on $(r_i - r_{0i})/r_p$, the ratio of excess return to the portfolio return. A general criterion that accounts for transaction costs is

$$\min_{w} \delta(w - w_0)' \Omega (w - w_0) + \sum_{i=1}^{m} k_i E[\tau_i(r_{i,} w)] \tag{6.72}$$

with the controlling parameter, $\delta$, on the tracking error. By explicitly incorporating the trading cost in the criterion, it is shown that the total cost of the transaction is reduced.

Moorman (2014) [271] considers three portfolio strategies: A levered-momentum strategy, a zero-cost-decile momentum strategy and equally-weighted naïve strategy. Because the momentum strategies, as noted earlier, adapt to changing market conditions swiftly, they require frequent rebalancing and can reveal the drawbacks of any method to reduce transaction costs. First there is a no-trade region where the distance between target and the current weight is not significant enough to justify rebalancing. Write the rebalanced weight as,

$$w_t^* = \alpha_t w_{0t} + (1 - \alpha_t)\, w_t, \tag{6.73}$$

where '$\alpha_t$' is the rate of rebalancing; if $\alpha_t = 1$, then there is no rebalancing. It is easily seen that the current month's unbalanced weights are related to prior month's rebalanced weights as,

$$w_{0t} = w_{t-1}^* \left( \frac{1 - r_t}{1 + r_{pt}} \right), \tag{6.74}$$

where $r_{pt}$ is the return on the portfolio before-transaction cost. Note

$$r_{pt} = \sum_{i=1}^{N_t} w_{it}^* r_{it} - c_{it} |w_{it}^* - w_{0t}|, \tag{6.75}$$

where $c_{it}$ is the cost of transaction. The '$\alpha$' is chosen using the following objective function that results from constant relative risk aversion utility:

$$\max_{\alpha} \frac{1}{T} \sum_{t=0}^{T-1} \frac{(1 + r_{pt+1})^{1-\gamma}}{1 - \gamma}, \tag{6.76}$$

where '$\gamma$' is the coefficient of relative risk-aversion. The parameter '$\alpha$' is chosen via grid search, outside the no-trade zone. An empirical investigation of various distance and rebalancing methods reveals that transaction cost reduction is most successful for the lower momentum and equally-weighted market portfolios but more difficult for the momentum portfolios because of higher turnover.

## 6.5 Transaction Costs, Shorting and Liquidity Constraints

It is clear, based on the discussion in this chapter that the managers tend to mimic better performing portfolios or use information that they have about future values of the assets in the construction and rebalancing of their portfolios. But at the end of the investment horizon, the performance evaluation of the selected portfolio depends on the average returns after adjusting for the risk premia. Consider the $k$-factor models, (6.14), $r_{it} = \alpha_i + \beta_i' f_t + \epsilon_{it}$ and the equation that relates the average return to risk premia ($\lambda_k$), $\mu_i = r_f + \beta_i' \lambda_k$ in (6.15). The Equation (6.14) can also be written more compactly as,

$$r_t = \alpha + B f_t + \epsilon_t, \tag{6.77}$$

where $\alpha$ is a $m \times 1$ vector and $B$ is the $m \times k$ matrix of factor premia. Thus the portfolio return,

$$r_{pt} = w' r_t = w' \alpha + w' B f_t + w' \epsilon_t = \alpha_p + \beta_p' f_t + \epsilon_{pt}. \tag{6.78}$$

Because '$w$' is random, note

$$E(r_{pt}) = E(\alpha_p) + E(\beta_p' f_t) = \alpha' E(w) + \text{Cov}(\beta_p, f_t) + E(\beta_p') E(f_t). \tag{6.79}$$

Note the first term is the result of asset choice, the second term is the result of the factor model theory and the last term reflects the risk premia.

**Choice of Assets:** The screening of candidate assets relies upon assessing the underlying factors for return behavior of a portfolio. Factors are generally grouped into fundamental factors and macroeconomic factors, although there can be additional considerations such as if investing in certain assets is considered to be socially responsible, or environmentally safe, etc. Fundamental factors are related to asset characteristics and other information such as analysts' forecasts and the macroeconomic factors relate to those activities in the economy that are likely to be closely associated with the asset behavior. These may be industry factors or sentiment factors. These factors can be standardized and an aggregate score can be used to sort out the assets. Several sorting methodologies are available in the literature. Using the factor scores, sort the stocks and go long on stocks with high scores and go short on stocks with low values. This method is easy to implement and it diversifies away risk of individual stocks. Some disadvantages are that it may be difficult to infer at times, which variables provide unique information and portfolios may pick up different undesirable risks. Some of the commercial risk models based on the choice of these variables are discussed in the last section.

Patton and Timmermann (2010) [284] test for a positive relation between ex-ante estimates of CAPM beta and the resulting returns using daily data. Stocks are sorted at the start of each month into deciles on the basis of their betas using the prior one year daily data and value-weighted portfolios are formed. Returns for these portfolios are recorded for the following month. If the CAPM holds, one should expect the average returns should increase with the betas. This has been empirically confirmed. The authors also examine if the post-ranked betas of portfolios ranked by ex-ante beta estimates are monotonic, confirming the predictability of the betas. Sorting the stocks based on their betas and forming decile portfolios, the performance is evaluated in the subsequent month. Specifically from the CAPM model, $r_{it} = \alpha + \beta_i r_{mt} + \epsilon_{it}$ for $i = 1, 2, \ldots, 10$, the following hypothesis, $H_0 : \beta_1 \geq \beta_2 \geq \cdots \geq \beta_{10}$ is tested. The estimates range from 1.54 to 0.6 and the hypothesis is confirmed. This approach is extended to two-way sorts; firm size and book-to-market ratio or firm size and momentum. It is concluded that the size effect in average returns is absent among loser stocks. Also momentum effects are stronger for small and medium size firms.

**Alpha and Beta Strategies:** Portfolio managers have traditionally developed strategies to maximize '$\alpha$' in the factor model. One of the performance measures, information ratio defined in Section 5.2 is taken as an index for the portfolio's performance. But after the 2007–2008 financial crisis, there is a great deal of interest in smart-beta (referring to the coefficients of the factor model) strategies. This is a broad term that refers to anything from ad-hoc strategies to a mix of optimized simpler portfolios and investments in non-traditional asset classes. The construction is generally rule-based and is transparent. The commonality among them is in their exposure to factors that investors are comfortable to choose from. Kahn and Lemmon (2015) [225] provide a framework to understand the beta strategies along the decomposition of return in (6.79) into three returns. To summarize:

- "The smart-beta return arises from static (i.e., long-term average) exposures to smart-beta factors.
- The pure alpha consists of three pieces:
  (a) The average security selection return (beyond smart-beta factors).
  (b) The average macro, industry and country returns (beyond smart-beta factors)
  (c) The return due to smart-beta timing."

The authors advocate an optimal blend of active smart-beta and index products. The optimality depends on the understanding of expected returns and risks. Other considerations include how the impact of the underlying factors will change and how the volatility and correlations among the factors are likely to change.

Frazzini and Pedersen (2014) [144] present a model with leverage and margin constraints that can vary over time and can be different for different investors and formally study the implications which are verified by the empirical analysis as well.

The CAPM model assumes that an investor is interested in maximizing the expected return for a given leverage on the investor's risk tolerance. But some investors may be constrained in their leverage (such as insurance companies that are required to maintain a certain amount of reserves, etc.) and may over-invest in risky securities. The growth of ETFs indicates that some investors cannot use leverage directly. The inclination to invest in high-beta assets requires lower risk-adjusted returns than low-beta assets that require leverage. The authors set out to study how an investor with no leverage constraint can exploit this and other related issues.

Consider the two period model where agents trade '$m$' securities that each pay '$\delta_t$' as dividend and '$W$' shares are outstanding. An investor chooses a portfolio of shares, $w = (w_1, \dots, w_m)'$ and the rest is invested at the risk-free rate, $r_f$. Denoting '$P_t$' as the $m \times 1$ price vector, the following utility function criterion is optimized.

$$\max w' \left( E_t(P_{t+1} + \delta_{t+1}) - (1 + r_f)P_t \right) - \frac{\gamma}{2} w' \, \Omega \, w \tag{6.80}$$

subject to the portfolio constraint

$$m_t \cdot w' P_t \leq W. \tag{6.81}$$

The quantity '$m_t$' is a leverage constant and if $m_t = 1$, it implies that the investor simply cannot use any leverage. If $m_t > 1$ it would imply the investors have their own cash reserves. Solving above with $\psi_t$ as the Lagrangian multiplier for the constraint in (6.81), we obtain:

$$\begin{aligned} &\text{Equilibrium Price: } P_t = \frac{E_t(P_{t+1} + \delta_{t+1}) - \gamma \Omega \, w^*}{1 + r^f + \psi_t} \\ &\text{Optimal Position: } w^* = \frac{1}{\gamma} \Omega^{-1} \left( E_t(P_{t+1} + \delta_{t+1}) - (1 + r^f + \psi_t) \, P_t \right) \\ &\text{Aggregate Lagrangian: } \psi_t = \sum_i \left( \frac{\gamma}{\gamma^i} \right) \psi_t^i \end{aligned} \tag{6.82}$$

Some key implications are:

(i) An asset's alpha in the market model is $\alpha_t = \psi_t(1 - \beta_t)$; thus alpha decreases as beta increases.

(ii) Sharpe ratio increases with beta until beta is unity and then decreases.

(iii) If portfolios are sorted out in terms of betas, expected excess return is $E_t(r_{t+1}) = \frac{\beta_t^H - \beta_t^L}{\beta_t^L \beta_t^H} \cdot \psi_t$, where the superscripts denote high and low beta portfolios.

These findings are useful for studying how investment decisions can be made. Investors who are constrained in their access to cash tilt toward riskier securities with higher betas. Empirically it has been shown that portfolios with higher betas have lower alphas and lower Sharpe ratios than portfolios of low-beta assets. This is currently an active area of research.

## 6.6 Portfolio Trading Strategies

The discussion in this chapter mostly centered around optimal ways to construct portfolios. The trading strategies generally center around rebalancing the portfolios if the tracking error exceeds certain thresholds. Strategies related to execution depend on the price impact of rebalancing transactions and will be commented on in Chapter 9. The transaction costs consist of brokerage commissions, market impact and the spread; the spreads and market impact vary between assets and are generally higher for less liquid assets with lower market capitalizations. With portfolio trading, the optimal strategy will depend on the current returns and the future expected returns of all assets in the portfolio as well as on the risk and return correlations along with their transaction costs. As mentioned in Gârleanu and Pedersen (2013) [157]: "Due to transaction costs, it is obviously not optimal to trade all the way to the target all the time." They argue that the best portfolio is the combination of efficient portfolios: Current portfolio to avoid turnover and associated transaction costs and the future optimal portfolio. The best portfolio is thus forward-looking. The critical factor appears to be alpha decay (mean-reversion) of returns. Assets with slower decay will get more weight. Results are empirically tested out using commodity futures using the data on returns for various durations. The suggested approach requires continuous trading and rebalancing as opposed to periodic rebalancing followed in practice.

The basic model along the lines of (3.37) can be stated as follows:

$$r_{t+1} = cf_t + a_{t+1}, \tag{6.83}$$

where $r_t$ is a $m$-dimensional excess return vector and $f_t$ is a $n \times 1$ vector of known factors and $a_t$'s are a white-noise sequence. Further, it is assumed that

$$\Delta f_{t+1} = -\Phi f_t + \epsilon_{t+1}, \tag{6.84}$$

where $\Delta f_{t+1} = f_{t+1} - f_t$ and $\Phi$ is a $n \times n$ matrix of mean-reversion coefficients of the factors. The factors could be macroeconomic or asset or industry related factors that have some predictive ability. Assuming '$x_t$' is the number of shares vector, then the transaction cost associated with trading $\Delta x_t = x_t - x_{t-1}$ can be stated as:

$$\text{TC} = \frac{1}{2} \cdot \Delta x_t' \Lambda \Delta x_t, \tag{6.85}$$

where $\Lambda$ is a positive definite matrix reflecting the level of trading costs and can be taken to be proportional to risk. Assuming '$\rho$' $\in (0, 1)$ is a discount rate, $\gamma$ is the risk-aversion coefficient, the objective function can be stated as:

$$\max_{x_0, x_1, \ldots} E_0 \left[ \sum_t \left( (1-\rho)^{t+1} \left( x_t' r_{t+1} - \frac{r}{2} x_t' \Sigma x_t \right) - \frac{(1-\rho)^t}{2} (\Delta x_t' \Lambda \Delta x_t) \right) \right]. \tag{6.86}$$

The above is solved via dynamic programming. Various solutions are offered. We summarize them below.

- Optimal trading rate is higher if the transaction costs are smaller and if the risk aversion is larger.
- Target portfolio downweights the faster decaying factors and more weight is given to persistent factors.
- With a persistent market impact, the cost of more frequent trading is smaller.

There are some practical considerations that must be taken into account while implementing the ideas proposed here. It is important to develop reasonable predictive models in (6.83) and (6.84). The factors '$f_t$' are usually taken to be some functions of past returns. Generally, the multivariate version of (6.83) should yield better predictions than the univariate models as observed from the commonality literature.

**Commercial Risk Models:** Many of the trading outfits have some version of factor models, where exposure to factors is tailored toward the preferences of individual investors. We briefly discuss Axioma's Robust Risk Model and the Barra Equity Model for Long-Term Investors here. As discussed earlier, factors used in multi-factor models can be broadly grouped as:

- Industry and country factors
- Style factors related to the equity itself
- Macroeconomic factors that are closely related to asset's universe
- Statistical factors that are observed but cannot be easily interpreted

For international investments, currency factors also can be included as many domestic companies trade in foreign markets as well. Some features of the modeling procedure suggested in Axioma are as follows:

- Use robust models to fit (6.21) so that the influence of outliers in the model estimation is minimal.
- Develop common factor models, common to certain stocks as there may not be any factors unique to an asset.
- In the factor model, it is possible that the dimension of factor vector, '$f$', if observed factors are considered, can be larger than the dimension of the return vector, '$r$', and so it is suggested to use asymptotic PCA which is based on the singular value decomposition of $R'R$ rather than $RR'$, where $R$ is a $m \times T$ returns matrix.
- Hybrid solutions where fundamental and statistical factors are combined optimally are suggested.
- Industry that has thinly distributed (measured by Herfindahl's Index) firms is usually aggregated with similar industry to get reasonably reliable estimates.

These notes are taken from the Axioma Robust Model Handbook (dated June 2010). For the description of the Barra Model, refer to Bayraktar, Mashtaser, Meng and Radchenko (2015) [35]. The model provides factors by investment horizon: short, mid and long horizons. Some distinguishing characteristics of this model are:

- Implied volatility adjustment to improve risk forecasts.
- Volatility regime adjustment methodology to calculate market volatility levels.
- Sixty industry factors based on the Global Industry Classification Standard.
- Statistical factors include momentum and reversal indicators.

Some of the specific measures that are worth noting are:

**Historical Beta:** $\beta = (1 - w)\beta_s + w\beta_{\text{ind}}$, $w = \dfrac{\sigma(\beta_s)}{\sigma(\beta_s) + \tau\sigma(\beta_{\text{ind}})}$, where '$\beta_s$' is obtained from regressing excess returns on the cap-weighted excess returns, '$\beta_{\text{ind}}$' is from regressing excess cap-weighted industry returns on the cap-weighted excess returns and $\tau$ is the tuning parameter.

**Downside Risk:** Consider the returns on a lower quartile, then Lower Partial Moment = $\dfrac{1}{|I|} \sum_{i \in I} (r_i - r_v)^2$. For other variations of this measure, refer to the citation.

**Liquidity:** LIQMA $= \dfrac{1}{T} \sum_{t=1}^{T} \dfrac{|r_t|}{v_t/\text{ME}_t}$, where $v_t$ is the traded dollar volume and $\text{ME}_t$ is the market value.

## 6.7 Exercises

1. Using the historical log returns of the following tickers: CME, GS, ICE, LM, MS, for the year 2014 (in file `investment_data.csv`), estimate their means $\mu_i$ and covariance matrix $\Sigma$; let $R$ be the median of the $\mu_i$'s.

(a) Test if the returns series are random via the autocorrelation function.

(b) Solve Markowitz's minimum variance objective to construct a portfolio with the above stocks, that has expected return at least $R$. The weights $\omega_i$ should sum to 1. Assume short selling is possible.

(c) Generate a random value uniformly in the interval $[0.95\mu_i, 1.05\mu_i]$, for each stock $i$. Resolve Markowitz's objective with these mean returns, instead of $\mu_i$ as in (b). Compare the results in (b) and (c).

(d) Repeat three more times and average the five portfolios found in (a), (b) and (c). Compare this portfolio with the one found in (a).

(e) Repeat (a), (b) and (c) under no short selling and compare the results obtained with short selling.

(f) Use the LASSO method to construct the portfolio and compare it with the result in (e).

2. Suppose that it is impractical to construct an efficient portfolio using all assets. One alternative is to find a portfolio, made up of a given set of $n$ stocks, that tracks the efficient portfolio closely—in the sense of minimizing the variance of the difference in returns.

Specifically, suppose that the efficient portfolio has (random) rate of return $r_M$. Suppose that there are $N$ assets with (random) rates of return $r_1, r_2, \ldots, r_n$. We wish to find the portfolio whose rate of return is $r = \alpha_1 r_1 + \alpha_2 r_2 + \cdots + \alpha_n r_n$ (with $\sum_{i=1}^{n} \alpha_i = 1$) by minimizing $\mathrm{Var}(r - r_M)$.

(a) Find a set of equations for the $\alpha_i$'s.

(b) Another approach is to minimize the variance of the tracking error subject to achieving a given mean return. Find the equation for the $\alpha_i$'s that are tracking efficient.

(c) Instead of minimizing $\mathrm{Var}(r - r_M)$, obtain $\alpha$'s that result from minimizing mean-squares of the difference $(r - r_M)$. Note the mean squares is the sum of variance and bias squared.

3. The file `m_logret_10stocks.csv` contains the monthly returns of ten stocks from January 1994 to December 2006. The ten stocks include Apple, Adobe Systems, Automatic Data Processing, Advanced Micro Devices, Dell, Gateway, Hewlett-Packard Company, International Business Machines Corp., and Oracle Corp. Consider portfolios that consist of these ten stocks.

(a) Compute the sample mean $\hat{\mu}$ and the sample covariance matrix $\hat{\Sigma}$ of the log returns.

(b) Assume that the monthly target return is 0.3% and that short selling is allowed. Estimate the optimal portfolio weights by replacing $(\mu, \boldsymbol{\Sigma})$ in Markowitz's Theory by $(\hat{\mu}, \hat{\boldsymbol{\Sigma}})$.

(c) Do the same as in (b) for Michaud's resampled weights described in the text using $B = 500$ bootstrap samples.

(d) Plot the estimated frontier (by varying $\mu_*$ over a grid that uses $(\hat{\mu}, \hat{\boldsymbol{\Sigma}})$ to replace $(\mu, \boldsymbol{\Sigma})$ in Markowitz's efficient frontier.

(e) Plot Michaud's resampled efficient frontier using $B = 500$ bootstrap samples. Compare the plot in (d).

4. The file `portfolio_data.csv` contains historical monthly returns for one set of 6 portfolios and another set of 25 portfolios, formed based on Size and Book-to-Market ratio (BM). The data is obtained from French's data library. The portfolios are formed as the intersection of size (or market equity, ME) based portfolios and book equity to market equity ratio (BE/ME) based portfolios (2×3 forming the first set of 6 portfolios and $5 \times 5$ forming the second set of 25 portfolios). These portfolios are discussed in the 1993 paper by Fama and French.

In this exercise we will only work with the first set of 6 portfolios, which are contained in the columns named beginning with "PF6," with the rest of the column names following French's naming convention about the size and BM of the corresponding portfolios – SML contains small size + low BM, SM2 contains small size + medium BM, SMH contains small size + high BM, BIGL contains big size + low BM, etc.

Finally, the last 4 columns of the data set contain the Fama-French factors themselves along with the risk-free rate: MktMinusRF contains the excess return of the market over the risk-free rate, SMB contains the small-minus-big size factor, HML contains the high-minus-low BM factor and RF contains the risk-free rate.

(a) Using the entire sample, regress the excess returns (over the risk-free rate) of each of the 6 portfolios on the excess market return, and perform tests with a size of 5% that the intercept is 0. Report the point estimates, $t$-statistics, and whether or not you reject the CAPM. Perform regression diagnostics to check your specification.

(b) For each of the 6 portfolios, perform the same test over each of the two equi-partitioned subsamples and report the point estimates, $t$-statistics, and whether or not you reject the CAPM in each subperiod. Also include the same diagnostics as above.

(c) Repeat (a) and (b) by regressing the excess portfolio returns on all three Fama-French factors (excess market return, SMB factor and HML factor).

(d) Jointly test that the intercepts for all 6 portfolios are zeros using the $F$-test statistic or Hotelling's $T^2$ for the whole sample and for each subsample when regressing on all three Fama-French factors.

(e) Are the 6 portfolios excess returns (over the risk-free rate) series co-integrated? Use Johansen's test to identify the number of co-integrating relationships.

5. The file `portfolio_data.csv` also contains returns on 25 size sorted portfolios along with the risk free rate, excess market return along with two Fama-French factors. The monthly data spans 1926 to 2012.

(a) Fit CAPM to the 25 portfolios. Give point estimates and 95% confidence intervals of $\alpha$, $\beta$, the Sharpe index, and the Treynor index. (Hint: Use the delta method[1] for the Sharpe and Treynor indices.)

[1] If $\hat{\theta} \sim N(\theta, \Omega)$, then $g(\hat{\theta}) \sim N(g(\theta), g'\Omega g)$, where $g(\theta)$ is a one-to-one function of $\theta$. Use the diagonal elements $g'\Omega g$ to obtain the standard error of $g(\hat{\theta})$.

(b) Test for each portfolio the null hypothesis $\alpha = 0$.

(c) Use the multivariate regression model to test for the 25 portfolios the null hypothesis $\alpha = 0$.

(d) Perform a factor analysis on the excess returns of the 25 portfolios. Show the factor loadings and the rotated factor loadings. Explain your choice of the number of factors.

(e) Consider the model

$$r_t^e = \beta_1 \mathbf{1}_{t<t_0} r_M^e + \beta_2 \mathbf{1}_{t \geq t_0} + \epsilon_t$$

in which $r_t^e = r_t - r_f$ and $r_M^e = r_M - r_f$ are the excess returns of the portfolio and market index. The model suggests that the $\beta$ in the CAPM might not be constant (i.e., $\beta_1 \neq \beta_2$.) Taking February 2001 as the month $t_0$, test for each portfolio the null hypothesis that $\beta_1 = \beta_2$.

(f) Estimate $t_0$ in (e) by the least squares criterion that minimizes the residual sum of squares over $(\beta_1, \beta_2, t_0)$.

(g) Fit the Fama-French model and repeat (a)–(c).

6. Portfolio Rebalancing: Consider the 10 stocks in the file `m_logret_10stocks.csv` from Exercise 3: Divide the duration of the data into four time intervals.

(a) At the end of each interval, compute the optimal portfolio weights using the risk-aversion formulation; choose $2 \leq \lambda \leq 4$ (evaluate at $\lambda = 2, 3$ and 4). Comment on how the portfolio weights have changed and why.

(b) For each interval, construct factor models; sort the stocks based on the first factor. Follow 130/30 (short up to 30% of the portfolio value and use the resulting funds to take long positions on the better performing stocks) strategy and evaluate the allocation procedure.

7. The file `exchange_rates_1.csv` contains exchange rates between US$ and 25 major currencies; the daily data spans January 3, 2000 to April 10, 2012. Consider the data from January 3, 2002 for the following currencies: EUR, GBP, DKK, JPY and CNY. Compute $r_t = \ln p_t - \ln p_{t-1}$, returns for each series.

(a) Estimate the mean vector and the covariance matrix of the returns.

(b) Construct a portfolio with Markowitz's minimum variance objective with expected returns at least R, the median of the elements in the mean vector. The weights must be positive and should sum to 1.

(c) Repeat (a) and (b) for each year and comment on the variation in the composition of the portfolios.

8. Consider the exchange rate data in the file `exchange_rates_3.csv` for 24 currencies; consider the returns on these currencies along with the market return and the risk free rate. The daily data spans 3/1/2005 to 10/4/2012. We will use the data from 2005 to 2010 to construct a portfolio and the data from 2011 to 2012 to evaluate the portfolio.

(a) Compute the mean $\hat{\mu}$ and sample covariance matrix, $\hat{\Sigma}$ of the log returns.

(b) Compute the shrinkage estimate $\hat{\Sigma}^*$ of the covariance matrix, under a one-factor model.

(c) Estimate the optimal weights, both with and without short selling under the regular estimate of $\Sigma$ and as well as under the shrinkage estimate. Also consider an equal weighted portfolio and the portfolio weighted by the inverse of variances. Compute the optimal weights for all combinations and discuss them.

(d) Estimate the optimal weights using LASSO both for short and no-short positions.

(e) Evaluate the different weighting schemes in terms of their performance during the validation period, 2011–2012. Summarize your findings and offer some intuitive explanations.

9. Consider the data in file `exchange_rates_3.csv` in Exercise 8.

(a) Perform a PCA on the 24 exchange returns; use the first two principal components on factors in a two-factor model for $F$, estimate $F$.

(b) Using the estimated $\hat{F}$ in (a) as the shrinkage target, compute a new shrinkage coefficient and the new shrinkage estimate of $\Sigma$.

(c) Compare the efficient frontier corresponding to this estimate with those obtained in (1.c).

(d) Group the currencies in terms of their performance during 2005–2010 into four quantiles. Construct a portfolio based on the top performing quantile and another portfolio based on the bottom quantile; evaluate their performance using the data for 2011–2012.

10. Consider again the data in the file `portfolio_data.csv`. Consider the returns on 25 size sorted portfolios along with the risk free rate and Fama-French factors. The monthly data spans 1926 to 2012. We will use the data from 1926 to 2000 to construct a portfolio and the data from 2001 to 2012 to evaluate the portfolio.

(a) Compute the mean $\hat{\mu}$ and sample covariance matrix, $\hat{\Sigma}$ of the log returns.

(b) Compute the shrinkage estimate $\hat{\Sigma}^*$ of the covariance matrix, under a one-factor model.

(c) Estimate the optimal portfolio weights, both with and without short selling under the regular estimate of $\Sigma$ and as well as under the shrinkage estimate. Also consider an equal weighted portfolio and the portfolio weighted by the inverse of variances. Compute the optimal weights for all combinations and discuss them.

(d) Evaluate the different weighting schemes in terms of their performance during the validation period, 2001–2012. Summarize your findings and offer some intuitive explanations.

11. Consider the data in file `portfolio_data.csv` in Exercise 10.

(a) Perform a PCA on the 25 portfolio returns; use the first two principal components on factors in a two-factor model for $F$, estimate $F$.

(b) Using the estimated $\hat{F}$ in (a) as the shrinkage target, compute the new shrinkage estimate of $\Sigma$.

(c) Compare the efficient frontier corresponding to this estimate with those obtained in (10.c).

# 7

# *News Analytics: From Market Attention and Sentiment to Trading*

## 7.1 Introduction to News Analytics: Behavioral Finance and Investor Cognitive Biases

The determinants of large sporadic movements in stock prices that are not justified by fundamentals have been speculated and studied by many economists, starting with Keynes (1937) [230]. The early empirical studies on this (see Cutler, Poterba and Summers (1989) [96]) linking stock news to stock prices did not find any significant relationships between the macroeconomic events and stock performance, particularly in terms of large returns. The rational financial theory that assumes the market price of a stock must be equal to the present value of future expected cash flows is generally unable to explain these patterns. So alternative explanations are sought from the behavioral finance perspective. It is argued that investors make decisions based on their sentiment, that may not be related to data at hand about the stock performance. Generally, competing with these sentiment-driven investors can be costly. Several market crashes can attest to these premises.

There are a number of studies finding evidence to support the hypothesis that the content of news media can predict stock market activities. News communication channels have evolved over time, from slow print media, where the information is screened and edited to almost instantaneous social media, where participation is high but information is of varying quality. The web, which is in-between, can be slower than social media, but a certain degree of quality can be normally expected. Antweiler and Frank (2004) [22] find evidence for the relationship between internet message activities that can be characterized into "buy," "sell" or "hold" recommendations and trading volume and volatility.

Tetlock (2007) [312] provides an excellent summary of theoretical models of the effect of investor sentiment on stock prices. These models generally assume that there are two types of traders, noise and rational traders. Noise traders hold random beliefs about the future stock performance and rational traders hold informed beliefs possibly based on the past performance of the stock and expected future potential. The difference between the two beliefs, when measured right is taken to reflect the investor sentiment. Changes in the stock behavior can be due to a number of other factors as well, such as risk aversion, inventory management, etc. The so-called media pessimism can be a proxy for such stock-related, or in general market-related, factors as well, but its timing is used to distinguish sentiment-related influence on the stock behavior. Lower sentiment generally leads to downward pressure on prices resulting in low returns at short horizons, but will be reversed in the long run. If it is due to true information about the stock, the trend in returns can be expected to persist. Investor sentiment is also likely to influence the trading volume with "irrational" traders trading heavily over "rational" traders with shorter horizon in mind. It is assumed that informed traders have access to information prior to it becoming widely known. The empirical research clearly demonstrates the following:

- The firm's return on a news day positively predicts its returns after the news. The gradual dissipation of the liquidity shock after the news, leads to return momentum.
- Returns on high-volume news days are better (positive) predictors of post-news returns than returns on low-volume news days.
- The contemporaneous correlation between the firm's trading volume and the magnitude of its price changes temporarily increases around news days.
- The price impact of informed trading in the firm's stock temporarily decreases as news reduces information asymmetry.

Baker and Wurgler (2006) [27] investigate how investors' beginning-of-period sentiment affects the cross-section of future stock returns. A number of proxies for sentiment are considered. These are, closed-end fund discount measured by the difference between the net asset value of closed-end stock fund shares and their market prices, NYSE share turnover, the number and average returns of the IPOs, the equity share in the new issues and the dividend premium which is measured by the log difference of the average market-to-book ratios of payers and non-payers. These six proxies for investment are used to form a sentiment index via principal component method. The first principal component is then shown to be related to future stock returns after adjusting for the stock and firm characteristics such as firm size, age, profitability, book to market equity, etc. The predictability of investor sentiment is stronger for high volatility, non-dividend paying, extreme growth and distressed stocks. Because the turnover measure is highly volatile, it is dropped from the index and the revised index values that are available are now based on the other five variables.

In an extended article, Baker and Wurgler (2007) [28] consider some additional proxies as well and these are able to better discriminate the two factors that could lead to mispricing, a change in sentiment of the irrational traders and a limit to arbitrage from the rational investors. Market turnover, defined as the ratio of trading volume to the number of listed shares, market Volatility Index (VIX), equity issues as a ratio of total new issues that includes debt issues as well and a measure of insider trading patterns are all considered to be other proxies for sentiment. The constructed index is orthogonalized to a set of state variables: Real growth in durables, nondurables and services consumption; growth in employment and the National Bureau of Economic Research recession indicator. The time-varying index constructed from these is shown to match up well with the anecdotal accounts of the timing of bubbles and crashes.

It must be noted that the sentiment index constructed by Baker and Wurgler (2006, 2007) [27, 28] is based on monthly data, as the underlying variables are somewhat slow-moving. At low frequency, the index does provide a predictive direction for the stock performance. The index given in Figure 7.1 is updated and posted in the authors' website.[1] Yu and Yuan (2011) [329] and Stambaugh, Yu and Yuan (2012) [305] show how the index can shed light on well-documented anomalies. These anomalies include unexpected deviations in performance during times of financial distress; the firms with high probability of failure have lower, not higher, subsequent returns, which is not expected by the standard asset pricing models. Another one is the momentum effect, which refers to empirical observation that high past returns generally forecast high future returns. Also it is noted that higher past investment predicts abnormally lower future returns.

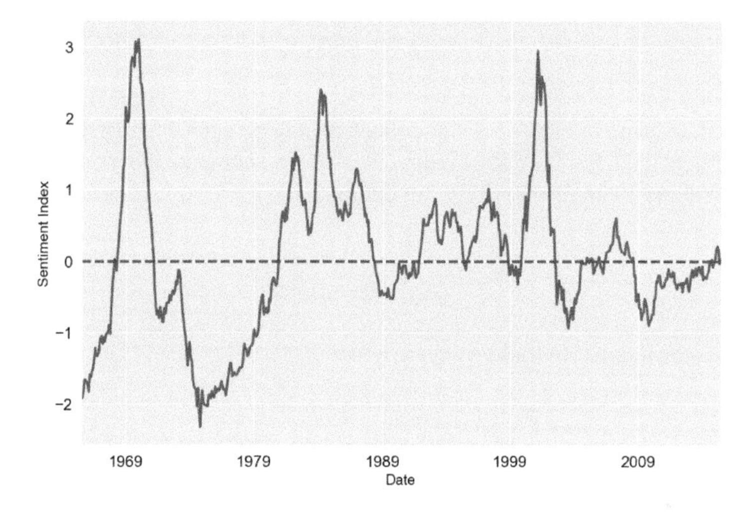


Figure 7.1: Sentiment Index.

While an extensive discussion of these results at the stock level is not possible here, we want to present the descriptive statistics of key performance indicators and other associated characteristics at the market level. The goal of this exercise is to show how conditional on the sentiment scores, the averages over different key characteristics differ. The data on the Fama-French factor model, see Fama and French (2015) [134], is taken from Ken French's website.[2] It consists of (details are directly from the site):

- SMB (Small Minus Big) is the average return on the nine small stock portfolios minus the average return on the nine big stock portfolios.

[1] https://www.dropbox.com/s/ip2812eich83tw2/Investor_Sentiment_Data_20190327_POST.xlsx?dl=0

[2] https://mba.tuck.dartmouth.edu/pages/faculty/ken.french/Data_Library/f-f_factors.html

- HML (High Minus Low) is the average return on the two value portfolios minus the average return on the growth portfolios.
- RMW (Robust Minus Weak) is the average return on the two robust operating profitability portfolios minus the average return on the two weak operating profitability portfolios.
- CMA (Conservative Minus Aggressive) is the average return on the two conservative investment portfolios minus the average return on the two aggressive investment portfolios.
- $R_m - R_f$, the excess return on the market, value-weighted return of all CRSP firms incorporated in the US and listed on the NYSE, AMEX, or NASDAQ that have a CRSP share code of 10 or 11 at the beginning of $t$, and good return data for $t$ minus the one-month Treasury bill rate (from Ibbotson Associates).

The daily data on the above factors for the same duration as the duration of the sentiment data given in Figure 7.1 was collected and aggregated to match the monthly level sentiment data. For the excess return, $R_m - R_f$ we compute the monthly averages and as in Yu and Yuan (2011) [329], we compute the realized variance of the excess return as,

$$\hat{\sigma}_t^2 = \frac{22}{N_t} \sum_{t=1}^{N_t} (R_{mt} - R_{ft})^2, \tag{7.1}$$

where $N_t$ is the number of trading days in a month. The aggregated factors are plotted in Figure 7.2.

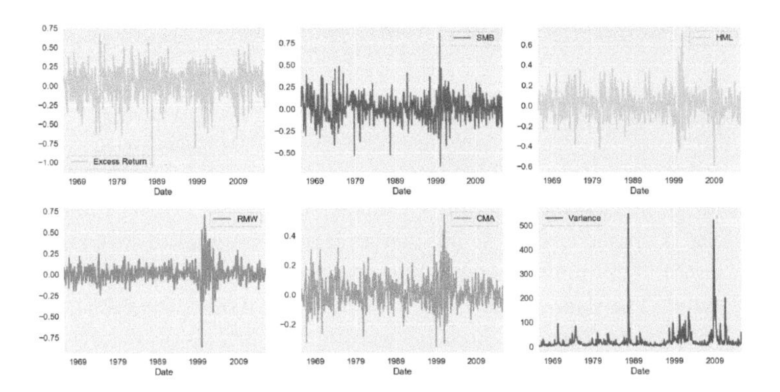


Figure 7.2: Time Series Plot of Fama-French Factors.

To classify the sentiment into low and high, Baker and Wurgler (2006) [27] simply use negative and positive scores. In order to allow for natural variation in sentiment scores, we classify the score below the first quartile (−0.5) as low sentiment and the score above the third quartile (+0.5) as high sentiment and in between as normal sentiment. Table 7.1 provides the summary of averages of key factors. The contrast between the low and high sentiment periods is quite evident.

Table 7.1: Fama & French Factors Across Sentiment Levels

| | Low | Normal | High | All |
|---|---|---|---|---|
| N | 152 | 283 | 168 | 603 |
| Sentiment | -1.297 | 0.031 | 1.120 | 0.000 |
| Excess Return | 0.0127 | 0.044 | -0.004 | 0.023 |
| SMB | 0.031 | 0.009 | -0.010 | 0.009 |
| HML | 0.014 | 0 | 0.052 | 0.018 |
| RMW | -0.001 | 0.007 | 0.035 | 0.013 |
| CMA | 0.011 | 0.003 | 0.040 | 0.016 |
| Realized Variance | 23.89 | 24.58 | 16.83 | 22.25 |

Some observations noted from the descriptive statistics follow: While the pattern in excess return is consistent with what is observed in Yu and Yuan (2011) [329], the realized variance in a high sentiment period is lower than in other periods, but this is consistent with the mean-variance relationship as expected by the asset pricing models. The SMB, which measures the difference in returns between small stocks and large stocks, is negative during high sentiment periods indicating the portfolio of small stocks does not do well in these periods. The HML, the difference in returns between high and low B/M stocks, tends to be positive and higher during high sentiment periods. The RMW, which measures the performance difference between robust and weak profitability portfolios, is also higher during high sentiment periods. Finally, CMA, which measures the difference in performance between conservative and aggressive firms, continues to do well in high sentiment periods.

Da, Engleberg, and Gao (2011) [97] review studies on investor attention and its impact on stock performance. The proxies such as extreme returns, trading volume, news and headlines, advertising expense, etc., were used as indirect measures of attention. Because some of these proxies can be driven by factors unrelated to investor attention, it must be noted that they, at best, are proxies. Instead it is suggested that search frequency of key words related to a stock in a search engine is a more direct measure of the attention as the users who search directly or indirectly express their intent and thus their interest.

Google publishes the search volume index (SVI) for key words and, except for rarely-searched tail terms, the index tends to be fairly reliable due to the significant market share of Google in terms of search volume. Da et al. (2011) [97] find SVI is a leading indicator of extreme returns and news, and abnormal SVI matches with known return anomalies. Generally, we must emphasize that what is meant by an anomaly is an empirical deviation from what is postulated by economic theory. A positive abnormal SVI generally predicts higher stock prices in the short term and price reversals in the long term. The SVI is correlated with other proxies of investors, particularly retail investors, attention and is available in a timely fashion.

Da, Engleberg, and Gao (2015) [98] extend this idea to study the market-level sentiment and its correlation to aggregate market returns. The market-level index is based on the top thirty search terms that relate to the general public's concern, such as unemployment, recession, bankruptcy, etc. The index is shown to predict short-term return reversals, increase in volatility and flows from equity funds to bond funds. The change in the index is demonstrated to be correlated with certain sentiment indicators, of Baker and Wurgler (2006) [27].

Stambaugh et al. (2012) [305] consider several asset-pricing anomalies that may be attributed to financial distress, momentum, asset growth, etc. For each anomaly, the strategy that goes long in the highest-performing decile and goes short in the lowest-performing decile is examined. However as Huang et al. (2015) [207] observe, "...whether investor sentiment can predict the aggregate stock market at usual monthly frequency is still an open question." It must be noted that the market is efficient for the most part and any information gets impounded rather immediately at the higher frequency level. Thus there is a great deal of interest to mine news, attention and sentiment as they occur and information from social media has come to play a greater role. We present some evidence to that effect in the next section.

## 7.2 Automated News Analysis and Market Sentiment

As noted in Chapter 1, trading now mostly occurs at a relatively high frequency level and hence any news related to a stock gets processed almost instantaneously. The literature reviewed in the last section refers to low frequency sentiment data and hence its influence on returns is not easy to notice. New technologies that can handle automatic news collection, extraction and categorization of relevant information are fast-emerging. Quantitative models that incorporate this information in investment decisions are being actively researched and automation of these activities can greatly help traders shorten their reaction time to emerging news feeds. There has been a surge of academic interest in this topic (see Mitra and Mitra (2011) [270] and Peterson (2016) [285]). Several technology companies (iSentium, SMA, MarketPsych, RavenPack, etc.) have sprung up in the last decade offering the service of aggregating the web information or sentiment, related to stocks.

Financial news can be classified into two main groups. Some financial news related to a stock is released on a regular basis, such as earnings statements and models to analyze such events are well structured. Earnings expectations are derived from stock-related performance indicators, and when actual earnings differ from traders expectations, they lead to adjustment of the a priori positioning. Other news, related to the product or personnel of the company, tend to be unscheduled and may also affect trading decisions. Although this news is unstructured, it can also be easily processed. Usually the early exploiters of this type of information can expect to gain some advantage over others. These are generally termed news-based event strategies and have been fairly well-studied in the finance literature.

Unstructured news streams do not come at regular intervals, tend to be qualitative and are usually in the form of a text. In order to correlate this information with the stock performance, they need to be appropriately quantified. Because this information originates from numerous sources such as in the case of social media, it needs to be properly aggregated. Without any aggregation and smoothing of news flows, it may be difficult to extract the signal from the noise. Some common steps involved in making the news usable for trading are:

- Identify the unique, relevant and authentic news on a timely basis.
- Quantify the textual information into informative scores, whose variation would cover negative to positive sentiment/connotation.
- Aggregate the scores from different sources.
- Develop a baseline model adjusting for seasonality of the sentiment scores.
- Relate the abnormal deviation of the sentiment scores to abnormal stock price movements.

As many of these concepts and related issues are discussed in the references cited in this chapter, we will be brief in our discussion. Das and Chen (2007) [102] show how it is possible to capture the net sentiment from positive and negative views on message boards using statistical natural language processing techniques. The emotive sentiment is elicited using different classification algorithms and the results are pooled via majority voting. Opinions are extracted from message board postings and they are classified into three types: Bullish, bearish and neutral. The classification algorithms include parts of speech tagging and support vector machines. The algorithms are initially trained using humanly pre-classified messages and are designed to learn and apply these rules out-of-sample. Relating the sentiment to the performance of stocks in the Morgan-Stanley High-Tech Index, it is observed that there is no strong relationship to individual stock prices but there is a statistical relation to the aggregate index performance. There is also a strong relationship between message volume and volatility consistent with Antweiler and Frank (2004) [22]. These examples clearly illustrate that investor sentiments expressed publicly via media have some impact on stock behavior.

We want to briefly provide some details on a few examples of web analytics providers; interested readers should consult these providers' websites for further details.

- **Ravenpack (NewsScore):** The key information on entities (company, organization, currency, commodities and place) from major news sources are gathered and processed for their relevance and are assigned event sentiment score. The scores are aggregated on a rolling window basis. In addition, the aggregate count as event volume, the event novelty score, event novelty elapsed time, etc., are provided. For equity markets, composite sentiment score and the impact projections of the news are also available. The main news providers are Dow Jones Newswires, The Wall Street Journal (all editions) and Barron's.

- **Thomson Reuters (News Analytics):** Data fields include relevant score of the news item to the asset, number of sentiment words or tokens, sentiment classification, novelty of the content, feed volume, etc. The sources are Reuters, PR Newswire, etc.

- **Thomson Reuters (MarketPsych Indices):** News and social media information in real time is delivered as data series; it is categorized into three types of indicators: Emotional indicators, Macroeconomic metrics and Buzz metrics on the asset level. The data is updated on a minute basis. The social media data includes blogs, internet forums and finance-specific tweets.

- **iSentium (iSense):** Extracts sentiment signals using natural language processing architecture on Twitter data; the data feed includes retweets, if the author has a finance-related bio, number of followers, etc. The impact is measured by the number of retweets for each author's tweets.

- **Social Market Analytics (SMA):** The data set provides a snapshot of sentiment factors at a 15 minute interval sourced from Twitter/StockTwits messages. The raw scores are adjusted for the average and volatility. A measure of unusual volume activity is also provided.

In the next section, we will summarize studies that use the information provided by these agencies for developing trading strategies. To provide the theoretical underpinning of the effect of investor sentiment, two assumptions are made (see Delong, Shleifer, Summers and Waldmann (1990) [103]): The market consists of two types of traders, noise and rational, and they are both taken to be risk averse. Tetlock (2007) [312] defines sentiment as the level of noise traders' random beliefs to rational traders' updated beliefs; it is argued that media (collective) pessimism is a proxy for either the sentiment or risk aversion. It is expected that high media pessimism will predict low returns at short horizons and at longer horizons, a reversion to fundamentals will occur.

## 7.3 News Analytics and Applications to Trading

We have reviewed select studies that have clearly documented the influence of news on stock market behavior. In this section, we will outline first the key methodologies used in relating the data on the inflow of news, to key characteristics of stock trading and stock performance indicators. A simple framework that we will follow is from Gross-Klussmann and Hautsch (2011) [170] and is depicted in Figure 7.3.

The news arrival can generally be modeled as a point process but to remove any noisy friction in the data, both in the newsfront and in the stock price movement, it

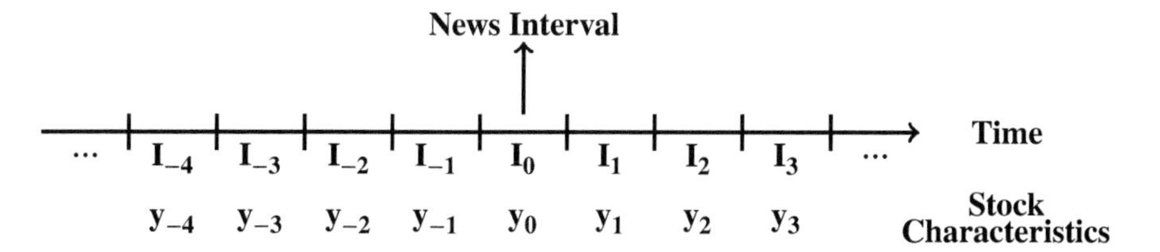


Figure 7.3: News Interval.

may be advisable to aggregate the information to short time intervals. The stock characteristics can include the return as well as volume, volatility, imbalance in demand/-supply, etc., during the interval. In addition to time aggregation, we also need to consider aggregation over different variables or dimensions of sentiment as in Baker and Wurgler (2006) [27]. We describe these in two contexts below.

**Aggregation over Dimensions**

In the use of monthly data from Google, Da et al. (2015) [98] aggregate the daily change in a related search term, $i$, over the top thirty searched terms to come up with an index called, FEARS (Financial and Economic Attitudes Revealed by Search) as

$$\text{FEARS}_t = \sum_{i=1}^{30} R^i(\Delta\text{ASVI}_{it}), \tag{7.2}$$

where $\Delta\text{ASVI}_{it}$ is an adjusted (winsorized, deseasonalized, and standardized) daily change in search volume,

$$\Delta\text{ASVI}_{it} = \ln(\text{SVI}_{it}) - \ln(\text{SVI}_{it-1}). \tag{7.3}$$

Here, $R(\cdot)$ denotes the ranking function. Generally search volume data exhibits non-stationarity as well as strong weekday seasonality. It has, for instance, been found that web searches tend to be higher on Monday through Wednesday than on other days. The above adjustments to search volume take care of both concerns. Equation 7.3 is essentially a differencing operator that addresses the stationarity. A similar differencing for the weekday effect can be easily incorporated. These ideas were discussed in Section 2.7.

The sentiment index in Baker and Wurgler (2006) [27] is a composite index based on the five standardized sentiment proxies mentioned earlier. Because some variables take longer to express the same sentiment, the proxies are lagged. The series are smoothed over the prior twelve months and the index uses the $t-12$ value.

In the construction of this index, the principal component of the five variables and their lags are used. Thus the first stage index has ten loadings; using the correlation between the index and current and lag values of the variables, the time index that is most appropriate for each of the variables is selected. In the original construction, the variables RIPO and PD-ND are lagged but the rest are used as of current time. The sentiment index is also further refined to adjust for common business cycle components, measured through growth in the industrial production index, growth in consumer durables, nondurables and services and an indicator variable for National Bureau of Economic Research (NBER) recessions. The resulting regression residuals, after adjusting for the time series effects, are argued as better, cleaner measures of sentiment. These are then used in the principal component analysis. Thus mathematically from Section 3.2, if $x_t = (x_{1t}, x_{2t}, \dots, x_{5t})'$, the first principal component $l'x$ is obtained by

$$\max_{l} \ l'\Sigma_{xx}l \qquad \text{subject to} \quad l'l = 1. \tag{7.4}$$

Because the variables here are standardized, it should be noted that the covariance matrix, $\Sigma_{xx}$ is also the correlation matrix. The first linear combination associated with the eigenvector that corresponds to the largest eigenvalue of the matrix, $\Sigma_{xx}$, gives the sentiment score,

$$s_t = \frac{l'x_t}{\sqrt{\lambda}} \tag{7.5}$$

in Figure 7.1. Here $\lambda$ is the largest eigenvalue of $\Sigma_{xx}$ so that the sentiment score, $s_t$ has unit variance. Because it is a standardized measure (as the eigenvalues of $\Sigma_{xx}$ add up to five) both the eigenvalues and eigenvectors are easy to interpret.

## Relating Sentiment to Stock Performance

The general framework relating the sentiment scores ($s_t$) to performance indicators ($y_t$) is simply through lagged regression. For example, the returns $r_t(y_t)$ are related to ($s_t$) and other control variables and can take the form:

$$r_{t+k} = \beta_0 + \beta_1 s_t + \gamma' z_t + u_{t+k}, \tag{7.6}$$

where $z_t$ is a set of stock-related control variables, such as Fama and French factors. The left side can also be (annualized) realized volatility, calculated by sampling the price over '$N$' times during a day.

$$rv_t = 250 \sum_{d=1}^{N} r_{t,d}^2. \tag{7.7}$$

Because of the persistence of volatility, some adjustments via differencing should be considered when relating sentiment scores to volatility. Finally, it is possible that the effect of positive sentiment on stock returns could be different from that of negative sentiment. Fitting models similar to (7.6) for both scenarios (positive and negative sentiment regimes) separately would be a worthwhile exercise.

**Aligned Investor Sentiment**

In the above set-up, the construction of the sentiment scores (Equation 7.5) or FEARS (Equation 7.2) is done first and then it is related to–for instance–the returns via (7.6) subsequently. Huang, Jiang, Tu and Zhou (2015) [207] combine the two steps, together arguing the linear combination in (7.5) is constructed better in referencing to the stock characteristics and not on its own. They demonstrate that the new combination of predictors based on the PLS method described in Section 3.2 has much greater predictive power for the aggregate stock market and its predictability is both statistically and economically significant. Because it is a supervised learning procedure, one should expect it to fare better.

**Social Media/Twitter Data:** Over the past years, significant development has been made in sentiment tracking technologies that can extract indicators of public mood from social media such as blogs and Twitter feeds. Although each tweet is of limited number of characters, the aggregate of tweets may provide an accurate representation of public sentiment. This has led to the development of real-time sentiment tracking tools (Opinion Finder, GPOMS (Google Profile of Mood Status), etc.) to track the public mood and relating it to economic indicators. Crowd–sourced data generated by social network sites such as Twitter/StockTwits is being increasingly used by quantitative researchers to generate trading signals. The wisdom of the crowd surprisingly appears to outperform the wisdom of the experts in a variety of instances. For a recent application relating sentiment data to earnings announcement, see Liew, Guo and Zhang (2017) [243]. We now illustrate a simple trading algorithm using the iSentium data.

iSentium expertise lies in providing market sentiment indicators. Market–related texts from Twitter are processed through Natural Language Processing techniques and the contents are assigned sentiment scores in the range of $-30$ to 30 with positive score indicating an optimistic view and negative score, a pessimistic view. For a specific keyword the data contains the following information: Time when the tweet is recorded, sentiment Score, total number of tweets since Jan 1, 2012, whether it is a retweet, whether the author has a finance–related bio, number of followers of the author, number of tweets posted by the author, and average number of retweets for each of the author's tweets, measuring the author's impact.

In addition to per–tweet data, the company also provides binned data, binned over various time durations. Interestingly tweet volume is particularly high for popular ETFs and large cap technology stocks. To illustrate, we consider the per-tweet data from Jan 1, 2012 to September 10, 2014 for the **Nasdaq 100** ETF (ticker symbol: **QQQ**). The typical data is listed in Table 7.2.

It is quite apparent from the short list of entries in Table 7.2, the same author of a tweet can have varying sentiments in a short duration of time. It is also likely that a few authors can dominate the volume of tweets in a short duration ($\approx$90 minutes in the sample data in Table 7.2). Consequently in the simplified application we describe here, the daily summaries of sentiment scores are calculated and matched with the daily performance of Nasdaq 100 ETF. Thus, the sentiment data is smoothed so that

Table 7.2: Typical Elements of iSentium Data

| Date | Time | SS | Retw | Retwld | authid | FinBio | nfollow | ntweets | impact |
|---|---|---|---|---|---|---|---|---|---|
| 1/1/2012 | 3:04:41 | 9 | 0 | −1.00E + 00 | 202647426 | 1 | 222 | 16 | 0 |
| 1/1/2012 | 3:10:06 | 9 | 0 | −1.00E + 00 | 23059499 | 1 | 10501 | 27 | 0 |
| 1/1/2012 | 3:19:46 | −11 | 0 | −1.00E + 00 | 202647426 | 1 | 222 | 25 | 0 |
| 1/1/2012 | 3:43:52 | 0 | 0 | −1.00E + 00 | 202647426 | 1 | 222 | 30 | 0 |
| 1/1/2012 | 3:50:01 | 0 | 0 | −1.00E + 00 | 23059499 | 1 | 10501 | 48 | 0 |
| 1/1/2012 | 4:10:01 | −5 | 0 | −1.00E + 00 | 23059499 | 1 | 10500 | 89 | 0 |
| 1/1/2012 | 4:11:02 | −5 | 0 | −1.00E + 00 | 207429198 | 0 | 51 | 2 | 0 |
| 1/1/2012 | 4:12:10 | −5 | 1 | 1.53E + 17 | 16704290 | 1 | 1170 | 3 | 0 |
| 1/1/2012 | 4:26:07 | −10 | 0 | −1.00E + 00 | 207429198 | 0 | 51 | 9 | 0 |
| 1/1/2012 | 4:30:03 | −10 | 0 | −1.00E + 00 | 23059499 | 1 | 10500 | 104 | 0.0625 |

extreme values are weighed in appropriately. Note that twitter activity occurs every day but the markets are open only on weekdays. No attempt is made to account for the cumulative effect of weekend sentiment scores on the beginning-of-the-week Nasdaq 100 ETF performance. The Nasdaq 100 ETF data contains daily price bars (with open, high, low and close) and the volume of transactions.

The time series plot of the sentiment related variables and the Nasdaq 100 ETF trade related variables are given in Figure 7.4. Here the variables are defined as:

- RetO= $\ln(P_{t\cdot 0}) - \ln(P_{t-1.0})$ based on opening price
- Volume is logged
- Volatility= $0.5(H_t - L_t)^2 - 0.386(C_t - O_t)^2$, where $H_t = \ln(P_{t\cdot H})$, the high price observed during the day, etc.
- SSAav is the average sentiment score for the day
- FBav is the proportion of tweets by the tweeters with a finance bio
- Retav is the average number of retweets per day

While the focus is on sentiment score, the other two sentiment variables indicate the intensity of activity by the twitter participants in the finance field. It can be seen from Figure 7.4, that the sentiment average (SSAav) exhibits some variation over time but it is generally stationary.

There is some memory in the series as the first few autocorrelations are found to be significant. All the other sentiment related series also exhibit some memory. The same can be noted for the volume and volatility series as well. One of the key tools to connect two series in terms of lead-lag relationships is the cross-correlation function defined in Chapter 3 (see Section 3.3). For the sentiment data ($s_t$) to be of any predictive use, it should be correlated to lead return ($r_{t+k}$) and lead volatility ($V_{t+k}$). Figures 7.5 and 7.6 capture the cross-correlations, but one can observe that there is some concurrent correlation between $s_t$ and $V_t$. This may be due to the fact that the

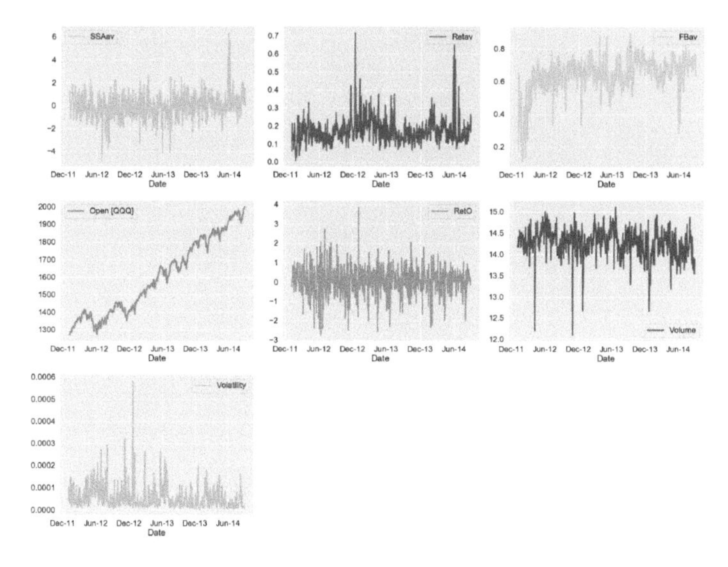


Figure 7.4: Sentiment Score and Nasdaq 100 ETF.

aggregation of data was done over the entire day, which may be too long a duration for aggregation of high frequency data, as traders may use any signal fairly quickly. These relationships can be confirmed in Table 7.3 with stepwise regression model fit. It can be observed that all three variables: Return, volume and volatility, are fairly well-explained by the sentiment data.

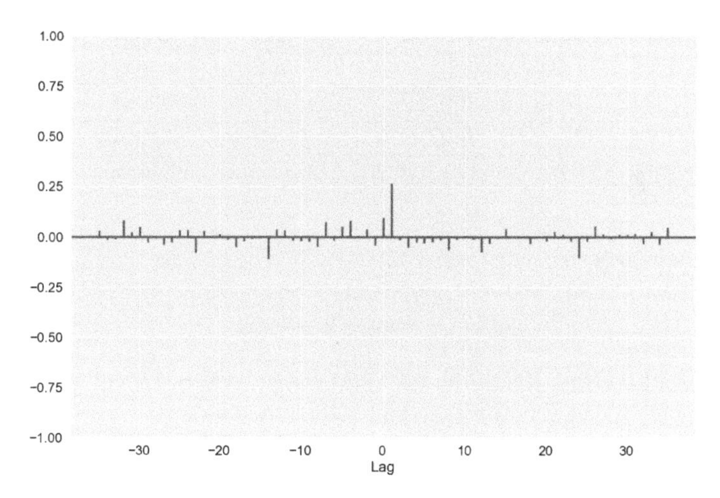


Figure 7.5: CCF between SSAav and RetO.

The techniques described in Chapter 5 (see Section 5.3) can easily be applied to develop strategies leveraging sentiment data. For illustration, we propose two naïve strategies for trading using such data; both in the spirit of the discussion related to

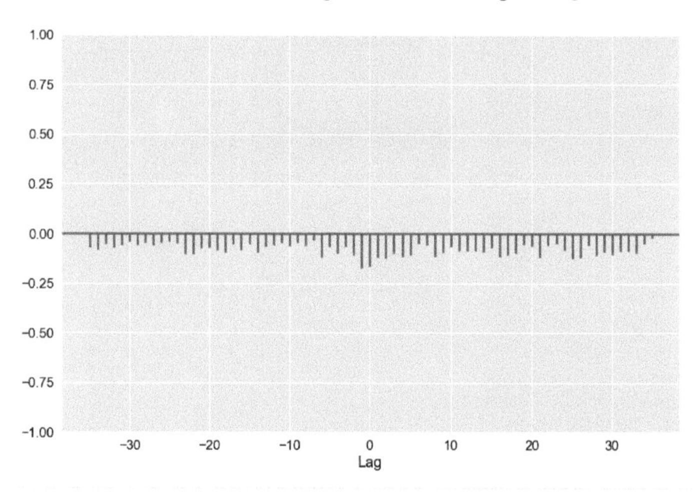


Figure 7.6: CCF between SSAav and Volatility.

Table 7.3: Regression of Nasdaq 100 ETF variables on sentiment factors

| | Response Variable | | |
|---|---|---|---|
| Predictors | Return | Volume | Volatility |
| SSAav | — | −0.0482 | −0.00005 |
| SSAverlag | 0.1895 | −0.0291 | −0.00006 |
| AuHTAV | — | −0.000017 | −0.000001 |
| RetAV | — | — | −0.000063 |
| NfollowAV | — | — | −0.000001 |
| AVetW | — | — | 0.000047 |
| Constant | 0.0759 | 14.41 | 0.000085 |
| $R^2$ | 0.0732 | 0.1154 | 0.1357 |

∗All indicated factors are significant at 0.05 level

sentiment data by Baker and Wurgler (2006) [27]. The first one develops rules based on the sentiment scores and the second one is based on the more direct regression relationship between returns and the sentiment scores:

1. **Crossover Strategy:** If the difference between 3-day EWMA and 30-day EWMA of sentiment scores is positive, go long; otherwise go short.

2. **Open-to-Open Strategy:** Regress open-to-open returns on 10-day EWMA and its lag to predict the return; if positive, go long, hold the position until the next day's open.

It is easy to note that the crossover strategy is similar to the moving average oscillator rule, discussed in Chapter 5, comparing the short term trend with the long term trend. The strategy exploits the deviation between the two. Both strategies can be evaluated on a moving window basis. Here we present results for the cross-over strategy only. As noted from Figure 7.4, the Nasdaq 100 ETF price has been steadily going up and thus a simple buy and hold strategy over time would be profitable. So we compare performance of the strategy with the buy and hold approach in Table 7.4.

Table 7.4: Descriptive Statistics (Averages): Cross-over strategy

| | SSAve | RetAve | FBAC | Reto | Volume | Volatility | $N$ |
|---|---|---|---|---|---|---|---|
| Short | −0.5223 | 0.1783 | 0.6539 | −0.0866 | 14.327 | 0.000061 | 333 |
| Long | 0.4398 | 0.1793 | 0.6477 | 0.2206 | 14.220 | 0.00044 | 309 |
| All | −0.0592 | 0.1788 | 0.6509 | 0.06125 | 14.28 | 0.000053 | 642 |

The negative signal from sentiment score matches with the negative return and higher volatility. The averages are in the right direction. The equity curve in Figure 7.7 highlights the result of the strategy.

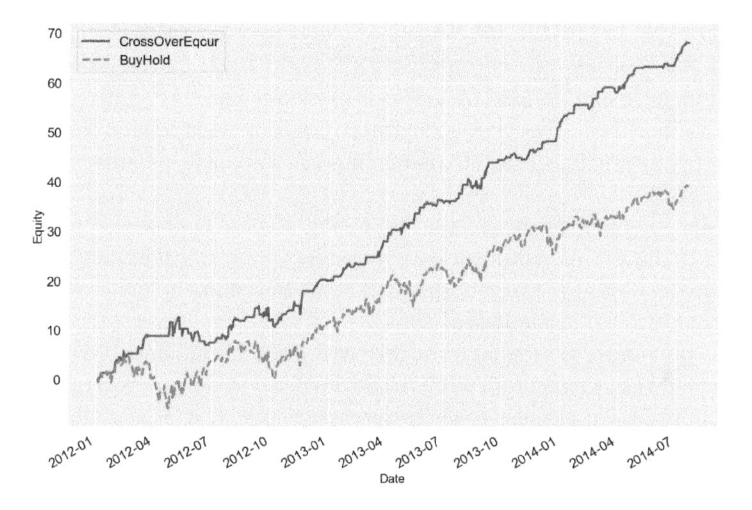


Figure 7.7: Equity Curve for Nasdaq 100 ETF.

**Bloomberg News and Social Sentiment Data:** The news agency applies statistical machine-learning techniques to process the textual information related to equities and quantifies the sentiment both at the (news) story-level and at the equity-level. The scores indicate at the story-level if the sentiment is positive, negative, or neutral with a confidence level; these are aggregated to provide the company-level sentiment. The computation is based on both news and Twitter feeds with a rolling window of half-an-hour. The predictive strength is tested out using the performance of the equity at a daily level as well as through changes in the stock price during the day. The approach is similar to cross-sectional momentum strategy that is outlined in Chapter 5. It can be shown that the stocks in the top decile of the day's sentiment scores tend to do better during that day.

Sentiment can also be used to predict the nature of earnings reports. Event driven strategies are still a significant part of trading strategies and it can be shown that sentiment scores prior to earnings releases can augment these strategies. The intraday strategy is based on using the arrival of information during an interval of pre-defined duration and making trading decisions based on the sentiment scores and their confidence. Generally traditional news combined with a twitter feed can generate stronger signals. A methodology used to relate sentiment scores whether from traditional news or from Twitter is to first construct an index,

$$I_t = \ln\left(\frac{1 + \|BV_t\|}{1 + \|BE_t\|}\right) \tag{7.8}$$

that adjusts for the baseline activities, here $BV_t$ denotes the number of bullish tweets and $BE_t$ denotes the number of bearish tweets. Defining the intraday return, $r_t = \ln(P_t^{\text{close}}) - \ln(P_t^{\text{open}})$, where $P_t$ refers to the stock price on a given time unit. Then forming, $y_t = (I_t^N, I_t^{TW}, r_t)'$, the VAR model that was discussed in Chapter 3 is used to identify the lead-lag relationships; here $I_t^N$ is the index for the news sentiment, such as Google search volume for relevant keywords and $I_t^{TW}$ refers to the index based on Twitter data. The VAR model provides a broader framework to consider other information such as market volume, market volatility, etc., that can be easily incorporated.

We want to conclude this section by noting some recent research that further demonstrates the use of sentiment data in trading. First Constantinos, Donkas and Subrahmanyam (2013) [85] establish a link between sentiment and momentum. The news diffuses through the actions of various 'newswatchers' and creates momentum, which was previously attributed to mainly past price movements. Reactive trading gets corrected when momentum positions are reversed as the difference between fundamental information and the news becomes apparent. It is argued that those who follow news closely will under-react if the information contradicts their sentiment. This means that bad news among loser stocks will diffuse slowly where sentiment is optimistic. The sentiment measure that is used is the widely recognized Conference Board's Consumer Confidence Index. The central hypothesis that sentiment impacts momentum via cognitive dissonance is generally supported by empirical analysis.

The statistical arbitrage rules developed in Chapter 5 are widely used in the industry to exploit market anomalies, whether they are due to fundamental news or sentiments, which can lead to temporary mispricing of assets. The efficacy of technical analysis and how it is related to investor sentiment is also studied in Smith, Wang, Wang and Zychowicz (2016) [303] by focusing on its usage by the sophisticated group of hedge fund managers.

Hedge fund managers integrate both fundamental and technical analysis in their decision-making process on market timing. Sentiment analysis is shown to be an important catalyst in this regard. It is documented that the performance of technical analysis users is better than nonusers in high-sentiment periods but technical tools are less valuable during low sentiment periods. The sentiment data is found to be very useful for market timing. The idea of using sentiment data for return movements has been extended to other dimensions as well.

**Manager Sentiment Index:** It should be expected that corporate managers sentiment could play a role in investment decisions as they may have an informational advantage over outside investors. Jiang, Lee, Martin and Zhou (2019) [218] construct a manager sentiment index from the textual tone of conference calls and financial statements related to the firm as it reflects the management subjective views. The measure is the difference between the positive and negative words standardized by the total word count on a monthly basis. This new index is shown to predict future aggregate stock returns consistent with theory and its predictive power is shown to be greater than other macroeconomic predictors discussed earlier. The proposed index can act as complementary to other investor sentiment measures.

## 7.4 Discussion / Future of Social Media and News in Algorithmic Trading

We have presented in this chapter summaries of studies that document the efficacy of statistical arbitrage in high-sentiment periods when short-sale constraints can discourage the elimination of overpricing that may be due to sentiment. The benefits seem to disappear in low-sentiment periods. These results are quite robust and stand true in different volatility regimes as well. The momentum rule results in profits when investors are optimistic which happens during high sentiment periods.

In a recent article, Zhou (2018) [331] discusses open issues related to measuring and using investors sentiment. The first one is related to aggregating various proxies of available sentiment measures. The aggregation by PCA captures only the total variance in the proxies but does not necessarily relate to the investment characteristics. An alternative is to use PLS or canonical correlation methods to find the optimal aggregation of sentiment measures but this can change over time. The problem is somewhat related to combinations of forecasts, discussed in Section 4.6.3. In the use of iSentium data, we dealt with the time aggregation of individual sentiment scores by simply averaging them as each tweet contains other information such as retweets, number of followers and whether the tweeter has a finance bio, etc. All these additional details may capture the reliability of a sentiment score and so deserve further attention.

The second issue mentioned is to carefully study the relationship between technical rules-based sentiment (which results in momentum or contrarian strategies) and other sentiment measures. One indicator that is based on technical analysis is

$$\text{Williams } \%\, R = \frac{\text{highest high} - \text{closing price}}{\text{highest high} - \text{lowest price}} \times 100. \tag{7.9}$$

The above is calculated based on the past ten days and indicates if a stock is overbought or oversold, which may be the result of executives optimism or pessimism. If the above measure is less than 20, it is taken to indicate that it is 'overbought' and if it is greater than 80, it is taken as 'oversold'. The idea is similar to the oscillator index discussed in Chapter 5.

Finally, which sentiment features are useful to forecast asset returns? What does the volatility in sentiment mean? How long does the memory in sentiment last? These are open issues that need further investigation. There is also a legitimate growing concern about the quality of news or information as we move from print media to web to social media, and the dissemination speed increases.

Notwithstanding these challenges, the influence of news media on the stock market behavior is likely here to stay. A recent study of von Beschwitz, Keim and Massa (2019) [320] documents how high-frequency traders rely on news analytics for directional trading on company-specific news.

# Part IV

# Execution Algorithms

In this part of the book, we cover a core topic in Algorithmic Trading: Execution Strategies. Once again, we will strive to give, from the practitioner's perspective and context, the necessary building blocks to grasp the subject and finally, the quantitative underpinnings. Our hope is that in reading this part, our readers would have enough understanding of the topic to be able to develop a basic implementation of an execution product and be able to create supporting Pre- and Post-trade analytics.

We start in Chapter 8 by discussing the modeling of trade data. We focus on several fundamental topics that are critical to practitioners but are often not given enough attention by other treatments of the subject, including details on some of the analytics used in practice. Chapter 9 then explains one of the fundamental subjects of execution: Market Impact (MI). We bring in the intuition on the inner workings of market impact and build the functional form of standard MI models from first principles. We then go into more details through a review of select literature to guide the reader that may be interested in more depth. Finally, Chapter 10 reviews the core details of the topic by covering Execution Strategies and their components. Here again we give a practical perspective with a hope of setting the reader on a more solid understanding of the issues faced in the industry.

# 8

# *Modeling Trade Data*

Many articles and books have covered the topic of trade data modeling. Our goal in this chapter is to review modeling approaches commonly used in the industry, and as in the rest of this book, we aim to provide some context on the role these models play in practice, the challenges and considerations that are involved in dealing with trade data, day in and day out, and then finally provide the quantitative foundation the reader needs to progress further. This chapter follows the same pattern as used in other parts of this book. Much of the discussion here is based on intuitive ideas. We discuss the importance of normalization analytics and their role in execution algorithms. We then provide a review of common microstructure signals that are widely used in High Frequency Trading (HFT) and execution as a foundation for more advanced alpha signals. Finally we conclude with a more in-depth quantitative treatment on some of the sophisticated models discussed in applied and academic research areas.

## 8.1 Normalizing Analytics

A challenging aspect of execution, unlike other areas in algorithmic trading which usually retain the optionality to trade when they desire, is the need to develop strategies that are consistent across market conditions and apply to a large number of stocks. The strategies will need to perform for very liquid instruments (that trade easily at the minimum tick size) as well as for illiquid stocks that trade only infrequently and exhibit wide bid-offer spreads.

To complicate matters further, these liquidity characteristics do not remain constant during the day. Intra-day trading variations are well documented and are firmly rooted in institutional investors' behavior. Immediately after the open there is a lot of trading activity as investors incorporate the information from overnight news into their trading decisions. The higher uncertainty on the stock valuation is reflected by a much higher volatility and spread in the early trading hours. Once the price discovery period is over (usually within 15–30 minutes in US markets), the trading behavior settles in, activity is reduced and so do spreads and volatility. Higher activity then resumes towards the end of the day, in particular due to the fact that many funds are benchmarked to the day's closing price. European markets have similar predictable "bumps" in and around the US market open as the market absorbs the information content of the reaction of US investors.

As mentioned earlier, there are also important sources of inter-day variability with the impact of special days. For example, the trading activity during futures/options expiration is heavily tilted towards the close, since the settlement prices are related to the closing price on that day. On Fed announcement days trading activity slows ahead of the announcement as investors await any surprise decision. Activity jumps immediately after that, incorporating any relevant information.

The complexity of dealing, in a systematic way, with all the above idiosyncrasies is a daunting task, and it would be unwieldy and prohibitively expensive to build and maintain models for each event. Hence, the practitioner's approach to address this is to parametrize many of the trading decisions using normalizing features, variables that help adapt the trading decision to a particular situation.

The next section provides a brief review of the most commonly used analytics in the industry and gives pointers to some modeling approaches. We will only provide a relatively coarse treatment of the subject from a practitioner's perspective. It should be sufficient as a starting point for anyone to build the necessary analytics for a basic execution algorithm. But this subject can benefit from more systematic academic discussion.

### 8.1.1 Order Size Normalization: ADV

Assume a trader needs to execute an order for one million shares of a stock. If this order is large it has to be carefully managed, while for a tiny order the trader can simply "fire and forget." Intuitively then, from a trading perspective, due to the large differences in stock characteristics and liquidity, order size should be viewed as a relative measure. The most commonly used order size normalizer is the Average Daily Volume (ADV). While the term ADV has a simple connotation, it is actually a very important measure for order sizing, position limit, etc. There are some other considerations as well: Which center measure to use—average, median, some other measure robust to outliers? Over what horizon? How to treat special days? The answer is likely to be application-specific and there is very little consensus or systematic study on the use of different measures.

For Amazon stock, the volatility of the volume data can be seen from Figure 8.1. One can also observe that the volume is strongly autocorrelated, with periodic predictable spikes for special days and unpredictable 'surprises' that depend on the day's information flow. As a baseline metric, the 30-day mean or median daily volume is probably the most commonly used metric in the industry. That said, other window lengths such as 66-day can present additional advantages as this time horizon covers a full quarterly earning cycle and can be considered as incorporating all the seasonal effects of the daily volume. The median is more stable and smooth while the average is more reactive to changes (see Figure 8.2). The prediction of ADV plays an important role in order execution strategies as normalized order size is one of the main features used in the scheduling of the trade.

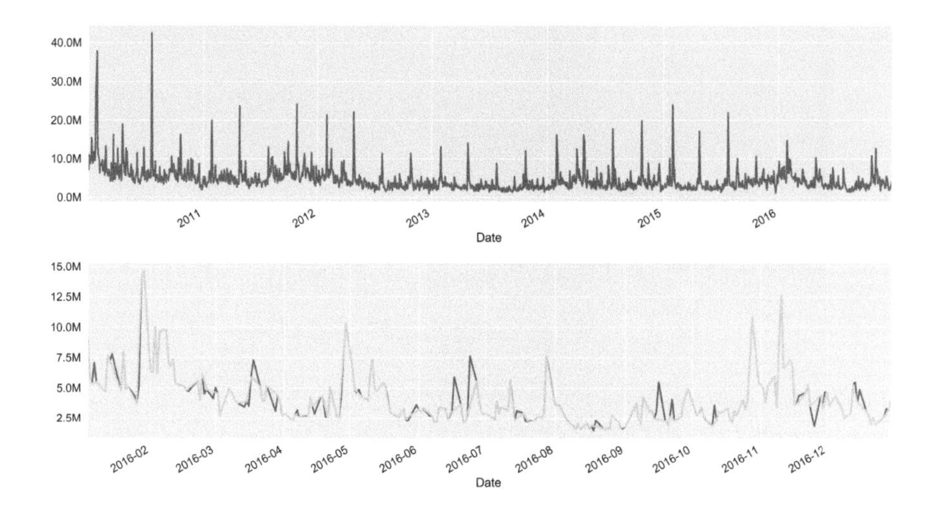


Figure 8.1: Amazon Daily Volume.

Another common use of relative order size, as we will see in Chapter 9, is in the calibration of a market impact model. One may argue that because the calibration is done ex-post, we should use the realized daily volume instead of the ADV as a normalizer. However, empirical research has shown that this leads to somewhat poorer predictions. It is possible that the large noise-to-signal ratio present in impact estimation is better adjusted by a more stable smoothed data in place of a noisy but more precise one. For consistency reasons because we use ADV as normalizer in the calibration it is often suggested to use the same normalizer for market impact estimation instead of the best prediction of daily volume. This can be problematic in particular on days where the volume is predictably higher. For a discussion on volume weighted average price (VWAP) modeling and prediction refer to Brownlees et al. (2011) [62] which was discussed in Chapter 5.

### 8.1.2 Time-Scale Normalization: Characteristic Time

Trading dynamics have an intrinsic time scale that varies significantly across the stock universe. The specifics of any particular stock microstructure determine how fast information is incorporated in its stock price. It is important to take this into consideration in a trading algorithm. For instance, when evaluating a decision of posting vs. taking liquidity, one needs to estimate the likelihood the order would get executed before the allotted time to obtain a fill expires. For instance, we would not post passively if, on average, it takes three minutes for an order placed at the bottom of the queue to get executed and we only have thirty seconds to complete a trade. However posting passively would be appropriate if it takes fifteen seconds on average for the queue to turn over. There are several measures used in practice:

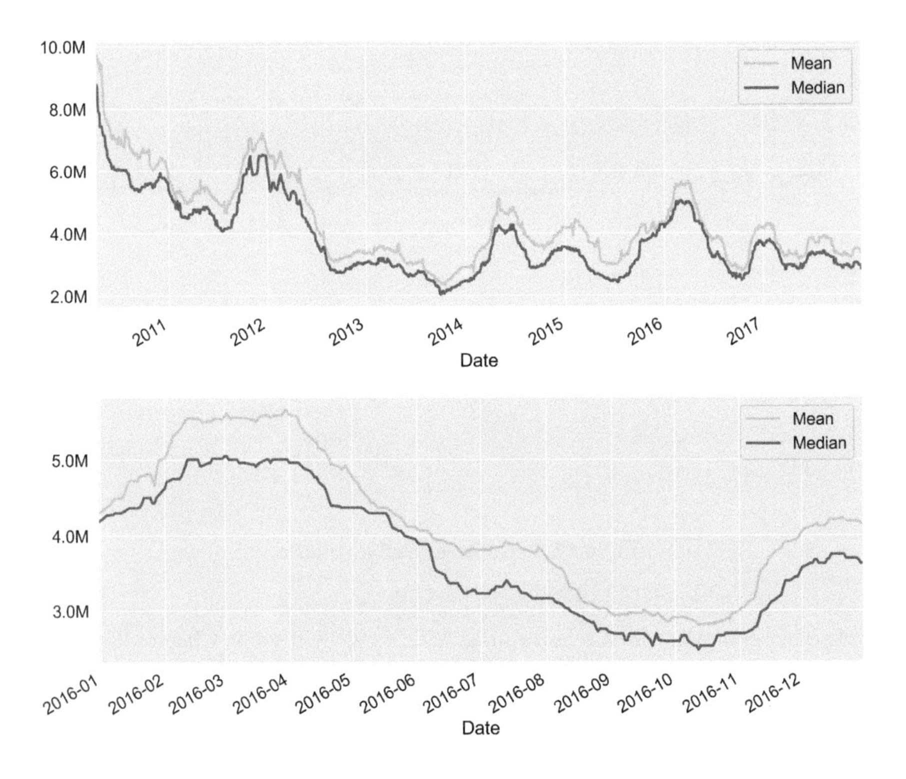


Figure 8.2: Amazon ADV based on Rolling Stats.

- **Straight Clock Time:** A surprising number of traders completely ignore this and just use the same time interval for all stocks. Simple to use, but it is definitively sub-optimal.
- **Event/Tick Time:** The normalization is done by counting a certain number of events such as quote changes, e.g., ten ticks. Often used in HFT, it is probably the most effective method, but requires to keep track of every single tick change which can be computationally intensive. It is also not as easy to conceptualize as the natural clock time as ten ticks could represent microseconds for some heavily traded stocks or could be of the order of seconds or minutes for certain illiquid instruments.
- **Volume Time:** This is the time it takes for a certain specified amount of trade volume (e.g., 1000 shares) to accumulate. This is coarser than tick time, but easier to implement as most algorithms keep track of how much volume has been traded. It also suffers from challenges similar to tick time as the difference between liquid and illiquid instruments can be substantial.
- **Turnover/Characteristic Time:** This measures the time it takes for a quote to move on average (more often median). In a simple setting, a quote moves when

all liquidity is removed from a level. This would be the time it takes for a passive order to get executed, if it is placed in the order book queue. It is a simple measure since this can be calculated in advance from historical data. It also, in a way, represents a natural time scale which makes it easy to conceptualize.

In Figure 8.3, we show a simple example of characteristic times for various stocks calculated as one minute average of daily medians in October 2016. One can see that the differences can be quite large.

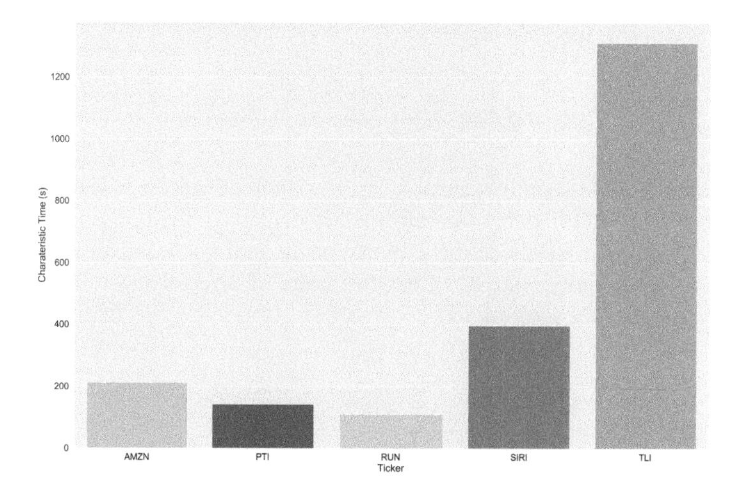


Figure 8.3: Characteristic Time for Various Stocks in October 2016.

### 8.1.3 Intraday Return Normalization: Mid-Quote Volatility

Time scale normalization is a 'horizontal' normalization (in a price chart the $x$-axis is time) and for an effective normalized approach we also need a 'vertical' normalization, i.e., in price space. For instance if we decide to layer the order book as a way to control adverse selection, child orders should be spaced in a way that exhibits a consistent pattern of behavior across different stocks. Placing the orders at the same distance for all stocks would cause volatile stock to always execute at multiple levels and for stable stocks, the layering would be useless as the deeper orders would rarely get executed. A commonly used normalizing variable is mid-quote volatility in ticks, estimated as the average number of ticks the mid-quote moves in a certain time period stated in terms of a characteristic time.

It must be noted that the characteristics of microstructure noise in mid-quote prices are somewhat different from actual transaction prices. It is argued that mid-quotes are less impacted by bid-ask bounce than the transaction prices and are probably closer to the underlying 'efficient' price. The local discrepancies between the observed and the efficient prices may be due to delays in acquiring and processing relevant information. Anderson, Archakov, Cebiroglu and Hautsch (2017) [16] postulate an elegant model extending Roll's model (4.27). Taking '$p_t$' as the log quote-mid-point, the model for price dynamics is stated as:

$$\begin{aligned} p_t &= p_{t-1} - \alpha(p_{t-1} - p^*_{t-1}) + a_t \\ p^*_t &= p^*_{t-1} + \epsilon_t; \quad 0 < \alpha < 2, \end{aligned} \tag{8.1}$$

where '$p^*_t$' is the efficient price. Note the first equation, when $\alpha = 1$ is the same as Roll's model if '$p_t$' is taken to be the transaction price. In this formulation, $\mu_t = p_t - p^*_t$ is taken to be the mispricing component. From (8.1), note that the return variance, because $r_t = -\alpha\mu_{t-1} + a_t$,

$$\mathrm{Var}(r_t) = \frac{2\sigma_a^2 + \alpha\sigma_\epsilon^2}{2-\alpha} = \sigma_\epsilon^2 \, \frac{2\lambda + \alpha}{2-\alpha}, \tag{8.2}$$

which is the mid-quote volatility, and $\lambda = \sigma_a^2/\sigma_\epsilon^2$ denotes the noise to signal ratio. The last term can be used to isolate the information content in the mid-quote volatility. In Figure 8.4 below, we provide the estimate of the mid-quote volatility for various stocks. From the diagram, one can infer the range of price changes from liquid to illiquid stocks.

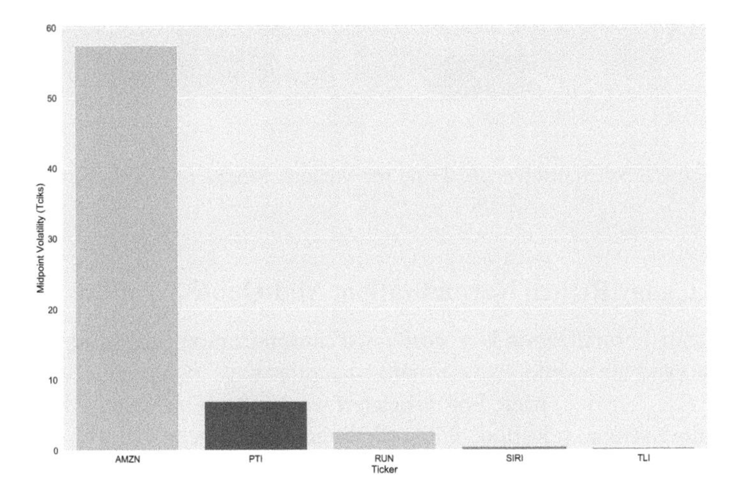


Figure 8.4: Mid Quote Volatility for Various Stocks in October 2016.

### 8.1.4 Other Microstructure Normalizations

A common aspect of naïve trading strategies is trading in an obvious and predictable way. Some high frequency trading (HFT) strategies are trained to spot the presence of large institutional orders, so that they can build out their inventory early and can then sell (buy) back at higher (lower) prices. Since an execution strategy has to trade, it is impossible to remain completely invisible to the watchful eye of other sophisticated market participants, but with some modest ease its impact can be minimized. A common approach used to appear less conspicuous is to normalize decision so as to avoid changing the distributional characteristics of various microstructure metrics that are usually closely monitored. Some examples are:

- **Number of Trades per Second/Period:** This is used as a normalizing variable to control how often one crosses the spread for instance. A noticeable increase in the number of trades per unit of time in a certain direction could easily be interpreted by other participants as a sign of urgency.
- **Average Trade Size:** An unusually large print on an exchange or a larger than average block in a dark venue would be easily spotted and may give away the presence of the strategy in the market. The average trade size also can be used as a normalizer to decide on the size to send to take liquidity. This measure would be different for different types of venue (lit venues, gray venues, dark venues, etc.).
- **Average Quote Size:** Once the decision to trade passively is made, it is also important to size the order appropriately. Placing an order that would noticeably change the displayed size from the average inside quote size would be interpreted as a directional signal especially if the size creates a strong imbalance in the order book (see later on order book imbalance).

Another usage of microstructure normalizers is to support decision making by comparing current conditions with historical averages. A couple of illustrative examples would be:

- **Spread:** When the decision is to cross, in that if the spread has to be crossed to take liquidity, and if the current spread is meaningfully larger than its longer term historical average, it may be an incentive to wait a little longer in the hope that the spread closes in; if the spread is smaller than usual, all else being equal, the spread can be crossed immediately before the spread returns to its historical level.
- **Quote Size:** Certain popular liquidity seeking strategies look at the far side visible quantity and compare it to the historical distribution. If the displayed size is "rare enough" (usually that depends on the urgency) the strategy would cross the spread to capture it before it disappears.[1]

### 8.1.5 Intraday Normalization: Profiles

Incorporating intraday periodicity is a critical component of any execution strategy. Intraday periodicity arises from primarily behavioral and practical reasons

---

[1]We want to mention the predictive power of order book imbalance. One would think that if a larger than usual size appears at the far side it would create a strong imbalance that would push the price in more favorable direction. That is true and the soundness of the approach could be questioned. On the other side if the displayed size is rare enough it could imply an error or a naive trading decision and the quantity available could be large enough to put the algorithm ahead of schedule which would allow it to back off and take advantage of better prices without being forced to catch up to its schedule.

around trading discontinuities such as the open and the close (and the lunch break period in certain Asian markets) and other scheduled events (e.g., the US open for European markets). Information dissemination tends to occur at certain time clusters and so does the trading activity. The intraday variability of the measures affected by this periodicity is so strong that not considering them would lead to sub-optimal decisions. Thus, standard practice is to incorporate intra-day periodicity using normalized profiles, so that the deviations from the expected value are stated in relative terms. While most microstructure variables discussed above showcase some of this seasonality, the most used profiles are:

- Volume Distribution, the percentage of the expected daily volume that will trade.
- Volatility Distribution, the percentage of the expected daily return volatility that will be observed.
- Spread Multiplier, the percentage deviation from the average spread.

Of particular importance is the Volume Distribution (or daily volume profile) as it remains the key ingredient to create a VWAP strategy. For such a strategy, the volume profile is directly used as the trading schedule. This will ensure that the strategy trades proportionally; a larger slice of the order when more volume is expected to trade thus reducing the risk of trading at prices away from the period VWAP.

Volume distribution during the course of a day is most generally shaped like a flattened U (often called the Volume Smile). Volatility and spread distributions are both shaped vaguely exponentially. Volume is generally clustered around the opening auction and the first several minutes after the market opens. This makes intuitive sense as the market is re-opening after a trading halt and all new information accrued overnight needs to be incorporated in the trading decisions. Spread and Volatility are also large during this period signifying the increased uncertainty around the stock valuation. After this period of heightened activity, trading normally settles in and volume flattens out only to pick up again in the period before closing. Mutual funds and other investment firms often use the closing price to determine the net asset value (NAV) for creation/redemption purposes and thus prefer trading around the closing auction to minimize the risk of large negative deviation, while still striving to minimize price impact.[2] With the recent strong push away from active investing into passive instruments, in particular such as ETFs, the shape of the volume profiles has become increasingly back-loaded. It is estimated that the last hour of trading accounts for 25–30% of the daily volume. The Volume Smile is thus turning into a Volume Smirk! Volatility and spread also are at maximum at the open to account for the uncertainty around the market reaction to overnight news. As soon as price discovery takes place, their values quickly settle and gradually decay to minimum usually at the end of the day.

---

[2]In this section, we ignore the fact that certain markets have lunch breaks. This clearly complicates but does not change the overall picture or approach and it's a relatively trivial extension to the treatment of profiles.

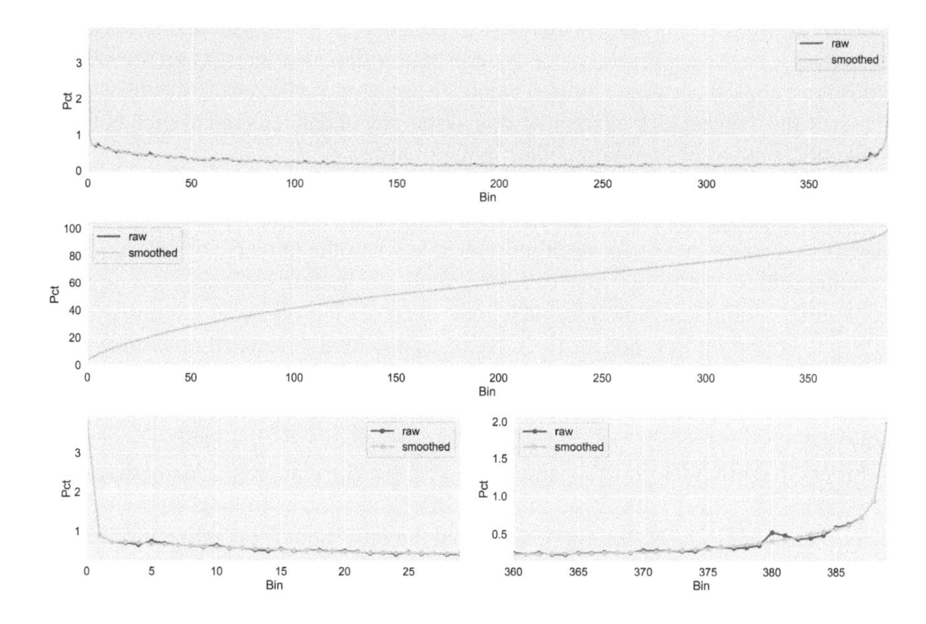


Figure 8.5: Amazon Profile, Cumulative Profile and Smoothing Details.

Having accurate profiles is of critical importance in execution trading and in the next section we delve into some of the practical decisions and considerations on how one would create these profiles. Figure 8.5 provides an example of profile for Amazon.

**Building Intraday Profiles:** Building stable profiles presents some non-trivial challenges at multiple levels. These are both logistic related and data and modeling related. It is a trade-off between availability of data and cost to build and maintain a database. We provide some valuable pointers in building the profiles. This treatment is primarily focused on intraday volume models. Similar approaches can be extended to other variables with strong intraday periodicity.

## Modeling Approaches

As a supervised learning model, intraday periodicity is definitively not trivial to model and until recently these problems have received little coverage. As a consequence, the proposed models are generally complicated and take a lot of effort to create and maintain. While more sophisticated approaches are now being developed, the most common approach is to use a discrete piece-wise linear profile with uniform time bins and creating a baseline model by using a form of per-bin averaging over a variable length of history. With that as a starting point some additional processing is done to improve the predictive performance of the profiles as discussed below. For the reader interested in more quantitative approaches, one can start with Bialkowski et al. (2006) [41] and Kawakatsu (2018) [226].

**Number of Bins:** As previously mentioned the most common approach for developing profiles is to use a discrete, piece-wise linear profile. The decision on the number of bins is a trade-off between required precision and granularity during the most active periods of the day and lack of stability due to sparsity of data/change in each bin. One minute bin is the most common choice in the industry.

**Per Stock Profiles vs. Clustered Profiles:** One of the key decisions to be made is whether to create a profile for each individual stock in the universe or use some sort of group profiles. With a universe of several thousand trading instruments it might be impractical to build and maintain stable individual stock profiles. The stability of the profiles is important in order for their use to be effective. Overly illiquid and slowly moving stocks are usually good candidates for clustered profiles.

A common approach is to use a statistical clustering approach using an appropriate distance measure based on trading characteristics and to identify the number of clusters that provides good separation. Given that the market has many thousands of instruments to deal with, large numbers of clusters, eight to nine hundred is not uncommon. The result of this approach is that the most liquid and active instruments will likely end up in their own cluster, while lower liquidity stocks will form their own clusters.

**Length of History:** Another open question is the length of history that should be used. The more data is used, the more stable the profiles. However unless great care is given to various seasonal effects and continued changes in market microstructure, there is the risk for the profile of not properly incorporating more recent shifts. This is a particular concern now, as the explosion of ETF trading has been consistently shifting more and more volume toward the close. The length of history can be used as a hyper-parameter to be trained on a per-cluster basis.

**Volume Profile Smoothing:** Intraday volume profiles tend to be extremely noisy and require smoothing to be effective. Without smoothing the volume profile performs worse than a flat profile. Care must be taken to avoid smoothing over real discontinuities (e.g., there is a predictable spike in volume in European profiles when the US market opens). These spikes usually are preceded by a period of reduced trading activity, as traders wait for the event. Smoothing these spikes will have the effect of spreading the spikes over previous bins and thus reducing the precision of the profile. One common approach is to remove statistically significant spikes and replace them with a local average for the purpose of smoothing and then bring the spike back. It has been observed that this approach improves the overall predictability of the profile. As far as smoothing algorithms there are a lot of choices available. Kernel smoothing methods discussed in Chapter 2 have been proven effective. However, care must be given at the boundaries (beginning and end) of the profile.

**Special Days:** Special days present additional challenges due to limited data (e.g., triple witching occurs only four times a year). So using larger clusters (one cluster as a limit) is often the only option. One way to handle this would be to use these separately calibrated profiles across all special-day clusters. See Figure 8.6 for an example of FOMC (Federal Open Market Committee) special day. However this would lose the granularity of the regular day clustering. In order to maintain some of the regular cluster level characteristics a common approach is to use a "difference" profile created by a normalized per-bin ratio between an average regular day profile and the special day profile. This captures the special day systematic shift that can then be multiplied back to the cluster profile essentially mixing the two components. While in most cases this has proven effective we would recommend out-of-sample testing at the cluster level to see in which profile this approach is effective.

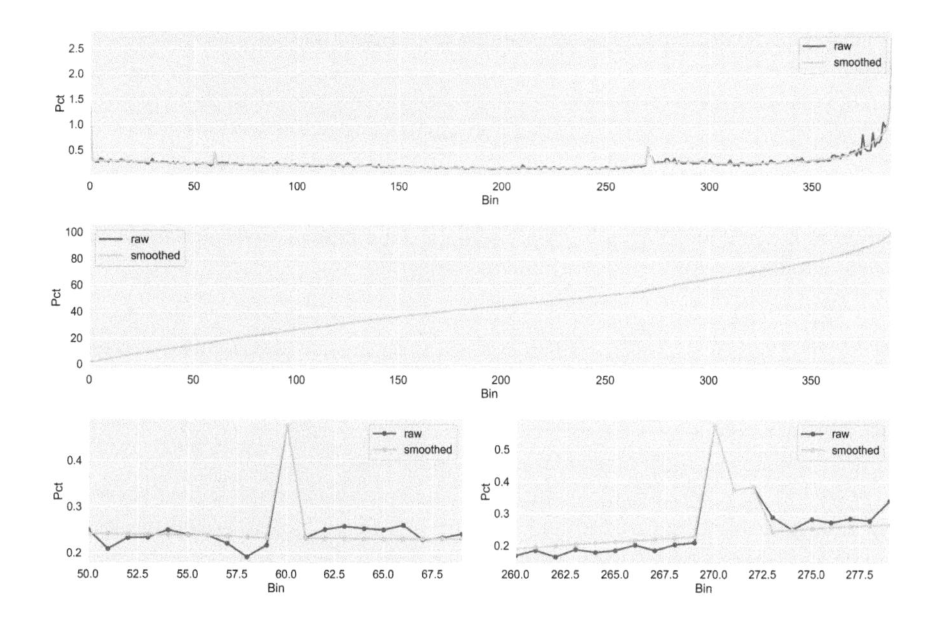


Figure 8.6: FOMC Profile, Cumulative Profile and Smoothing Details.

**Dynamic Profiles:** Volume distributions are relatively stable but for reasons we have discussed above they are subject to large distortions due to some 'surprise' events such as unexpected news. In such cases the volume can spike dramatically and the overall realized distribution will end up being highly skewed. To address this issue, dynamic models that take into account the clustering of volume during surprise events skewing the rest of the day's profile more towards the event must be developed. Most models used in the industry are proprietary and often heuristic and some academic research is called for. They also require a sophisticated real-time analytics infrastructure, and they should be able to instantly measure and apply these adjustments while the algorithm trades.

As a side note on a practical matter, unexpected volume spikes and the related distortion of the volume profile are likely still the most common reason why the execution coverage desks get calls from clients complaining of their poor performance. This is because the volume spikes usually happen in conjunction with a price dislocation.

A strategy trading against a static profile would under-trade around the event thus leading to a significant deviation from the VWAP benchmark.

### 8.1.6 Remainder (of the Day) Volume

A related measure that is of real importance in execution is related to estimating the remainder of the day's volume; how much volume is still expected before the market closes. This is important for sizing the orders to be sent so as to avoid excessive impact. As this will trade into the closing auction, it is important to plan how many shares should be reserved for the auction. We will discuss more of this in the next section.

Denote the expected cumulative volume from time $t$ to $T$ as,

$$rdv_t = \mathrm{E} \sum_{j=t}^{T} V_j. \tag{8.3}$$

The naïve approach is to take the estimated volume for next day as $\tilde{V}_{T-1}$ based on the data up to the previous day and simply remove the cumulative volume up to time $t$, to estimate the rest of the day's volume to come,

$$\widehat{rdv_t} = \hat{V} - \sum_{j=0}^{t} V_j. \tag{8.4}$$

Note this approach ignores the effect of same day events and, considering how noisy the '$\hat{V}$' estimate can be, it leads to relatively poor prediction.

A second approach uses the historical volume profile for each discrete period and gets an aggregate estimate of $\hat{V}$ and the current volume to extrapolate the full day volume and subtract that to the current volume. Let $\tilde{X}_j$ be the proportion of the daily volume expected in time bin $j$. Then

$$\widetilde{rdv_t} = \frac{\sum_{j=0}^{t} V_j}{\sum_{j=0}^{t} \tilde{X}_j} - \sum_{j=0}^{t} V_j. \tag{8.5}$$

This is commonly used and works reasonably well after the noon hour when the activity level is somewhat established and stable. However in the first two hours of the day, the estimates turn out to be quite poor.

The natural model is to combine the two methods (8.4) and (8.5) with per bin shrinkage factor $\lambda_t$. Thus,

$$\widehat{rdv_t}^{*} = (1-\lambda_t)\widehat{rdv_t} + \lambda_t \widetilde{rdv_t}. \tag{8.6}$$

The shrinkage factor $\lambda_t$ is then calibrated at the bin level to give the best mix of static and dynamic weights. We will not discuss the calibration of this model here but only point out some interesting results. As we would expect $\lambda_t$ is small at the beginning of the day, raising quickly to seventy-eighty percent by noon continuing to increase to one. This is reached by construction near the close. What may be a bit surprising is that the slope after the initial jump around mid-day is relatively small implying that the estimate of the daily volume continues to provide stability and a more accurate prediction up until the close. This implies that there is some form of "reversion to the mean" in later parts of the day after some intraday volume surprise. There is some literature to support the view that the trading in the first two hours after market opens sets a tone for the rest of the day. For a recent study of market intraday momentum, see Gao, Han, Li and Zhou (2018) [156].

### 8.1.7 Auctions Volume

It is somewhat surprising that not much academic research is available on such an important aspect of trading. Even for practitioners handling the execution of the opening and closing auctions, this is probably a bit more than an afterthought. This is usually handled as part of the volume profiles estimation, in which the opening and closing auctions are respectively the first and last bin of the profile. As per our earlier discussion of volume profiles this ends up being nothing more than an average of the proportion of daily volume. The estimates can be updated intraday using both dynamic volume profiles and rest of day's volume models as discussed above. More sophisticated approaches leveraging the auction imbalance information that is published by the exchanges (around the last 5–10 minutes of the trading day depending on the exchange) can be used as well.

As mentioned, the current trend towards passive investments and ETFs has driven the liquidity more and more towards the close. This means that the closing auction is becoming increasingly the center of liquidity management. It is quite likely that no other area of research is more relevant and impactful than focusing on better understanding price and volume dynamics during the closing period. In particular in understanding the dynamics in relation to create-redeem activities, across asset classes.

## 8.2 Microstructure Signals

Although the focus of the book is somewhat on alpha signals, no coverage of the subject would be complete without some discussion on the topic of microstructure signals. Because any successful research in this space can be instantly utilized in practice, not much is publicly available. Here we give a brief overview of common signals considered as a starting point for the interested researcher/practitioner. Some of these may have been covered earlier already in the book.

**Mid-Price:** This is taken to represent 'fair' price in the market. With $p^b$ the highest bid price and $p^a$ the lowest ask price, mid-price is

$$M = \frac{p^a + p^b}{2}. \tag{8.7}$$

This changes relatively infrequently and so it is taken to be a low frequency signal.

**Quote Imbalance:** Arguably the most used predictive microstructure signal by practitioners is the quote imbalance. The simplest measure of quote imbalance can be defined as:

$$\text{qimb} = \frac{bs - as}{as + bs}, \tag{8.8}$$

where 'bs' is the bid side quote and 'as' is the ask side quote. The intuition behind why quote imbalance is predictive of the next quote movement is simple: If more buyers enter the market before crossing the spread, they are likely to post at the best bid. The more quotes pile up, the more unlikely it is they will find a seller and eventually a buyer might run out of patience and cross the spread. If the opposite size is small (i.e., the quote imbalance is heavily slanted toward the buy side) it is likely the trade will clear the price level leading to a price move.

**Book Imbalance:** More sophisticated measures of the imbalance consider not only the inside quote but the size distribution over the entire book. A simple version can be stated as a weighted average across multiple levels:

$$\text{obimb} = \frac{1}{k} \sum_{i=0}^{k} w_i (bs_i - as_i), \tag{8.9}$$

where $w_i$ would be a function of the distance of the $i$ price point (possibly normalized by the quote volatility) and $k$ is some appropriate depth of the book. Most academic studies set $k = 5$ or 10 on either side of the order book. With market fragmentation, one may consider the imbalance in the superbook.

**Weighted Mid-Price (or Microprice):** A common approach to incorporate quote imbalance in a 'fair' price is to quote the adjusted mid-point (See example Figure 8.7). This is defined as

$$p^w = wp^a + (1 - w)p^b, \tag{8.10}$$

where $w = bs/(bs + as)$. This has the following intuitive characteristics:

- For zero imbalance this is equal to the mid-price.
- For positive imbalance this tends toward ask price implying that the fair price is higher than the mid-price.
- For negative imbalance this tends to move toward bid price implying the fair price is lower than the mid-price.

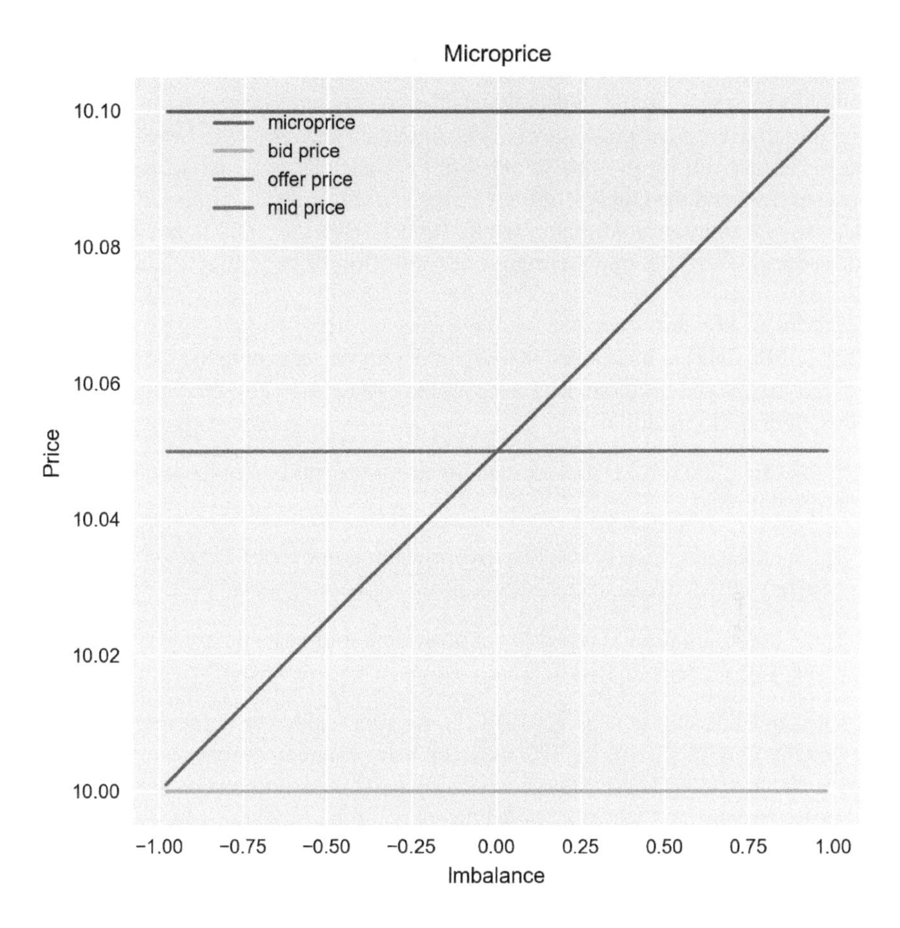


Figure 8.7: Microprice for Different Levels of Imbalance.

One can easily devise other measures of microprices that incorporate order book imbalance as well as the impact of the last traded price. Stoikov (2018) [307] provides a measure of microprice as

$$p^{\text{micro}} = M + g(w, S), \tag{8.11}$$

where $S$ is the spread, $S = P^a - P^b$ and the function '$g$' is applied on the imbalance and the spread to move from mid-price to fair price. The function '$g$' is estimated from past order book data and is calibrated toward efficient price. It is verified that $P^{\text{micro}}$ is a better predictor of future price.

**Trade Imbalance:** For every trading transaction there is always one side that is the initiator of the trade, and in general the counterpart that runs out of patience that decides to pay the spread. This may be due to trading by an informed trader or by a trader who has some urgency to trade. In either case an imbalance between the amount of buyer-initiated volume versus seller-initiated volume would likely be instructive. While this metric is simple to use, the challenge is to categorize a trade as buyer or seller initiated. For most practitioners, who might not have access to Level III message data and have to rely on quotes and trades, this is not trivial as there is a limited amount of research around the efficacy of the various classification algorithms. For Level III data, this information is available for lit venues. However, this is not the case for dark venues, where the trade direction classification of large blocks could be quite valuable.

The most popular approach for trade classification was proposed by Lee and Ready (1991) [241] as mentioned in Chapter 5 with subsequent revisit on this topic by Chakraborty, B. and Moulton, P. C. and Shkilko, A. (2012) [71]. The simplest version of this algorithm is as follows:

- If the execution price is greater than the prevailing mid-price the trade is taken as buyer-initiated.
- If the execution price is less than the prevailing mid-price the trade is classified as seller-initiated.
- If the execution price is equal to the prevailing mid-price the trade is marked the same way as the previous trade.

As noted in Chakraborty et al. (2012) [71], for short sales, the algorithm has proven only moderately predictive to 70% accuracy using contemporary quotes and trades. This is due to various causes. One such cause is the asynchronous nature of the trade and quote streams and where a quote change could be time stamped slightly before the trade that affected it, leading to mis-classification. The classification is shown to improve if the quotes are lagged.

## 8.3 Limit Order Book (LOB): Studying Its Dynamics

A comprehensive study that includes all the key measures discussed in the last section is desirable. Before we formally present the existing models, we want to provide a review of how the analysis of LOB data has been approached. Biais, Hillion and Spatt (1995) [40] study the trading activity in the Paris Bourse (a fully automated limit order market), the dynamics of order flow and how the order flow varies with the state of the order book and marketplace events relevant to the asset. It is found that the conditional probability of a limit order placement is larger, when the bid-ask spread is larger and when the order book is not deep. If a market order has been placed, the chance that the next order posted provides liquidity is generally higher. The placement of orders also follows a pattern with new orders coming in the morning when the price discovery occurs and cancellations and large orders occur in the evenings. The durations between trades do indicate that the intensity of trade varies during the course of a day. The main tool used in these calculations is a contingency table which provides both marginal and conditional probabilities. With market fragmentation, the calculation of these probabilities has become quite complex.

### 8.3.1 LOB Construction and Key Descriptives

It is important to be able to construct the order book from the Trades and Quotes data. Here we present the example and the procedure as discussed in Cao, Hansch and Wang (2009) [67] which is quite elegant. The buy and sell side orders are represented as step functions.

- The height of the first step of the book is the mid-price. For step '$i$', $i = 1, 2, \ldots$, the height on the buy side is the difference in price, $p_i - p_{i-1}$.
- The length of a step '$i$' is the aggregate number of shares, $Q_i$ at price '$p_i$'.
- Similar procedure is carried out in the sell side.
- To make the order book comparable for different stocks, heights are normalized by the price gap between the tenth price and the mid-price.

To illustrate we consider the same data used in Cao et al. (2009) [67, Table II] which is given below in Table 8.1 (partially):

Table 8.1: Descriptives of the Shape of the LOB

| | Length (%) | | Height (%) | |
|---|---|---|---|---|
| Steps | Buy | Sell | buy | sell |
| 1 | 11.97 | 11.94 | 5.07 | 5.23 |
| 2 | 13.23 | 12.67 | 8.55 | 8.97 |
| 3 | 12.06 | 11.56 | 9.25 | 9.59 |
| 4 | 11.04 | 10.77 | 9.84 | 10.12 |
| 5 | 10.59 | 10.49 | 10.24 | 10.45 |
| 6 | 9.91 | 10.16 | 10.63 | 10.62 |
| 7 | 9.55 | 9.87 | 10.95 | 10.71 |
| 8 | 9.06 | 9.58 | 11.28 | 10.88 |
| 9 | 8.80 | 9.52 | 11.71 | 11.38 |
| 10 | 3.80 | 3.43 | 12.48 | 12.06 |

Figure 8.8 below presents the shape of the limit order book. With the presence of high frequency trading it must be noted that this figure changes quite dynamically.

**Slope of LOB:** To begin with, we can consider the slope of the order book both on the demand side and on the supply side, using the quotes on both sides of various depths. This provides information on order book imbalance that can be helpful for the trader to decide on the optimal time to enter or to exit the market. It is observed from Figure 8.8 that the bid-ask spread is at least twice the difference between quotes at successive depths. Thus the slope of the book is steeper closer to the best quotes.

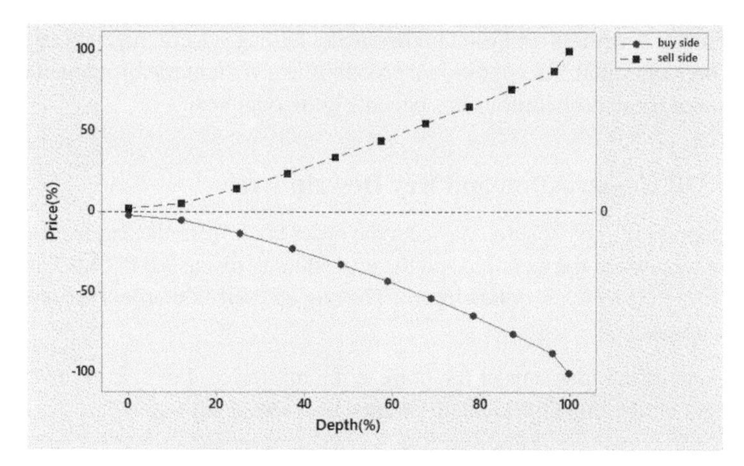


Figure 8.8: The Limit Order Book.

**Order Flow:** The orders can be characterized as buyer or seller initiated and how aggressive they are. The aggressiveness is quantified by the size of the order. Most of the orders are small, indicating that they could be part of a large parent order. Both placement and cancellation of orders seem to decrease away from the best quotes. As the intra-day trading pattern follows typical diurnal pattern, both the depth and order flow must be adjusted for the time of the day. Some observations (stylized facts) worth noting are:

- Large (small) trades on one side of the market tend to be followed by large (small) trades on the same side. The same thing could be said about new limit orders and cancellations as well. It is speculated that this may be due to traders reacting similarly to the same events or due to parent order splitting and automatic placement of child orders.

- Cancellations on the buy side of the book are more frequent after market buy orders and the same holds for sell orders. This may be due to the fact that large sell orders tend to convey negative information about the stock while large buy orders convey positive information. Cancellations may also occur because of placement of some market orders which are intended to probe the presence of hidden orders.

- Conditioning on the state of the book by size of bid-ask spread and the depth, it is observed that trades are more frequent when the spread is tight but new orders inside the quotes are more frequent when the spread is large. The changes in spread are mainly due to liquidity shocks.

- Frequency of trades tends to be clustered because competing traders who monitor the market closely are likely to place the orders when they read the market in their favor. It is observed that the trading activity is more intense with the information flow. Also observed are the following stylized facts: When the spread is large after liquidity shocks, traders place their orders quickly to take advantage of time priority in the queue reforming itself. Thus the spread tends to reverse back to its original level. The expected time interval is generally lower after large trades than after other trades, whether the spread is large or not.

The study by Biais et al. (1995) [40] provides a set of concrete measures that could be used to study the dynamics of the limit order book. We will now delve into model formulations.

### 8.3.2 Modeling LOB Dynamics

**Review of Early Models:** We begin with one of the earlier models to appear in the econometric literature (Lo, MacKinlay and Zhang (2002) [250]); they used the data from Investment Technology Group (ITG) that contained time stamped information on each limit order from submission to cancellation or execution. With three possible paths for each submitted order, the associated (a) time-to-cancellation/modification, (b) time-to-first fill and (c) time-to-completion are tracked and models are developed for each path on both sides of the order book. Because ITG largely dealt with institutional investors, the results apply only to a limited segment of traders. The modeling approach is to get the execution times as the first-passage time to the limit price assuming that price follows a geometric Brownian model with a drift:

$$dP(t) = \alpha \cdot P(t)\, dt + \sigma \cdot P(t)\, dW(t), \tag{8.12}$$

where $W(t)$ is a standard Brownian[3] motion. A buy limit order with a price '$P_l$' is executed in the time interval $[t_0, t_0 + t]$ only if $P_{\min} \leq P_l$ and the probability is

$$P_r(P_{\min} \leq P_l \mid P(t_0) = P_0) = 1 - \left(1 - \left(\frac{P_l}{P_0}\right)^{2\mu/\sigma^2}\right) \Phi\left[\frac{\ln(P_0/P_l) + \mu t}{\sigma\sqrt{t}}\right], \tag{8.13}$$

where $\mu = \alpha - \frac{1}{2}\sigma^2$ and $\Phi(\cdot)$ is the cumulative distribution function of the standard normal.

Note (8.13) can be alternatively stated as, if '$T$' denotes the limit order execution time, $F(t) = P(T \leq t \mid P_0)$ for buy orders and similarly for sell orders, $F(t) = P_r(P_{\max} \geq P_l)$. To see if the model is appropriate for the data at hand compare the theoretical c.d.f. of $F(t)$ to empirical c.d.f. using the well-known result that these c.d.f.'s are uniformly distributed. Fix '$\tau$' as the fixed sampling interval and with $r_t = \ln(P_t) - \ln(P_{t-1})$, the estimates of $\mu$ and $\sigma^2$ are:

$$\hat{\mu} = \frac{1}{N\tau}\sum_{j=1}^{N} r_j, \quad \hat{\sigma}^2 = \frac{1}{N}\sum_{j=1}^{N}\frac{(r_j - \hat{\mu}\tau)^2}{\tau}, \tag{8.14}$$

where '$N$' is the number of observations in the sample.

---

[3] Also known as the Wiener process, $W(t)$ is continuous and $W(t) - W(s) \sim N(0, t-s)$ and is modeled by a random walk.

The above First Passage Time (FPT) model was used for filled orders without taking into account that there were other orders that were canceled or were modified in that duration. The model's other limitations include not accounting for time priority, not including other relevant variables such as price volatility, spreads, etc. The model's performance using the empirical cumulative distribution function is shown to be lacking. The model (8.13) is expanded as

$$F(t) = p_r(T_k \leq t \mid X_k, P_{lk}, S_k, I_k), \tag{8.15}$$

where $T_k$ is the execution of time of $k$ orders, $P_{lk}$ is the limit order price, $S_k$ is the size and $I_k$ is the side indicator. An alternative function, the hazard rate (briefly discussed in Chapter 4)

$$h(t) = \frac{f(t)}{1 - f(t)} = \frac{f(t)}{S(t)} \tag{8.16}$$

where $f(t)$ is the density function, is found to be useful to operationalize and explain the modeling of survival rate of time-based events; Note $S(t)$ is called the survivor function. The censoring information ($\delta_i = 1$ if observation '$i$' is censored) can be incorporated with $t_i$, $i$th realization of the random variable, $T$. It is assumed that the censoring mechanism is independent of the likelihood the limit order is executed. Using the generalized Gamma (defined in Chapter 4) for $f(t)$, the models are estimated.

The following variables are used as explanatory variables:

- Distance between mid-quote and limit price.
- Prior trade indicator for buyer or seller initiated.
- Measures of liquidity.
- Measures of market depth.
- Measures that reflect change in trading activities.

It is shown that this model does fare better than the FPT model to capture the execution times. But an important limitation is "...censoring as a result of prices moving away from the limit price would be a violation of the underlying assumption since prices at the time of censoring are not included in $X_i$." We have observed in our analysis of Level III data, a key determinant for cancellation of an order is price moving away from the limit price that the trader desired.

**Other Early Models:** There are a number of studies that have looked into order book dynamics using models that arise from point processes that were briefly discussed in Chapter 4. Bouchard, Mezard and Polters (2002) [51] investigate the interaction between order flow and liquidity; it is observed that the distribution of incoming limit order prices follow a power law around the current price and the overall shape of the book in terms of volume on both sides is rather symmetric. Smith, Farmer, Gillemot and Krishnamunthy (2003) [304] and Farmer, Gillemot, Lillo, Mike and Sen

(2004) [140] study how the order flow influences price formation; the price impact of market orders is a function of touch price depth and spread size.

To have a more complete and realistic picture of how the LOB evolves, we must also consider the presence of 'hidden liquidity'. Almost all exchanges allow traders to hide all or portions of their orders. The so-called, "iceberg order" is executed into several smaller parts and is queued along with the other orders, but only the displayed quantity is visible and is part of the market depth. When the order reaches the front of the queue only the displayed order is executed. While in practice hidden liquidity represents a significant portion of exchanges liquidity; in the models discussed below, we will not address the hidden liquidity issue. This will be commented on in a later section.

Modeling of LOB dynamics has drawn on ideas from economics, physics, statistics and psychology. The approach taken in economics literature is to focus on the behavior of traders and the dynamics is modeled as sequential games (Rosu (2009) [294]). Others, such as researchers from physics, have treated order flows as random and statistical mechanics techniques are used to study the dynamics. Other studies, such as Parlour and Seppi (2008) [282] and Bouchaud et al. (2009) [49] are useful references for studies that come from an economics point of view. We will review some models that are of recent origin.

**Recent Models:** Our coverage here is by no means complete as we discuss only select models. In our choice, we mainly went with empirically verified models.

**Cont, Stoikov and Talreja (2010) [88]:** In this model of the limit order book, the number of limit orders at each price level in the book is taken to be a continuous-time Markov[4] chain. The evolution of the order book through market orders, limit orders and cancellations is represented as a counting process[5]. The model replicates the evolution of the events using independent Poisson processes. The study views LOB as a system of queues subject to order book events whose occurrences are modeled as a multidimensional point process. Before we state the questions and the models implications, we want to define certain useful quantities.

It is taken that the price grid is $P(n) = \{1, 2, \dots, n\}$ multiples of a price tick. The number of outstanding orders in the book $|X_t^i|$ at price $i$, collectively, $X_t \equiv (X_t^1, \dots, X_t^n)$ is taken to be a continuous-time Markov chain; if $X_t^i < 0$, then there are $-X_t^i$ bid orders at price '$i$' and if $X_t^i > 0$, then there are $X_t^i$ ask orders at price $i$. Now we have,

$$\begin{aligned} \text{Best Ask Price: } p_A(t) &= \inf\{i : X_t^i > 0\} \\ \text{Best Bid Price: } p_B(t) &= \sup\{i : X_t^i < 0\} \\ \text{Mid-Price: } p_M(t) &= (p_A(t) + p_B(t))/2 \\ \text{Spread: } p_S(t) &= p_A(t) - p_B(t) \end{aligned} \tag{8.17}$$

[4]This just means that the current state of the order book depends only on the previous state.

[5]This process keeps track of the number of various events over time.

The depth of the order book is stated relative to best bid and best ask on either side of the book. This way the model can be applied to the order book at any moment of time. Define

$$Q_i^B(t) = \begin{cases} X_{p_A(t)-i}(t), & 0 < i < p_A(t) \\ 0, & p_A(t) \le i < n \end{cases}$$

and (8.18)

$$Q_i^A(t) = \begin{cases} X_{p_B(t)-i}(t), & 0 < i < n - p_B(t) \\ 0, & n - p_B(t) \le i < n \end{cases}$$

Thus, $Q_i^B(t)$ and $Q_i^A(t)$ denote the number of buy orders at a distance '$i$' from ask and the number of sell orders at a distance '$i$' from bid, respectively. This representation highlights the shape (or depth) of the book relative to the best quotes on either side as in Figure 8.8.

The following assumptions are made on the order arrivals: Market buy or sell orders arrive at independent exponential times with rate, $\mu$. Limit buy or sell orders arrive at a distance of '$i$' ticks, from the opposite best quote at independent exponential times with rate $\lambda(i)$. The cancellations of limit orders occur at a rate proportional to the number of outstanding orders, $\theta(i)X$. All these events are assumed to be mutually independent.

How the limit order book is updated with the inflow of above events can be described as follows: For a state $x \in \mathbb{Z}^n$ and $1 \le i \le n$, let $x^{i\pm1} = x \pm (0, \dots, 1, \dots, 0)$, where '1' denotes the change in the $i$th component. For example, a limit buy order arrival at price level, $i < p_A(t)$ increases the quantity at level '$i$', from $x$ to $x^{i-1}$ with rate $\lambda(p_A(t) - i)$. Similarly, a limit sell order arrival changes the order book from $x$ to $x^{i+1}$ with rate $\lambda(i - p_B(t))$ for $i > p_B(t)$. A market buy order decreases quantity at the ask price $i$ and hence $x \to x^{p_A(t)+1}$ with rate '$\mu$' and a market sell order at the bid price, '$i$' will change the book status, $x \to x^{p_A(t)+1}$ with rate $\mu$. Cancellation of a buy order at price level, '$i$', will decrease the quantity at the rate $\theta\left(p_A(t) - i\right)|X_p|$ for $i < p_A(t)$ and cancellation of a sell order will decrease the quantity at the rate of $\theta\left(i - p_B(t)\right)|X_p|$ for $i > p_B(t)$.

It is assumed that the limit order arrival rate $\lambda(i)$ follows a power law

$$\lambda(i) = \frac{\kappa}{i^\alpha}, \tag{8.19}$$

which is confirmed by several empirical studies. The empirical estimates of the parameters are based on the following quantities: $s_m$ is the average size of market orders, $s_l$ is the average size of limit orders, and $s_c$ is the average size of canceled orders. Also let $N_l(i)$ be the total number of limit orders that arrived at a distance '$i$' from the opposite best quote and $T_*$ is the total trading time in minutes and $N_m$ is the number of market orders during the same time. Then

$$\hat{\lambda}(i) = \frac{N_l(i)}{T_*}, \quad 1 \le i \le 5 \tag{8.20}$$

where $\hat{\lambda}(i)$ is extrapolated beyond five positions using the power law in (8.19) simply by minimizing $\sum_{i=1}^{5}(\hat{\lambda}(i) - \frac{\kappa}{i^{\alpha}})^2$, over '$\kappa$' and '$\alpha$'. The arrival rate is estimated as:

$$\hat{\mu} = \frac{N_m}{T_*} \cdot \frac{s_m}{s_l} \tag{8.21}$$

The cancellation rate as noted earlier, is defined as proportional to the number of orders at that price level,

$$\hat{\theta}(i) = \begin{cases} \dfrac{N_l(i)}{T_* Q_i} \cdot \dfrac{s_c}{s_i}, & i \leq 5 \\ \hat{\theta}(5), & i > 5 \end{cases} \tag{8.22}$$

It is understood that the cancellation is not due to executing market orders.

While these descriptives are easy to compute and are intuitive, the main interest in modeling high frequency dynamics of the order book is to predict short-term behavior of various key quantities, that were described earlier and might help in algorithmic trade executions. Some relevant questions are as stated before: Given the state of the order book, what is the probability that mid-price will move up, what is the probability of executing both buy and sell orders at the best quotes before the price changes, etc.

We present some key results in Cont et al. (2010) [88]. These results generally make good intuitive sense. The probability of queue going up when there are no orders in the queue, for $1 \leq d \leq 5$, given that the best quotes are not changing is

$$p_{\text{up}}^{d}(m) = \begin{cases} \dfrac{\hat{\lambda}(d)}{\hat{\theta}(d)m + \hat{\lambda}(d) + \hat{\mu}}, & d = 1 \\ \dfrac{\hat{\lambda}(d)}{\hat{\theta}(d)m + \hat{\lambda}(d)}, & d > 1. \end{cases} \tag{8.23}$$

Other questions such as the probability that the mid-price goes up when the spread $\geq 1$, etc., are based on the following quantities:

- Probability of a market buy order:

$$\frac{\mu^a}{\mu^b + \mu^a + \sum_j (\lambda_B(j) + \lambda_A(j) + \theta(j)Q_j^A(t) + \theta(j)Q_j^B(t))} \tag{8.24}$$

- Probability of a limit buy order $d$ ticks away from the best ask:

$$\frac{\lambda_B(d)}{\mu^b + \mu^a + \sum_j \left(\lambda_B(j) + \lambda_A(j) + \theta(j)Q_j^A(t) + \theta(j)Q_j^B(t)\right)} \tag{8.25}$$

- Probability of a cancel buy order $d$ ticks away from the best ask:

$$\frac{\theta(d)Q_d^B(t)}{\mu^b + \mu^a + \sum_j \left(\lambda_B(j) + \lambda_A(j) + \theta(j)Q_j^A(t) + \theta(j)Q_j^B(t)\right)} \tag{8.26}$$

The model was evaluated using data sets from the Tokyo Stock Exchange, and is shown to capture realistic features of the order book profile.

The use of Poisson processes to model the flows of limit orders, market orders and cancellations makes this method analytically tractable. This model captures the steady-state shape of the order book, but it may not be useful for predicting the short-term behavior of LOB which is more relevant from the traders' point of view. The assumption of Poisson process results in that the intensity of arrivals and cancellations does not depend on the state of the order book, which is somewhat unrealistic given the feedback loop relationship between the behavior of the market participants and the history of the order book. Recent works show that the memory-less property may not be empirically valid. Orders tend to be clustered due to dependencies between liquidity taking and liquidity providing orders and also due to algorithmic execution of parent orders that are split into numerous child orders. Also, the assumption that all orders are of the same size is also too simplistic.

Models based on marked point processes are proposed when the mark besides price, represents order types, size, etc. The joint modeling assumes that the order flow is self-exciting, i.e., new orders or cancellations occur in an orderly point process with an intensity function that depends on the history of the limit order book. These models are somewhat difficult to estimate and do not yet fully capture the empirical results such as volatility clustering, etc.

**Queue-Reactive Model:** Most market participants consider their target price with reference to market price, which depends on the order flows and the resulting changes in the state of the LOB. Huang, Lehalle and Rosenbaum (2015) [206] study the LOB dynamics via a Markov queuing model when the market price remains constant and how the market price changes. The change in market price can occur when the execution occurs at the best price or a new order appears within the spread. This dual split of the model is termed 'queue-reactive model'. In the framework considered, three specific scenarios are possible: bid and ask sides are independent, bid and ask sides are independent except for the first two positions, bid and ask sides are independent on either side and cross dependence between bid queue and ask queue is possible.

We will not delve into extensive details here, but briefly indicate the main gist of the model. Going for '$k$' positions on both sides of the order book, $X(t) = (q_{-k}(t), \dots, q_{-1}(t), q_1(t), \dots, q_k(t))$ is modeled as continuous time Markov jump process. The key quantities in the calculations are the intensity rates of arrivals of limit orders, cancellations and market orders. Under the independence assumption, the following behaviors are observed:

- Limit order arrival in the first position is approximately a constant function of queue size. At other positions away from the best positions, the intensity is a decreasing function of queue size.

- Intensity of order cancellation is an increasing concave function of $q_{\pm k}$, not linear as in Cont et al. (2010) [88]; but after the first position, intensity is much lower as they are likely to move to the top of the book.

– The rate of market order decreases exponentially with the available volume in the best positions. For the second position the shape of the intensity function is similar to the first position but the intensity decreases after that.

In the case of dependency where bid side and ask side are correlated to each other, the above observations include the consideration of queue sizes on the other side as well. In a paper with similar ideas, Abergel and Jedidi (2013) [2] present a model that captures a stylized description of order book with the assumption of independent Poissonian arrival times. They also show that the price process converges to a Brownian motion.

### 8.3.3 Models Based on Hawkes Processes

**An Introduction:** Hawkes (1971) [186] introduced a model for self-exciting and mutually exciting point processes which capture the property that occurrence of an event increases the chance of further events occurring. Before we describe its applications in LOB modeling, we want to briefly discuss the main features of the model. If we let $\lambda(t)$ be the conditional intensity measured through the change in the number of events that occur in the time interval $(0, t)$ given the information available, then the self-exciting process has the intensity

$$\lambda(t) = \mu + \int_0^t \gamma(t-u)\, dN(u) = \mu + \sum_{T_i<t} \gamma(t - T_i), \tag{8.27}$$

where $0 < T_1 < T_2 < \cdots < T_n < \cdots$ are the times when events occur. When $\mu > 0$ and $\gamma(u) = 0$ this provides a Poisson base level for the process. The function $\gamma(u) \geq 0$ is called the exciting kernel. Each event will increase the intensity which will then decrease until the next event occurs again when the intensity will go up again. Some of the kernels suggested in the literature are:

$$\begin{aligned} &\text{Exponential kernel: } \gamma(u) = \alpha \cdot \beta \cdot e^{-\beta u}, \qquad \mu > 0 \\ &\text{Power law kernel: } \gamma(u) = \frac{\alpha\beta}{(1+\beta u)^{1+p}}. \end{aligned} \tag{8.28}$$

Here '$\alpha$' represents the overall strength of excitation and '$\beta$' controls the relaxation time.

Hawkes (2018) [187] provides a review of these processes in financial applications. Two extensions that are found to be useful are: One related to marked point processes, where associated marks with the event can trigger the excitement and the second relates to processes where there are different types of events and how they can influence each other:

$$\begin{aligned} &\text{Marked Hawkes Process: } \eta(\lambda_t) = \mu(t) + \sum_{T_i<t} \gamma(t - T_i, \xi_i), \qquad T_i < t \\ &\text{Mutually Exciting Process: } \lambda_{i,t} = \mu_i + \sum_{j=1}^{D} \sum_{T_{j:r}<t} \gamma_{ij}(t - T_{j:r}). \end{aligned} \tag{8.29}$$

Here '$\xi_i$' are marks such as 'volume' and '$D$' are the different types of events with their own point processes.

**Models for Volatility Clustering:** The features of volatility clustering and the significant autocorrelations in durations between order arrivals and significant cross-correlation of arrival rates across various event types are better captured by the multi-dimensional Hawkes process. For a given $M$-dimensional point process, let $N_t = (N_t^1, \dots, N_t^M)$ denote the associated counting process and the Hawkes process is characterized by intensities, $\lambda^m(t), m = 1, \dots, M$ as

$$\begin{aligned} \lambda^m(t) &= \lambda_0^m(t) + \sum_{n=1}^{M} \int_0^t \sum_{j=1}^{P} \alpha_j^{mn} e^{-\beta_j^{mn}(t-s)} dN_s^n \\ &= \lambda_0(t) + \sum_{n=1}^{M} \sum_{t_j<t} \sum_{j=1}^{P} \alpha_j^{mn} e^{-\beta_j^{mn}(t_j - t_i^n)}, \end{aligned} \tag{8.30}$$

where the number of exponential kernels, $P$, is fixed and $t_i^n$ is the $i$th jumping time of the $m$th variate. The scale and decay parameters, $\alpha^{mn}$ and $\beta^{mn}$, express the influence of the past events $t_i^n$ of type '$n$'. The baseline model in (8.30) does not incorporate the effect of bid-ask spread on order flow. It is shown in the literature that $\alpha$ and $\beta$ parameters do depend on the bid-ask spread. A simulated two-dimensional Hawkes process is given in Toke 2011 [313]. As noted earlier, the main characteristic of the Hawkes process is that intensity goes up at each event and decays exponentially between events.

The evolution of the order book is driven by different intensity functions, but the model changes with the value of the spread. The base intensity and the decay rates are all assumed to be functions of the spread. The model is quite complex and requires the estimation of a large number of parameters, which in turn requires a vast amount of data for calibration.

We will briefly discuss the likelihood estimates of the model in (8.30). First, we want to observe for a Poisson process, the intensity function $\lambda(t) = \lambda_0$, is a constant. The log-likelihood of the model,

$$\ln \mathcal{L}(\{t_i\}_{i=1,2,\dots,N}) = \sum_{m=1}^{M} \ln \mathcal{L}^m(\{t_i\}), \tag{8.31}$$

where

$$\ln \mathcal{L}^m(\{t_i\}) = T - \sum_{i=1}^{N} \sum_{n=1}^{M} \frac{\alpha^{mn}}{\beta^{mn}} (1 - e^{-\beta^{mn}(T-t_i)}) + \sum_{t_i^m} \ln[\lambda_0^m(t_i^m) + \sum_{n=1}^{M} \alpha^{mn} R^{mn}(l)]$$

and

$$R^{mn}(l) = \sum_{t_\kappa^n t_l^m} e^{-\beta^{mn}(t_l^m - t_\kappa^n)}$$

is the cumulative decay function. The Multiplicative Random Time Change Theorem states that with a certain transformation of the variables, the Hawkes process can be changed to a unit rate homogeneous Poisson process. Transforming the time variable '$t$' into '$\mathcal{T}$', where

$$\mathcal{T} = \int_0^t \lambda(s)\, ds \tag{8.32}$$

Hawkes process becomes a Poisson process with unit rate. Thus,

$$\mathcal{T}_i - \mathcal{T}_{i-1} = \Lambda(t_{i-1}, t_i) = \int_{t_{i-1}}^{t_i} \lambda(s)\, ds \tag{8.33}$$

is exponentially distributed.

More generally in the '$M$' dimensional multivariate case,

$$\Lambda^m(t^m_{i-1}, t^m_i) = \int_{t^m_{i-1}}^{t^m_i} \lambda^m_0(s)\, ds + \int_{t^m_{i-1}}^{t^m_i} \sum_{n=1}^{M} \sum_{t^n < t^m_{i-1}} \alpha^{mn} e^{-\beta^{mn}(s-t^n)}\, ds$$
$$+ \int_{t^m_{i-1}}^{t^m_i} \sum_{n=1}^{M} \sum_{t^m_{i-1} < t^n < s} \alpha^{mn} e^{-\beta^{mn}(s-t^n)}\, ds$$

will follow an exponential distribution. The $\Lambda^m(t^m_{i-1}, t^m_i)$ are known as compensators and it can be verified empirically if they follow an exponential distribution.

**An Application of Hawkes Processes:** We want to illustrate the model with an application. We first look at the statistical properties of the data and then look at the results of fitting one and two dimensional Hawkes processes to it. Our findings show that a two dimensional model based on Hawkes processes performs considerably better than a model based on two independent Poisson Processes. This provides strong evidence of the fact that the two key features of Hawkes processes (path dependency of the order flow and the possibility of modeling the interaction between the dimensions) are capable of capturing and replicating some of the key features of the empirical data. Moreover, there is evidence of significant and asymmetric interplay between the buy and the sell sides for the Limit Orders and evidence of limited interplay between the two sides when it comes to Market Orders. Finally, a study of the statistical properties of the orders showed a complex of alternating symmetric and asymmetric behavior of the LOB.

The INET data for XOM (Exxon Mobil) for the 1st of September 2010 is used here. Figure 8.9 shows the submission frequency plot and we can see how in the first hour and a half and in the last half an hour of the trading day there is a much higher submission frequency (with peaks of 25/30 orders submitted per second) than during the rest of the day (around 5–10 orders per second). This means that if we choose to look at the limit order flow with a model that does not allow for a varying baseline intensity, we will have to drop these two periods of increased trading activity and keep the data for only the middle portion of six and a half hours of the trading day.

We then look at the data for the central portion of the six hours of the trading day and find that market orders are just a fraction (4%) of Limit Orders. Similarly, there is an imbalance in the side of the order submission (see Table 8.2). This shows how,

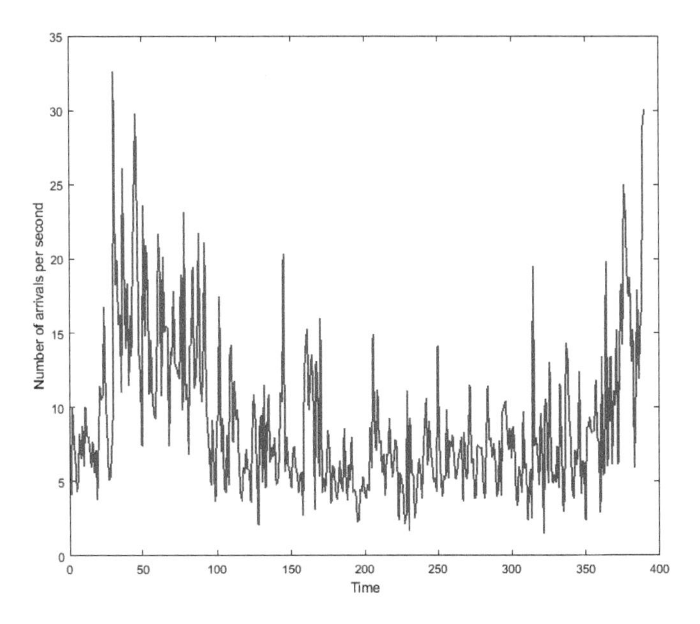


Figure 8.9: Frequency of Limit Order Submission for 09/01/2010 (time in minutes after market opening).

Table 8.2: XOM data for 09/01/2010

| | Limit Orders | Market Orders | Total |
|---|---|---|---|
| Buy | 48,541 | 2,107 | 50,648 |
| Sell | 41,194 | 1,882 | 43,076 |
| Total | 89,735 | 3,989 | 93,724 |

on this trading day, most of the "action" was happening on the Buy Side of the book which could mean that market participants were confident in an increase of the price of XOM (in line with the historical performance of the stock for that day).

The next step was to re-construct the Limit Order book from the order flow so as to look at the life of the Limit Orders (LO) after their submission. From Table 8.3, only about 6.5% of all submitted LO get at least partially filled with the percentage dropping to only about 6% for those LO that actually get completely executed. Also, out of the LO that are completely filled, nearly 90% of them are filled in just one execution. This allows us to say that the size magnitude of the LO is comparable to that of the Market Orders, given that nearly all of them require only one Market Order to be fully filled.

It can be observed that very few limit orders are actually submitted close to the market (hence have a chance to get executed) or that many of them do not stay in the book long enough to get executed. This led us to look into the life-time of the limit orders and results are given in Table 8.4. In this particular case, limit orders stay in the

Table 8.3: LO executions and percentage breakdown by number of fills required

| Number of LO | % at least partially fill | | % completed in one fill | | | % completely fill |
|---|---|---|---|---|---|---|
| 89,735 | 6.57 | | 5.35 | | | 6.15 |
| # fills | 1 | 2 | 3 | 4 | 5 | 6 | ≥ 7 |
| | 87.14 | 10.83 | 1.34 | 0.37 | 0.13 | 0.06 | 0.13 |

Table 8.4: Mean LO cancellation and execution times (in seconds)

| No Fills | | Partial Fills | |
|---|---|---|---|
| Buy Orders | Sell Orders | Buy Orders | Sell Orders |
| 177.8 | 250.5 | 46.1 | 63.0 |
| First Fill | | Completion | |
| 25.3 | 56.2 | 25.6 | 57.6 |

book for much longer if they are on the sell side, than if they are on the buy side. This life-time difference for limit orders on the two opposite sides of the book is consistent with what we saw earlier in the order imbalance of the limit order and market order submissions. In fact, in this case as well we see how the buy side appears to be the more active side of the book. The shorter execution time for the orders on the buy side could be caused by either more aggressive pricing or by the price movements of the markets while the shorter lifespans and the higher number of orders on that side seem to indicate the use of more active trading strategies. However, there is another interesting observation that can be made. We see that on average a sell limit order that is filled for the first time (but that does not get completely filled), is canceled after about 7 seconds, compared to 20 seconds that it would take had the order been on the buy side. This means that on average it takes three times longer for a partially (and never completely) filled order to be canceled if it is on the buy side (which is the more active side) that if it is on the sell side (which is the less active side). This asymmetry between buy and sell sides at the aggregated level highlights the different dynamics between liquidity demanders and providers.

The next step was to observe changes in the behavior of the book when moving further away from the market. Observe the order submission and cancellation frequencies as a function of distance from the market price. Figure 8.10 and Figure 8.11 show that most limit orders are submitted and canceled at the market (respectively, 40% and almost 30%) and that the rates drop sharply with each price level further away from it. This supports the conclusion that the very small percentage of executed orders is not caused by the fact that most limit orders are submitted far away from the market and hence do not get the chance to get executed. We can also see how approximately 70% (96%) of all Limit Orders are placed in the top 5 (20) levels of the book and that 69% (96%) of all cancellations also occur in the top 5 (20) levels. This is a first indication of the fact that Level III data offers more information than Level II (with the top 5 levels of the book) we are only able to capture around 70% of all the occurring events. However, this also indicates that there is no need to use all the information provided in Level III data because by looking at only the top 20 levels of the book we can capture nearly all of the occurring events. Both plots show a very similar behavior for the buy and sell side which led us to assume an almost symmetric behavior of the book. This assumption was later tested and confirmed at a 99% confidence level with a two-sided Kolmogorov-Smirnov (KS) Test.[6]

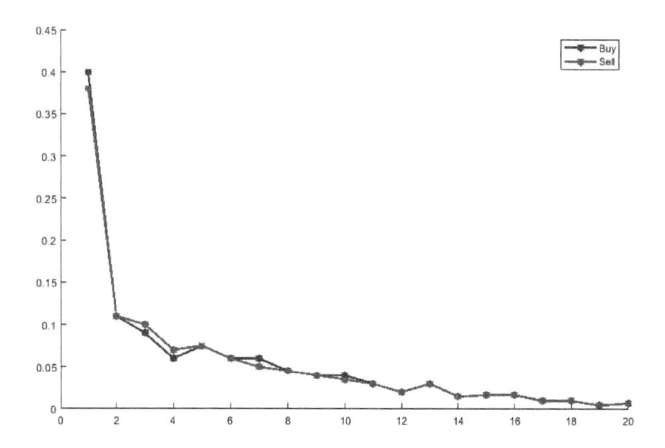


Figure 8.10: Submission Frequencies–Distance near touch price
KS Test Statistics= 0.61 → fail to reject null at 1% sig. level.

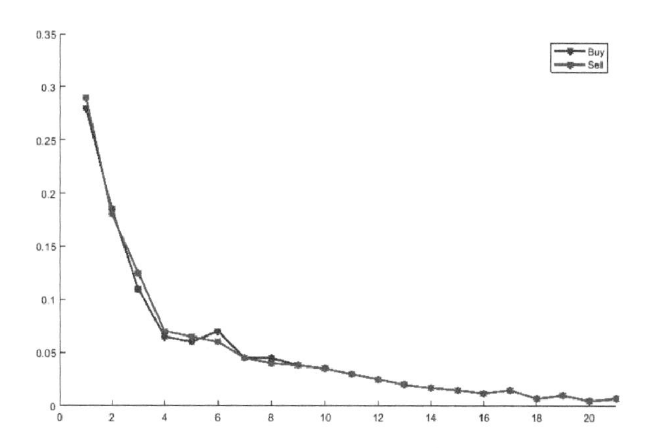


Figure 8.11: Cancellation Frequencies–Distance from near touch price
KS Test Statistic= 0.83 → fail to reject null at 1% sig. level.

[6]The KS test is based on the distance between two empirical cumulative distribution functions, $F_1(x)$ and $F_2(x)$; for additional details refer to a book on mathematical statistics.

A better understanding of what happens to the limit orders once they enter the book is key in developing a more accurate way of choosing when to submit a limit or a market order. For this reason, we examined the cancellation frequencies as a function of position in the queue and we found that (Figure 8.12) most of the cancellations occur when the limit orders are in fourth or fifth position in the queue. In this case as well, we carried out a test of the assumption of a symmetric behavior of the book which confirmed previous similar findings. However, it would be more interesting to look at the cancellation frequencies as a function of overall distance from the market (as a measure of distance from possible execution).

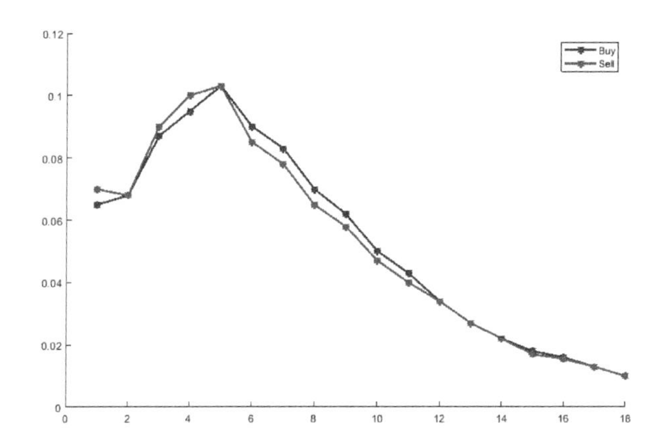


Figure 8.12: Cancellation Frequencies—position in queue.

With two more trading days (3rd and 9th of September 2010) data available, we are able to study the robustness of the above results. The submission and cancellation frequencies as a function of relative price, which were symmetric for the 1st of September, were not symmetric anymore on the 3rd of September. In this trading day, the peak of activity on the buy side occurred at twelve price levels away from the market. This seems to suggest that on the 3rd of September the markets were betting that the price of XOM would drop in the foreseeable future, possibly reacting to some news event. A closer investigation of that time period revealed in the previous days, the announcement by Exxon Mobil of the plan of closing a refinery, which then led to a drop in the price shortly after 09/03/2010.

A second interesting difference was in the amount of volume on the two sides of the book. Across the three days, there was a significant volume imbalance with considerably more volume on the sell side than on the buy side but the amount of imbalance between the two sides changed significantly across days. Overall there is no explanation as to why there would be such a strong volume imbalance in favor of the sell side other than some kind of temporary anomaly. These findings seem to point out the fact that the shape and behavior of the LOB change dramatically across different days and that, in order to capture a more general/average behavior of the book, many more days of data are required. This preliminary analysis is simply meant to illustrate the complex behavior of limit order book dynamics.

**One-Dimensional Models:** We begin with a one dimensional model for the Limit Orders submitted on the buy side and one for those submitted on the sell side. Similarly, we build a one dimensional model for the Market Orders submitted on each side of the book. In the one dimensional case, the intensity function driving the Hawkes process is of the form

$$\lambda(t) = \lambda_0 + \alpha \sum_{t_i<t} e^{-\beta(t-t_i)} \tag{8.34}$$

with $\lambda_0$ being the baseline intensity, $\alpha$ the jump in the intensity after the occurrence of an event and $\beta$, the rate of exponential decay of the intensity after the occurrence of an event.

Table 8.5 and Table 8.6 show the results of the MLE estimation of the parameters of the one dimensional models for the Limit Orders on the two sides of the book. The difference in the values of the estimates of the parameters for the two sides of the book (strongest on the 9th and weakest on the 1st of September) indicates that there is a level of asymmetry between the two sides. Also, even though the magnitude of the values of the parameters is the same for all three days (on each side of the book) there is still some variation between them, especially when we look at the model for the buy side. This seems to suggest that the behavior of the book can change significantly from day to day and that a considerable amount of data might be necessary in order to capture any kind of general behavior.

Table 8.5: 1-D Limit Order Model (buy side)

| | $\lambda$ | $\alpha$ | $\beta$ |
|---|---|---|---|
| 09/01/2010 | 0.008108431 | 1.421767060 | 1.435094466 |
| 09/03/2010 | 0.006127003 | 1.290012415 | 1.299564704 |
| 09/09/2010 | 0.002341542 | 1.023410145 | 1.026757239 |

One of the advantages of having chosen an exponential decay for the intensity function of the process, becomes apparent when we interpret these parameters. Using a well-known property of exponentially distributed variables, we can tell the half-life of the excitation effect caused by the occurrence of an event. For example in one-dimensional models, we find that, on average, half of the excitation effect caused by

Table 8.6: 1-D Limit Order Model (sell side)

| | $\lambda$ | $\alpha$ | $\beta$ |
|---|---|---|---|
| 09/01/2010 | 0.002055786 | 1.071071139 | 1.073989940 |
| 09/03/2010 | 0.002342963 | 0.983678355 | 0.986999404 |
| 09/09/2010 | 0.006123845 | 0.946420813 | 0.955774394 |

the occurrence of an event is dissipated after about 0.7 seconds ($\ln 2/\beta \approx$ half-life). Also, the small estimated values for $\alpha$ indicate that the occurrence of an event has a very small impact on the likelihood of another event occurring. This seems to suggest that there is a weak self-excitation effect in the limit order process, when modeling it with a one-dimensional model.

We then build, in a similar fashion, the two one-dimensional models for the flow of buy and sell Market Orders for all three days and obtain the following parameter estimation (see Table 8.5 and Table 8.6). The first thing that is obvious is the asymmetry

Table 8.7: 1-D Market Order Model (buy side)

| | $\lambda$ | $\alpha$ | $\beta$ |
|---|---|---|---|
| 09/01/2010 | 0.0238743 | 7.3187478 | 19.8477514 |
| 09/03/2010 | 0.01450873 | 3.38168999 | 7.70581048 |
| 09/09/2010 | 0.0267254 | 4.8818834 | 13.5427254 |

Table 8.8: 1-D Market Order Model (sell side)

| | $\lambda$ | $\alpha$ | $\beta$ |
|---|---|---|---|
| 09/01/2010 | 0.02083999 | 9.38842281 | 24.55777564 |
| 09/03/2010 | 0.01616096 | 6.64370599 | 16.68090519 |
| 09/09/2010 | 0.0239992 | 8.2180945 | 25.0115980 |

between the buy and the sell sides. Nevertheless, for the Market Orders the asymmetry appears to be much more significant than it was for the Limit Orders, given the bigger differences in the values of the estimated parameters. Also, for Market Orders we see that the occurrence of each event seems to have a much stronger impact, on the likelihood of occurrence of future events than was the case for Limit Orders. In fact, the values of $\alpha$ in Table 8.7 and Table 8.8 are considerably higher than those we saw in Table 8.5 and Table 8.6. Similarly, the values of $\beta$ are also much larger in the Market Order models than they were in the Limit Order models which means that the half-life of the excitation effect is much shorter for Market Orders (0.03 seconds vs. 0.7 seconds) than for Limit Orders. These differences in the results of the estimations are not that surprising if we consider what we saw in Table 8.2 for order submissions. Market Orders are submitted in much fewer numbers than Limit Orders, which explains why even though it occurs, such self-excitation has a very short lifetime. The significant increase in the chances of another Market Order occurring given that one has just occurred (even though for a very short time), might seem to contradict the fact that very few Market Orders actually occur. However, what can be inferred is that Market Orders depend strongly from the occurrence of other Market Orders and that they tend to cluster around submissions.

**An Extension:** An important limitation of this basic Hawkes Process model is that each event has the same impact on the intensity function. This implies that each event carries the same amount of information in the process. However, it is easy to imagine that a large LO would carry much more information about future price changes than a small one and this should be accounted for in the model. We include order size in the intensity function by scaling the size of the jump $\alpha$ in the value of the intensity by the ratio $\frac{w_i}{\overline{w}_i}$, where $w_i$ is the size of the LO and $\overline{w}_i$ is the average LO size up to that point. The intensity function for this "extended" one-dimensional model becomes

$$\lambda(t) = \lambda_0 + \alpha \sum_{t_i<t} \frac{w_i}{\overline{w}_I} e^{-\beta(t-t_i)}. \tag{8.35}$$

We can then compare the performance of this "size-adjusted" model with the "basic" one by looking at the Q-Q plot of the compensators for the two models. But the results show how in both cases (buy and sell Limit Orders) adjusting for size does not improve the fit of the model. This result is somewhat counter-intuitive given more than 70% of all Limit Orders are for 100 shares. This would imply that any deviation from this size doesn't have an impact on traders ability to conceal their intentions, and that there is no significant information left in the size of LO once orders are submitted.

**Two-Dimensional Models:** The next step is to build two-dimensional models that would simultaneously model the buy and sell sides of the book and with two separate (yet interacting) intensity functions. The intensity functions for this kind of model become

$$\begin{aligned}
\lambda^B(t) &= \lambda_0^B + \int_0^t \alpha^{BB} e^{-\beta^{BB}(t-u)}\, dN_u^B + \int_0^t \alpha^{BB} e^{-\beta^{BS}(t-u)}\, dN_u^S \\
\lambda^S(t) &= \lambda_0^S + \int_0^t \alpha^{SB} e^{-\beta^{SB}(t-u)}\, dN_u^B + \int_0^t \alpha^{SS} e^{-\beta^{SS}(t-u)}\, dN_u^S
\end{aligned} \tag{8.36}$$

and we have five parameters for each dimension of the model for a total of ten parameters across two dimensions (buy and sell). The two additional parameters in each dimension account for the cross-excitation effect between the two dimensions. In our case, $\alpha^{BS}$ tells us the increase in the likeliness of a LO occurring on the buy side after the occurrence of a LO on the sell side, while $\beta^{BS}$ tells us the rate of decay of the buy side intensity after the occurrence of a LO on the sell side (and vice-versa for $\alpha^{SB}$ and $\beta^{SB}$). We then used MLE to estimate the parameters of the model and can make a few remarks about the results (Table 8.9).

Table 8.9: Two-Dimensional Limit Order model

| | | | | |
|---|---|---|---|---|
| $\lambda^B = 2.480 \cdot 10^{-5}$ | $\alpha^{BB} = 14.386$ | $\beta^{BB} = 31.517$ | $\alpha^{BS} = 0.303$ | $\beta^{BS} = 0.464$ |
| $\lambda^S = 2.677 \cdot 10^{-5}$ | $\lambda^{SS} = 15.657$ | $\beta^{SS} = 37.299$ | $\alpha^{SB} = 0.149$ | $\beta^{SB} = 0.312$ |

There is some asymmetry (see Figure 8.13) in the cross-excitation effects with the occurrence of an event on the sell side having a stronger impact on the likeliness of occurrence of an event on the buy side rather than the other way round ($\alpha^{BS} > \alpha^{SB}$). On the other hand though, since $\beta^{BS} < \beta^{SB}$, the effect on the sell side of an occurrence on the buy side is more persistent than that of an event on the buy side on the

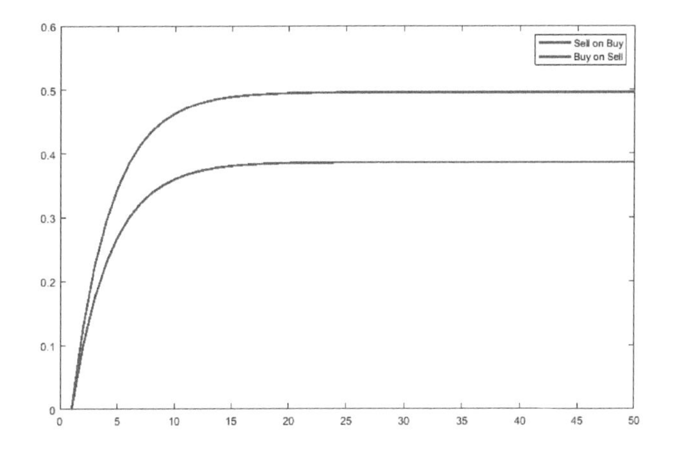


Figure 8.13: Asymmetry in Cross-Excitation Effect.

sell side (2.22 seconds vs. 1.49 seconds). The self-excitation components of the model are rather similar for both dimensions and it is interesting to point out the very short persistence of the self-excitation effect in both cases (about 0.02 seconds).

In order to evaluate whether the two key features of a model based on Hawkes processes (path dependency of the order flow and the possibility of modeling the relation between the various dimensions) are really helpful in improving the performance of the model, we can compare the performance of the above model to that of a bi-dimensional model based on two independent Poisson Processes (note that Poisson process does not account for any dependency in the order flow and by construction this assumes independence between the two dimensions). The results proved to be very encouraging with a remarkably better fit to the empirical data for the Hawkes Process based model.

Following a similar procedure, we build a bi-dimensional model for the Market Orders and performed MLE to estimate the parameters. The results (Table 8.10) showed very weak and symmetric cross-excitation between the dimensions which, however, turns out to be very persistent (141 seconds for the half-life of the effect of buy on sell and 215 seconds for that of sell on buy). The self-excitation effect is comparable across the two dimensions and is also very short lived (with a half-life of about 0.02 seconds).

Finally, we compared the performance of our model to that of the bi-dimensional Poisson Process. Similarly to the LO case, the model at hand significantly outperforms the Poisson Process one for both dimensions.

Table 8.10: Two-Dimensional Market Order model

| | | | | |
|---|---|---|---|---|
| $\lambda^B = 2.701 \cdot 10^{-5}$ | $\alpha^{BB} = 11.793$ | $\beta^{BB} = 36.153$ | $\alpha^{BS} = 0.0037$ | $\beta^{BS} = 0.0049$ |
| $\lambda^S = 3.451 \cdot 10^{-5}$ | $\alpha^{SS} = 12.454$ | $\beta^{SS} = 35.063$ | $\alpha^{SB} = 0.0017$ | $\beta^{SB} = 0.0032$ |

**Limitations and Findings:** There are some important considerations to be made from this analysis. The first issue is related to the cleaning of the data that was necessary in order to adapt it to model fit. In fact, Hawkes Processes are "simple" processes that do not allow for more events to occur simultaneously. Unfortunately, in the INET data it is quite common to encounter clusters of 10/20 events reported with the same timestamp. This is typical of high frequency data due to the close proximity with which events are often reported and to the obvious limitation of the chosen time scale (e.g., milliseconds). In our case we decided to keep only the first event of each cluster, dropping all the others. A possible alternative used in the literature is to simply uniformly distribute the events from the clusters across the time interval given by the occurrence of the cluster and that of the next event. However, this approach could introduce bias in the parameter estimation phase. In fact, the clustering of events by itself is of some interest.

Also, an important limitation of the bi-dimensional Hawkes Process model is that it is very computationally intense. Evaluating the five parameters for one of the two intensity functions using data from only one trading day took almost two days. It becomes clear that in order to evaluate the parameters of a model using data for a longer time period or for evaluating the parameters for a model with higher dimension (hence with more parameters) will require considerably more computational power.

In this section, we have provided a brief discussion of the Hawkes process for modeling the limit order book. As demonstrated here with an example, the work requires further attention on empirical modeling.

## 8.4 Models for Hidden Liquidity

Investors have the option to place limit orders that are not visible to other traders but are observable to exchange officials. These are called 'hidden' or 'iceberg' orders. The invisible order retains price but not time priority. When the visible portion gets executed, another fraction of the hidden order becomes visible. On the NYSE, an order improving the quotes can be fully hidden. The posting of hidden orders is a second stage decision by the traders after choosing between market and limit orders. The risk of placing limit orders has been discussed earlier, and an important element of it is the transparency or exposure risk that can signal to other market participants the intentions of the trader. It is also speculated that orders are hidden when there is increased participation of informed traders and to minimize price impact when the execution probability is fairly high.

De Winne and D'Hondt (2007) [324] study the choice between hidden and visible limit order placement via a logit model. The predictors include characteristics related to exposure and picking-off risks, such as order size relative to order book depth, the competitiveness of the price and the imbalance of the order book. Traders who monitor the order book closely can infer the presence of hidden orders and their depth by repeated posting on the opposite side. Using the detailed (more than level III) data from Euronext, orders are classified for their aggressiveness based on how much liquidity is taken out from the best opposite quote. It is observed that generally hidden orders are less aggressive than other limit orders. In modeling the data, because the exposure risk is larger toward the market close, the likelihood of more hidden orders at the near end of a trading session is also taken into account and so is the trading intensity that varies during the day.

The predictors of the models and the sign of estimated coefficients are listed below:

- Spread (+)
- Order size as the ratio of depth on the same side and on the opposite side (+)
- Market imbalance based on best five visible quotes (−)
- Time left before the market close (+)
- Order aggressiveness measured by the five orders submitted prior to incoming order (−)
- Price aggressiveness, measured by the ratio of distance between the order price away from the best price on the same side and the price away from the opposite side's best price (−)

It is concluded that the relationship between the decision to hide and the order aggressiveness on the opposite side is ambiguous. All other estimates generally agree with what was postulated. When order aggressiveness is used as the response variable, it is shown that the traders adjust their order submissions when they see a signal of hidden order on the opposite side. The overall conclusion is that hidden orders appear to be posted by non-informed traders.

Using the data from the Spanish Stock Exchange, Pardo and Pascual (2012) [280] study the market reaction to the presence of hidden orders discovered during the trading process. The high frequency data reveals that traders on the opposite side become aggressive once they detect the hidden volume, but the price impact of hidden volume is purely temporary.

To illustrate the presence of hidden orders, we consider the Level III data whose description is given in Section 1.4.2 (Table 1.10). With this, we can construct the order book in real time (Figure 8.8). For modeling, some key elements considered are:

- The depth (in number of shares) at that price level before and after the order is submitted.

- The relative price equal to 1 'at the market' and 2 at '1 price level away from market, etc.

If the event is execution or cancellation,

- The distance in cents from opposite best price in the book.
- The duration the limit order has been in the book.

Some that are common to all events (excluding Hidden Executions) are:

- Whether the Super Book (the market) is Locked, Crossed or in a normal state.
- The relative spread.
- The number of shares at the Top of the book on that side, and on the opposite side of the book.
- The number of shares at the Top 5 price levels on that side, and on the opposite side of the book.
- The weighted average price of the shares on the Top 5 levels on that side, and on the opposite side of the book.
- The relative spread of the weighted prices.

If the event is a hidden order execution:

- The price and the size at which the Hidden Order is executed.
- An indicator of what side is the trade; in fact, given there is no BUY or SELL indicator for hidden trades, the Lee-Ready Rule is applied.
- The distance in cents from opposite best price in the book.

To illustrate, we use CISCO data for a single day; CISCO stock is heavily traded and thus captures typical intense market activity. To avoid the bias in results due to excessive trading activities around the opening and closing of the market, we consider only the duration between 9:35 a.m. to 3:55 p.m., thus discarding data outside these time limits. Table 8.11 below provides how the orders submitted on both sides (Buy and Sell) are dealt with. It is clear a majority of orders gets canceled with only 13%

Table 8.11: Distribution of Orders

| | Buy | Mid-point | Sell | Total |
|---|---|---|---|---|
| Canceled | 166975 | — | 175693 | 342648 (86.9%) |
| Executed (Visible) | 21041 | — | 23595 | 44636 (11.3%) |
| Executed (Hidden) | 1702 | 3864 | 1394 | 6190 (1.8%) |
| Total | 189718 | 3814 | 200662 | 394194 |

of orders getting executed.

It is interesting to note that although only 1.8% of the trades are hidden, they seem to occur and be inter-dispersed through the trading day. Figure 8.14 indicates that hidden orders are clearly embedded with the visible orders. Two useful observations follow from Table 8.12 below: The hidden orders are generally executed by orders

Table 8.12: Type of Executed Orders Versus Side (entries are median price and depth 1)

| Type of Order | Buy | No Action | Sell | Overall |
|---|---|---|---|---|
| Visible | 20.50 | — | 20.51 | 20.50 |
| | 39863 | — | 39682 | 39800 |
| Hidden | 20.50 | 20.52 | 20.54 | 20.51 |
| | −150 | 50 | −50 | 50 |
| All | 20.50 | 20.52 | 20.51 | 20.50 |
| | 36047 | 50 | 36606 | 32765 |

placed closer to the top of the book and they seem to result in better price.

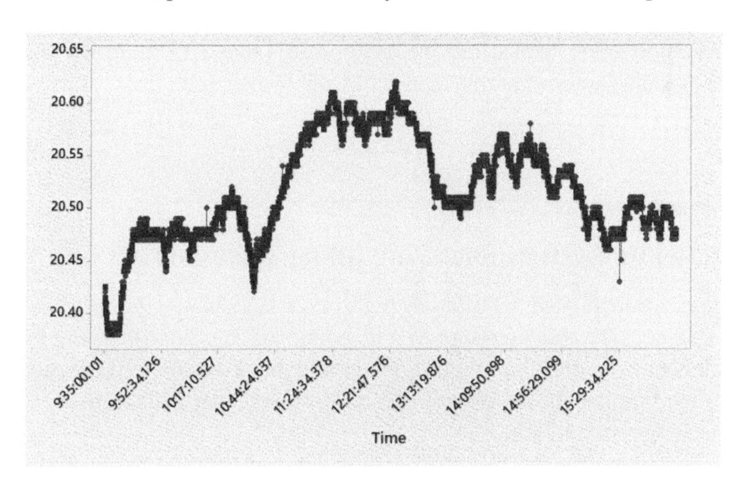


Figure 8.14: Price Plot (Red: Hidden; Blue; Visible).

Recent work by Bessembinder, Panayides and Venkataraman (2009) [38] evaluates the strategies for order exposure using Euronext-Paris Bourse data; the hypotheses are stated below:

- Order exposure increases execution probability and decreases time-to-completion.
- Order exposure decreases execution probability and increases time-to-completion (supported).
- Hidden orders are used primarily by (uninformed) traders to mitigate the option value of a standing limit order.

- Hidden orders are used primarily by (informed) traders to protect themselves against defensive trading strategies (not true).
- Hidden orders usage is more for larger orders, aggressively priced orders and when adverse selection risk is high.

The discovery and estimation of hidden depth is of interest to traders; while discovery has been discussed in great detail in the context of dark pools, we want to comment on the existence and execution of hidden orders in the presence of a vast majority of visible orders. We study this via transition matrix; probability that the next trade is a hidden trade given the status of the current trade, whether it is visible or hidden:

Table 8.13: Transition Matrix

| | | Current Trade | |
|---|---|---|---|
| | | Visible | Hidden |
| Next Trade | Visible | 0.95 | 0.35 |
| | Hidden | 0.05 | 0.65 |

This information along with depth information in Table 8.13 indicates that hidden orders are executed somewhat together at the top of the book.

## 8.5 Modeling LOB: Some Concluding Thoughts

With the availability of voluminous high frequency data, as we saw in this chapter, there is a considerable literature to understand market microstructure issues and order book behaviors. In a thorough study on order book dynamics, Hall and Hautsch (2006) [171] formulate the key questions faced by market participants as the following hypotheses:

(1) An increase of the depth on the ask (bid) side

- increases the aggressiveness of market trading on the bid (ask) side,
- decreases the aggressiveness of limit order trading on the ask (bid) side,
- increases the probability of cancellations on the ask (bid) side.

(2) Past price movements are

- negatively (positively) correlated with the aggressiveness of market trading on the bid (ask) trade,
- positively (negatively) correlated with the aggressiveness of limit order trading on the ask (bid) side,
- negatively (positively) correlated with the probability of cancellations on the ask (bid) side.

(3) A higher volatility decreases the aggressiveness in market order trading and increases the aggressiveness in limit order trading.

(4) The higher the bid-ask spread, the lower the aggressiveness in market trading and the higher the aggressiveness in limit order trading.

It is empirically demonstrated that it is important to take into account the state of the order book to better understand the submission and cancellation behaviors. It is also clear based on the INET data analysis provided earlier that, while the Hawkes process tends to do better than other approaches, there is still a lot of room to improve on the models presented in this chapter.

# 9

# *Market Impact Models*

## 9.1 Introduction

Market Impact (MI in short) is likely the single most important concern in trading today. It is also complex and the least understood. It is something every trader experiences, but it cannot be easily measured a priori (only the execution costs are directly observed); and it can only be estimated ex-post and even then far from any accuracy. We can broadly identify where it comes from and we intuitively sense that we can, at least partially, control it through our trading approach, but we do not fully understand its dynamics and how to alter it as many, many players are involved in a market.

To complicate matters further, large scale trading decisions most often arise from an expectation of price momentum. This means that trading will often occur in the presence of market drift (alpha) and the realized execution cost is a combination of alpha and the price impact. In ex-post analysis, decomposing the alpha from the expected impact is extremely arduous, if at all possible, and it is particularly a constant concern of any quantitatively driven (or not) trading investment as it is critically important to achieve an optimal execution. For instance, if we find that most of the cost is due to price drift, the best option would be to accelerate the trading and pay more in impact to capture the more attractive prices. Conversely, if the cost is mostly driven by impact then it would behoove us to slow down our trading to minimize the impact.

Additional complications arise from the complexity of the market structure with its different trading phases (continuous trading, auctions, etc.) and types of venues (lit venues, gray pools, dark pools, etc.). This requires an ability to measure the differential impact of trading in a particular way in a particular venue, if one hopes to optimally access that liquidity.

Unlike many other practical topics outlined in this book, MI or price impact (PI) has been extensively studied by academics for decades on its own or as part of the more general problem of Optimal Execution which is covered in the next chapter. Unfortunately, the research is somewhat abstract, stylized and is mathematically cumbersome, such that it has only marginally (and very slowly at that) impacted (pun intended) the consideration by traders and portfolio managers. More than thirty years has passed since the seminal paper by Kyle (1985) [234] that launched a large body of work on market impact in the industry. The most often used transaction cost model (model estimating the expected cost of an order implicitly incorporating market impact) is the Bloomberg Transaction Cost Analysis (TCA), a simple (as far as anybody can tell) model of the form:

$$\mathrm{TC} = 0.33\,\sigma\sqrt{\frac{Q}{ADV}} + 0.5\,\mathrm{ba}_{\mathrm{sp}}, \tag{9.1}$$

where $Q$ is the parent order size, ADV is the average daily volume, $\sigma$ is the stock volatility and $\mathrm{ba}_{\mathrm{sp}}$ is the average bid-ask spread.

This chapter cannot hope to cover this important topic in any great detail, but it will strive to build an intuition of the drivers of MI that the reader can expand upon. It will then introduce the reader to the most common approaches to market impact modeling used in practice with some insight on how to calibrate them. The chapter will end with a historical review of the most relevant theoretical and empirical research in the area of market impact to benefit the readers. Empirical studies in market impact modeling have been very few. This chapter has an illustration with a large data set based on the authors' work.

## 9.2 What Is Market Impact?

There are several definitions of market impact, some are more insightful than others. One that is general and more intuitive is:

**Market Impact is the cumulative market-wide response of the arrival of new order flow.**

This definition incorporates the effect of both trades and order submissions (or cancellations). It also includes the effects that, while changing the overall system "hidden state," do not immediately lead to a price change. There are several likely components to this response:

- The first is a purely "mechanical" one. The removal of liquidity, either via a trade or via a cancellation, can lead to change in the prevailing mid-price.
- The second is "economic related." Any change in the shape of the order book alters the supply/demand sides leading to a re-establishment of local equilibrium price.
- Another component can be called "speculative." A price move may lead participants to expect the initiation (or continuation) of a momentum trend, triggering some "copycat" trading in the same direction.
- Then there is an "arbitrage" component where participants try to step in to provide liquidity in prevision of the price move and immediately get out of it at a better price.

These effects are then propagated through the system with various levels of feedback loops leading to very complicated "chaos-like" macro market dynamics. There is also evidence that these effects are amplified by the "novelty" of the price move meaning that the behavior can be highly path dependent.

## 9.3 Modeling Transaction Costs (TC)

As discussed, MI is a quantity that cannot be directly observed but can be inferred. What is observed are the prices that one pays to execute an order of a certain size, for a particular instrument, over a specified time period. These prices incorporate different components that can be broadly grouped into specific categories:

1. **Explicit Costs:** These costs, in theory at least, are known in advance. These include exchange fees, clearing costs, taxes, etc. They also include all the fixed costs for connectivity, for accessing market data, etc. Part of these costs are independent of how the order is traded and depend only on the number of shares traded and total notional value (e.g., clearing fees, taxes). Other costs vary with implementation because they depend on the choice of venue (different venues have different fee schedules) and the order aggressiveness (in most venues one is charged to take liquidity, while receiving a rebate for providing liquidity).

2. **Spread Costs:** It is often necessary to cross the spread to take liquidity and the price achieved is thus higher than the prevailing price (usually defined as the midpoint). If the order is executed passively, the spread cost could be negative.

3. **Exogenous Price Evolution:** At the risk of oversimplifying, this is the theoretical price path that would have happened had we not traded.

4. **Market Impact**, as defined in the previous sections.

We begin with a review of some commonly used TC models: Average TC models. These are macro level models that estimate the cost broadly as a function of characteristics related to order and execution strategy. We ignore the explicit costs and exogenous price evolution. This is an oversimplification but can be rationalized with a few assumptions:

- We are interested in an average transaction cost, averaged over many orders.
- Price behaves, to a large extent as a random walk.
- We assume that the arrival of orders is uncorrelated to any specific price trajectory on the day of trading.

Under these assumptions the price evolution accounts only for uncorrelated noise in the cost estimate and can be ignored. Thus in our simplified framework for transaction cost we have two remaining components: Market impact and spread costs.

$$\mathrm{TC} = f(\mathrm{mi}, \mathrm{sc}) = \mathrm{mi} + \mathrm{sc}. \tag{9.2}$$

In most cases, an additional simplification is made in that the two components are additive.

We start with the simpler component: Spread Costs. This cost is due to crossing the spread on a portion of the order. If the spread is crossed every time the trade occurs, this component could be modeled as half the average bid ask spread $\mathrm{ba}_{\mathrm{sp}}$. In general, during the execution of an order, we may trade aggressively, incur the half spread and when we trade passively, we are paid the half spread. Since execution strategies are by definition liquidity demanding, the ratio will be tilted towards aggressive trading. Thus, the most basic spread cost can be modeled as a portion $\beta$ of spread. It should be noted that the spread cost depends on other features also. For example, it seems intuitive that if there is more urgency, the more often the trader will be willing to cross the spread. Thus, the spread cost could also be a function of an urgency parameter, which is often proxied by the average trading rate. Once we incorporate bid-ask spread and the trading rate, we can, as a first approximation, assume that the spread cost is not dependent on other stock and order characteristics. Thus,

$$\mathrm{TC} = \mathrm{mi} + \beta \mathrm{ba}_{\mathrm{sp}}. \tag{9.3}$$

We turn our attention to the average market impact. Clearly, the size of the parent order $Q$ is the primary driver of market impact. As we discussed in Chapter 8 size must be adjusted for the liquidity level of a stock. So a normalized order size $\frac{Q}{\mathrm{ADV}}$ is usually used. Additionally it is well known that, all else being equal, the more volatile the stock is the more the expected cost. The intuition is that volatility (and spread to some extent) is a measure of uncertainty around the "true" value of the stock, and so one would then expect that the effects of "sensing" a large order in the market would be magnified. Note that volatility is correlated to spread, and so higher volatility stocks will have a higher spread cost.

A more general average TC model is

$$\mathrm{TC} = f\left(\frac{Q}{\mathrm{ADV}}, \sigma\right) + \beta\, \mathrm{ba}_{\mathrm{sp}}. \tag{9.4}$$

The functional form of most models assume a multiplicative power law relationship for both relative size and volatility:

$$\mathrm{TC} = \alpha \sigma^{\delta} \left(\frac{Q}{\mathrm{ADV}}\right)^{\gamma}. \tag{9.5}$$

It is well established in MI research that the exponent $\gamma$ for relative size is less than one (relatively) well approximated by 0.5 and thus (9.5) is known as a square root law. This is an empirical fact and there are several theories for why this is the case. See Gatheral (2010) [161] for a discussion about this. The square root formula implies that the cost of trading is independent of the time it takes to liquidate. If ADV and $\sigma$ are fixed, the price impact depends only on the trade size.

For the exponent $\delta$ of volatility there is some debate and the data seem to indicate that the model is somewhat mis-specified. On one hand, there is some evidence that a stock twice as volatile is twice as expensive to trade, so many models suggest $\delta = 1$. On the other hand when we compare average costs in different volatility regimes, it has been shown that if the whole market were to dramatically increase in volatility, like at the height of the 2008 financial crisis, the average cost did not increase by the same amount, indicating that $\delta < 1$. One way to account for that would be to use two measures of volatility: Overall market volatility, $\sigma_M$, with an exponent less than 1 and relative volatility $\frac{\sigma}{\sigma_M}$ that has exponent of 1. The most common approach is to re-calibrate the model over time fixing $\delta = 1$ and allowing varying $\alpha_t$ to absorb the effect of any regime change. We thus come to the "standard" functional form for a TC model:

$$\text{TC} = \alpha\sigma\sqrt{\frac{Q}{\text{ADV}}} + \beta\,\text{ba}_{\text{sp}}. \tag{9.6}$$

For many years, traders have used this simple sigma-root-liquidity model described for example by Grinold and Kahn (1994) [169]. The model (9.3) stated in another form:

$$\text{TC} - \text{Spread Cost} = \alpha \cdot \sigma\sqrt{\frac{Q}{\text{ADV}}}. \tag{9.7}$$

Thus the relative size (RS), $\frac{Q}{\text{ADV}}$ is the key factor in the pre-trade cost estimation and the impact is proportional to volatility and RS. The square root of relative size is expected because risk capital should be proportional to the square-root of the holding period. In Toth, Lemperiere, Deremble, Lataillade, Kockelkoren and Bouchaud (2011) [314], the authors present an argument that if latent supply and demand are linear in price over some plausible range of prices, which is a reasonable assumption, market impact should be square-root.

The square-root formula as stated in (9.7) refers only to the size of the trade relative to daily volume. It does not refer to, for example, the rate of trading, how the trade is executed and the market capitalization of the stock and other stock characteristics. If the trading is quite aggressive, the square-root formula tends to break down. Moro, Vicente, Moyano, Gerig, Farmer, Vaglica, Lillo and Mantegna (2009) [272] observe that the price path during the execution follows a power law:

$$p_t - p_0 = \alpha \cdot \left(\frac{t}{T}\right)^{2/3}, \tag{9.8}$$

where $T$ is the duration for the execution of the parent order. Immediately after completion of a parent order, the price begins to revert. Models used in industry are mostly functions of the determinants identified above in Equations (9.7) and (9.8).

The reader may notice that MI and thus TC are highly dependent on the trading approach but standard models do not seem to incorporate anything that accounts for the type of trading strategy. An important reason for this has to do with data availability. To calibrate such a model, we need data that will allow us to calibrate the average effect of the use of different strategies to trade the same amount of stock over the same amount of time. Such experimental data is not readily available to most researchers.

## 9.4 Historical Review of Market Impact Research

As we had discussed at the beginning of this chapter, one of the main focuses of Algorithmic Trading is to minimize execution cost. When a large order is executed in the market, it causes market impact (MI), which is commonly defined as the deviation of the post-trade market price from the market price, that would have prevailed had the trade not occurred (Fabozzi, Focardi and Kolm (2006) [131]). To alleviate MI and the implied transaction cost (TC), a trader may use an algorithmic trading strategy to slice a large "parent" order into a sequence of "child" orders. While a slicing strategy is typically motivated by the liquidity constraint associated with executing a large order in one print, it is usually balanced against the risk and opportunity costs associated with the execution of a sequence of child orders over a longer duration.

In asset management, with the competitive focus of the industry on cost management and with the diminishing returns of crowded strategies, understanding and managing MI has become a crucial component of a systematic investment process. For an estimate of its size among the transaction costs that include explicit costs such as brokerage commissions and fees, see Table 9.1 from Borkovec and Heidle (2010) [47] for cost estimates. Managing MI has important implications on three aspects of algorithmic trading: Pre-trade TC estimation, post-trade TC analysis and the design of optimal trading strategies (Almgren (2008) [9]). The post-trade analytics provide a necessary diagnostics for a trader to reflect on the choice of trading strategies as well.

Table 9.1: Empirical Transaction Cost Estimates[1]

| Type | Fixed | Variable |
|---|---|---|
| Explicit | Commissions (7.4bp)<br>Fees | Bid-ask spreads (1.9 bp)<br>Taxes |
| Implicit | | Delay Cost (9.5bp)<br>Market Impact Cost* (35.6 bp)<br>Timing Risk & Opportunity Cost (11.7 bp) |

*Major Component
1: Readers should note changes in the market that have taken place since these numbers were reported. The commissions appear to be expensive for current trading. Most developed markets have commissions less than 1 bps, while E.M. markets are in the range of 2–5 bps for e-trading. The bid-ask spread makes sense if it is half a spread of liquid US stocks. For other markets, it can be much higher.

The study of price impact dates back to Bagehot (1971) [25] who postulated that the market consists of heterogeneously informed traders. The specialist cannot distinguish between informed and uninformed (liquidity) traders and fixes a spread that can balance the trading cost. The asymmetric information models based on this notion have been extensively studied in finance literature. The traders do convey information and Hasbrouck (1991) [180] provides empirical evidence that the market maker can infer information from the characteristics of the sequence of trades. The seminal work by Kyle (1985) [234] and Huberman and Stanzl (2004) [209] indicate that an order of large size (parent order) must be sliced into a number of small order (child orders) so that the volume per se will not signal informed trading. The hypothesis that time between the trades could convey the intensity of information flow (Easley and O'Hara (1992) [113]) and its impact on price is studied in Dufour and Engle (2000) [109]. Long durations are usually associated with no news and the variations in trading intensity are taken to be positively correlated with the behavior of the informed traders.

**Origins of Market Impact:** Most conventional definitions of the MI of a trade, equate it with the deviation of the post-trade market price from the market price that would have prevailed had the trade not occurred. Figure 9.1 shows an idealized MI picture of $n$-share sell order from Fabozzi, Focardi and Kolm (2006) [131] but it is augmented to include other information:

(i) The state of the order book that establishes a pre-trade equilibrium.

(ii) How this equilibrium is disturbed by a market order to sell.

(iii) As the sell order depletes the bid order book, it obtains an increasingly lower trade price and results in the trade print, and,

(iv) How over time the price gradually moves up back to recover some of the price drop.

Accordingly, the difference between (iv) and (iii) is called temporary impact, whereas the difference between (iv) and (i) is called permanent impact. If we are looking to model the MI effect of sequential trades, as shown in Figure 9.1, the temporary and permanent impacts will be superpositions of those of individual trades. The permanent impact is often assumed to be immediate and linear in the parent trade size, and the post-trade equilibrium to be the same for the order of '$n$' shares or a sequence of '$m$' orders of '$n/m$' shares each (Almgren, Thum, Hauptmann and Li (2005) [11], Fabozzi et al., (2006) [131]). By contrast, the temporary impact is usually a function of how the "parent" trade is split into smaller "child" trades.

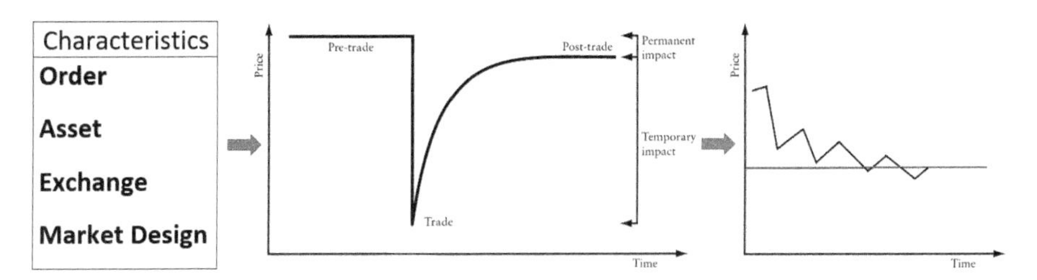


Figure 9.1: Idealized MI Model for a Sell Trade with Some Determinants.

The mechanism responsible for the permanent and temporary impacts has been a subject of significant interest. One common view links price changes accompanying large trades to the information cost and liquidity cost (Chan and Lakonishok

(1995) [74]; Holthausen, Leftwich and Mayers (1990) [199]) potentially attributing it to any transaction. In terms of information cost, the market response to the fact that a market participant has decided to sell (buy) shares of a particular stock could be perceived as conveying new (private) information about the fundamentals of a security, e.g., firm's management, debt conditions, and thus its future prices. Consequently, it results in a permanent price change with no price reversals. In terms of liquidity cost, the market response to the fact that liquidity was removed out of the order book, can be perceived as the effect of distorting the liquidity demand/supply equilibrium. The consequence is a temporary price impact that dissipates soon after the trading occurs, at a speed that depends on the market ability to absorb liquidity demand. The temporary component is thus associated with liquidity provision.

Independent of the market microstructure responsible for MI, two central concerns to much theoretical and empirical work are the functional form (shape) of the MI and the key determinants of this form. As seen in Figure 9.1, MI is influenced by trade-related factors (e.g., relative trade size, time and speed of trade), asset-specific factors (e.g., market capitalization, shares outstanding, bid-ask spread) and exchange- and market-related factors (e.g., liquidity, trading volume, institutional features). Models that incorporate these factors tend to be more complex but seem to result in better predictive power of market impact, that is useful in pre-trade cost estimation.

## 9.5 Some Stylized Models

Numerous stylized MI models offer some insights into possible shapes of the permanent and temporary impact functions. Some of these models focus only on the permanent impact, while other models attempt to capture the effect of temporary impact.

**Equilibrium Permanent Impact Models:** Models of one class, including Kyle's (1985) [234] and Huberman and Stanzl's (2004) [209], rest on the efficient market hypothesis positing that all available information gets impounded in prices. These models use a time-independent framework to establish an equilibrium price after a trade is executed. This means that trades are assumed to have only a permanent price impact (Huberman and Stanzl (2004) [209, p.1260]). Under these settings, it is shown that, if no quasi-arbitrage possibilities exist (to ensure viable markets), the permanent impact must be linear in the trade quantity and is symmetric between buys and sells. Linearity in trade size has been established by Kyle (1985) [234] as well. In fact, Huberman and Stanzl (2004) [209] add that "Nonlinear, time-dependent price-update functions may assume "chaotic" shapes without giving rise to price manipulation. Only when additional assumptions are made on the shapes of these functions does the analysis become meaningful."

Almgren et al. (2005) [11] develop a different market impact model, in the context of finding an optimal execution strategy for a large trade [see the discussion in Section 10.4.2] that is sliced into several smaller trades. They describe the random process satisfied by the stock price, using an arithmetic Brownian motion that depends directly on a permanent impact function that is linear in trade size and therefore has no effect on the optimal execution strategy. However, they recognize that if the total number of units traded is sufficiently large, the execution price may steadily change between trades, in part because the supply of liquidity is exhausted at each successive price level. They assume that this effect is short-lived because liquidity returns back to a level after each period and a new equilibrium price is established. They model this effect using a temporary price impact function that is non-linear in trade size. They then postulate that both the permanent and temporary impact functions follow power laws based on the empirical results (Loeb (1983) [251]; Lillo, Farmer and Mantegna (2003) [244]).

**Hybrid Impact and Propagator-Style Models:** Models of another class assume that trading agents have zero intelligence (instead of being fully rational) and take random decisions to buy or to sell, but that their action is interpreted by all the other agents as containing some potential information. The mere fact of buying, or selling, typically leads to a change of the ask $p^a$, or bid $p^b$, price and hence to a change of the mid-point $p = \frac{p^a+p^b}{2}$. The new mid-price is also expected to follow a random walk (at least for sufficient large times), if it is immediately adopted by all other market participants as the new reference price around which new orders are launched.

Madhavan et al. (1997) [256] develop a price formation model postulating that the (bid-ask) mid-price '$p$' changes because of unpredictable public information shocks (news) and microstructure effects that include statistical effects of order flow fluctuations (e.g., autocorrelation) and trading frictions (e.g., asymmetric information, dealer costs). This postulate automatically removes any predictability in the price returns and ensures market efficiency. If all trades have the same volume and the surprise component of the order flow at the $k$th trade is given by $\epsilon_k - \rho\,\epsilon_{k-1}$, where the signs of trades $\epsilon_k$ are generated by a Markov process with correlation $\rho$,[1] we can write the following evolution equation for the mid-price as:

$$\Delta p_{k+1} = p_{k+1} - p_k = \theta[\epsilon_k - \rho\epsilon_{k-1}] + \eta_k, \tag{9.9}$$

where $\eta$ is the shock component and the constant $\theta$ measures the size of trade impact. Empirical estimations of their model are not aimed at investigating the shape of impact functions.[2]

Others extend the model of Madhavan et al. (1997) [256] to gain insight into the shape of impact functions. Bouchaud et al. (2009) [49] use (9.9) to compute several

---

[1]This means that the expected value of $\epsilon_k$ conditioned on the past only depends on $\epsilon_{k-1}$, given by: $E(\epsilon_k \mid \epsilon_{k-1}) = \rho\,\epsilon_{k-1}$.

[2]First, both information flows and trading frictions are important factors in explaining intraday price volatility in individual stocks. Second, information asymmetry decreases steadily throughout the day, however, dealer costs increase over the day (possibly reflecting the costs of carrying inventory overnight) so that bid-ask spreads exhibit intraday patterns commonly noted in previous research studies.

important quantities. Principally, they write the lagged return impact function (for time points $n$ and $n+l$), denoted $R_l$ as

$$R_l = p_{n+l} - p_n = \theta \sum_{j=n}^{n+l-1} [\epsilon_j - \rho\epsilon_{j-1}] + \sum_{j=n}^{n+l-1} \eta_j, \tag{9.10}$$

and then show that the lagged impact function is constant and equal to:

$$R_l = \theta(1-\rho^2), \quad \forall l. \tag{9.11}$$

Now define the "bare" impact of a single trade taken at time $l$, denoted $G_0(l)$, which measures the influence of a trade at time $n-l$ on the bid-ask mid-price at time $n$. Written in terms of $G_0(l)$, the mid-point process is expressed as:

$$p_n = \sum_{j=-\infty}^{n-1} G_0(n-j-1)\,\epsilon_j + \sum_{j=-\infty}^{n-1} \eta_j. \tag{9.12}$$

Then it can be shown that $G_0(0) = \theta$ and $G_0(l) = \theta(l-\rho)$ for $l > 0$. Bouchaud et al. (2009) [49] also observe that "The part $\theta\rho$ of the impact instantaneously decays to zero after the first trade, whereas the rest of the impact is permanent. The instantaneous drop of part of the impact compensates the sign correlation of the trades."

Bouchaud et al. (2004) [50] have developed an earlier price evolution model, similar to (9.10), where the price at time $n$ is written as a sum over all past trades of the impact of one given trade propagated up to time $n$:

$$p_n = \sum_{n'<n} G_0(n-n')\,\epsilon_{n'} \ln V_{n'} + \sum_{n'<n} \eta_{n'}, \tag{9.13}$$

where $V_n$ denotes the volume of a trade at time n and $G_0()$ is assumed to be a fixed non-random function that only depends on time differences. The $\eta_n' s$ are also random variables, assumed to be independent from $\epsilon_n$, and are used to model all sources of price changes not described by the direct impact of the trades. The authors then consider constraints imposed on the shape of response function, $G_0$, by three empirical results they discuss: (a) the mid-price price process is close to being purely diffusive, even at the trade-by-trade level; (b) the temporal structure of the impact function first increases and reaches a maximum after some number (100 to 1000) of trades, before decreasing back with a rather limited overall variation; (c) the sign of the trades shows surprisingly long-range, power-law correlations. These empirical results are reconciled with (9.13) by assuming that the response function $G_0()$ must instead also decay as a power-law in time, with an exponent precisely tuned to ensure simultaneously that prices are nearly diffusive and that the response function is nearly constant. Gatheral (2010) [161], assuming that the trading costs should be non-negative, demonstrates that the exponential decay of market impact is compatible only when the market impact is taken to be linear. Thus the debate on the functional form of MI is ongoing.

## 9.6 Price Impact in the High Frequency Setting

Markets are now highly fragmented and exchanges use a fee structure to attract order flows. The earlier models are based on relatively low frequency asset information flow. With high frequency market makers playing a bigger role in the trading activities, the liquidity provision is now focused on short horizons. Parent orders are sliced into child orders when sent to the market resulting in a correlated sequence of trades. In this setting, the estimation of market impact has to be considered for a short duration, particularly the temporary aspect of market impact. For an elegant description of these issues related to high frequency market microstructure, see O'Hara (2015) [276].

Conrad and Wahal (2020) [83] examine realized spreads and price impact in all common stocks from 2010 to 2017. In this section, we cover the results in that paper as they are the most relevant to the high frequency setting. The measure of market impact is still the bid-ask spread. As mentioned in Section 1.5.1, Ho and Stoll (1981) [196] discuss a model where the market maker's objective is to maximize profit while minimizing the chance that it may not happen, by determining the optimal bid-ask spread. The key aspect of their model is the consideration of trade size of market maker's time horizon. Execution costs are measured directly by the quoted spread, effective spread, and realized spread, and indirectly by the Roll's (1984) [291] implied spread (see Exercise 6, Section 2.12). Huang and Stoll (1996) [205] suggest a decomposition of price changes into permanent and temporary components and provide a summary of studies in various markets.

First for the sake of completeness, we define the spreads:

**Quoted Half-Spread:** $(p_t^a - p_t^b)/2$. Because the spread represents the cost of trades, half-spread is taken to be a measure of cost for a single trade. It reflects the order processing cost, inventory cost and anticipated loss to traders with superior information. This measure is based on the assumption that trades occur at the quotes. To compare the transaction cost, this half-spread is standardized by the mid-quote as $(p_t^a - p_t^b)/(p_t^a + p_t^b)$.

**Effective Half-Spread:** $|p_t - m_t|/m_t$, where '$m_t$' is the quote mid-point existing at time of trade. The effective spread is less than the quoted spread as the trades take place inside the spread and is generally considered to be a better measure.

Market impact is not a static concept and from the way it is defined, it is focused on the impact of the current trade on future trades. It is possible that some traders have adverse information, the price can move against the expected direction which may result in losses for the trader. To measure the market impact, it is suggested that actual post-trade prices be used. This measure is the realized spread.

**Realized Spread:** $[(p_{t+\tau} - p_t)/[p_t = b_t]]$ for trades at bid and for trades at ask, $-[(p_{t+\tau} - p_\tau)/[p_t = a_t]]$. Here '$\tau$' is the length of time after the trade. This measure is also known as price reversal as earnings are realized only if prices reverse after a trade.

The difference between the effective spread and the realized spread is taken to be the amount lost to informed traders. We will adapt the expressions given in Conrad and Wahal (2020) [83], to summarize the empirical findings:

$$\begin{aligned} \text{Price Impact:} p_{i_{t\tau}} &= q_t(m_{t+\tau} - m_t)/m_t \\ \text{Realized Spread:} rs_{t\tau} &= q_t(p_t - m_{t+\tau})/m_t \\ \text{Effective Spead:} es_t &= q_t(p_t - m_t)/m_t \end{aligned} \tag{9.14}$$

The time horizon '$\tau$' must be chosen optimally. The span since the trade should be long enough to cover observing offsetting trades. If it is too short it may not reflect price reversal and if it is too long, the trade behavior cannot be simply attributed to the current trade. Huang and Stoll (1996) [205] suggest using $\tau = 5, 30$ minutes. They compare the results for NASDAQ (dealer market then) and NYSE (auction market) and show that in all spread measures, dealer market had higher cost.

Conrad and Wahal (2020) [83] study the distribution of the spreads in the high frequency context and argue $\tau = 5$ minutes is too long in the current high-frequency setting. Before we describe their results, observe $p_{i_{t\tau}}$ measures the change in the fundamental value of a stock and $rs_{t\tau}$ can be taken as the cost of trading. They quantify the probability in two time scales, clock time and trade time, measured by aggregate dollar realized spreads adjusted for total trading volume. This seems to decline sharply over time. Here are some key findings:

- $\tau = 5$ minutes is outdated; a horizon of no more than 15 seconds for large stocks and 60 seconds for small stocks is recommended. Use of daily return reversals is not accurate in a high frequency setting.

- Speed matters. Liquidity providers face stiff competition in this arena; to read, generate and send orders quickly and to extract liquidity require machinery of high speed.

- The order book imbalance, a measure used for reading future price movements is likely to play a more significant role than changes in fundamental volume.

- Stock volatility does not seem to matter much to the price of liquidity provision; what matters is non-diversifiable risk.

## 9.7 Models Based on LOB

It has been noted that the models developed to study market impact generally do not consider situations that may arise from strategies where price impact may depend upon the state of the order book. The second generation models consider the behavior of the limit order book. In the last section, we summarized recent work in this area. Here we consider joint modeling of several key order book measures. Hautsch and Huang (2012) [185] consider a time-aggregated model. The core activities in the limit order book are captured in Table 9.2 and the market depth is kept to the three best quotes on both sides of the market. Let $Y_t$ be a $K$-dimensional listing of the variables

Table 9.2: Model (9.15) Variable Definitions

| Variable | Description |
|---|---|
| $p_t^a$ | Logarithm of the best ask after the $t$-th event. |
| $p_t^b$ | Logarithm of the best bid after the $t$-th event. |
| $v_t^{a,l}$ | Logarithm of the market depth at the $l$-th best ask after the $t$-th event. |
| $v_t^{b,l}$ | Logarithm of market depth at the $l$-th best bid after the $t$-th event. |
| $\text{BUY}_t$ | Dummy equal to one if the $t$-th event is a buyer-initiated trade. |
| $\text{SELL}_t$ | Dummy equal to one if the $t$-th event is a seller-initiated trade. |

in Table 9.2; the co-integration model in Chapter 3 is fit for the data:

$$\Delta Y_t = \mu + \alpha\beta' y_{t-1} + \sum_{t=1}^{p-1} \Gamma_i \Delta Y_{t-i} + u_t. \tag{9.15}$$

The co-integrating $(K \times r)$ matrix, $\beta$ has first two columns restricted to have entries as zeros except for buy and sell indicators. To study the market impact, the model is written in the reduced-form to capture the impulse-response coefficients as:

$$Y_t = \mu + \sum_{i=1}^{p} A_i Y_{t-i} + U_t, \tag{9.16}$$

where $A_1 = I_k + \alpha\beta' + \Gamma_1$, $A_i = \Gamma_i - \Gamma_{i-1}$ for $1 < i < p$ and $A_p = -\Gamma_{p-1}$. The VAR($p$) model in (9.16) can then be written in the form of VAR(1) as

$$Y_t^* = \mu^* + A^* Y_{t-1} + U_t^*, \tag{9.17}$$

where $\mu^* = (\mu', 0', \ldots, 0')'$, $Y_t^* = [Y_t', Y_{t-1}', \ldots, Y_{t-p+1}']$, $U_t^* = [U_t', 0', \ldots, 0']'$ and $A^* = \left[\begin{array}{c|c} A_1 \cdots A_{p-1} & A_p \\ \hline I_{K(p-1)} & 0 \end{array}\right]$. It can be shown that

$$Y_t = J M_t + \sum_{i=0}^{t-1} J A^i J' U_{t-i}, \tag{9.18}$$

where $J = [I_k, 0, \cdots, 0]$ is a $k \times kp$ selection matrix and $M_t = A^{*t} Y_0^* + \sum_{i=0}^{t} A^{*i} U_{t-i}^*$.

Recall that the market impact measures the impact of current trade on the future trades; it is measured by the impulse response function

$$f(h, \delta_h) = E[y_{t+h} \mid y_t + \delta_y, t_{t-1}, \ldots] - E[y_{t+h} \mid y_t, y_{t-i}, \ldots] = J A^{*h} J' \delta_y, \tag{9.19}$$

where $\delta_y$ is the change to $Y_t$, due to the current trade. The long-run effect, $f(\delta_y) = C\delta_y$, where $C = \beta_\perp(\alpha'_\perp(I_K - \sum_{i=1}^{p-1}\Gamma_i)\beta_\perp)^{-1}\alpha'_\perp$. Here $\alpha_\perp$ and $\beta_\perp$ are orthogonal to $\alpha$ and $\beta$ in (9.15).

Hautsch and Huang (2012) [185] estimate the market impact using the data from Euronext Amsterdam for heavily traded stocks. The main findings are, besides the existence of co-integrating relationships between quotes and depths, limit orders have long-term effects on quotes, the price movement is influenced by the order size and the decrease of spreads after the arrival of a limit order is reverted back asymmetrically soon after. The between-stock variations in market impact are shown to be related to trading frequency of the stocks. These empirical observations are immensely useful for pre-trade transaction cost estimation.

Cont, Kukanov and Stoikov (2014) [87] consider the dynamics of market liquidity that is a function of limit orders, market orders and cancellations to model the price impact function. The outstanding limit orders provide a measure of depth in the market on both buy and sell sides and thus also provide a measure of order flow imbalance (OFI). By setting the number of shares (depth) or price levels beyond the best bid and ask equal to $D$, define the price change in the interval $(t_{k-1}, t_k)$ as,

$$\begin{aligned}\Delta P_k^b &= \delta\left[\frac{L_k^b - C_k^b - M_k^s}{D}\right]\\ \Delta P_k^s &= -\delta\left[\frac{L_k^s - C_k^s - M_k^b}{D}\right],\end{aligned} \tag{9.20}$$

where $L_k^i$, $C_k^i$ and $M_k^i$ for $i = b, s$ denote arrival of the limit orders, cancellations and market orders in the defined interval. Here $\delta$ is the tick size. If $P_k = \dfrac{P_k^b + P_k^s}{2\delta}$, that is mid-price normalized by tick size, for any short interval, $(t_{k-1,i}, t_{k,i})$, it is postulated that

$$\Delta P_{k,i} = \beta_i \mathrm{OFI}_{k,i} + \varepsilon_{k,i}, \tag{9.21}$$

where $\mathrm{OFI}_k = (L_k^b - C_k^b - M_k^s) - (L_k^s - C_k^s - M_k^b)$ and $\Delta P_k = \frac{\mathrm{OFI}_k}{2D} + \epsilon_k$. Here it is important to note that '$\beta_i$' is taken to be price impact coefficient and is taken to be inversely proportional to market depth:

$$\beta_i = \frac{C}{D_i^\lambda} + v_i. \tag{9.22}$$

The models in (9.21) and (9.22) can be regarded to represent instantaneous price impact. The model as it can be seen from the set-up assumes that all activities in the limit order book have an average linear impact, $\beta_i$, in the $i$th interval.

Cont et al. (2014) [87] using the TAQ data for a month in 2010 for fifty stocks estimate the model in (9.21) over ten-second intervals. The model is generally confirmed by the data with $\hat{\beta}_i$ significant in most of the samples and $\hat{\alpha}_i$ only significant in less than twenty percent of the samples. It must be noted that a significant number of trades posted gets canceled within a few seconds of submission. Various theories that are postulated (see Hasbrouck and Saar (2009) [182]) are yet to be fully empirically tested. Hence instead of using OFI, trade imbalance ($\text{TF}_k = M_k^b - M_k^s$) is used in Equation (9.21). Results are still reasonable, but not as strong as using OFI in the model. Finally the relationship between $\hat{\beta}_i$ and $D_i$ (price impact coefficient and depth) is studied; the results indicate for model (9.22), $\hat{C} \sim 0.5$ and $\hat{\lambda} \sim 1$.

Some interesting conclusions that could be useful for trading are drawn. The impact coefficient varies more with the amount of liquidity provision (OFI) than the trade imbalance (TI). It can be also used to monitor adverse selection. If a limit order is filled after a positive OFI, price is likely to go up and for a limit sell order, a positive change would imply that the order was executed at a loss. These conclusions need to be firmed up with further studies in this area.

## 9.8 Empirical Estimation of Transaction Costs

As noted earlier, a key aspect of MI modeling is to provide information about expected pre-trade costs. The studies that have been reviewed on empirical analysis thus far can help us understand how the price formation and the market impact function evolve during a trading period. But most participants who trade large quantities of equities want to know the trading costs a priori. Here we review the relevant studies and augment with our own research in this area. First we review the factors that are found to influence these costs in block trades. Although block trades differ in fundamental ways to algorithmic trades, they can be treated as all equivalent to a parent trade except that no splitting into child orders occurs.

**Determinants of Cost**

- **Trade Size.** Many studies observe that the (relative) size of trades positively relates to the magnitude of MI (Keim and Madhavan (1997) [228]), consistent with the prediction of stylized model (Kyle (1985) [234]; Huberman and Stanzl (2004) [209]).

- **Market Capitalization.** Keim and Madhavan (1996) [227] find that the asymmetry of trading costs between buys and sells is more pronounced among smaller stocks. This is consistent with what is broadly known to be the effects of market capitalization on MI (Lillo et al. (2003) [244]; Chan and Lakonishok (1997) [75]). From a practical point of view, market cap itself may not be a true factor but it can be a proxy for some liquidity characteristics.

- **Price Volatility.** Price volatility has been found to be positively related to MI (Chiyachantana, Jain, Jiang and Wood (2004) [78]). One explanation is that elevated volatility is associated with greater dispersion in beliefs, with risk averse traders less inclined to participate in markets, thereby resulting in larger price concessions or MI (Domowitz et al. (2001) [107]).

- **Trading Activity.** Dufour and Engle (2000) [109, p. 2467] find that the price impact of trades increases, as the time duration between transactions decreases, and more broadly that the price impact is especially large due to increased trading activity, as measured by high trading rates. In their view, "times when markets are most active are times when there is an increased presence of informed traders; we interpret such markets as having reduced liquidity." Likewise, Yang (2011) [328, p.91] finds that "a trade shortly after the previous trade results in higher price impact than one after a long period." He also offers "evidence that increased trading activity (as measured by short durations between trades) is associated with larger price impact, therefore implying a higher degree of information-based trading."

- **Time of Day.** It is reported that the MI for buy trades decreases as time passes during the day, with the highest impact in the first trading hours (Frino, Jarnecic and Leopone (2007) [150]).

At the higher frequency levels other determinants to consider are:

- Intraday volatility on a 5-min basis
- Depth of the order book
- Spread
- Intraday Price Dynamics
- VIX level and return

### 9.8.1 Review of Select Empirical Studies

While there is some discussion about the theoretical models for market impact, published studies based on large scale data are somewhat rare especially when it comes to models useful for pre-trade cost estimation, Almgren et al. (2005) [11] took US stock trade orders processed by Citigroup for Dec 2001 to June 2003 and estimated a model of the market impact. For two select stocks, IBM and DRI, the power law model for market impact are estimated. The temporary impact is defined as $\widetilde{p}_k - p_{k-1}$, for $k$th execution. The impact cost depends on relative size (Parent order/Average daily volume), turnover ratio (ADV/Shares outstanding), the duration of trading and volatility.

While the square root formula in (9.7) is quite robust and holds even beyond equity markets, recent empirical studies confirm the need for more general power law functions; the quotient of RS varies depending on other factors that get included in the model. Zarinelli, Treccani, Farmer and Lillo (2015) [330] show that logarithmic functional form fits the data better for the large order data that they consider.

Limited access to broker-proprietary algorithmic trading data has stifled detailed academic inquiry in this area. While the impact of algorithmic trading is extensively discussed in the finance literature, formal empirical results are rare. The few exceptions we are aware of are: Engle, Ferstenberg and Russell (2012) [121], who use proprietary algorithmic trading data from Morgan Stanley to study the risk-return trade-off associated with increasing the number of child trades and duration of parent trade; Domowitz and Yegerman (2005) [108], who use ITG's Transaction Cost Analysis Peer Group Database to compare execution costs for a set of 40 buy-side clients; and Hendershott and Riordan (2011) [194], who use algorithmic trading data from firms listed on the Deutsche Boerse DAX to measure the contribution of algorithmic trading to price discovery. A few academic studies resort to using proxies based on publicly available TAQ data. Two examples are: Hendershott et al. (2011) [192], who use NYSE electronic message traffic data as a proxy for algorithmic liquidity supply to study how algorithmic trading effects liquidity and bid-ask spreads; and Chaboud et al. (2014) [70], who use FX quote data (highest bid and lowest ask) from an electronic limit order book platform called EBS to construct mid-quote series for studying the effect of algorithmic trading on FX rates' volatility. The use of mid-point prices, in the lack of actual arrival price and completion price for a parent trade, has been criticized for failing to capture the possible asymmetry of the bid and ask price impacts of trading, which has been observed to exist for block trades. In any event, none of these studies deals with the estimation of MI models per-se.

A recent study by Frazzini, Israel and Moskowitz (2018) [143] presents results on trading costs and price impact function using a large number of trade executions over several countries, but the data is from a single large trader. It is argued that the cost estimates based on effective spreads ignore the price impact. The results confirm that trade size is a critical component of trading cost function; but the cost function produces lower cost estimates than estimates based on TAQ data which is a market wide data set representing traders with different trading skills. The price impact measure as a function of trade size on a ratio to average daily volume can be modeled via square root function. More volatile stocks have higher transaction costs and costs are higher during times of high market volatility.

**Our Study:** Velu, Gretchika, Benaroch, Nehren and Kuber (2015) [319] address these issues using a large algorithmic trading data set. The data contains "parent" algorithmic executions, including the initial arrival price and fill price as well as the child trades' average fill price (computed as the share-weighted average executing price of the child orders). The algorithmic executions are generated using Volume Weighted Average Price (VWAP) and Percent of Volume (POV) strategies. Recall the volume weighted average price (VWAP) strategy schedules to trade some constant percentage of the expected market volume regularly throughout the day, where the proportion traded is adjusted based on historical market volume profiles and the predicted volume. Hence, in VWAP, the trading schedule is predetermined based on a prediction of the daily market volume distribution, although the timing of executions can be randomized so as to make the trading pattern less predictable. In any case for efficient execution, it is essential to have good predictive models for volume.

By contrast, the percent of volume (POV) strategy trades a fixed percentage of the current market volume where the trading schedule is dynamically determined based on a historical analysis of volume profiles used to anticipate trading volumes. Unlike implementation shortfall strategies, VWAP and POV do not explicitly take into account the expected MI in generating their algorithmic executions. Moreover, although firms typically implement proprietary versions of these strategies with parameters reflecting their experience and clients' profiles, VWAP and POV nonetheless generate relatively standardized algorithmic executions. In other words, from the perspective of MI, these two strategies can be safely considered "plain vanilla" in the sense that firm-to-firm differences in their behavior and performance tend to be insignificant. The measures employed, therefore, do not depend on any unique traits of the strategies and so they do not have any influence on the cost estimates per se. The unique aspect of the data is that if the direction of the meta order (on the buy side or sell side) is explicitly provided.

The data set allows us to study the effect of parent orders characteristics on the market impact and the implied transaction cost (TC). Specifically, we estimate the parameters of a family of power-law models, investigate the explanatory power of different determinants of market impact and TC, and develop insights useable for pre-trade cost estimation. Our model estimation effort is informed by existing stylized market impact models and empirical findings about the market impact of block trades. These stylized models are reviewed earlier in this section. In addition to studying the effects of buy and sell trades, order size, price volatility and market cap, we examine the effect of the number of child trades and duration of the parent trade as well. Finally, we also summarize two extensions of our model estimation effort that pertain to the effect of the bid-ask spread on market impact and to the market impact of ETFs traded via the algorithms mentioned above.

This study has some important findings. A central novel insight that emerges from the analysis is that market impact can be substantially decreased by reducing the number of child order trades. There could be several explanations, but a simple plausible one is that reducing the number of child trades reduces the information leakage about trading intentions. This explanation is in line with the main premise of Farmer, Gerig, Lillo, and Waelbroeck (2013) [139] that, in an efficient market, market impact is determined by the information that is disseminated through trades. It is also consistent with studies showing that the price impact is larger as the duration between trades decreases and there is increased trading activity (Dufour and Engle 2000 [109]; Yang 2011 [328]), and that the number of trades may drive prices more than the actual trade size (Jones, Kaul and Lipson (1994) [223]).

Another finding is the nuanced difference in market impact and TC that is related to buy and sell trades. On the overall level, this difference is statistically insignificant and this presumably contradicts the buy-sell market impact asymmetry commonly reported for block trades. However, a deeper examination reveals a more nuanced difference between buy and sell trades. Visual inspection of the TC vs. trades' relative size and the TC vs. stocks' market cap shows that in some sub-ranges there is a difference between buys and sells. After dividing trades into 24 bins based on their relative size and testing the TC difference within each bin we observe that the difference is statistically significant in six bins; six out of 24 exceeds the proportion expected when examined through multiple hypothesis tests at a 5% significance level. Consequently, separate market impact and TC models for buy and sell trades are estimated, and the following observed about the difference between the coefficients of buys and sells. This difference is significant in models containing a subset of the explanatory factors (e.g., relative size alone, or relative size and volatility), but it is within a two standard errors threshold in models containing all the factors. Our findings notwithstanding, it is necessary to investigate further whether the underlying causes may be linked to the specifics of trade flow, the logic behind many of the trading algorithms.

It is important to mention two caveats. One relates to the role of permanent versus temporary impact in our analysis. The data set does not contain information about the price dynamics after the completion of the parent trades, that is, only completion price for the last child order is available. Therefore, it was not possible to determine what portion of the measured market impact is permanent and what portion of it is temporary. Another caveat relates to the high variability in the data. To obtain a better signal out of the noisy data, the data was smoothed through binning and extraction of key descriptives. The models are based on these descriptives. Nevertheless, for robustness testing all models were estimated using the raw unsmoothed data and the results are qualitatively similar, albeit the $R$-square values are understandably much lower.

**Select Empirical Results of Our Study:** The market impact is modeled as a non-linear function of several stock-related and trade-related factors. The power law form is generally recommended for non-linear shape. The data analyzed in Velu et al. (2015) [319] consists of over 250,000 parent orders fully executed over a one-year period, Oct 1, 2009–Sept 30, 2010. All trades were completed within a single trading day. Each parent order was sliced into multiple child trades; each child trade was executed in one or several partial fills. Only parent trades were executed through widely used VWAP (volume-weighted average price) or POV (percent of volume) strategies. This ensures that the participation rate is well-defined and stays roughly constant over the duration of the parent trade. We define

$$\text{Market Impact (MI) in bps } = \text{side} \cdot \left(\frac{\text{Completion Price} - \text{Arrival Price}}{\text{Arrival Price}}\right) \cdot 10000$$

$$\text{Transaction Cost (TC) in bps} = \text{side} \cdot \left(\frac{\text{Average filled Price } - \text{Arrival Price}}{\text{Arrival Price}}\right) \cdot 10000,$$

where the arrival price is the price when the first child order is executed and completion price is the price when the last child order is executed. Every trade in this data is explicitly identified as buy or sell. This contrasts with studies that use TAQ data that do not have the same explicit identification. Using Lee and Ready's (1991) [241] heuristic rule to determine the trade direction can be sometimes inaccurate (see Asquith, Oman and Safaya (2011) [23] and Chakrabarty, Moulton and Shkilko (2012) [71]). To make the analysis robust, we binned the explanatory variables and obtained smoothed averages; the analysis is based on binned data.

Results start with a visual inspection of the relationships. The graphs clearly indicate that the relationship between MI and RS is non-linear, and it remains non-linear with binning of other factors as well. The relationship between MI and relative size is monotonic till the relative size reaches 20% or so (Figure 9.2). From Figure 9.3 we can infer the following: trades with longer durations clearly have a lower MI; trades for higher volatility stocks, too, can be read to have a higher MI; trades on larger market cap stocks have a smaller MI; lastly, MI becomes higher as the number of child trades increases. This last relationship is important and deserves special attention particularly when designing execution strategies. There is a clear trade-off between low "immediate" liquidity impact of small child orders and long term information leverage of the parent order level when hitting the market too frequently. But the power law model is more general and can accommodate interaction effects as well. These graphs for various factors besides RS can be collapsed into a master curve, with appropriate normalization, as given in Lillo et al. (2003) [244].

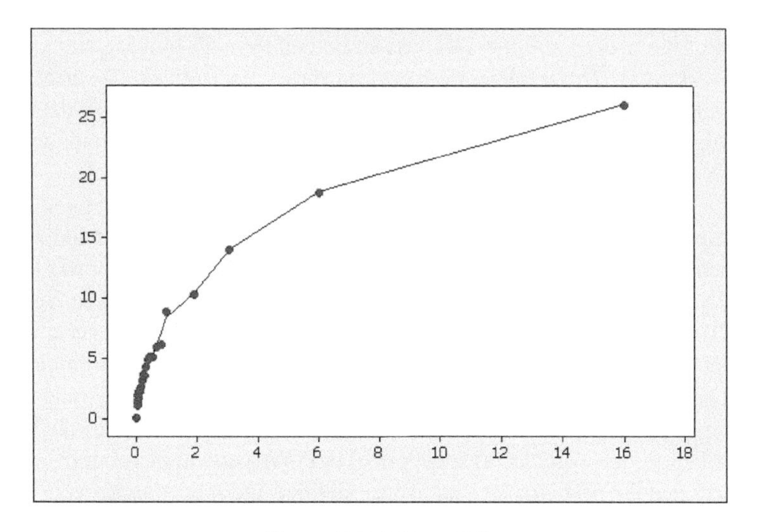


Figure 9.2: MI vs. RS.

Table 9.3 shows the results of power-law models ; the standard errors are shown in parentheses below the estimated regression coefficients. All coefficients in the models presented in Table 9.3 are statistically significant. This is consistent with Almgren et al. (2005) [11], except that in their model the market-cap-to-volume ratio is not significant. It should be noted that the values of $R$-square are significantly higher than other published studies, mainly due the fact that the analysis is based on smoothed data. The response variable is not an individual parent trade cost, but rather the average cost for trades in a bin representing a combination of various levels of predictors. Thus, the models developed here are intended to predict the average MI for a combination of predictors. Building empirical models for noisy data almost always requires some smoothing of raw data. While there are excellent methods such as kernel smoothing, smoothing in a higher dimension is not easy due to the 'curse of dimensionality' problem. Thus, we chose this simple approach of binning.

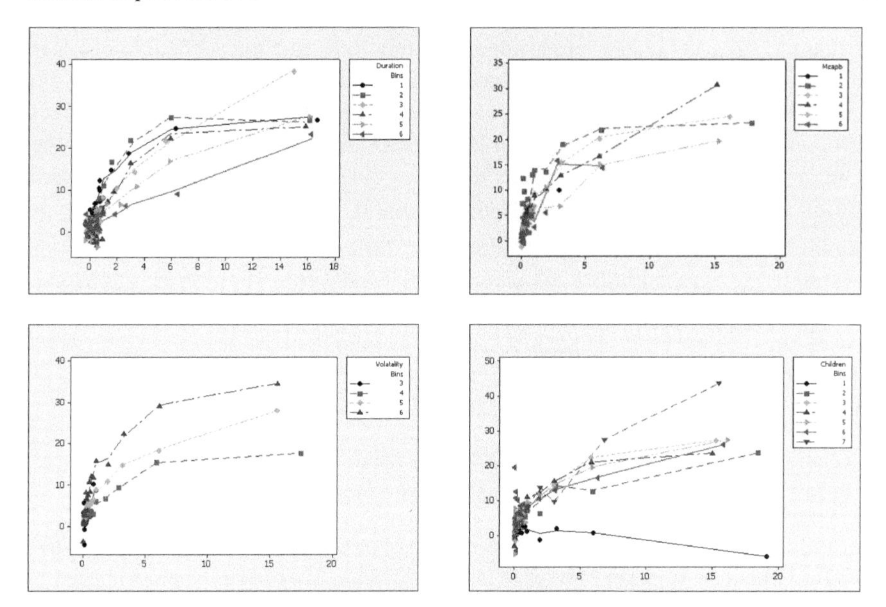


Figure 9.3: MI vs. RS with Other Factors.

Table 9.3: Response Variable is MI (bps) ($\hat{\sigma}^2 = 225, n = 2,211$ bins)

| Predictor | Model 1 | Model 2 | Model 3 | Model 4 | Model 5 | Model 6* |
|---|---|---|---|---|---|---|
| Rel. Size | 0.53 | 0.51 | 0.56 | 0.63 | 0.43 | |
| | (0.02) | (0.02) | (0.02) | (0.02) | (0.02) | |
| Volatility | | 1.12 | 0.92 | 1.17 | 0.53 | |
| | | (0.09) | (0.09) | (0.11) | (0.1) | |
| Duration | | | −0.24 | −0.25 | −0.47 | −0.27 |
| | | | (0.02) | (0.02) | (0.3) | (0.1) |
| Market Cap | | | | −0.09 | −0.12 | −0.04 |
| | | | | (0.02) | (0.02) | (0.07) |
| No. of Children | | | | | 0.53 | 0.55 |
| | | | | | (0.04) | (0.16) |
| Constant | 9.6 | 19.6 | 15.9 | 9.4 | 2.7 | 1.3 |
| | (0.34) | (1.07) | (0.9) | (1.3) | (0.46) | (1.3) |
| $\hat{\sigma}^2_{error}$ | 143 | 136 | 125 | 124 | 114 | N/A |
| $R^2$ | 0.37 | 0.40 | 0.45 | 0.45 | 0.50 | 0.03 |

*Response Variable = $\frac{\text{MI(bps)}}{(\text{Vol})(\text{RS})^{0.6}}$

Several relevant observations can be made about the coefficients in the estimated models. For example, Model 3, which is traditionally well cited in the literature, can be written as:

$$\text{MI(bps)} = 15.9 \cdot \sigma^{0.92} \cdot (RS)^{0.56} \cdot T^{-0.24}.$$

Note that the volatility coefficient is approximately 'one' and the coefficient of RS is about 0.5, which has indeed been observed in the literature. It is consistent with the folklore that it takes one day's volatility to trade one day's volume. The duration effect is also consistent as most of the parent trading occurs in less than two hours in the data set, not always in real life. This has led us to consider an additional model where $\frac{\text{MI(bps)}}{(\text{Vol})(\text{RS})^{0.6}}$ is used as a normalized response variable. This variable can be taken as the multiplicative residual term of Model (2) in Table 9.1. The regression results of this variable on other variables are given under Model (6*). Two observations can be made: The duration and the number of child trades are still significant factors. However, the increase in $R$-square is somewhat smaller than the unrestricted versions of the same model, labeled as Model (5) in Table 9.1.

Other key findings can be summarized as follows:

- Relative size (RS) and stock volatility ($\sigma$) consistently exhibit positive influences on MI; their impact varies depending on other factors in the model.
- Duration ($T$) has an inverse effect. Its use in the model must be carefully calibrated as the execution of parent trades using child trades can be highly concentrated.
- Stocks with higher market capitalizations tend to have smaller price impact for the same relative size.
- The number of child trades, which is usually determined by the trading algorithm, does reveal a positive effect on MI.

There are several practical considerations that may limit the use of these models for pre-trade cost estimation. These challenges in MI calibration are due to the following facts:

- Not enough data for certain combinations of the factors involved.
- MI is a noisy measure which is hard to calibrate ex-post.
- Sometimes, the market moves at a much larger magnitude than MI.
- Use of raw data results in a poor predictive models; thus a need for aggregation. But aggregation obviously ignores the wide variation within the aggregated unit.

# 10

# *Execution Strategies*

We finally turn our discussion to one of the central themes of this book. Execution strategies are arguably the core means by which capital flows in and out of the financial markets. As discussed in Chapter 1, large money managers, mutual fund companies and hedge funds have substantial holdings (some have assets under management in the trillions of dollars). They are in constant need to trade, in order to re-balance their portfolios to achieve their investment objectives. These trades are usually of large volume that is not instantaneously available in the market at reasonable prices. This forces the trader to slice the order in smaller sizes that can be more easily absorbed by the market; They also need to skillfully trade these slices to minimize any footprint that may be left in the market under the form of price impact. Started as nothing more than utilities tools to implement the simpler, smaller trades, the field has ballooned into a multi-billion dollar industry where a large number of sell-side brokers as well as fintech-driven companies are fiercely competing. The execution has started in a somewhat chaotic way. Albeit slowly, the field has matured and is now starting to focus on the real core aspects of the execution problem. With the better understanding of the behavior of the limit order book and with the advancement of communication and computing technologies, we expect the field to evolve substantially in the coming years.

In this chapter, we start with a practitioner view of the field, briefly describing the core components, and the approaches being used in practice. We will walk through a brief history and underlying ideas behind the current product lines. We strive to dispel some of the misconceptions, that may still exist in the industry. This may be considered controversial by some of our peers and clients alike, but that we feel our description is closer to reality. The goal is not to stir any controversy but we hope to provide the reader with an uncluttered approach to the field to better focus on things that really matter and on the problems that still need to be solved.

The second part will take the treatment to a more formal level providing a review of the academic research area named "Optimal Execution" and our view on the modeling approaches that will comprise the next generation Execution products.

## 10.1 Execution Benchmarks: Practitioner's View

Execution generally refers to the aspect of Algorithmic Trading that is concerned with implementing, as efficiently as possible, an investment decision of a portfolio manager (PM). The main elements of that decision are: The instrument to transact, the direction (buy or sell) and the quantity (how many shares or contracts). It also often includes additional trading instructions such as how to trade a particular order; for example, the maximum (minimum) price the trader is willing to buy (sell).

As such, unlike other algorithmic trading strategies we have discussed in previous chapters, WHAT to trade is exogenous to the strategy, which is only concerned on HOW best to trade the order. In addition to the PM directed parameters, the strategy will also require: Start Time, and End Time, providing lower and upper bounds around when the strategy is allowed to transact. Finally, there is a set of additional parameters that must be considered: Some generic (e.g., maximum participation rate, limit price) and some strategy-specific (e.g., Urgency). These add constraints the strategy will need to abide by, thus essentially limiting the "Solution Space" that can be explored. As stated before, in most relevant cases, the order to be executed is too large for the market to absorb as a whole so the order is sliced down in one form or another. Execution strategies provide various approaches to this slicing issue, so as to achieve the trader's specific objective.

We begin with an alternative view. The usual approach would typically introduce various benchmarks (mostly some form of price or a participation rate) that essentially codify the objective of the strategy and then describe the algorithmic trading strategy that evolved to minimize the distance to that benchmark. This is not, in our opinion, how algorithmic trading has mostly evolved and a misconception that still mires the industry. Traders do not have different objectives when executing a trade such as achieving a certain benchmark. The objective is always to get the "best price" possible for any order consistent with the constraints spelled out by the PM. The problem is that this "best price" is an unknown ex-ante and worse than that, it is unknowable even ex-post. Unknowable because the choice of the trader will most likely change the path of the price and volume evolution; thus, there is no way to know how that price would have evolved had the trade not occurred. Thus execution strategies have evolved not in order to achieve a particular benchmark but as a set of approaches, with increasing sophistication. Additional parameters provide the trader different ways to maximize their chance of achieving that elusive 'best' price.

**Benchmarks:** The standard benchmark to measure a trading outcome is the "Implementation Shortfall" (IS). This is simply the execution price compared to the "prevailing" price at the time the trader received the order.[1] Formally,

$$\text{IS} = \text{side} \cdot \frac{p_{\text{exec}} - p_{\text{arrival}}}{p_{\text{arrival}}}, \tag{10.1}$$

where 'side' is the side multiplier: 1 for buys, $-1$ for sells, $p_{\text{exec}}$ is the execution price achieved and $p_{\text{arrival}}$ is the prevailing price at order arrival. This is usually the mid-quote at time of arrival, but sometimes, the last traded price before the arrival of the order is used instead.[2] Note $p_{\text{exec}} = \frac{\sum_{i=1}^{n} p_i x_i}{X}$, where $x_i$ is the child order volume traded in the $i$th duration and $X$ is the total volume of parent order traded. With child orders split equally, $\frac{x_i}{X} = \frac{1}{n}$.

Implementation Shortfall is the most sensible choice because it is related to Profit and Loss (P&L) achieved during the trade. It suffers however from several shortcomings:

- This benchmark while measuring the effects of the trading approach chosen (i.e., market impact) also captures the exogenous price dynamics, due to embedded alpha in the investment decision and the activities of other participants. In many instances, these exogenous sources have larger impact than the trading approach. Even on the endogenous side, the more significant components of the realized outcome are not under the control of the strategy and are instead dependent on the order characteristics as well as the choice of other constraints applied. The exact same strategy trading the same name and quantity may beat the benchmark by many basis points on one day and miss it by just as much the next day, depending on the prevailing market conditions. This makes Implementation Shortfall somewhat not well suited to understand how well a strategy is working.
- The IS benchmark implicitly attributes special meaning to $p_{\text{arrival}}$, as the anchor price. This has led to a widespread belief that, in order to minimize the slippage against arrival price, one should adjust the trading approach as price deviates either in favor or against the $p_{\text{arrival}}$. We argue that, lacking any additional information and under a weak market efficiency assumption, this is clearly erroneous as the strategy would more likely make a decision based on spurious noise leading to a much wider distribution of outcomes. That being said if there are exogenous factors that make the $p_{\text{arrival}}$ informative, like a PM or traders view strongly that the particular price is an attractive stable price, then leveraging that additional

[1]From a PM perspective, the Implementation Shortfall is a measure against the prevailing price at the time the decision was made. In this book we take the trader's perspective.

[2]These choices while seemingly inconsequential are actually somewhat tricky as they can lead, in some fringe cases, to significant measurement outliers. For example, if right before the order arrival, the spread gaps increase for a short moment leading the mid-point to be very different from what one would call the prevailing price. Alternatively for a less liquid stock that does not trade often, the last traded price may be several minutes old and the prevailing price may have moved significantly.

information would lead to a better outcome. Finally there are other behavioral reasons for the large usage of "price adaptation" with respect to the $p_{\text{arrival}}$. Often traders are more interested in not looking bad on a trade by trade basis and are willing to accept inferior prices but minimizing the chance of being called out for missing the benchmark.

- This benchmark tells us little about how close the trade was to the "best price" and thus, on its own, does not provide any diagnostic guidance on how to improve a particular trading approach

For this reason early practitioners started using auxiliary benchmarks meant to help measure how well the strategy implemented a particular approach of trading. The strategy usually took the name of the benchmark it tries to achieve. We cannot talk about the benchmark without discussing, at a high level, the approach taken by the eponymous strategy.

This approach, while sensible, had some unintended consequences that, to some extent, we are still facing today. Early adopters and sellers of these algorithms started equating the benchmark and the strategy used to achieve it, as the actual ultimate objective of the strategy. This led to the belief that these strategies should not be used if a trader has an IS benchmark in mind. Our claim is that all these strategies are in some sense IS strategies. When no better information around price direction/dynamics exist, the focus of the strategy is on minimizing the footprint, by managing the realized trading rate which, as we know, is the best understood driver of market impact.

Let us review the basic benchmarks and related strategies.

## TWAP

The Trade (Time) Weighted Average Price (TWAP) is the simplest of all the benchmarks and is simply the simple average price of '$N$' trades over the time horizon when the parent order is traded.[3]

$$p_{\text{TWAP}} = \frac{1}{N} \sum_{i=1}^{N} p_i. \tag{10.2}$$

While conceptually simple, this benchmark compares the execution price (here $p_{\text{exec}} = \frac{1}{n} \sum_{i=1}^{n} p_i$, where '$n$' is the number of child orders) with the average price achieved by all participants over the same time horizon allotted. This measure may be consistent with TWAP execution strategy where the parent order is divided equally and traded periodically over a certain time horizon (Figure 10.1). Although technically, to achieve the best trade (weighted) price, one would want to trade more when

[3]The literature universally expands the acronym as Time Weighted Average Price. We believe that to be misleading. TWAP evolved as a juxtaposition with the VWAP benchmark where every trade is weighted by the volume associated. The average price over the period is simply the trade weighted (i.e., unweighted) equivalent. Thus Trade Weighted Average Price seems a more accurate name.

there are in general more trades. While as a benchmark it is essentially a measure of mediocrity, it is stable and corrects for the exogenous price moves. That being said, it also has some drawbacks. As a benchmark, TWAP is unknown until the end of execution (each trade in the market changes the value of the benchmark, contrary to the arrival price). It is also directly impacted by the trader's own trades (they count toward the TWAP market price) and thus not really an independent benchmark and could be manipulated. It is hard to argue that there is anything optimal in this approach but lacking any other information and knowledge about the behavior of price and volume, it may be the only alternative when completing the order is a requirement. We will see below an arguably better approach (POV) if order completion is not a requirement.

While very simple TWAP is still actively used, in particular, in some less developed markets, it is also used by some systematic quantitative strategies, where the joint trading schedule of a long-short portfolio is optimized as part of the investment process and then each individual short term slice (5–15 m) is sent to a TWAP algorithm. This helps to maintain the dollar neutrality of the portfolio.

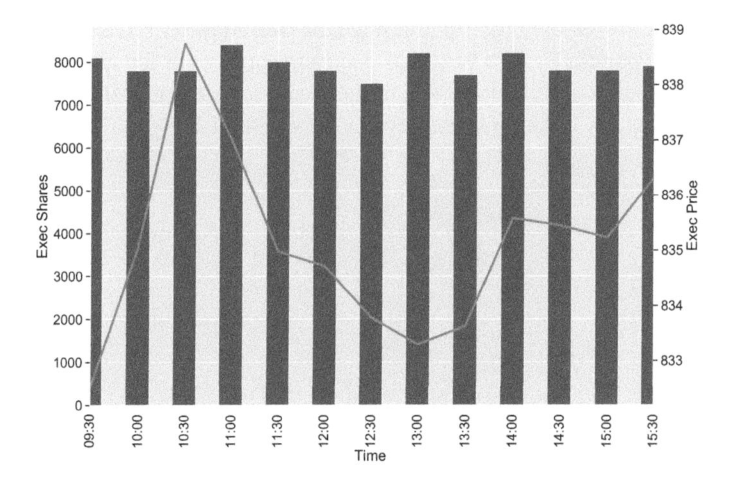


Figure 10.1: Example of TWAP Trade for AMZN.

**VWAP**

The Volume Weighted Average Price (VWAP) is the next simplest benchmark that is commonly used. Simply weigh every market trade by the total trade size over the duration.

$$p_{\text{VWAP}} = \frac{\sum_{i=1}^{N} p_i V_i}{V}. \tag{10.3}$$

This benchmark is arguably a better measure of the fair market price over the trading period (trading times and prices that are closer to the times and prices of other market participants—who might have a better assessment of the short term price-trade).

Other arguments for and against the VWAP benchmark are the same as the ones discussed for TWAP.

The VWAP strategy, in order to achieve its goal, takes advantage of the relative stability of the volume distribution over the day we discussed in Section 8.1.5. Instead of trading equally throughout the time period, the strategy splits the parent order proportionally to the volume profile re-normalized over the time horizon.

The VWAP strategy is conceptually superior to TWAP as a market impact reduction approach since by trading more (less) when more (less) volume is expected to trade it does a better job at minimizing the realized trading rate. That being said if the volume profile is, for some special circumstances, extremely noisy or unpredictable, trading VWAP might actually be inferior to TWAP since the error in our volume profile prediction might cause the strategy to achieve a higher trading rate than the simpler strategy. Thus, the need for good volume forecasts for various discrete time intervals is crucial.

There is an argument to be made that, reiterating the lack of any price dynamic/prediction information, VWAP is the best approach to trade an order when the order needs to compete. It is used by quantitatively driven funds in which the investment strategy correctly sizes the order to minimize the overall impact. Generally, while in recent years the popularity of VWAP has decreased in favor of the more flashy "Liquidity Seeking" strategies, it is still one of the most used strategies in the market. We will discuss Liquidity Seeking strategies in a later section.

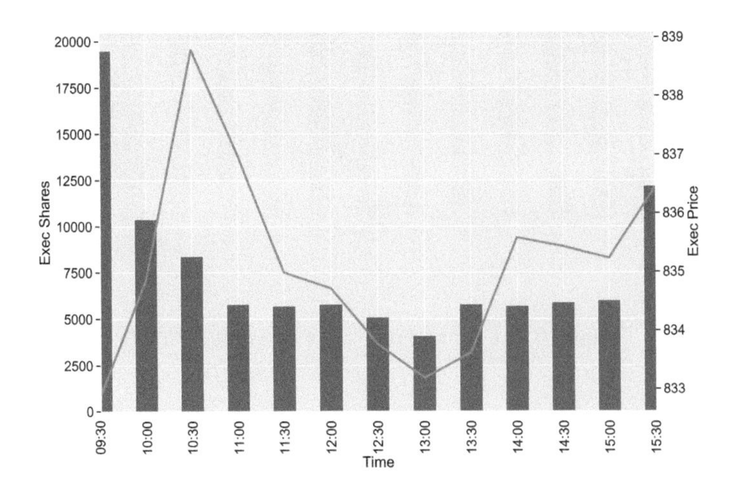


Figure 10.2: Example of VWAP Trade for AMZN.

**Inline with Volume, a.k.a. POV**

The previous paragraph hinted that one of the limitations of VWAP (and TWAP) as a market impact controlling tool in that they do not take into consideration the high degree of variability in the daily volume. If, in a particular day, the actual volume is significantly less than expected (due to lack of activity or large distortion in the volume profile) the VWAP strategy will just trade at a higher rate than originally envisioned by the trader leading to a higher total cost. Conversely, if trading a multiday order on a day where more volume is traded than expected, the trader would likely trade a smaller portion of the order than the market would have been able to absorb for the same expected impact cost.

If completing the order is an absolute hard necessity, arguably, there is little that can be done. However if there is some flexibility around completing the order then another approach can be taken. This is where the Inline strategy comes into play. The Inline or alternatively called POV (Percentage of Volume) strategy has a target trading rate as a benchmark and is not directly concerned with price. The aim of trading at a fixed rate is a way to directly control market impact. While seemingly straightforward, following a certain participation rate in real time is not an easy task and these strategies can have some severe drawbacks unless carefully implemented.

- Early implementation was purely reactive, meaning observe the volume for a certain period and then trade the prescribed portion of that volume in the next iteration. Due to noise in the volume dynamics this could lead a strategy to oversize the market after a large volume spike. In particular, large block prints, if not correctly managed, can cause a strategy to severely over-trade.
- Even when a volume forecast is used there is still room for error in the estimates and any shortfall needs to be correctly managed and possibly adjusted over multiple periods to avoid excessive impact.
- The POV benchmark can be extremely noisy, particularly at the beginning of the order due to the granularity of trading. This may result into no trading at the beginning until enough volume is traded and then immediately chasing that volume aggressively, or may lead to falling behind the target volume.

Because of these drawbacks and complexities Inline strategies have somewhat lost their popularity and the more flexible Liquidity Seeking approaches have taken their place.

The POV benchmark does not provide a way to measure the effectiveness of the strategy as an impact prevention approach, in particular because of the issues discussed above. The PWP $X$ (Participation Weighted Price at $X\%$ of the market) price is used as a measure of execution quality for the Inline benchmark. It represents the price resulting from trading if $X\%$ of the market volume was traded until completion of the order (in practice, the VWAP of the period from order start until ($1/X\times$ Order Size) has traded in the market).

**Target Close**

As previously discussed many quantitative strategies leverage time series data based on closing prices. As a result, some PM/traders favor this price point as a benchmark for the execution strategies. This is also the case for passive indexers whose fund

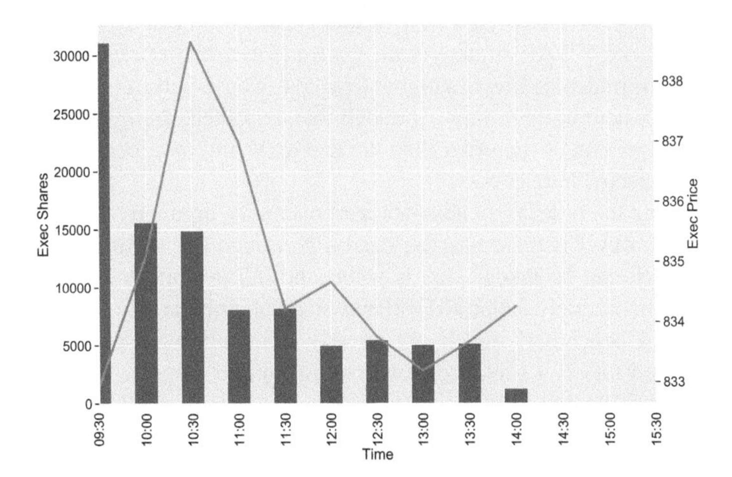


Figure 10.3: Example of 5% POV Trade for AMZN.

tracking error is computed based on closing prices. This clearly places a particular importance to the closing price and thus the desire to devise strategies for minimizing the risk from large negative deviation from that price.

It is important to note that in general even Target Close is an IS strategy, as it is still trying to achieve the elusive best price possible. If the ultimate price was unimportant the strategy would be trivial: Place the whole quantity as a Market-on-Close (MOC) order into the closing auction no matter how big the order is. The closing price could be negatively affected by the excess imbalance but the order WILL achieve the closing price benchmark. Instead the objective is to limit market impact while trading at times closer to the end of the day, limiting the risk of large deviations from the close price. Most strategies attempt to forecast the amount of closing volume and size the MOC slice accordingly, thus limiting the chance of negatively affecting the close price and then trade the rest in the latter part of the continuous phase accelerating the trading rate toward the close.

### Auxiliary Benchmarks

In addition to these standard primary benchmarks, several additional metrics are often used to evaluate execution performance. Here is a non-exhaustive review:

- **Spread Normalization:** As a way to make performance more comparable across the stock universe, various benchmarks are often normalized by the prevailing spread. Intuitively the expected slippage of the same VWAP strategy trading two orders: One very liquid and with 4 bps spread and the other illiquid with a 20 bps spread is likely to be very different. But as a percentage of the period spread they would be comparable.

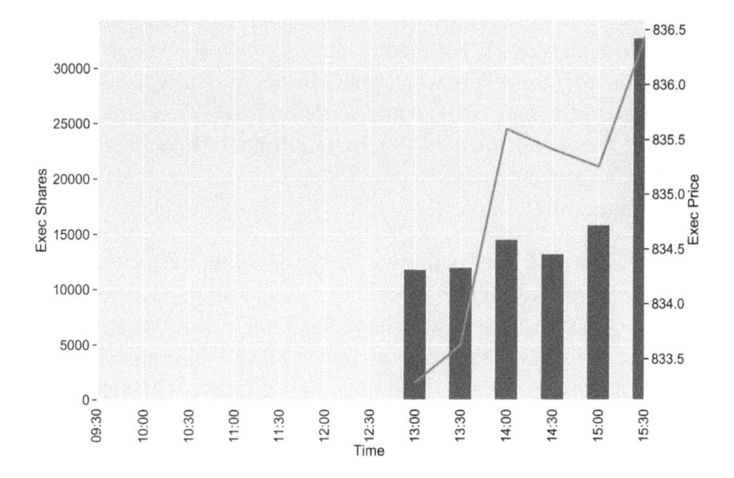


Figure 10.4: Example of a Target Close Trade for AMZN.

- **Completion Rate:** Balancing low slippage versus arrival price, may be considered as a desirable outcome. But if only a fraction of the parent order gets executed, the trader is left with a potentially significant opportunity cost, as the expected alpha is not fully captured. In order to avoid rewarding strategies that achieve low implementation shortfall, by being extremely passive and stop trading when prices become unfavorable, most practitioners add a penalty term for the unexecuted portion of the order. There are several approaches, the simplest is to apply a slippage to the unexecuted quantity. The slippage is usually the difference between the arrival price and the closing price of the day.

- **Close:** Even though it is not the central benchmark discussed above the closing price being the last accessible price of the day is often used to compute the opportunity cost for unexecuted quantity. There are different variations of this benchmark; most notably the previous day's close (for strategies that select their target holdings through an overnight process using the latest available close price as an input to their model) or the open price to reflect the fact the previous close is a price point that was not accessible to the strategy.

- **Make/Take Ratio:** This ratio provides some indication of the execution efficiency. How much of the liquidity was sourced on the passive side of the spread (at the bid or lower for a buy order), versus how much was the algorithm required to pay the spread? This metric highlights the quality of the order placement of the execution strategy through its ability to source liquidity, while achieving its desired trading rate. It is worth noting, however, that more passive fills do not guarantee overall better performance with regard to the primary benchmark. For instance, if a security exhibits strong intraday adverse momentum, a very passive strategy favoring passive fills would likely result in lower participation rates. This will delay the total execution and expose the traders to unfavorable prices later in the day. Additionally a well-performing strategy would leverage short term signals to trade more aggressively when prices move away and cancel passive orders to avoid adverse selection. This would score unfavorably on this benchmark. In general, the reader is encouraged to remain critical of benchmarks and metrics, and should strive to create more sophisticated versions that more accurately capture existing trade-offs.

- **Mean vs. Variance of Performance:** While absolute performance tends to be the main focus of execution benchmarking, a trader might also want to control its variability. In practice, traders with large amounts of flow on a daily basis often focus more on the attained average as their risk of being adversely affected by outliers is somewhat mitigated by the number of orders. Conversely, occasional traders are often willing to accept a slightly less optimal average performance if it is accompanied by less variability.

- **Reversion:** An additional way to analyze market impact or short term opportunity cost is to study the stock price trajectory after the execution is completed as noted in Chapter 9. A sustained high demand for liquidity results in a displacement of the supply-demand equilibrium, as liquidity providers manage their inventories, which may lead to an adverse price move (market impact). Empirical evidence (as mentioned in Gatheral (2010) [161]) suggests that the temporary component of price impact reverts once the execution stops, and the impact function follows a power law decay function (see Chapter 9 for more details). Hence, studying reversion at a time window commensurate with the order duration informs the trader of the impact cost that may have been incurred due to a certain participation rate. This rate might not be correctly adapted to the liquidity that the market was able to provide at the time of the transaction.

## 10.2 Evolution of Execution Strategies

In the previous section, we described the main approaches used by the original algorithms that still account for the majority of the trading flow going through execution products around the globe. These strategies are conceptually simple and overall quite effective. While over time they evolved in sophistication with better analytics, order placement, more nuanced handling of the schedule, etc., the approach has largely remained the same. As the business evolved, practitioners started looking for better approaches, with some quantitative underpinning that would move beyond the naïve methods that are the basis of these historical algorithms. We briefly look at how these efforts have evolved and the drivers behind this evolution. More rigorous details will be presented later in this chapter.

**Implementation Shortfall**

At the turn of the century, things changed with the seminal paper by Almgren and Chriss (2000) [10] that proposed a formal approach to "optimal execution." The approach was based on a mean-variance optimization framework similar in spirit to the Markowitz optimal portfolio allocation setup.

We begin with a high level discussion. The optimal schedule proposed, trades-off minimizing impact with the volatility of the overall cost. Like in Markowitz, the trade-off between mean and variance is governed by a hyper-parameter, '$\lambda$', which was meant to incorporate the "risk aversion" specific to a particular trader (see Section 6.1.1 for the mean-variance formulation). Because the overall volatility cost depends on the remaining quantity, the variance is highest at the beginning of the order and thus the optimal schedule will trade-off more impact cost and thus a higher trading rate to reduce that variance. As the position decreases, the risk decreases and thus the optimal schedule reduces its trading rate. The optimal schedule displays a very typical front loaded shaped trajectory that vanishes to zero at the trading end time (Figure 10.5).

This approach had generated a great deal of interest in the industry and brokers increased their investment in quantitative personnel who could understand and implement these ideas leading to the evolution of a new type of quant: The execution quant. Most algorithm providers implemented a version of this algorithm and popularity of the approach led to a dramatic increase in the interest of the academic community to find better approaches to the Optimal Execution problem. Because the approach is fundamentally based on the dynamics of market impact, this renewed interest from academics has resulted in significant research on market impact models.

The approach had its critics. Many traders were highly skeptical. They did not believe that a completely static approach, would provide an optimal way to trade, without taking into account what is happening in the market. These traders would want that a real IS strategy be dynamic, "picking spots," backing off when prices are not favorable and thus taking advantage of prices. Additionally, the dramatic increase of dark liquidity that promised the ability to trade in large positions with zero (limited) impact, further challenged the idea of a static schedule to achieve the best price possible. Finally, the actual implementation was somewhat problematic because it did not accommodate the possibility of the order being amended. Essentially because historically the infrastructure of most algorithms did not maintain any state and when a parameter was changed, they reacted by creating a brand new parent order. The IS optimization would then re-evaluate the strategy and front-load again the schedule creating further impact on the stock. For traders that use tight limits, this created real havocs.

Some buy-side quants were also skeptical. They correctly argued that the IS schedule, by being a mean-variance trade-off, is not actually providing the best price. The optimization trades off some expected cost for a reduction in the uncertainty of the outcome (i.e., standard deviation). But to large trading operations, the average shortfall is really the only thing that matters. Some would take additional volatility if a better price can be achieved. In this setting their risk aversion is zero. What happens when, in this elegant quantitative model, one sets a zero risk aversion? In this case the reduction in market impact is the only component in the optimization and the Almgren-Chriss' model reduces to VWAP! So a large subset of users shrugged off the revolution and continued to use VWAP as the main optimization approach.

Execution quants struggled to adapt the IS framework to incorporate any feedback. The restrictive mathematical framework made it complex to evolve. Several providers kept the core ideas but abandoned the original formulation; others tried to incorporate more dynamic approaches by moving to a stochastic control framework. Some still kept the original formulation fixed and bolted on a set of heuristics.

New academic work expanding on the framework is now available; however, the solutions often appear non-intuitive such as: "The optimal trading approach is to trade a large portion of the order at the beginning and the end of the time horizon with only moderate trading in the middle." These new results are yet to be adopted by practitioners as they are considered to be somewhat unrealistic. While the debate is still ongoing, there are a few IS implementations that follow the original framework. The approach is slowly losing out in favor of a more dynamic set of algorithms.

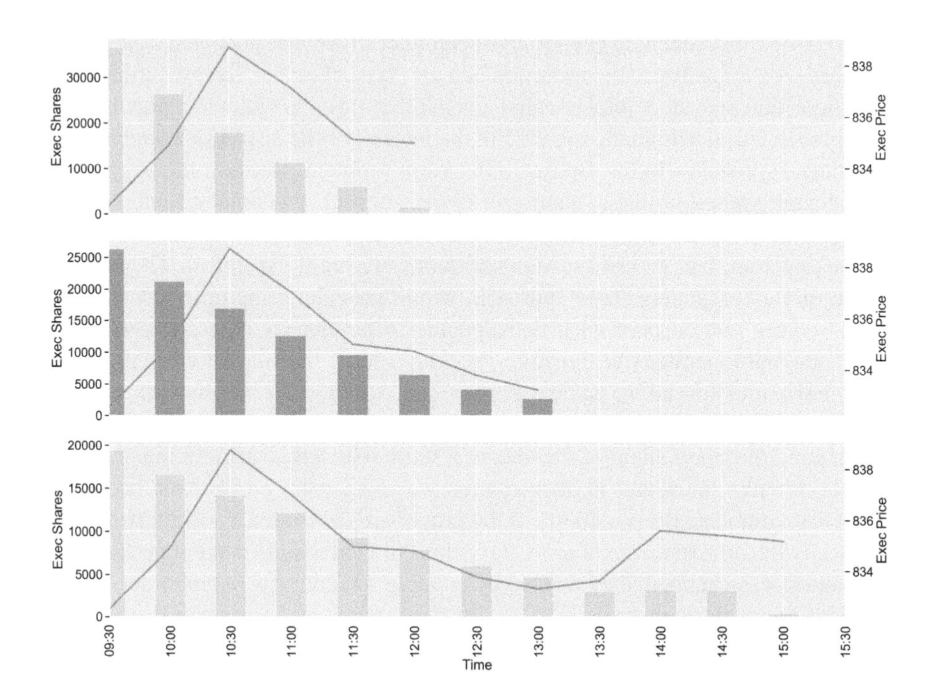


Figure 10.5: Almgren & Chriss' Solution for Different Risk Aversions.

### Liquidity Seeking Algorithms

The next generation of algorithms, broadly termed Liquidity Seeking, are to some extent a rejection of schedule-based approaches that had dominated for a while. This came because these approaches at times have to make sub-optimal decisions just to maintain a schedule. These decisions have significant impact on the performance of a strategy. By going for a completely dynamic approach, it is hoped that the excess impact due to "schedule catch-up" behavior would be reduced.

Liquidity Seeking algorithms encompass a range of approaches and urgencies that vary from very aggressive, taking as much liquidity as possible up to a certain maximum limit price, to strategies that post all liquidity in dark pools, to highly opportunistic strategies that pounce at particular situations in the market such as far-side outsized liquidity (i.e., when the visible quantity at the far side is unusually large),[4] and to everything in-between.

The downside of these approaches is that, because there is not a clear schedule to follow, the behavior can be highly non-deterministic and it implicitly excludes any timing risk component. The order that does not find enough liquidity would remain, possibly to a large extent, incomplete. That order would then have to be traded the next day. To address that, traders started demanding that the algorithms maintain a minimum participation rate thus reintroducing a scheduling component. Another common issue, in particular in the early days, was the tendency to maintain a large portion of the order hidden at the mid-point, which resulted in a potential for severe adverse selection. If the residual of a large order is fully executed at the mid-point, then there is a high degree of probability that the price immediately after the trade would have been more attractive, and thus creating a regret. This led some traders to demand that even in the dark pools, the order should be "paced," again reintroducing the concept of a schedule.

In the last few years, innovation on execution trading styles has somewhat tapered off. This slowdown is partly due to the push of many buy side trading desks towards a more predictable and systematic approach toward the selection of strategy and algorithm providers. The so-called "Algorithm Wheels" are becoming more popular in large buy side institutions as their requirements for demonstrating best-execution, call for simpler and homogenized trading approaches. They want to efficiently use as much of the sample set to make data-driven decisions.

Competition is as fierce as ever, due to recent regulatory changes driven by Europe's MIFID II regulation. However, the focus has mostly been on improving the various underlying components, seeking more sophisticated approaches that can further squeeze out performance from existing strategies. To some extent, sophisticated traders have started to move away from strategies and towards more abstract objectives and constraints, allowing brokers to pick the best approach to achieve them. They now engage with the algorithm providers with unprecedented openness in their effort to further minimize the overall trading costs. The benchmarks have not changed and can be still confusing and elusive; however, relative performance is easy to measure, thus at least workable. This new freedom to experiment and single-minded focus on

---

[4]Far side outsized liquidity is essentially similar to quote imbalance which, we discussed, points to the price going in the favor of the order. For this reason taking that liquidity at that price would seem counter-intuitive and, while this is a common practice, there is little to no evidence that the approach is highly beneficial.

performance is leading to a renewed energy and innovation that could signal a new "Golden Age" of electronic trading.

## 10.3 Layers of an Execution Strategy

We turn our attention to how the strategies discussed are actually implemented. To a large extent the approach to all strategies is somewhat homogeneous and can be decomposed into a set of trading concerns, that are well understood and can be looked at individually. In this section we review these layers providing some high level concepts around standard and more modern approaches.

### 10.3.1 Scheduling Layer

This layer is responsible for actually implementing any particular strategy. Often described as the macro trader of the strategy, it is in charge of decisions spanning the order duration, such as allocating the parent order quantity through the life of the order. In short, it answers the question: "how much should I trade in each period?" The scheduler is aware of the overall trading intention, total quantity to transact (size of the parent order), strategy parameters, constraints (maximum POV, limit price), and order duration. Taking the VWAP algorithm as an example, the scheduler is the component in charge of following the prescribed volume curve between the start and end times prescribed by the user. To derive the allocation, most schedulers discretize the trading horizon in more or less granular bins (generally one to five minutes), each bin considered as a distinct trading interval, to be handled by the child order placement module.

The scheduler will also decide which market phases should the algorithm interact with, if not specified by the user; for instance, whether to participate in auctions or not. For illiquid stocks, call auctions can be a significant source of liquidity, so participating in them can limit the need for excessive trading in the continuous session. However, trading too much in the open auction—for example—can create an anchoring effect leading to a more permanent market impact and price displacement for the remainder of the continuous session, and thereby negatively impacting follow-up executions. As often in electronic trading, there are not always absolute solutions, and these trade-offs must be given due consideration.

The scheduling layer plays an important role in implementing liquidity sourcing preferences, such as the level of participation in volatile periods of price discovery when spreads are wide (e.g., in the first few minutes after the market open), or allocating volume between lit and dark venues (where one swaps certainty of execution for lower price impact). As discussed in Chapter 8, algorithms are to focus on leveraging many normalizing variables, in order to handle different characteristics of the stock universe both on regular and on special days. Modern scheduling layers also should respond to realtime feedback loops from the market and adjust their execution as necessary such as front-loading the execution if the market is expected to move unfavorably, or back-load it in case of expected favorable move or adjust the balance across liquidity sources (lit or dark markets).

In summary the scheduler's role is to organize, implement and adjust the overall plan of action for the strategy and it is clearly the most important component of an algorithmic trading strategy.

### Implementing a Scheduler

The above description is too broad and does not provide much in the way of details on how to build a Scheduler in practice. It is not easy to cover all the intricacies and details of a real world implementation, but we will provide some ideas used in practice. There is no standard approach to build a Scheduler and this treatment may be too simplistic and unsophisticated. But the details may be useful.

**Naïve Approach:** We start with the simplest possible implementation. As discussed, a schedule is often represented by a vector of parameters in equally spaced discretized bins. As an example, for a schedule-based strategy,[5] the vector will contain the quantity traded, the price, etc. In each time period, we submit the information to the order placement layer on the exact quantity required for the particular bin. We expect that the order placement will ensure completion of the bin quantity by the end of the allotted time. If the order placement cannot fulfill the request due to limit price or POV constraints, the scheduler simply ignores the order and moves on to the next bin.

Clearly, this implementation could lead to severe under-trading and is likely to result in excess impact, due to immediate "catch up" at the end of a time bin. Less often traded stocks are not likely to fully fill the required liquidity within each bin, exacerbating the amount required to catch-up. One could normalize the time bin by a multiple of characteristic time that would somewhat harmonize the behavior across the stock universe but it may not address the other limitations.

A better approach would be to allow the order placement layer to engage in fulfilling the desired quantity without "forcing" a hard catch-up. Any unfilled quantity would simply be added to the next bin. This will limit the excess impact but the tendency for the order to fall way behind the desired schedule may still prevail. Clearly a trade-off is needed. An additional drawback to this approach is that in some cases the scheduler will be behind schedule most times.[6]

**Trading Bands:** A common approach to address the drawbacks mentioned above is via the trading bands. These are essentially auxiliary schedules above and below the desired schedule. They represent how much behind and ahead the scheduler is allowed

---

[5]For POV strategy, this could be based on the previous bin volume or a forecast of next bin volume.

[6]For POV orders, it is possible to be ahead of schedule when using volume forecasts which at time will overestimate the expected bin volume.

to deviate from the desired target. Figure 10.6 illustrates a visual representation of a VWAP schedule with its trading bands.

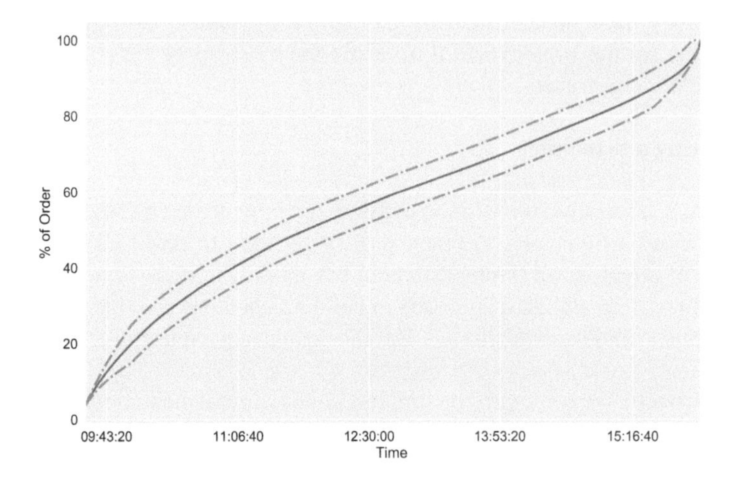


Figure 10.6: Example VWAP Schedule with Bands.

Using the trading bands, the scheduler can determine if and how much it needs to catch up to the schedule so as to not fall too much behind or in general end up with the risk of catch-up.[7] The scheduler can then determine how much shortfall it needs to make up and the opportunistic quantity it has available because it is allowed to be ahead of schedule. It can then pass this information to the order placement so that the scheduler can be more nuanced about managing the required trading needs. Figure 10.7 gives a visual representation of the information available to a general scheduler.

At the beginning of every time bin, the strategy will assess four main state variables:

1. **Target Quantity:** Where the strategy should be at time $t_0$ ( point A )
2. **Current Quantity:** Where the strategy is (point B)
3. **Next Quantity:** Where the strategy should be at time $t_1$ (point C)
4. **Opportunistic Quantity:** Where the strategy can (at a maximum) be at time $t_1$ (point D)

Another important consideration is at what time, if no trading happens, will the strategy hit the lower band and thus will need to catch up (point L). The difference between the Target Quantity and the other states provides the necessary information

[7]It is well known that arbitrary, unconditional, catch-up is one of the significant drags in performance. Also many traders are extremely risk averse and tend to cancel an order that is too far behind schedule.

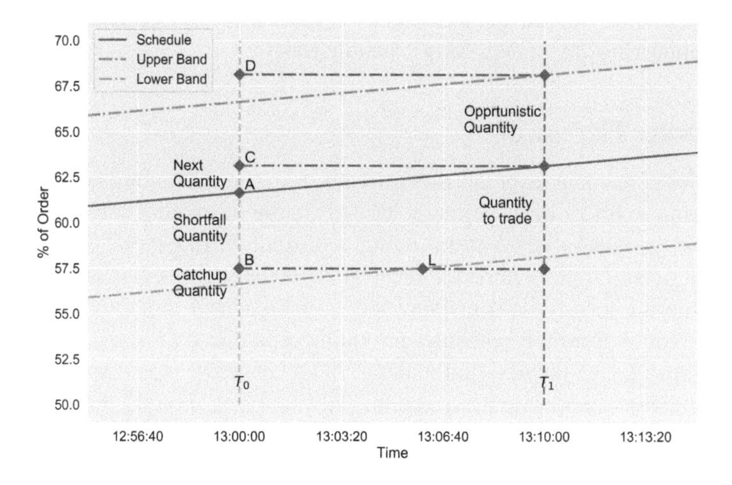


Figure 10.7: Schedule Information Set.

for the strategy to deliver the required instructions to the order placement layer to maximize the chances to execute the required shares. These instructions can include, how to stay on schedule and how much leeway it has to stay behind, etc. Providing the order placement with a quantity range has additional benefits, because it allows the execution to be ahead of schedule when conditions are favorable leading to reduced catch-up behavior and thus resulting in reduced impact and better performance.

The distance between the bands and the schedule can be made more stock-specific allowing to further refine the behavior across the trading universe. It can also be made client-specific as it also represents the traders' risk aversion from schedule deviation. The wider the bands, the less risk of unilateral catch-up but larger the risk of significant deviation from the desired schedule.

**Stateless Approach:** An improved approach is not to use time bins at all, but work in a more flexible set-up. The scheduler evaluates the same state variables on a continuous basis (every few seconds for example) and adjusts the instructions to the order placement to refine the trading policy. The schedule can also use several time points in the future to provide a term structure to the policy and thus plan ahead. This is useful, in particular, when the schedule starts having a significant "curvature" toward the end of an order.

**Systematic Deviation:** A final refinement to this approach, is the usage of predictive signals (a.k.a. alpha) within the schedule. If a prediction of short term price move is possible, it can leverage that information by making a systematic decision to get ahead or stay behind schedule and take advantage of that price information. This has to be done carefully to avoid incurring any additional market impact or risk of catch-up that will cancel out any possible benefit of this move. The schedule also needs to have a component that will move the strategy back on schedule if the price prediction is not possible. This will prevent the strategy to remain ahead (behind) for longer than necessary, incurring any unnecessary schedule risk.

### 10.3.2 Order Placement

The order placement layer sits downstream from the scheduling layer and can be termed the micro trader of the strategy. It is in charge of making decisions spanning the next few seconds or next several minutes of trading. In general this layer is not aware of the overall trading intention received by the scheduler. It is only responsible for implementing locally its decisions. More sophisticated methods leverage additional instructions from the scheduler providing a picture on the macro state of the order that may nuance the decision making. In any case, we can simplify the role of the micro trader as answering the question: "at what price should I trade the quantity allocated for the upcoming period to complete the order at the best possible price?" Now we provide the historical evolution of Order Placement techniques.

**Naïve Order Placement:** The simplest approach is to send the given quantity as a market order with essentially no uncertainty in the outcome. The aggregated order-book (Superbook) can provide a good, conservative (because it does not include the possibility of mid-point hidden orders), estimate of the average price that would be achieved. Depending on the stock and order characteristics the size to trade may still be larger than what can be absorbed by the market without "eating" into the order-book. This could simply be remedied by decreasing the schedule interval.

The first order placement modules were not that much different from this naïve approach. While for highly liquid instruments with tight spread, this approach may achieve decent results, as we go down the liquidity spectrum, this approach does not work well. One needs to consider the additional complexity that comes with leveraging non-marketable orders.

**Peg and Cross:** The next approach is often termed "Peg and Cross." It consists of placing a limit order for the scheduled quantity at the near side. If the market moves away, the order can be repriced (alternatively one can rely on a peg order type to do this) and size updated if necessary. If the order is not fully executed by the time the schedule reaches the catch-up zone, an aggressive order can be sent to get back on schedule.

Variations of this approach are still the norm in many order placement products across the industry. An advantage of the approach is that if the order is executed passively on a maker-taker exchange, it will earn a rebate which makes the overall cost of trading lower. On the downside, this technique will consistently be adversely selected with most of the passive executions happening, when the price is "coming through" and is likely to be more attractive right after the order is filled.

One approach to minimize the adverse selection is to "layer" the order book with some of the available quantity so as to capture execution at better prices when they come through. Ideally, with a fast infrastructure, if short term signals like order book imbalance are used to predict when the probability of price drop may reach a critical threshold, the trader can cancel the top order, minimizing the chance of adverse selection. This is a common approach of HFTs as they want to be at the top of the book to capture a fill right before the price moves away.

A common variant of the Peg and Cross technique is to leverage additional price points, in particular the mid-price. As the need for liquidity increases, the order placement can either ping the mid-price of various exchanges or have a quantity hidden at the mid-price to increase the probability of getting filled.

**Beyond Peg and Cross:** More sophisticated approaches are possible and are used in some advanced implementations. These models are proprietary and the authors do not claim to have seen them all. We give a few additional pointers.

One direction in which these models evolved is to leverage Reinforcement Learning techniques (see Section 4.6.2). These models are notoriously hard to calibrate and require a good order book simulation environment. They are also somewhat limited in scope as most simulators require what is called a "small order" assumption, meaning that the quantity traded does not meaningfully alter the state of the market and thus does not require to inject any strong assumptions in the simulator on how the market will react. This is definitively an interesting space for research and could lead to valuable breakthroughs in order placement strategies in the future.

Other approaches use more heuristic methods around liquidity interaction. The focus is more on speed, low latency and alpha signal utilization to trade aggressively, when the price is ready to move away and "stepping out" of the way, when the market is about to come through.

An area of interest is to develop an approach to order placement that is uniform across stocks, apart from the usual normalization practices. As an example, for a subset of instruments with very volatile and thin order books, where posting liquidity does not really add any value, one could benefit from a more opportunistic approach. Stocks with ultra-long queues could use different methods to maintain "density" within the order book by canceling orders when no liquidity is needed. Finally, low liquidity stocks likely require a completely new approach. The focus would be on generally illiquid markets in Europe and APAC.

### 10.3.3 Order Routing

The output of the order placement layer can be modeled as an array of price levels and associated quantities such as $[[18.85; 600], [18.84; 400], [18.81; 200]]$ which is then passed on to the order router for execution or quoting in the market.

The order routing layer sits downstream from the order placement layer and is in charge of the ultra high-frequency decisions (of the order of microseconds to milliseconds). Given the volume and complexity of events to process in a fragmented market environment and the sensitivity to latency, the routing decisions are often delegated to the Smart Order Router (SOR) rather than being handled by the algorithmic strategy itself. While it is also generally unaware of the overall trading intention received by the scheduler, it is responsible for implementing the decisions made in the order placement module. As such, the SOR answers the question "to which exchange(s) should I route the child order in order to maximize the probability of getting a fill at or better than the desired price?" This component is of particular importance in fragmented markets, and may be absent in execution strategies for single markets (Emerging Markets Equities, Futures, etc.).

While probably the simplest component of the execution stack from a quantitative perspective, the SOR implements rather complex decisions to achieve several objectives:

- **Sourcing Liquidity Efficiently:** This is the primary objective of the component in charge of handling the fragmentation complexity. But it can take different forms depending on the instructions received from the order placement module. For passive orders, sourcing liquidity efficiently means maximizing the probability of fill and minimizing the time to fill. For aggressive orders, this means maximizing the fill rate.

- **Optimizing Price:** The instructions received by the SOR are in the form of a limit order and a Time-in-Force. For Immediate or Cancel order (IOC) instructions meant to remove liquidity from the order book, the limit price is a maximum (minimum) for a buy (sell) order. This gives the router the optionality to eventually find a better price and provide price improvement to the order placement module.

- **Managing Execution Fees:** Venues have various pricing models (flat price, maker-taker, taker-maker) and charge different prices to access the liquidity. Thus, the SOR can optimize order routing to minimize fees (or maximize rebates), at the potential expense of lower fill rates.

In practice, these objectives are assigned different weights to reflect the traders' preferences because of the implicit trade-off they represent. There are also some factors that directly influence routing decisions:

- **Relative Price Target of the Child Order:** Near touch, far touch or mid-point. This is the main input influencing routing decisions. The handling of other features described below depends on the price being targeted.

- **Algo Parent Order Urgency or Liquidity Demand:** Low urgency parent order algorithms target low percentage of volume over the course of the execution, as a result of low risk aversion or low perceived alpha. As a consequence, they do not have to access all available liquidity at all points in time and can focus on quality over quantity. On the other hand, high urgency algo strategies are used by traders with fast-realizing alpha and favor liquidity extraction over quality considerations.

- **Algo Child Order Urgency:** Irrespective of the longer scale urgency of the parent order, child orders also have different degrees of urgencies depending on the reason why they were generated. Passive orders are associated with a time horizon by the end of which they are expected to be filled. The longer the horizon, the deeper in the book they can be posted. Aggressive orders also have varying degrees of urgency, from low if the algorithm is crossing the spread to maintain a target execution schedule (for instance after hitting the lower band surrounding the schedule due to insufficient passive fills), to high if the algorithm is crossing the spread due to a trading signal that may indicate a short lived opportunity to capture advantageous liquidity (being due to its size or price).

- **Child Order Size:** For aggressive orders, the relative size of the child order compared to available far touch size offers some optionality. If the order is small compared to the available liquidity, the SOR can search for price improvement, without getting exposed to the opportunity cost, resulting from cancellation of far touch orders. Conversely, oversized orders require fast routing in order to maximize fill rate before the liquidity fades away.
- **Intended Liquidity Destination:** Lit or dark markets.
- **Execution Cost:** Venue specific cost can also vary for aggressive and passive orders, and some strategies may explicitly target an execution cost to remain profitable.

**Implementing a Smart Order Router**

The first consideration in the implementation of a SOR in a fragmented HFT landscape is speed. This is a requirement for the processing of both order instructions and inbound market data that is used for making routing decisions. A modern SOR is expected to receive direct feeds from the exchanges and process the market data to reconstruct the composite order book in real time, which in many cases is achieved through hardware-based solutions. From an implementation point of view, for time sensitive decisions, the optimal part is generally modeled offline and then stored as a collection of routing tables. This will allow the SOR to swiftly implement the desired routing sequences based on the characteristics of the top level algorithm and its local intent. For passive orders, targeting the near touch or mid-point, the SOR can afford to solve the problem online and determine optimal allocation across eligible venues.

As described above, the objectives of the router can differ across orders and across time. They take into consideration multi-dimensional features, resulting in a large number of scenarios. For illustration purposes, we provide a few examples of opportunities and challenges the SOR faces for different target price points.

- **Near Side Orders:** When submitting passive orders, the main objective is to optimize the queue position so that the probability of obtaining a fill[8] is maximized, before the market moves away. There are a few factors that directly influence the fill probability of orders routed to the near side:

[8]The SOR can be either cost-agnostic and favor fill rates and time to fill, or cost-aware and try to maximize rebates in markets that have a maker-taker pricing model. This would, obviously, result in different routing strategies.

- **Price:** The closer to the top of book, the higher the probability of fill, and lower expected time to fill, all else equal.
- **Queue Length:** Long queues, in particular when price volatility is low, exhibit significant position competition on the passive side, resulting in longer time to fill with generally lower probability of fill. The decrease in fill probability is more pronounced at deeper levels of the book as they barely trade beyond the touch.
- **Resting Time:** The longer an order rests in the order book, the higher the chance for its fill as it progressively accrues queue priority in price-time order books, following executions and cancellations.
- **Size:** Smaller orders (with respect to the available liquidity) tend to enjoy higher fill rates than the bigger orders.

In a fragmented market, these factors take different values on different venues giving the router an opportunity to dynamically optimize liquidity-sourcing as volume shifts from exchange to exchange. This is essential for achieving best execution as naïvely sending a passive child order to a venue, resting it there for a period of time without getting any execution while transactions occur in other venues, might incur opportunity cost if the market moves away and if the child order ends up being traded at a less favorable price later on.

Naïve posting strategies allocate shares evenly across venues, or proportional to the market share of venues. Others utilize simple routing tables, where venues are ranked in decreasing order of preference (based on market share or fees) and shares are allocated sequentially up to an allowable quantity before moving on to the next venue. More quantitative routing decisions for lit passive posting can be expressed as a constrained optimization problem: Given '$X$' shares to execute, $[n_1, n_2, \ldots, n_Y]$ shares displayed on the book of '$Y$' lit venues with expected trading rates $[v_1, v_2, \ldots, v_Y]$ over the desired execution horizon, what combination of shares should be sent to different venues to maximize the overall fill probability and minimize time to fill? Some recent academic research ideas in this area are discussed in Section 10.5.

- **Mid-Point Orders:** For these orders, the destination is usually dark pools. This presents other inherent challenges. While lit executions and order books are public and can be observed in market data feeds, dark order books are invisible (making it difficult to estimate the fill probability a priori, at the time of routing). Dark executions tend to be reported in aggregate (for instance, under a single condition code), thus making attribution of actual execution to a venue more challenging. From solely observed market data, dark executions in the US cannot be directly attributed to any venue. However, the trader, through execution fills received, is in possession of censored observations of the available—otherwise invisible—liquidity. That is when receiving a fill for '$x$' shares one knows there were at least '$x$' shares available to trade on the other side in a particular venue and can use this information as an input for the next round of decisions.

Additionally, not all dark venues are made equal, and in practice, the quality of their liquidity is usually assessed as a function of:

- **Average Order Size:** The larger the better as it indicates the presence of natural liquidity that most execution algorithms favor dealing with.
- **Uniqueness:** There is no precise rule here, but a fill can be considered unique if there is no other dark trade happening on the same symbol close to the moment preceding and following the print. More unique prints are also generally indicative of the presence of natural liquidity as market makers (considered to be non-natural liquidity) tend to quote in multiple venues at once.
- **Short Term Reversion:** Measured as the trajectory of the mid-point immediately following an execution (usually at time-frames ranging from a few milliseconds to a few seconds). Large, adverse, moves are often indicative of a certain toxicity of the venue.

These metrics directly influence routing decisions. While a venue may be executing large quantities, the quality of its liquidity may not always make it an ideal destination for other types of executions. For low urgency executions (low participation rate, low expected alpha), the router can be more restrictive in the choice of venues, focusing on venues with larger—more unique—liquidity. For high urgency orders (demanding both large liquidity and immediacy due to superior alpha), it may step up to access all venues in order to maximize liquidity extraction.

Here too, naïve posting strategies allocate shares either evenly across venues, or they utilize simple routing tables as mentioned earlier. For an aggressive order meant to sweep dark pools quickly before performing another action, the main consideration is usually speed. As a result, the number of venues explored tend to be somewhat limited. Aggressive sweep tactics favor dark venues that are close in space (and time) and have the faster matching engines, to minimize the opportunity cost associated with the round trip time necessary to receive a fill or an IOC cancellation back. From a quantitative optimization perspective, because of the transient nature of dark liquidity and the inherent uncertainty, dark allocation is a classic multi-armed bandit problem: Given $X$ shares to execute in $Y$ venues with unknown available liquidity and unknown trading rates, what is the combination of shares to be sent to different venues that will maximize the fill probability and minimize the time to fill? The exploration versus exploitation dilemma favors incurring some opportunity cost by allocating shares to venues with low probability of execution, to obtain information about the potential presence of liquidity there at a particular point in time.

- **Far Side Orders:** For these orders, maximizing the fill rate is the primary objective; this is followed by obtaining price improvement if the child order is not particularly urgent. The challenge for aggressive take orders is to be able to capture all the liquidity that was visible at the time the child order was initiated. Market participants, in particular high frequency market makers, tend to display liquidity on multiple venues simultaneously. They are likely to cancel all other outstanding child orders once they receive execution confirmation from a venue, to avoid adverse selection. This will deprive an aggressive trader from capturing all the liquidity that is displayed.[9] Market makers may then replace their quotes deeper in the order book, as a reflection of their lesser appetite for additional inventory. Hence, the trading algorithm that tries to capture all visible liquidity on the far touch at a point in time, may end up getting only a partial execution and trade the remainder of the slice at a less favorable price. This rapid disappearance of liquidity immediately following a trade is known as a "fading" effect. This can be particularly costly over time for liquidity takers who may need to pay extra ticks to access the amount of volume they need. To guard against such fading, SORs now implement Time-on-Target coordination tactics.[10] Most advanced SORs now incorporate estimations of the (time) distance separating them from the matching engines of these venues, to optimize child order routing sequence so that all routed orders hit the various exchanges at approximately the same time. Another common tactic is to send large orders first so that the potential fading faced by later orders has less impact on total fill rate.

When the urgency of the child order is not high, and if the objective of an aggressive order is just to catch-up to a prescribed schedule with a low-frequency granularity (of the order of minutes), then the routing is not really time sensitive and can also leverage the SOR's ability to access both lit and dark venues. Even if the original order was intended to trade on a lit market, the low urgency of the desired action gives the SOR, an opportunity to source mid-point liquidity first—to get some price improvement—and complete the intended action with the residual quantity. This technique is generally described as "dark pinging." It sequentially scans a set of dark pools prior to performing low urgency spread crossing. The selection of which pools to scan through follows a similar approach as orders intended for mid-point routing in the previous paragraph: The trade-off between quantity and quality depends on the urgency of the overall order.

Similar to the top level trading algorithm, the performance of the SOR is evaluated using metrics such as fill rate on aggressive limit orders (higher fill rate meaning a more efficient routing less subject to fading), the ratio of routed versus executed quantity, and average time-to-fill on passive orders (less unsuccessful routing and shorter time-to-fill meaning less potential opportunity cost incurred by child orders due to inefficient routing), etc.

---

[9]This is possible due to the latency that exists when sending orders to multiple, physically distant, exchanges.

[10]Time-on-Target is a military term used to describe the coordinated firing or artillery by many weapons so that all munitions hit the target at the same time.

## 10.4 Formal Description of Some Execution Models

In the last two sections, we provided the practitioner's view of issues related to execution strategies. These strategies all have a common theme, that is to optimize dynamically where, how often, and at what price to trade, given the prevailing market conditions. In this section, we want to provide a brief review of the evolution of execution models. Given the complexity of the field, we will not be able to cover all facets of the issues, so interested readers should refer to some of the recent research as well as the original papers cited in this section.

We present below some select academic results in this area. The quantities that the trader can determine are the number of child orders ($n$) and their sizes ($x_k$) and the duration for the parent order trading ($T$). The relevance of this problem to the practitioners is also discussed in Chan and Lakonishok (1997) [73] and Keim and Madhavan (1997) [228]. Bertsimas and Lo (1998) [37] develop a model that suggests splitting the parent order to reduce the average trading costs. Almgren and Chriss (2000) [10] split the parent order to minimize the mean-variance of trading costs (implementation shortfall with risk) similar to Markowitz's portfolio theorem. Some recent works on this topic are also briefly mentioned in the following sections.

### 10.4.1 First Generation Algorithms

The performance of traders is generally evaluated by how well they execute in comparison to volume weighted average price (VWAP), because the VWAP strategy is considered to be closer to how a passive trader would trade. Madhavan (2002) [255] warns that the uncritical use of VWAP as a benchmark can promote non-optimal trading behavior. The choice of a benchmark obviously affects a trader's placement strategies. Recall the discussion earlier how traders are shifting the orders toward the market close as large number of trades occur during the closing hour. Using daily VWAP as a benchmark, leads to trades spread over the day and thus passive trading can significantly affect alpha capture. The VWAP strategy is designed so that order submissions follow historical volume patterns over the course of the day with reasonably fine time grids. This practical approach is expounded in this section.

We will follow the formulation of the optimization problem as given in Busseti and Boyd (2015) [64]. With $X$, the total number of shares to be traded over the time horizon, $(0, T)$ in '$n$' discrete time intervals, the problem is to decide how to divide the parent order $X$ into child orders $x_1, \ldots, x_n$ so that the executed average price $p_{\text{exec}}$ is closer to $p_{\text{VWAP}}$. It is assumed that the price, $p_t$, is a random walk and thus the return is white noise. It is assumed that the trading algorithm mixes optimally market and limit orders. Defining the cost of trade in time interval $t$ as $x_t\hat{p}_t$, where $\hat{p}_t$ is the effective price that depends on the instantaneous transaction costs such as bid-ask spread with '$s_t$' as the fractional bid-ask spread with a certain participation rate $x_t/V_t$, it can be shown that the effective price is

$$\hat{p}_t = p_t \left(1 - \frac{s_t}{2} + \alpha \frac{s_t}{2} \cdot \frac{x_t}{V_t}\right), \tag{10.4}$$

which gives in the end quadratic transaction costs which is a reasonable approximation as seen from Chapter 9. Here '$\alpha$' is taken to be the proportionality factor of market orders to total orders. Defining the proportional ratio

$$s = \frac{X p_{\text{VWAP}} - \sum x_t \hat{p}_t}{X p_{\text{VWAP}}} \tag{10.5}$$

as normalized slippage, where the numerator is the cash flow to the broker, the optimizing function is

$$E(s) + \lambda \operatorname{Var}(s). \tag{10.6}$$

Here '$\lambda$' is the risk-aversion parameter. Note that the quantities $p_t$ and $V_t$ are random variables and the decision variable is '$x_t$'. Note with some approximation, $p_t \sim p_{t-1} \sim p_{\text{VWAP}}$ and with applying the law of iterated expectation, the optimization in (10.6) simplifies to

$$\min_{x_t} \sum_{t=1}^{n} \left[ \frac{s_t}{2x}(\alpha x_t^2 k_t - x_t) + \lambda \sigma_t^2 \left( \left( \sum_{r=1}^{t} u_r / X \right)^2 - 2M_t \sum_{r=1}^{t-1} u_t \right) \right] \tag{10.7}$$

with constraints $x_t \geq 0$ and $\sum_{t=1}^{n} x_t = X$. Here the quantities $M_t = E\left[\sum_{r=1}^{t-1} V_t / V\right]$ and $k_t = E[\frac{1}{V_t}]$ refer to market intra-day volume distribution. Assuming that the spread is a constant which is generally true for actively traded stocks, the optimal static solution can be shown to be equal to:

$$x_t^* = XE\left(\frac{V_t}{V}\right), \quad t = 1, \ldots, n. \tag{10.8}$$

From the above discussion, it is clear that forecasting both the intra-day distribution of periodic volume, $v_t$, and the total volume for the day, $V$, is essential for achieving optimal results. The optimal result in (10.8) can be obtained also with slippage as defined in the previous section.

**Bertsimas and Lo Model:** The basis of the model discussed here is that trading affects not only the current prices but also the price that follows. Thus the problem involves dynamic optimization but the solution turns out to be static and elegant. Assume again that an investor wants to sell a parent order of $X$-shares in $(0, T)$, where $T$ is arbitrary, and let $x_k$ be the number of shares sold (child orders) with the realized price, $\widetilde{p}_k$. The cost of trading, also known as slippage is

$$s = X p_0 - \sum_{k=1}^{n} x_k \widetilde{p}_k, \tag{10.9}$$

where $p_0$ is the arrival price. The model proposed in Bertsimas and Lo (1998) [37] can be stated as

$$\min_{x_k} E_1 \left[ \sum_{k=1}^{n} x_k p_k \right] \text{ such that } \sum_{k=1}^{n} x_k = X. \tag{10.10}$$

The criteria in (10.10) call for determining the sizes of child orders a priori. This is the function of the Scheduler that was described earlier. To solve (10.10), the model for price, where the price impact and price dynamics are separated, is taken to be,

$$p_k = p_{k-1} - \theta x_k + \varepsilon_k, \tag{10.11}$$

where $\theta > 0$ represents the price impact. When $\theta = 0$, the model is simply a random walk model discussed in Chapter 2. Under (10.11) the best strategy for splitting the parent order, $X$, is shown to be

$$x_1^* = x_2^* = \cdots = x_n^* = \frac{X}{n}, \tag{10.12}$$

the equal split. What the trader needs to decide a priori is the number of splits, $n$. The allocation strategy in (10.12) can be taken as the time weighted average prices (TWAP) strategy in industry as mentioned in Section 10.1.

The dynamic programming solution to (10.10) via the Bellman equations turns out to be elegant and simple because the price impact ($\theta x_k$) does not depend upon the prevailing price ($p_{k-1}$) nor on the size of the unexecuted order ($X - \sum_{j=1}^{k-1} x_j$). The indirect assumptions here are that the volume curve is even over different times of the day and the volatility profile is flat. Bertsimas and Lo (1998) [37] also consider extensions of the model (10.11), where a serially correlated state variable such as market condition (S&P 500) is added. In this case it is shown that the best execution strategy at any point is a function of the state variable and the remaining unexecuted size.

As a point of comparison to the equal allocation strategy observe that the VWAP strategy is similar in the sense that it does not also depend explicitly on any price movement. Simply the allocation that is made in (10.8) can be intuitively justified as resulting from $\frac{x_i}{X} = \frac{V_i}{V}$. If the traders are restricted to only '$n$' time intervals, it is important to have a good forecast for the quantities, $V_i$ and thus, $V$, the total volume for the day.

### 10.4.2 Second Generation Algorithms

**Almgren and Chriss Model:** The seminal papers by Almgren and Chriss (2000) [10] and Almgren (2003) [12] consider the same problem taking into account the stock characteristics such as volatility via the mean-variance formulation of minimizing execution costs. This formulation treats the execution as a trade-off between risk and cost; the faster the execution, higher the cost but lower the risk. The objective again is to liquidate, $X$ units before time '$T$' at minimum cost, in time steps, $t_k = k\tau$ for

$k = 0, 1, \ldots, n$ where $\tau = T/n$ is the average length of the interval between successive child order submission times. If $X_k$ is the number of units held at $t_k$, the goal is to decide on the trading trajectory, $X_0, X_1, \ldots, X_n$ with $X_0 = X$ and $X_n = 0$. The number of units liquidated $x_k$ between $t_{k-1}$ and $t_k$ is called the trading schedule, with $x_k = X_{k-1} - X_k$. The trading strategy requires determining '$x_k$' in terms of information available at time $t_{k-1}$. Static strategies determine the trading schedule in advance of starting trade time, $t_0$, while dynamic ones depend on information available up to $t_{k-1}$. Almgren and Chriss (2000) [10] find trading trajectories that minimize $E(s) + \lambda\,\mathrm{Var}(s)$ for various values of $\lambda$.

The model for price impact is the same as in (10.11) but stated as

$$p_k = p_{k-1} - (\tau\theta)\frac{x_k}{\tau} + \sigma\tau^{1/2}\varepsilon_k, \tag{10.13}$$

where $v_k = x_k/\tau$ is the average rate of trading in $[t_{k-1}, t_k]$ and $\varepsilon_k$ are independent with mean zero and unit variance and '$\sigma$' is the volatility of the stock. The temporary impact represented by price per share received for sale in $[t_{k-1}, t_k]$ is

$$\begin{aligned} \widetilde{p}_k &= p_{k-1} - \eta\left(\frac{x_k}{\tau}\right) \\ &= p_{k-1} - c\,\mathrm{sgn}(x_k) + \frac{\eta}{\tau}\cdot x_k, \end{aligned} \tag{10.14}$$

where '$c$' can be taken to be the fixed cost of selling which is the sum of mid-spread and fees. As a simplifying assumption, the temporary impact component is not present in the next period. Note (10.13) can be written as,

$$p_k = p_0 + \sigma\tau^{\frac{1}{2}}\sum_{j=1}^{k}\varepsilon_j - \theta(X - X_k), \tag{10.15}$$

and thus the shortfall,

$$s = Xp_0 - \sum_{k=1}^{N} x_k\widetilde{p}_k = \sum_{k=1}^{N}\left(\frac{x_k}{\tau}\right)\eta - \sum_{k=1}^{N}[\sigma\tau^{1/2}\varepsilon_k - \theta x_k]X_k \tag{10.16}$$

is a function of the market volatility and the trade schedule, '$x_k$'. It captures the difference between the initial stock value and the weighted average of realized prices. From (10.16), it follows that

$$\begin{aligned} E(s) &= \theta\sum_{k=1}^{N} x_k X_k + \eta\sum_{k=1}^{N}\frac{x_k}{\tau} \\ \mathrm{Var}(s) &= \sigma^2\sum_{k=1}^{N}\tau X_k^2. \end{aligned} \tag{10.17}$$

If the trajectory is to sell at a constant rate, $x_k = \frac{X}{n}$ and $X_k = (n-k)\frac{X}{n}$, $k = 1, \ldots, N$,

$$\begin{aligned} E(s) &= \frac{1}{2}\theta X^2 + \varepsilon X + \left(\eta - \frac{1}{2}\theta\tau\right)\frac{X^2}{T} \\ \mathrm{Var}(s) &= \frac{1}{3}\sigma^2 X^2 T\left(1 - \frac{1}{N}\right)\left(1 - \frac{1}{2N}\right). \end{aligned} \tag{10.18}$$

The trajectory minimizes expected cost but variance increases with $T$. On the other hand if we liquidate the entire $X$ in the first time period,

$$
\begin{aligned}
E(s) &= cX + \eta \frac{X^2}{\tau} \\
\operatorname{Var}(s) &= 0.
\end{aligned}
\tag{10.19}
$$

If $n$ is large, hence $\tau$ small, $E(s)$ can be very large resulting in significant impact on price.

With these illustrations of extremes, the optimization problem can be restated as,

$$
\min_x \big(E(s) + \lambda V(s)\big), \tag{10.20}
$$

and if $\lambda > 0$, $E(s) + \lambda V(s)$ is strictly convex and thus has a unique solution. The solution to (10.20) can be written as a combination of exponentials. The results of the trading trajectory and the trade list are given below:

$$
\begin{aligned}
X_j &= \frac{\sinh(\kappa(T - t_j))}{\sinh(\kappa(T))} \cdot X, \quad j = 0, \ldots, N \\
x_j &= \frac{2\sinh(\frac{1}{2}\kappa\tau)}{\sinh(\kappa(T))} \cdot \cosh(\kappa(T - t_{j-\frac{1}{2}}))X, \quad j = 1, \ldots, n.
\end{aligned}
\tag{10.21}
$$

Here $t_{j-\frac{1}{2}} = (j - \frac{1}{2})\tau$, and $\kappa = \sqrt{\lambda\sigma^2/\eta}$ depends on the volatility and the risk aversion coefficient, '$\lambda$', which is termed the urgency parameter. The inverse of '$\kappa$' is termed the trade's "half-life"; the larger its value, the more rapid the depletion rate of the trade list. The solution given in (10.21) is shown to be adaptive at any point in time, '$j$'. Thus the initial optimal solution over the entire interval is also optimal over each subinterval. A fixed trajectory as in (10.21) was previously constructed by Grinold and Kahn (2000) [169]. The closed form solution in (10.21) is possible due to the linear structure of impact functions.

The value of '$\lambda$', the price of risk, depends on the trader's utility function. The suggested optimal strategy is just volume weighted average pricing (VWAP), that is trading at a constant rate, when $\lambda \to 0$, that is if the trader is risk-neutral. The trade schedule for various values of '$\lambda$' is given in Figure 10.8 and the projected execution schedule is given in Figure 10.9. Various extensions of the basic model (10.13) are considered by Almgren (2008) [9].

Some questions remain: What is the link between market impact and the dynamics of the limit order book, is the impact function linear, etc. The effect of a short term drift in prices and the serial correlation in the errors which is generally observed in high frequency trading can be incorporated easily in the model (10.15). Lorenz and Almgren (2011) [253] provide a Bayesian approach to the optimization problem. While the methodology yields elegant solutions, empirical use of these results requires a close monitoring of order book dynamics as the changes in demand and supply sides can be quite rapid. The general behavior of this adaptive strategy is aggressive; it depends on the prediction of short term price change; generally, if price goes up the tendency is to sell faster.

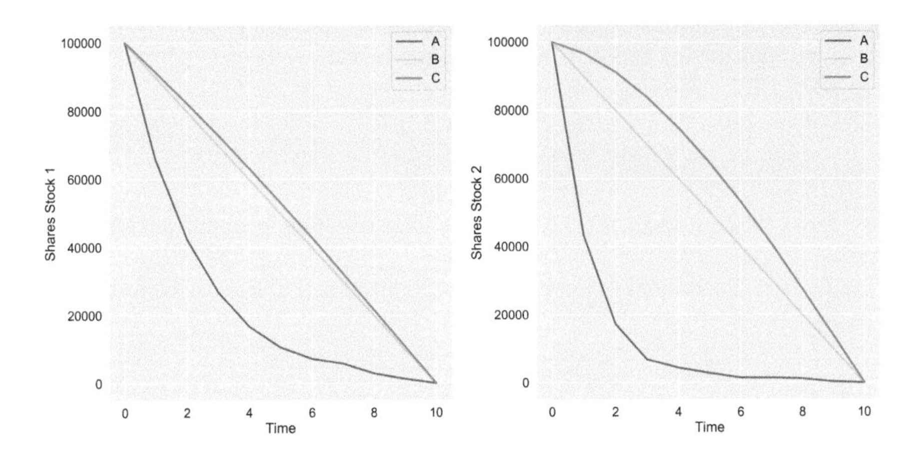


Figure 10.8: Optimal Trajectories for Two Securities; ($A$ : $\lambda = 2 \times 10^{-6}$; $B$ : $\lambda = 0$; $C = -5 \times 10^{-8}$; $B$ is the naïve strategy).

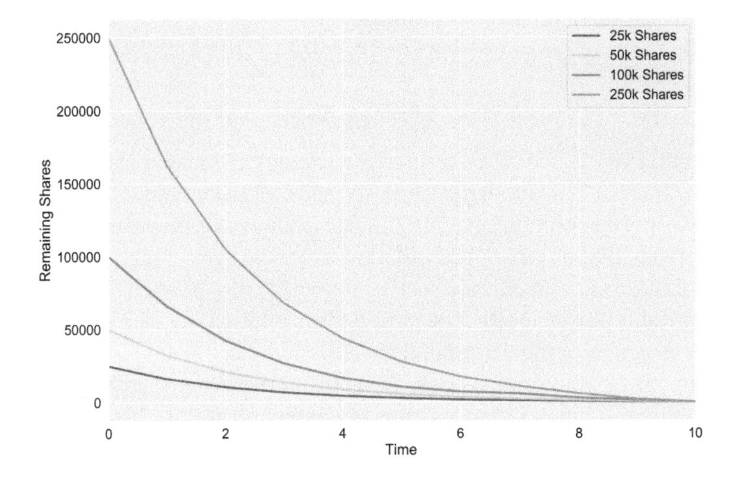


Figure 10.9: Projected Execution Schedules.

The above mean-variance formulation of execution cost is combined with the traditional mean-variance portfolio optimization in Engle and Ferstenberg (2007) [117] and Engle, Ferstenberg and Russell (2012) [121]. This is particularly useful in the context of portfolio rebalancing and the need for minimizing the transaction costs associated with all instruments involved in rebalancing. While the above methods can be taken as first generation methods that are purely based on slippage, more recent work focuses on adding other constraints such as incorporating the market VWAP and simultaneously determining the limit and market orders to achieve optimal results.

**Obizhaeva and Wang Model:** The models discussed in the previous sections assume a price impact function that is not sensitive to liquidity dynamics. Obizhaeva and Wang (2013) [274] develop a model that accounts for the dynamic properties of supply and demand, as observed from the limit order book. An advantage of this approach is that the resulting strategies are likely to be robust to different order book profiles. The model captures the so-called 'resilience'; that is, how the current trade affects the state of the future limit order book. The resulting strategy consists of a large trade initially aimed at moving the order book away from its steady state, followed by a number of small trades that will utilize the inflow from new liquidity providers. The speed at which the book replenishes itself is an important part of execution cost. The level of resilience depends on the level of hidden liquidity in the market. The price impact model considered here takes into account the dynamics of the limit order book.

It is assumed that the parent order $X$ will be traded in the interval $(0, T)$, '$n$' times at $t_1, t_2, \ldots, t_n$. To model the execution of a large order in the limit order book, it is assumed that the price of execution $(p_t)$ depends on the true value of the equity $(p_t^*)$ and the state variables $(z_t)$ such as past trades that may affect the book. Let $q(p_t^*, p_t; z_t)$ be the density of the limit orders on the other side of the book. The mid-quote $(\overline{p}_t)$ is generally taken to reflect the true price, $p_t^*$. If the trade is on the buy side, the initial large transaction, $x_1$, can push the ask price to a higher level, $p_1^*+(s/2)+(x_1/q)$, where '$s$' is the spread and the average execution price is then equal to $p_1^*+(s/2)+x_1/(2q)$. After the initial execution, the price converges to a new steady state, $p_t^* + s/2 + \lambda x_1$. Assuming that the limit-order book converges to its steady-state exponentially and at time '$t$',

$$q_t(p) = q \cdot I(p > A_t)$$

and (10.22)

$$A_t = \overline{p}_t + \frac{s}{2} + x_1 \cdot \kappa e^{-\rho t},$$

where $A_t$ is the asking price, $\kappa = 1/q - \lambda$ and '$\rho$' measures the resilience of the limit order book. If the most current ask price $(A_t)$ deviates from steady state level, $\overline{p}_t + s/2$, the new ask limit orders will come into the book at the rate of $\rho\, q\, (A_t - \overline{p}_t - s/2)$.

Given the above description of the limit order book dynamics, the optimal execution problem can be restated as follows:

$$\min_{x_k} E_1 \left( \sum_{k=1}^{n} [A_k + x_k(2q)]\, x_k \right), \tag{10.23}$$

such that $A_k = p_k^* + \lambda(X_1 - X_n) + \frac{s}{2} + \sum_{i=1}^{k-1} x_i \cdot \kappa e^{-\rho\tau(n-i)}$, where $p_k^*$, the true price is taken to follow a random walk.

The solution to this dynamic programming problem is given in Obizhaeva and Wang (2013) [274, p.14, Proposition 1] and is somewhat involved. Instead we state the strategy when $n \to \infty$ which is of practical interest as most parent orders are now divided into many, many child orders. The empirical market impact study that was presented in the last chapter attests to that. The solutions to (10.23) as $n \to \infty$ are:

$$\begin{aligned} \text{First and final orders} &: x_1 = \frac{X}{\rho T + 2} = x_n \\ \text{Spread of in-between trading} &: x_t = \frac{\rho X}{\rho T + 2}. \end{aligned} \tag{10.24}$$

The expected cost is determined as

$$\text{Expected cost} = \left(p_0^* + \frac{s}{2}\right) X_t + \lambda X_0 X_t + \alpha_t X_t^2 + \beta_t X_t D_t + \gamma_t D_t^2, \tag{10.25}$$

where $\alpha_t = \frac{\kappa}{\rho(T-t)+2} - \frac{\lambda}{2}$, $\beta_t = \frac{2}{\rho(T-t)+2}$ and $\gamma_t = -\frac{\rho(T-t)}{2\kappa[\rho(T-t)+2]}$. The initial and final orders are discrete in nature and the orders in-between are continuous. They make use of incoming orders with favorable prices.

Some comments are worth noting. The solution given in (10.24) does not depend on the market depth, '$q$' and the price impact, '$\lambda$'. It is shown that the price impact is not a factor, if the trade times are determined optimally and if they are not set a priori as in the other strategies. The reason for '$q$' not being a factor is due to the fact that the depth is taken to be constant at all times. This implies that there is enough liquidity in the market and the book gets replenished, albeit at a constant rate. The two factors that play important roles are the resiliency factor, '$\rho$' and the trading horizon, '$T$'. When $\rho = 0$ the execution costs are strategy dependent and when $\rho \to \infty$, the order book rebuilds itself faster. When '$T$' increases, the size of the first and final order decreases; if there is more time to trade, the trades are spread out to manage the execution cost. The net cost of this strategy is

$$\text{Net cost} = \frac{\lambda}{2} \cdot X^2 + \left(\frac{\kappa}{\rho T + 2}\right) X^2 \tag{10.26}$$

and it is shown to be smaller than the cost incurred if the strategy of constant rate trading is followed.

Obizhaeva and Wang (2013) [274] consider the extension of the optimization criterion in (10.23) to include the risk aversion as well, as in Almgren and Chriss (2000) [10]. Interested readers should refer to Section 8 of their paper. For a practical implementation of these methods, it is necessary to consider the trading that happens in multiple exchanges and how the replenishment patterns can differ over different exchanges, where 'liquidity' and the fee structure have become major considerations for the order flow.

In the models described so far in this section, it is assumed that the number of child orders '$n$' and the execution horizon, '$T$', are decided exogenously. Also the impact of a trade is modeled through the modified random walk model for the price with additional terms reflecting the permanent or temporary impact (Equations 10.13 and 10.14). The process of how price impact arises due to frictions in the liquidity access is an important part of market microstructure theory and this needs to be taken into account in determining the optimal execution. For example, a buyer in a seller's market can incur a lower cost of trading than a seller in a similar market. The Obizhaeva and Wang model is based on the arrival dynamics to the order book. Here we want to comment on another model briefly and the reader can refer to the cited papers for more details.

The approach taken by Easley, De Prado and O'Hara (2015) [111] is based on an asymmetric information model of the market maker's behavior. The key measure is the probability of information-based (PIN) trading that is estimated by the order book imbalance (see Easley, De Prado and O'Hara (2012) [110]). The PIN views trading as a sequential game between liquidity providers and liquidity takers, represented over the trading duration. The premise is that selling a large order in a market already imbalanced toward sell, will reinforce adverse selection from the other side and thus widen the bid-ask spread resulting in higher market impact. The optimal execution horizon (OEH) model in Easley et al. (2015) provides a framework for determining the trading horizon, '$T$', and does complement other earlier studies on execution strategies that minimize price impact. As the full treatment of this model requires developing many additional concepts, we do not comment any further here and leave further studies to the reader.

## 10.5 Multiple Exchanges: Smart Order Routing Algorithm

In highly fragmented equity markets, each venue maintains its own electronic limit order book, where orders are prioritized first based on prices and then at a given price level, according to the time of arrival. Exchanges publish information for each security in real-time but this information can change rapidly due to high levels of cancellations. The exchanges may differ with respect to best bid and offer price levels, market depth at various prices, etc. They may also differ in their fee structures. Under the maker-taker pricing model, exchanges offer rebates to liquidity providers and charge fees to takers of liquidity. These fees range from −\$0.001 to \$0.0030. Because the typical bid-ask spread for liquid instruments is \$0.01, the fees and rebates can be a fairly significant fraction of the trading cost, giving market participants an opportunity to optimize where child orders should be sent. The key features of the smart order router are given in Figure 10.10. The trade-off to address is the fact it may be expensive to route to all the exchanges, but by not routing to the right venue one might miss liquidity and hence incur greater market impact or opportunity cost. At any given time, the highest bid and the lowest offer among all exchanges comprise the National Best Bid and Offer (NBBO). Market orders are guaranteed to get the best price across all exchanges. Most traders use smart order routing algorithms with a goal to buy or sell the maximum number of shares in the shortest possible time, with the least market impact possible. But a key issue that should be considered is that many of the lit venues also have hidden orders (12–45%). How to estimate the size of the hidden orders and how to make routing decisions in the presence of hidden orders has been a focus of some recent research studies.

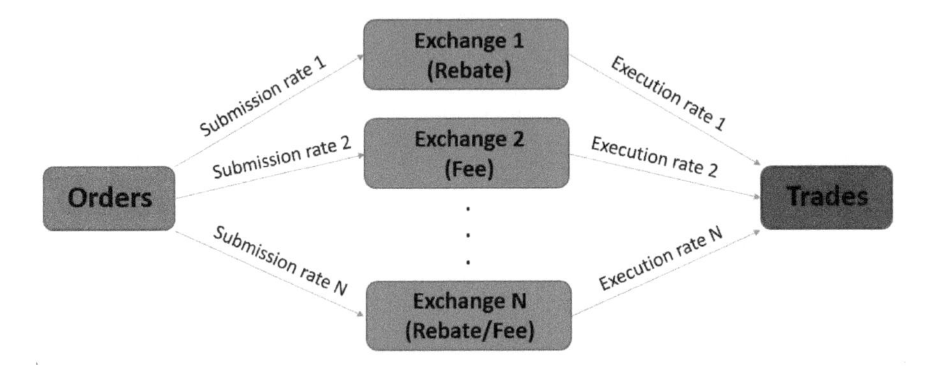


Figure 10.10: Features of the Smart Order Router.

While exchanges compete along many fronts, for example, rebates and fees, transparency and execution speed, the key variables driving efficiency remain liquidity and price improvement. As such, market structure is considered by market participants and regulators as the key determinant of the flow of liquidity. The coexistence of multiple exchanges as mentioned in Parlour and Seppi (2003) [281] raises specific questions that are of general interest and are stated below:

(a) Do liquidity and trading concentrate in a few exchanges?

(b) Do some market designs provide greater liquidity than others?

(c) Is the fragmentation of order flow desirable from a policy point of view?

(d) What is the constructive role for regulators to enhance liquidity?

Because exchanges operate under rules governing certain market designs, the performance and efficiency questions tend to be related to the types of markets existing, such as a pure limit order or a hybrid market (limit order book plus a specialist). In the hybrid market, a specialist can provide supplementary liquidity after a market order has arrived. Parlour and Seppi (2003) [281] conclude only under some conditions such as enforcement of time priority, is efficiency possible. They also conclude that merely increasing the number of trading venues may result in degradation of market quality, if the enforcement of time priority is not properly followed, as it might discourage traders from posting limit orders.

Foucault and Menkveld (2008) [142] consider the effect of fragmentation on two limit order markets. They examine the competition between Euronext and London Stock Exchange (LSE) in the Dutch market. The consolidated limit order book is found to be deeper after the entry of the LSE. General conclusions from the above study and others such as Hendershott, Jones and Menkveld (2011) [193] and O'Hara and Ye (2011) [177] are that, fragmentation of order flows improves the liquidity supply and protecting orders against trade-throughs is important.

The smart order router (SOR), generally works as follows. One can customize the use of the strategy by establishing a set of rules in splitting the parent order into child orders and then ensure that the orders are routed to the venue with the best price and liquidity (either in a pre-determined or adaptive manner). Successful SORs must be able to handle a variety of trading strategies for multiple venues, and they must deal with vast amounts of incoming market data. Publicly available information on the SORs used by major traders indicate that the order placement objective in general is to minimize client all-in shortfall with venue fees and rebates. All-in shortfall consists of shortfall on filled shares and cost of the clean-up trade for unfilled shares. The latter is taken to be an important component of order placement optimization, as it enables to quantify the effect of venue differences in fill rates, (see *Street Smart*, Issue 42, Jan 14, 2011). Other algorithms may choose to optimize the expected time to execute the client's overall order. Expected queue depletion speeds at various venues are usually predicted from recent trading data. Small orders tend to be placed in a single venue to minimize the queueing time whereas large orders will be placed in multiple venues to be exposed to maximum liquidity. With this approach, forecasting trading rates at different venues is crucial. The rates are modeled as a function of lagged rates, capturing the momentum and imbalances between supply and demand sides of the order book, but making it a non-trivial task in fragmented markets.

A recent empirical study by Battalio, Corwin and Jennings (2016) [33] confirms that brokers use both limit and marketable orders to execute trades. Past studies provide evidence that market orders are sent to venues with lower trading costs as the trading fees and rebates generally affect consolidated market depth. Maglaras, Moallemi and Zheng (2019) [257] show that limit orders are submitted to exchanges with high rebates and shorter waiting time for execution, while market orders are sent to venues that have lower fees and larger posted quote sizes. Cont and Kukanov (2017) [86] develop a model that is somewhat more realistic to the current practice by decoupling the 'order placement' decision from the scheduling decision and more importantly consider the option of placing limit orders on several exchanges, simultaneously. The model also accounts for the execution risk, the risk of not filling an order. Filling the unfilled portion may become costly and the allocation may shift toward market orders or toward overbooking, that is placing more orders than needed to refill.

To formalize the order routing problem, we assume a size of order $X$ is to be filled in the duration $(0, T)$. The decision is to split this order into a market order $M$ and '$K$' limit orders $L_1, \ldots, L_K$ with the same size to be placed in '$K$' exchanges that have queue sizes $Q_1, \ldots, Q_K$ ; thus the order allocation is summarized by the elements of the vector $X = (M, L_1, \ldots, L_K)$ that need to be optimally determined. If order cancellations in the duration $(0, T)$ in exchange '$K$' is represented by $\xi_K$, then the number of shares transacted can be written as

$$A(X, \xi) = M + \sum_{k=1}^{K} \left[ (\xi_k - Q_k)_+ - (\xi_k - Q_k - L_k)_+ \right], \tag{10.27}$$

where the terms in the parenthesis refers to the initial position and the final position of the queue outflows. The execution cost must account for fee $(f)$ and rebate structures (discussed in Chapter 9) at each exchange and the cost of adverse selection ($r_k$) and can be stated as follows:

$$C(X,\xi) = (h+f)M - \sum_{k=1}^{K}(h+r_k)\left[(\xi_k - Q_k)_+ - (\xi_k - Q_k - L_k)_+\right]. \quad (10.28)$$

Here $h$ is one-half of bid-ask spread.

The cost function in (10.28) can be modified to account for the cost of unfilled limit orders. These may be filled through market orders at a higher cost. Or it is possible that the prices have decreased resulting in additional adverse selection cost with penalty ($\lambda_u$) for falling behind and penalty ($\lambda_0$) for exceeding the target; Thus, the execution risk can be written as,

$$\text{ER} = \lambda_u(S - A(X,\xi))_+ + \lambda_0(A(X,\xi) - S). \quad (10.29)$$

To be more realistic, the market impact function can be considered as follows:

$$\text{MI} = \theta\left[M + \sum_{k=1}^{K} L_k + (S - A(X,\xi))_+\right]. \quad (10.30)$$

The total cost function that includes both implicit and explicit costs can be stated as:

$$V(X,\xi) = C(X,\xi) + \text{ER} + \text{MI}. \quad (10.31)$$

The random variable in (10.31) is $\xi$, the cancellations that occur at various exchanges. The minimization function is $E[V(X,\xi)]$, with some assumed distribution for $F$ of $\xi$. Under some reasonable assumptions such as the trader will not execute more than the target $X$ and market orders in the beginning of the duration $(0,T)$ are less expensive than at the end, when unfilled orders are converted to market orders, an optimal solution is shown to exist (see Cont and Kukanov (2017) [86] for details).

The analytical solution minimizing $V(X,\xi)$ is not easily tractable due to dimensionality issues. A numerical solution via gradient method is proposed. Random samples of $\xi$ are obtained and averaged to approximate $E(\xi)$. Let $q(X,\xi) = \Delta V(X,\xi)$ be the gradient of $V$. The following iterative algorithm is shown to converge:

- **Start with $\mathbf{X_0}$ and for $\mathbf{n = 1, 2, \ldots, N}$, do**
- $\mathbf{X_n = X_{n-1} - \gamma_N g(x_{n-1}, \xi^n)}$
- **End;** $\mathbf{X_N^* = \frac{1}{N}\sum_{n=1}^{N} X_n}$

The step size,

$$\gamma_N = \sqrt{k}S\left(N(h+f+\theta+\lambda_u+\lambda_0)_t^2 + N\sum_{k=1}^{K}(h+r_k+\theta+\lambda_u+\lambda_0)^2\right)^{-1/2} \quad (10.32)$$

can be seen as a function of all the costs associated with the execution of the order.

To summarize, recall that the algorithm needs the following input:

- **Trading Costs:**
  - One half of bid-ask spread ($h$)
  - Market order fee ($f$)
  - Effective limit order risks ($r_k$)
  - Market impact coefficient ($\theta$)
  - Penalties for overfilling or underfilling ($\lambda_0$, $\lambda_u$)
- **Market Variables:** Number of exchanges ($K$) and limit order queues ($Q_k$).
- **Execution Variables:** Time horizon ($T$) and target quantity ($S$).

While many of these quantities can be estimated using past transactions, the limit order queues ($Q_k$) and the cancellations ($\xi_k$) have to be obtained at the time of execution.

What we have presented in this section remains very limited as this area still needs significant academic research. So far, not much has been published based on empirics that would lead to the development of better stylized models.

## 10.6 Execution Algorithms for Multiple Assets

The execution of trading entry-exit decisions can be made using the methods described in the earlier section for individual assets. Most investors hold multiple assets in their portfolios and evaluate their investments as a composite rather than on an individual asset basis. In Section 6.6, we discussed trading strategies with the focus on rebalancing portfolios. The criteria used were in the traditional framework of mean-variance efficiency with some constraints on transaction costs. When it comes to execution, the focus is on execution costs and market impact. We will briefly outline the issues related to execution of multiple-security portfolios.

The first form of multi-asset execution we mentioned in this text was pairs strategies executing one side of a trade in proportion to the executions received on the other side. A natural extension of it is multi-leg algorithms where each side has multiple lines to be transacted over a certain horizon.

The implementation shortfall framework described previously relies on impact and risk dynamics of a single asset but can easily be extended to multiple assets. While the only real way to reduce risk in a single stock setup is to trade faster; in a portfolio scenario, it is possible to control risk by exploiting the correlation structure across stocks. Portfolio algorithms aim to solve the execution dilemma (lowering impact cost at the expense of increasing timing risk exposure or vice versa) faced by traders who want to trade large baskets, by offering two compelling sets of solutions:

- **At Comparable Cost:** Reduce the risk by neutralizing market or factor exposures.

- **At Comparable Risk:** Reduce the cost by allowing slower trading than a list of individual IS strategies.

Asset managers handle large trading lists when they periodically rebalance their holdings to tailor their exposure, or when they have inflows or outflows requiring to trade a sleeve of the portfolio. While this can present implementation challenges (such as managing cash neutrality between buys and sells across multiple markets), it also offers an opportunity for more efficient execution, if the basket is traded in a coordinated fashion rather than as a list of independent single lines. Based on the trader's view of what might be driving the short term risk of the portfolio during the execution window and the trader's risk aversion, the trading schedules can be optimized along different risk-mitigating dimensions:

- Minimizing Country / Sector exposure
- Minimizing Factor exposure
- Minimizing Beta/Delta exposure

Most implementations leverage QP solvers, and to create a flexible implementation for large portfolios that can be optimized in a reasonable time frame requires making numerous approximations.

From a pure trading implementation perspective, all the practical considerations mentioned with respect to single assets apply to portfolio trading algorithms as well, in particular when handling the relative liquidity of the buy and sell sides. The complexity that is quite unique to portfolio trading is the handling of dark liquidity in the trade schedule optimization. It presents both an opportunity and a challenge. Similar to a single asset execution, finding a large block of liquidity in the dark allows to considerably reduce timing risk, at cheaper trading cost than that is incurred in the lit market. That said, in a portfolio algorithm the trader relies on the correlation between names to balance the execution risk in order to slow down trading and minimize impact. So, receiving a large dark execution (or a series of simultaneous dark fills on the same side) at once, can reintroduce significant risk by creating an imbalance, leaving the trader exposed to adverse market movements. The algorithm is then left to adjust its positions rapidly and might incur large execution costs to get back to a risk balanced state. This has two practical consequences for developers of portfolio algorithms.

First, the names and quantities exposed to dark pools must be carefully selected. They should reflect the trade-off between cost reduction at the single line level, risk reduction at the portfolio level, and potential cost incurred to return to a risk balanced portfolio. The scenarios encountered are as diverse as one can imagine and make the formulation of the problem harder, once dark allocations are incorporated. To put things in perspective, we consider the two practical stylized examples as follows:

- A basket with a single position with a large percentage of ADV (i.e., large expected market impact) and small positions in liquid names (i.e., low expected market impact): In this scenario, the large percentage ADV order is the constraining factor for the execution speed of the entire portfolio. If that name can be traded as a block in the dark, it would allow the rest of the basket to trade faster (i.e., reducing exposure to timing risk) at little additional cost. This will lead to posting dark as much as possible.

- A basket with a few names with a larger marginal contribution to risk (MCR)—but not of overly large notional sizes—and positions in liquid and illiquid names: In this scenario, executing the large MCR names would result in a better outcome only if the names paired off with them on the other side of the basket from a risk perspective, are liquid enough to be executed rapidly at low cost. One possible way of assessing how much to post in the dark is to constrain the notional exposure with the available notional on the liquid names on the other side. That notional exposure can then be allocated among the large MCR names either naïvely (equal split or in descending order of MCR) or applying a more sophisticated approach: For instance an allocation as a function of the MCR and the conditional probability of fill given the size allocated.

Second, when the portfolio incurs a sudden change in risk profile—whether due to a dark fill or due to an abrupt market move in one direction—a reoptimization is necessary to return to a risk balanced position and to determine the new optimal trading schedule. This reoptimization can have different characteristics than the one that was performed at the beginning of the trade. Depending on the level of risk already exited up to that point in time, the trader has the ability to consider the temporary increase in risk as either something that needs to be reduced immediately in the next available trading bins, even if it generates larger trading costs, or something that can be reduced gradually in order to mitigate transaction costs. The optimal allocation depends on the liquidity of the names still available to trade on the other side of the imbalance.

**Cross-Impact:** Modeling the cross-impact of trading an asset on the trading of related assets has only been a recent area of interest. If a trader is liquidating simultaneously many assets as part of a rebalancing effort, one should expect some cross-impact. In the low frequency setting when the focus is not necessarily on 'execution' but on 'trading', this is studied under 'commonality' or 'co-movement' as discussed in Chapter 3, where the correlated changes are generally attributed to a general market factor that drives all the stocks. See Hasbrouck and Seppi (2001) [183]. In the high frequency setting, the problem is harder to investigate because of short-term market frictions. Almgren and Chriss (2000) [10] discuss the extension of their execution algorithm to multiple assets, but this approach has not been followed up. A recent study by Schneider and Lillo (2019) [298] extends the single-stock framework of Gatheral (2010) [161] to multiple stocks, but the formulation is in a continuous time framework.

As we generally followed a discrete time modeling approach in this book, we will present the results in that form, although a continuous-time presentation has its own elegance. The works cited earlier in the single asset case, Almgren and Chriss

(2000) [10] and Engle and Ferstenberg (2007) [117] consider multi-asset execution problems without cross-impact. A recent paper by Tsoukalas, Wang and Giesecke (2019) [316] provides details of dynamic portfolio execution and here we present only the essential summary. The set-up here is similar to Almgren and Chriss described in Section 10.4.2.

The execution window, $[0,1]$ is divided into '$N$' intervals indexed by $n \in \{0,1,\dots,N\}$ with $\tau = \frac{1}{N}$, the period length. Let there be '$m$' assets and the '$m$' dimensional buy and sell vectors of order sizes are denoted $x_n^+$ and $x_n^-$ and the aggregate vector as $x_n = \begin{bmatrix} x_n^+ \\ x_n^- \end{bmatrix}$. The net trade is thus given by $\delta' x_n = x_n^+ - x_n^-$, where $\delta' = [I_m, -I_n]$. If we assume that we begin with the total net trades to be traded as $z_0$, then

$$z_n = z_{n-1} - \delta' x_{n-1} \quad \text{and} \quad z_N = \delta' x_N. \tag{10.33}$$

It is assumed that the vector of bid and ask prices, $u_n$, follows a random walk,

$$u_n = u_{n-1} + \epsilon_n, \quad u_0 > 0 \tag{10.34}$$

with $E(\epsilon\epsilon_n') = \tau\Sigma$, where $\Sigma$ is a positive definite covariance matrix. Thus, the best ask and the best bid can be expressed as,

$$a_{i,n} = u_{i,n} + \frac{1}{2} s_i, \quad b_{i,n} = u_{i,n} - \frac{1}{2} s_i. \tag{10.35}$$

Recall that the temporary impact is instantaneous and the permanent impact is cumulative. If $x_{in}^+$ denotes 'buy' shares and $x_{in}^-$ denotes 'sell' shares, the ask and bid price changes are given as

$$a_{i,n}^* = a_{i,n} + \frac{x_{i,n}^+}{q_i^a}, \quad b_{in}^* = b_{i,n} - \frac{x_{i,n}^-}{q_i^b}. \tag{10.36}$$

The permanent price impact is denoted by

$$v_{i,n} = u_{i,n} + \lambda_{ii} \sum_{k=0}^{n-1} (x_{i,k-1}^+ - x_{i,k-1}^-), \tag{10.37}$$

which can be taken as the steady-state mid-price.

It is also assumed that the orders on both sides are replenished exponentially. While the above setup presents details for a single asset as the focus, we want to study cross-impact models, how a trade in one asset can impact the trade of another asset by shifting its demand or supply. We will present results for only the cost of two assets here, but the extension to multiple assets is rather straightforward. The Equation (10.37) is modified as,

$$v_{i,n} = u_{i,n} + \sum_{j=1}^{2} \lambda_{ij} \sum_{k=1}^{n} (x_{j,k-1}^+ - x_{j,k-1}^-), \quad i = 1,2. \tag{10.38}$$

The order book changes (reflecting demand/supply sides) can be captured by

$$d^k_{i,n+1} = e^{-\rho^k_i \tau}\left(d^k_{i,n} + \left[\frac{x^{\pm}_{i,n}}{q^k_i} \mp \sum_{j=1,2} \lambda_{ij}(x^+_{j,n} - x^-_{j,n})\right]\right), \quad k = a, b, \tag{10.39}$$

where $d^a_{i,n} = a_{i,n} - a^{\infty}_{i,n}$ and $d^b_{i,n} = b^{\infty}_{i,n} - b_{i,n}$ are termed displacements.

For the optimization description, we will set up the above quantities without the $i$th subscript. With the matrix $\Lambda = ((\lambda_{i,j})), e^{-\rho^k\tau} = \text{Diag}(e^{-\rho^k_i}),\ Q^k = \text{Diag}\left(\frac{1}{2q^k_i}\right)$, $\delta_a = (I_2; 0)$, $\delta_b = (0, -I_2)$, $k^a = 2Q^a\delta'_a - \Lambda\delta'$ and $k^b = 2Q^b\delta'_b + \Lambda\delta'$, we have

$$\begin{aligned} a_n &= u_n + \frac{1}{2}s + \Lambda(z_0 - z_n) + d^a_n \\ b_n &= u_n - \frac{1}{2}s + \Lambda(z_0 - z_n) - d^b_n \end{aligned} \tag{10.40}$$

and

$$\begin{aligned} d^a_n &= e^{-\rho^a\tau}(d^a_{n-1} + k^a x_{n-1}) \\ d^b_n &= e^{-\rho^b\tau}(d^b_{n-1} + k^b x_{n-1}). \end{aligned} \tag{10.41}$$

The manager's profit function can be written as,

$$\pi_n = x^{-\prime}_n(b_n - Q^b x^-_n) - x^{+\prime}_n(a_n + Q^a x^+_n). \tag{10.42}$$

Observe that the cumulative wealth equation up to the $n$th trade is $W_n = \sum_{i=0}^n \pi_i = W_{n-1} + \pi_n$. The state of the system is given as $y'_n = (1', z'_n, d'_n)'$. Now the dynamics can be aggregated as,

$$y_{n+1} = Ay_n + Bx_n, \tag{10.43}$$

where $A = \text{diag}(I_{m+1}, e^{-\rho^a\tau}, e^{-\rho^b\tau})$ and $B = (0, -\delta, e^{-\rho^a\tau}k^a, e^{-\rho^n\tau}k^b)'$. Thus, the manager's dynamic optimization problem can be stated as

$$\max_{x_n \geq 0} E_n(J_{n+1}) \quad \ni \quad \delta' x_N = z_N, J_{N+1} = e^{-aW_N} \tag{10.44}$$

with state dynamics $W_{n+1} = W_n - (u'_{n+1}\delta' + z'_0\Lambda\delta' + y'_{n+1}N)x_{n+1} - x'_{n+1}Qx_{n+1}$, $y_{n+1} = Ay_n + Bx_n$ and $u_{n+1} = u_n + \epsilon_{n+1}$. For methods to solve this problem, refer to the paper for complete details.

It is possible to simplify the optimization problem for the two assets case and derive some explicit expressions. As one can observe the link between the two assets comes in the off-diagonal elements of the '$\Lambda$' matrix. It will be of some use to compare the usefulness of these elements in estimating the total market impact of the two assets, traded as a portfolio versus traded as single assets.

## 10.7 Extending the Algorithms to Other Asset Classes

While the electronification of trading first took place in the equity space, recent years have seen a significant growth of electronic trading and market making in the fixed income world, opening new opportunities to quantitative trading strategies. We give a brief overview of the characteristics of some of these products.

**(a) Options:** Historically, equity options markets have presented unique challenges to the successful deployment of sophisticated algorithmic trading strategies. First, options markets are significantly more fragmented than their equity counterparts, not just in space (i.e., across multiple exchanges) but also in terms of instruments. For one equity instrument representing a company (or for one index), there is a grid of available options over multiple maturities and multiple strike prices. Similar to equities, options market data is disseminated in real time by the Options Price Reporting Authority (OPRA) which provides last sale reports and quotations from participant exchanges.[11] OPRA is a national market system plan that governs the process by which, options market data are collected from participant exchanges, consolidated, and disseminated. It also publishes other types of information with respect to the trading of options such as the number of options contracts traded, open interest and end of day summaries.

A key challenge that remains in the options space is the amount of data generated on a daily basis, and the significant bursts that can happen when underlying equity markets move sharply and force a rapid adjustment of a myriad of correlated instruments. As an example, the OPRA message rates statistics[12] for the fourth quarter 2018 were:

Peak Messages Per Second (millions): 19.8.
Peak Messages Per 100 Milliseconds (millions): 4.2.
Peak Transactions Per Day (billions): 45.9.

Automated options market making strategies were the first ones to be deployed, and we are now witnessing the emergence of execution strategies in the option space: For instance, targeting certain volatility levels (instead of price-based benchmarks). Given the natural relationship that exists with the underlying assets, options algorithmic trading strategies also require implementation of automated delta and gamma hedging execution strategies. It is also worth noting that most exchanges now support order types such as spreads making multi-leg strategies easier to implement.

**(b) Credit Derivatives:** When an entity, whether private or public, borrows money, the lender or debt holder bears some default risk until maturity. There is always a risk that the entity will not be able to repay the debt in full or per the agreed terms (coupons, maturity, ...). Credit derivatives allow debt holders to hedge all or part

---

[11] Options participant exchanges in OPRA: BOX Exchange, Cboe exchanges (BZX Options Exchange, C2 Options Exchange, EDGX Options Exchange, Options Exchange), Miami International Securities Exchange, MIAX PEARL, Nasdaq exchanges (BX, GEMX, ISE, MRX, PHLX, The Nasdaq Stock Market), NYSE exchanges (American, Arca).

[12] Source: `https://www.opradata.com/`

of that risk, as well as speculators to express their views on the creditworthiness of various entities.

A credit default swap (CDS) is a derivative contract allowing to buy or sell protection on a single reference entity. The protection buyer pays a fixed, running or upfront, premium in return for the right to receive a payment, should the reference entity suffer a credit event. Depending on the CDS contract, an eligible credit event might correspond to a payment default on the debt, a restructuring or a simple credit downgrade of the reference entity. In return for the premium, the writer of the CDS agrees to pay to the buyer ($1-$ recovery rate) times the notional of the contract upon the realization of an eligible credit event. Here, the recovery rate represents the fraction of the face value of the debt that can be recovered following a credit event.

By entering a default swap, the buyer is essentially purchasing an insurance against the risk of default of a borrower. However, that insurance being paid by the CDS issuer, bears counter-party risk (the ability of the issuer to pay the agreed amount in the event of a credit event affecting the borrower). Consequently, the valuation of a CDS depends, among other factors, on the default probability of the borrowing entity over the life of the contract, on the default probability of issuing counter-party, as well as the correlation between them.

The estimated CDS notional outstanding stands above $10 trillion, after having peaked at about $62 trillion at the end of 2007 (International Swaps and Derivatives Association (ISDA)), and north of 1 million trades are recorded per week. In the aftermath of the Global Financial Crisis of 2008, trading liquidity shifted from single name CDS toward CDS Indices, which are essentially baskets of single name CDS. Roughly half of the outstanding notionals originates from single reference entity contracts, while most of the other half emanates from credit indices (roughly $130,000$ trades per week in 2016).

Over-the-counter (OTC) derivatives markets are considered dealer markets, as they tend to be traded through a network of private dealers, who stand ready to provide liquidity in various instruments, while maintaining a relatively neutral net risk exposure. This, traditionally, is conducted one-on-one, off exchanges. However, the introduction of the Dodd-Frank Act of 2010 brought significant transformation to the US swap markets by mandating central clearing for standardized over-the-counter derivatives such as swaps, as well as forcing their execution on a swap execution facility (SEF).[13]

The term 'swap execution facility' defines a registered "trading platform in which multiple participants have the ability to execute or trade swaps by accepting bids and offers made by multiple participants" (Dodd-Frank Act, Section 733). Among the stated goals of mandating swaps execution on SEFs are: Promoting pre-trade transparency in the swaps market, as well as facilitating the real-time publication of trading information such as price and volume to enhance price discovery. As such the Dodd-Frank Act addressed, for swap markets, the three main challenges that had historically

---

[13]In Europe, the MiFID II directive adopted by the EU in 2014, with applicability on January 3rd, 2018, also introduces obligations for sufficiently liquid standardized derivatives contracts to be traded only on regulated platforms such as OTFs: Organized Trading Platform.

prevented a significant electronification of trading in fixed income markets: Lack of standardized instruments, lack of centralized trading platforms and lack of available transaction data.

SEFs usually offer multiple trading protocols, giving investors the flexibility to choose the most appropriate trading style for their needs. Existing trading protocols center around either an Order Book-like approach where all market participants have the ability to execute available bids or offers from multiple participants as well as leave their own limit orders; or a Request-For-Quote (RFQ)-like approach where participants request a single or two-sided market from multiple dealers during a real-time auction.

A variety of credit indices are fairly liquid and trading on SEFs, in particular US indices (CDX(c)) and European indices (iTraxx(c)). The CDX indices are further broken out by the type of debt covered such as Investment Grade (IG) and High Yield (HY) for the most liquid ones. Due to their liquidity, both CDX IG and CDX HY are examples of credit indices that lend themselves well to electronic market making activities.

However, the CDS Index market presents a certain number of idiosyncrasies compared to equity markets when it comes to market making. While there is an order book available to all participants, most of the volume still gets transacted via RFQ mechanism in which market makers are not always allowed to see the quotes. Consequently, market makers only have partial information regarding the true position of the market, when it is time to decide where to place their own orders. Using a mixture of historical trades and partial real-time information, market makers can reconstruct a theoretical mid-price of the market. They can set then the bid and ask quotes at an appropriate distance from that mid-price, accounting for the trade-off between their desire to obtain a fill and the risk associated with maintaining their inventory.

**(c) Corporate Bonds:** The corporate bond market is also an OTC market where market participants only have access to limited information prior to placing a trade. The major market participants are: Pension funds and insurance companies, who tend to have longer investment horizons; Hedge funds which tend to have more tactical trading allocations; and corporate treasury departments that have objectives and horizons that can span a wide spectrum.

Compared to other assets, it is also interesting to note that most of the secondary market transactions happen in the first few months following issuance, and after that a significant portion of the amount of bonds issued is held to maturity. Similar to other fixed income products, the key challenges for secondary trading of bonds and the electronification of its secondary markets historically have been a lack of centralized market place, lack of harmonized instrument characteristics and lack of transaction data.

The corporate bonds secondary market remains dominated by one-to-one privately negotiated trades, but electronic platforms allow transacting via similar execution protocols as what traders can use for CDSs: Limit order book type of executions as well as request-for-quotes (RFQ) on platforms such as Tradeweb, Bloomberg, MarketAccess among others.

In the US, for instance, under FINRA Rule 6730, Broker-dealer FINRA member firms have the obligation to report transactions in TRACE-eligible securities to the TRACE[14] database as early as practically possible, but no later than 15 minutes of the Time of Execution. FINRA Rule 6710 defines TRACE-eligible securities as USD denominated debt securities (whether issued by a US or foreign private issuer), and USD denominated debt issued or guaranteed by an Agency or Government-Sponsored Enterprise. Foreign sovereign, US Treasuries and money market instruments are specifically excluded from the eligible list. The TRACE database then disseminates price, size,[15] time stamp and direction of the trade to the public. While the rules require reporting in no more than 15 minutes, most reporting and public dissemination now happens within seconds or minutes of execution (with the exception of overnight trades being batch-reported the next morning), giving participants some transparency about the current market levels.

The combination of existing venues allowing for click-to-trade or RFQ protocols and trade events information facilitated the expansion of electronic trading for corporate bonds as well. A fifth to a quarter of the investment grade market is now transacted electronically, while on the high yield side, roughly 10% of the market is traded through RFQs.

Similar to credit indices, the first challenge for dealers in corporate bond electronic market making is to infer the value of the current "fair mid" for the market following each transaction. Given the relatively infrequent observations, even for bonds that are considered liquid, quantitative techniques employed tend to vary from the ones used in markets where higher frequency data are available such as in equities. Practitioners rely much more, for instance, on probabilistic state-space models to estimate unobservable state variables.

This type of approach obviously becomes increasingly more challenging as one moves down the liquidity spectrum. It is not uncommon for illiquid bonds to not trade for months, in which case assessing a fair mid-price cannot solely rely on prior transactions. Among possible solutions, modelers can use bonds of other maturities from the same issuers, or bonds from correlated issuers or comparable investment grades. Additionally, more liquid credit derivatives are also potential sources of information that can be leveraged for better valuations.

---

[14]Trade Reporting And Compliance Engine.

[15]The actual size disseminated back to the market is capped to prevent excessive information leakage: For investment-grade corporate bonds and agency debt securities, for any trade greater than $5 million, the par value is displayed as $5 MM+; for non-investment grade corporate bond, the displayed quantity is capped to $1 million par value.

# Part V

# Technology Considerations

Algorithmic Trading is primarily a technology endeavor. The complexity of an Electronic Trading operation is often mind-boggling and revolves around the orchestration and connectivity of numerous different systems. These systems are the core of any trading operation as they are in most cases used not only to trade client flow but also do all the internal facilitation and risk trading. The subject of most of this book is to provide a toolkit for creating and for performing profitable algorithmic and execution strategies. But from the perspective of the end users, who are usually the institutional asset managers, it is paramount to have platform stability and to be able to handle more mundane processes so that they are able to trade, receive executions and do end of day reconciliations, know their positions at any point of the day, etc. Shaving a few basis points by smart execution becomes relevant only when all the operational aspects of the trading interactions are handled flawlessly.

Designing and creating high performing algorithms is no simple matter and requires significant research work on the modeling and calibration of the strategy itself and the underlying execution. It also requires a good infrastructure: to measure and compare performance of the strategy in various settings, so that we can find and address sub-optimal behavior. Access to flexible and powerful research and developing transaction cost analysis (TCA) environments are very important aspects of the overall technology stack.

Finally, a critically important aspect of execution is to comply with (just citing the US example here) the Financial Industry Regulatory Authority (FINRA) and Securities and Exchange Commission (SEC) regulatory demands for any Broker Dealer including trade, reporting, Order Audit Trial Systems (OATS), National Market Systems (NMS) rules, etc. In particular since the Flash Crash event of 2010[1] and the spectacular trading error of Knight Capital of 2012,[2] the scrutiny and regulatory burden in the space has dramatically increased as regulators worried that a reckless and unchecked drive towards speed and automation could destabilize the equity markets.

This last part of the book gives a high level review of these important, often overlooked, but critical aspects of the subject. It can only be a cursory review as the subject is immense and probably deserves books of their own. But the discussion here should provide the reader with an understanding of what is entailed in running a large trading operation.

[1] https://en.wikipedia.org/wiki/2010_Flash_Crash

[2] https://www.bloomberg.com/news/articles/2012-08-02/knight-shows-how-to-lose-440-million-in-30-minutes

# 11

# *The Technology Stack*

There is a broad range of trading infrastructures and operations depending on the size and trading needs. None, however, is as broad and complex as a low touch trading operation of a large Broker-Dealer. For this reason we use that as a template for our exploration of the technology stack necessary to support such an operation. We start by reviewing the end to end flow of information and provide some detail on various components. As previously discussed most large Electronic Trading (ET) businesses also operate their own Alternative Trading System (ATS) so we will look at the technology setup needed to support that use case as well.

## 11.1 From Client Instruction to Trade Reconciliation

Figure 11.1 shows the end to end diagram of a hypothetical trading infrastructure. We can use this diagram to follow the full life cycle of a client order and the main components of the infrastructure involved.

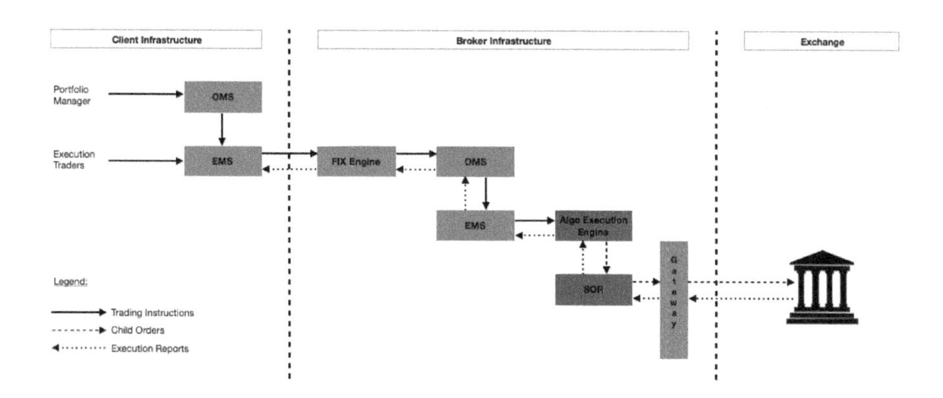


Figure 11.1: Sample End-to-End Trading Infrastructure.

**Client Side:** The trader enters an algorithmic order into the Execution Management System (EMS), a specialized trading infrastructure that integrates the internal systems and provides tools for institutional traders to manage day to day trading needs.

EMS most often refer to vendor products, Ez Castle[1], Portware[2] and FlexTrade,[3] are some of the most popular options. Many large buy-side trade organizations have more complex needs and often seek a more homegrown solution.

The trader normally selects from a drop-down menu, a particular provider and an Algo strategy and chooses the desired parameters. Brokers who want to expose their execution into an EMS need to certify with the EMS by providing a Financial Information eXchange (FIX) Specification document that highlights what strategies are available and what parameters are associated with each strategy and the validation information about the parameters. The EMS then integrates this into their system and exposes them in the front end for the trader to choose. FIX is a standard communication protocol specifically designed for financial applications.[4] Once the order is submitted, the EMS uses a preconfigured network connection to transfer the order to the broker. This session is created as part of the *Client Onboarding* step, a quite laborious process to set up a new client relationship and configure their financial limits and controls, billing, etc.

**Inbound Gateway:** We are now on the broker side where the order is received by an inbound network gateway. First, it is ensured that the message is a valid order and has all the necessary fields. Next, it will perform a set of risk and credit checks to ensure that the order is within the specified risk limits and the client's overall exposure, the maximum notional the client is allowed to trade and have in the market at any point in time. If any mismatch in these validation steps is detected the order is rejected and sent back to the client.

**Order Management System and Order Enrichment:** Once the order is validated, the order is created within the broker dealer infrastructure. The order life-cycle is quite complicated and more nuanced. It is critical that the state of the order is always up to date and transitions in states are to be carefully managed. The role of creating and managing the state of the order is fulfilled by an Order Management System (OMS). These infrastructures are at the heart of any trading operation and they are often built around a bus-based architecture, a software paradigm where loosely coupled components communicate with each other over a messaging middle-ware. The advantage of this approach is that other ancillary components within the infrastructure can "Listen in" on the messages and use the information to transmit to other systems or collect data for trade reporting and analysis.

One step often performed in this layer is Order Enrichment. This is a step that adds additional, lower level parameters that adjust and customize the execution behavior

[1] https://www.ezesoft.com/

[2] http://www.portware.com/

[3] https://flextrade.com/

[4] https://www.fixtrading.org/

for the specific client/algorithm/parameters triplet. This step also translates slight differences between the instructions the client sent and how the local execution system understands (e.g., Algo name could be sent as "Vwap" but needs to be "VWAP"). This is often accomplished by a specialized *Rules Engine* a software library that allows a set of rules to be applied to a structure. The rules are then reevaluated after any modification until no additional rules are triggered.

**Execution Strategy Stack:** As the famous quote goes: "Is where the magic happens!" The order reaches the software component where the Algo is actually executed. We delve into more details on the execution stack in the next section. So here, we assume that the strategy is initialized and starts executing. The strategy associated with the order contained in the OMS is often called the "Parent Order" and as consequence of the strategy logic one or more "child" orders are created in the OMS and sent forward for submission to one or more trading venues. Before the order is actually forwarded, it passes through an additional control layer, that ensures the strategy does not violate the risk limits and other "speed bumps," a general term used for limits that prevent the strategy from trading too fast or too aggressively or from sending too many child orders, etc.

**Outbound Gateway:** The child orders are received by another piece of infrastructure that is responsible for actually transmitting the orders to the market: The Outbound Gateway. All venues support additional protocols to communicate with market participants. Essentially all of them support a FIX protocol, but in most cases they also support a much faster "native" protocol that encodes the instructions in a compressed binary protocol. The role of the Outbound Gateway is to connect to the various venues and then act as a translation layer from the internal representation in the OMS to the external representation of the specific protocol implemented by the venue. The outbound gateway also listens in to the connection callback to capture the asynchronous events coming from the venue, like order insert/cancellation acknowledgments and executions and in turn updates the state of the child order representation in the OMS.

**Notifications to the Client:** As the strategy executes orders in the market, the OMS keeps the state of the parent order up to date with each execution and sends periodic updates via the inbound gateway back to the client's EMS that updates its own state and provides feedback to the trader that the strategy is executing, the average prices currently achieved, and other analytics necessary for the trader to understand how well the strategy is executing. The inbound control layer is also kept up to date so that the total state of all client orders is accounted for if and when a new order from the same client is received.

**Middle and Back Office:** We are now almost done. The step above completes the real-time feedback loop from the client through the executing strategy to the market and back. The rest of the processing is in most cases done offline by a set of infrastructures commonly referred as "Middle and Back Office Systems." These systems play a critical role in the business and regulatory side of trading. They are arguably the most important piece of the puzzle, without which no trading operation could function. As we previously discussed, performance of an algorithm is really just the cherry on the trading cake. Without a trader knowing what has been traded and what their current positions are, without the confidence that a once a buy order is executed, within the $T + 2$ settlement period (more on this below) the shares the trader just bought will be in the trader's account, nothing else would matter. Describing in detail the complexity and subtleties of the process that ensures that once an order is executed, the shares or the monies are in the right account at the right time is well beyond the scope of this book. But the authors feel it is important that the reader has at least some understanding and a high degree of respect for the critical role of these systems. The hard working people who operate them often spend nights and weekends handling the myriad of exceptions and trade breaks that are unfortunately all too common in the industry.

To start, let us quickly review the process by which share ownership is transferred between large institutional sellers and buyers. Stock ownership is almost never in paper form but is very likely stored as a book entry in a computer at a place called Depository Trust Company (DTC). This firm acts as the main custodian for all the shares that can be freely traded on an exchange. Brokers will have an account at the DTC that keeps custody of all the shares the broker holds on behalf of their clients. From the DTC perspective the broker is the owner of the shares. The process of transferring ownership goes through two steps:

- **Clearing:** It is the process of updating the accounts of the trading parties and arranging for the transfer of money and securities. In most cases, this step goes through yet another intermediary, a Central Clearing entity that provides various services; in particular, all trade aggregations so only the net shares across all trades are transferred.

- **Settlement:** It is the process of actual exchange of securities for cash. This in most cases (there are always exceptions in a complicated process) happens at $T + 2$.

The DTC then also handles other functions like the handling of all stock dividends and other corporate actions.

As discussed previously, in general, investors cannot trade directly with each other but only through an intermediary, a Broker. Many investment institutions often trade with multiple brokers but engage one (and sometimes more than one) as a Prime Broker where they centralize their balances. Prime brokers provide all sets of services, e.g., stock lending, financing, etc. The client will set up one or many accounts with the prime broker (sometimes thousands, depending on the complexity of the operations) where they record, which shares registered in DTC, do actually belong to the client. Some of the largest and most complex institutions, to limit credit exposure and for other competitive reasons, leverage multiple Prime Broker relationships in an even more complex Multi-Prime setup.

Let us finally turn to the main topic. In executing, broker middle and back office operations are segregated into two distinct parts and are managed independently:

- **The Market Side:** This deals with the interaction with the various exchanges and the clearing house and ensures that there is a perfect match between what the broker thinks has been traded and what actually was traded, at what price and how, i.e., taking vs. providing. This then is used to actually determine to whom the broker needs to pay, the various exchanges for their execution fees (or get paid by them if a rebate was earned).

- **The Client Side:** This handles the clearing and allocation of all traded shares to various accounts based on the client instructions (mostly in electronic format but sometimes still via email!) and ensures that their records match exactly with both the clients and the prime brokers. It also handles all reconciliation of the all too frequent "trade breaks" when something does not quite match exactly.

The two sides (Market and Client) are usually separated in a set of Middle office functions that handles the booking and allocations and the confirms with the client/exchanges and a back office that manages issues around settlement. The back office also handles the regulatory burden of managing the trade reporting process to the TRF/ACT and the OATS process that sends to FINRA, the full life-cycle of the order. As the reader can now appreciate, the role of this function is absolutely critical to the correct functioning of a trading operation.

## 11.2 Algorithmic Trading Infrastructure

This section details a somewhat generic approach for algorithmic trading infrastructure. Different providers have different approaches to this. For instance in some platforms, the core strategy is in one infrastructure but the order routing part is a separate (albeit similar) infrastructure. Others combine the two together, while some separate scheduling and order placement/routing components. The simplified discussion below assumes an all-in-one approach.

The core of an Algorithmic Trader infrastructure is a software framework generically called "Strategy Container." This facility sits in-between the low level OMS related functionality and the strategy, and provides a set of abstractions and interfaces to simplify the development of a trading algorithm. When a new order event is received from the OMS, the strategy container reads and validates the instructions like the name of the strategy and the necessary and additional parameters. If this is a valid new order instruction, it usually has to pass through an additional layer of controls to ensure that other variables such as order quantity or limit price are within the required boundaries. Once this validation step is completed the strategy container instantiates the code of a particular strategy, and initializes it with the specific parameters. It will also connect the algorithm to the necessary services that the algorithm needs for implementing the strategy. Let us look at some of the main services and related infrastructures (Figure 11.2).

**Static Data Services:** An Algorithmic strategy requires a nontrivial amount of reference data and other static data to operate. First and foremost, the particular instrument

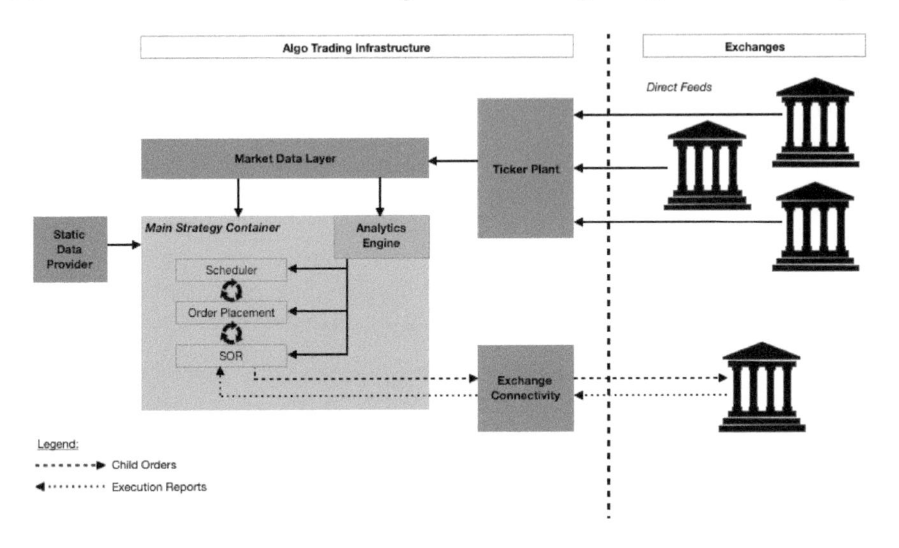


Figure 11.2: Sample Algo Trading Infrastructure.

the strategy is trading but also other information like what the primary exchange is, open/close time of the exchanges, etc. It will also require the normalizing analytics as well as any calibration parameters for various models. These static data services abstract the sources of the underlying data that could be databases, flat files and provide calls to other systems.

**Market Data Facility:** Access to real-time market data is the most important component of any trading algorithm. The strategy container is usually provided with two ways of accessing this data: via a data cache that is continuously kept up-to-date by an underlying thread, and via callback. The data cache also continuously updates additional core analytics used by the strategy such as total traded volume (filtered for specific condition codes) and others. A natural question is how do we actually get real-time market data? Connected to our market data facility is one of the most demanding and expensive pieces of infrastructure in the whole stack: The Market Data Plant. This is worth a small detour for a brief discussion.

**Market Data Plant:** Any serious trading operation, in particular in the post NMS and OPR era, requires to access protected quotes, to subscribe and deliver direct market feeds from all registered (thus protected) exchanges. For most use cases, top of book (Level I) data is not going to be sufficient. It is necessary to subscribe to at least Level II or even better Level III data. Each exchange family (and sometimes even within one) has its own proprietary multicast protocol. Also very important information like trade condition codes are proprietary as well and are to be normalized. For each Level III market data feed, a book building library is needed that interprets all the multicast events and updates the internal state of the book. Doing this in a very efficient way that can withstand market data spikes without dropping any multicast packet is not an easy task. It requires skill and an excellent networking and compute infrastructure, something that is quite expensive to build and maintain. For this reason trading operations often rely on third party vendors for this component that can provide a hybrid software and hardware solution to handle the entire operation. Some of the vendors in this space are: Reuters,[5] Redline Trading Solutions,[6] and Exegy.[7] Over and above the hefty cost of operations, exchanges have in the last decade significantly increased the cost of subscription for market data. This is now a significant component of the overall cost of running a trading operation (in the seven digit range!).

**Outbound Order Interface:** Now back to the discussion of the core services. The last core set of abstractions has to do with managing outbound child orders. This facility provides the strategy with the state of all outstanding child orders and ability to create, cancel and amend them in an asynchronous way and manage any exceptions like for example, canceling an order that has just been executed on exchange but the fill event has not yet reached the OMS.

**Main Strategy Loop:** We come to the *core of the core* of the whole infrastructure. While there are many ways to write a trading strategy a pattern has emerged over the years that can be described as somewhat as a standard approach. It is sometimes referred to as the Main Strategy Loop,[8] as described in Figure 11.3.

While the strategy container interacts with other working threads, and ensures that the state is always kept up to date, the core of the strategy flow is encoded in one stateless function. This function is called on either by a timer, any major event such as quote change/trade/execution, etc., or even on a tight loop. It then sequentially executes three separate stages:

1. **Processing:** Understand where the strategy is at any point in time by recovering the state from the state cache with all the information it needs. How many shares have been traded, and how many, based on the target strategy, should be traded and where should the trader stand in the near future.

2. **Evaluation:** Based on the full information set and the current state of the outstanding child orders in the market, determine which orders to cancel, which to amend, and which to submit.

---

[5] `https://financial.thomsonreuters.com/en.html`

[6] `https://www.redlinetrading.com/`

[7] `https://www.exegy.com/`

[8] Some practitioners for example believe that the right conceptual framework for a strategy is that of a State Machine: Based on what event just happen there is a "state transition" in the next state and writing a strategy is essentially to model and implement these state transitions. Conceptually elegant, in practice this approach is extremely hard to pull off as there are a lot of side effects and co-dependencies that makes it hard to cleanly decompose.

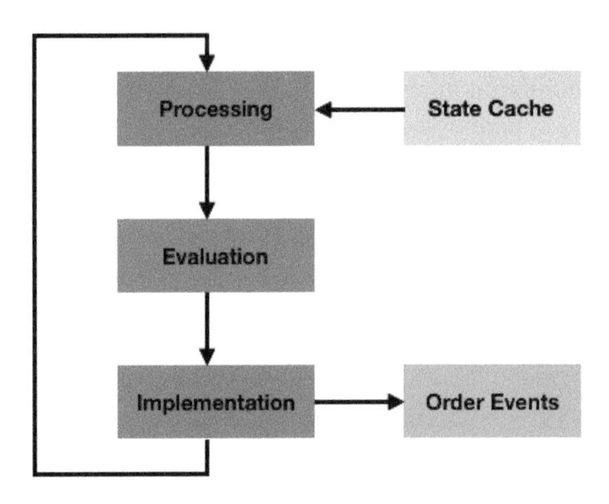


Figure 11.3: Sample Main Strategy Loop.

3. **Implementation:** Send these instructions to the Outbound Order Interface. In some cases Step 2 just calculates the correct exposure it wants in the market and it is the role of the implementation step to optimally decide if and how this should be implemented: Either by submitting additional orders or by amending the existing ones, etc.

After performing these three steps, the main loop will exit and repeat at some time in the future. Since the approach is Markovian, meaning only the existing state matters; if the strategy is called again immediately after, it will most likely make the same decisions it just made and determine in Step 2 that no action is necessary. Thus, the main loop simply exits.

**Additional Infrastructures:** Depending on the complexity and sophistication of the operation, the technology stack is often supplemented by additional components that serve certain specific roles. Let us look at some of the common ones.

**Realtime Analytics Engine:** More and more, algorithmic trading products have become highly sophisticated and dynamic and adapt as market conditions change. The normalization variables and signals are adjusted at every single tick or after every few trades to provide the most up-to-date picture of the immediate past and future market state. Conceptually these analytics could be built within the actual strategy engine but it is not only more cumbersome, it is also inefficient. In most cases one process cannot handle the computing load to run thousands of algorithms on all symbols. The infrastructure is thus "striped" in multiple processes handling a subset of symbols. Per symbol analytics could still be built within each process but many analytics, like for example computing the real-time price of an index mid-point, requires that the data on hundreds or thousands of symbols be computed in the same process. For each symbol, this would create large duplication, reducing the effectiveness of striping.

Additionally with more analytics, more load on the process is created that is used to handle the algorithmic decisions.

For these and other reasons of convenience needed in calculating real-time analytics, it is often delegated to a dedicated infrastructure. This makes it easier to build additional analytics and to combine existing ones with the more complex signals.

**Algorithm Switching Engine:** Most execution strategies have limited flexibility to change behavior in particular situations and opinionated traders may have strong beliefs that a different trading approach might be warranted in certain situations. Historically that trader may work with one of their favorite brokers to build a customized strategy for the purpose but that usually can take a long time and then the trader can use this approach with only one broker.

To provide simple dynamic customization, brokers began offering Algo of Algos functionality around what is commonly called an Algo Switching Engine. This Engine is a combination of a simple strategy container and a limited real-time analytics engine, where an Algo is represented by a simple set of if-then-else rules. The condition part of these rules will be simple combinations of the available signals and the action part will define a particular algorithm with associated parameters. Once the order is received by the switching engine, the rules are evaluated and an order for the specified strategy is sent to the strategy container. The rules are then re-evaluated every few seconds and if the conditions change the switching engine will send a cancel/replace message to the strategy container to change the underlying strategy.

With this simple approach, one can build very sophisticated dynamic algorithms thus offering almost unlimited capabilities. This has made algorithm switching engines very popular both with traders and with execution consultants that often differentiate themselves by smart usage of the switching engine. While an Algo Switching Engine is a very powerful tool, it has several meaningful downsides that one should consider:

- The rules that make up a custom strategy encode certain beliefs on the value of a particular triggering condition. If that condition had true predictive power it would be better integrated as an alpha signal and leveraged in a more nuanced way within the strategy rather than as a blunt instrument of an algo switching instruction. If the condition is actually, as most often the case, not predictive then the realized performance of the strategy will be, in general, made worse by the customization. Thus, while in principle, it is a good idea, in practice it should be discouraged unless its real value can be measured. This would make the algorithm switching engine ideal as a temporary exploratory tool rather than the overused tool as it currently is.

- Algorithms are usually not fully optimized for persisting state across a cancel/replace transition. This usually implies that the whole state of the algorithm is lost and a completely new order is created. Algos also tend to perform relatively poorly at the beginning of the order as various analytics and execution bands exhibit poor performance when initialized (e.g., participation bands unless handled correctly have a discontinuity at zero). Frequent use of switching transitions will have severe negative impact on algorithm performance.

- Once a trader is onboarded into a custom strategy, any proper systematic Post-Trade support by the broker is lost since the optionality embedded in the switching rules will largely drive the performance. Additionally, these orders cannot be used by brokers to measure and improve their own performance because the individual slices will be hopelessly biased by the switching rules.

As one can glean from the above treatment of an algorithm switching engine, the authors are not big fans of this mechanical approach and believe that this evolution has resulted in decreased quality of execution algorithms. The popularity of these approaches is still quite high and we believe currently the approach is overused. However, there are signs that this popularity is slowly waning as clients are starting to rely on the broker to fine tune the behavior of the strategies.

**Portfolio Algorithm Engine:** As we discussed in Chapter 10, some algorithmic strategies involve the coordinated trading of an entire portfolio of instruments. In this section, we briefly look at the technology infrastructure needed for these strategies and some related common issues.

The problems start immediately, at the interaction between the FIX infrastructure and the client's EMS. Until recently Basket trading was never well supported by various EMSs making the certification of a portfolio strategy a potentially very fiddly proposition as it requires supporting multiple approaches. Wherever basket trading is not supported, firms have resorted to simple ways like considering every portfolio order sent to that algorithm within a certain timeframe, or linked to a portfolio ID parameter, or a combination of both.

The portfolio algorithms infrastructure is not dissimilar to the one used in Algo Switching but with the added complexity of having to manage the schedule for multiple orders simultaneously. These algorithms will create a set of linked trajectories, often through some form of portfolio optimization process that implements the specific objective function. Once the trajectories have been generated the algorithm will, in most cases, use one or more existing algorithms possibly modified to adapt to the portfolio infrastructure. Certain algorithm implementations send short time slices to a TWAP algorithm (essentially transforming the trajectory into a piece-wise linear curve) while other implementations may use a modified VWAP style algorithm that supports a special parameter to overwrite the trajectory to be used. Once the portfolio trading starts, market conditions change or realized deviations of the portfolio shape (e.g., usually as an impact-control measure, the single slice will have maximum POV constraints applied, which if triggered, will distort the original intent) will often require the algorithm to re-optimize the joint schedule and the individual slices are amended to account for the changes.

This workflow is already quite complex but there are several additional mechanical and numerical issues that further complicate matters making portfolio algorithms really hard to pull off successfully. Here are some practical issues:

- Among the Portfolio IS style algorithms, the most common optimization approach is usually implemented as a Quadratic Programming (QP)[9] optimization problem where the schedule is discretized into bins representing the shares to be traded in a certain period. For large portfolios common in fund rebalancing and for relatively fine grained time bins, the number of variables that need to be used, explodes pretty fast (in particular formulating these problems requires use of several auxiliary variables to manage cost constraints, etc.). For example, a 1000 stock portfolio with 5m bins sent to a whole day Portfolio IS algorithm would have to solve a problem with more than 100k variables. This often results in a very slow optimization cycle that can take hundreds of seconds making it too slow for effective utilization (no client would want to wait several minutes to see the first trade!) and too intensive computationally for a scalable operation (e.g., typically a dozen such portfolios arrive at the beginning of the day). To solve this problem effectively requires nontrivial numerical methods that make the actual optimization problem very difficult to solve.

- Essentially, all such algorithms are discretized in time bins, and hence synchronizing the re-optimization steps to correctly line up with the executing algorithms is not easy and often leads to purely mechanical distortion of the realized schedule, even in very simple settings (e.g., the tendency of the orders to be behind schedule leads to the unexecuted quantity to be reallocated to the remaining bins causing a significantly stretched execution schedule).

- The portfolio level optimization problem requires a variance-covariance matrix for all instruments in the portfolio. As discussed, the universe of security that is traded with algorithms is usually very large and to have stable covariance matrix for such a universe requires the creation or purchase of a Factor Risk Model (either a fundamental factor model or maybe a simpler PCA style model, see Chapter 6). New issues and ticker changes create additional operational issues for these algorithms.

- Optimization linked schedules are very rigid and to avoid significant breaches of the provided constraints, the algorithm has to follow the schedule quite closely. This means that the algorithm usually cannot leverage additional opportunistic liquidity. In a market structure like the US where 30–40% of the liquidity is off-exchanges this makes these algorithms not very effective in sourcing liquidity. Some effort in incorporating dark pools into the optimization problem was attempted in Kratz and Schöneborn (2014) [233] but very little production algorithms have actually solved this problem effectively.

---

[9] `https://en.wikipedia.org/wiki/Quadratic_programming`

## 11.3 HFT Infrastructure

We have completed the whirlwind tour of the technology stack that one would find in a standard execution business. As mentioned, we did that because it is probably the most complex and broad use case. For other setups, some of the components may be very similar, but others may differ. There are probably a dozen different configurations and discussing all of the nuances in detail is well beyond the scope of this brief chapter but as a comparison we take a quick look at the other extreme, an HFT infrastructure.

In broad terms, HFT setup is in some ways a lot simpler while others may have technical complexities beyond what are usually found in execution. To begin with, HFT strategies are not driven by instructions; thus, there is no need for all inbound connectivity, OMS, etc. Additionally in order to achieve the maximum speed and the lowest "tick-to-trade" latency,[10] in general, the technology stack is condensed to run all in the same process. Direct market data feeds and direct exchange connectivity are just components of the same process reducing the complexities of a separate ticker plant, communication bus, and connectivity layer all together.

If the trading operation sits outside a broker-dealer, then the trade has to go through a third party broker-dealer via a sponsored access agreement. This usually requires that the trading system will connect to a broker provided appliance or software layer and through that connect to various venues via the native market access protocols. In most cases, the trading infrastructure is co-located in the same data center as the exchange(s) where that strategy will operate.

Finally, the other main difference is the need for the infrastructure to connect to a position keeping system to incorporate the current inventory as part of the decision making. Usually these positions are also computed internally, based on start of days positions in order to have an internal verification with additional controls, if the positions calculated internally deviate from the position calculated by the position keeping system.

In what other areas then do the technical complexities that exist beyond standard infrastructures lie? It is often driven by the pursuit of absolute speed and throughput. This implies having deep understanding of the networks, hardware and software, utilizing all the options in the book to squeeze additional nanoseconds. This includes specialized network switches, CPU overclocking, kernel modifications, use of assembly/machine language for the most latency sensitive parts of the codebase. It could also include setting up connectivity to far away exchanges, most notably the CME exchange where the S&P 500 Futures is traded, using proprietary networks like spread networks (now part of Zayo Group)[11] or via microwave towers like what is

---

[10]The total time it takes from a new market data event happening on exchange (tick), to that tick being received by the trading engine which in turn generates an order, to that order reaching the exchange.

[11]`https://www.zayo.com/private-dedicated-networks/`

provided by the ICE [12] and more recently also millimeter wave lasers like Anova[13]. More and more frequently, large portions of these ultra low latency setups leverage Field-Programmable Gate Array setups (FPGA)[14] and some vendors are even selling FPGA chips for all-in-one appliances where feeds, connectivity and strategy are programmed in hardware.

## 11.4 ATS Infrastructure

### 11.4.1 Regulatory Considerations

Alternative Trading Systems (ATS), are private US[15] securities trading platforms that meet the definition of "exchange" under federal securities laws but are exempt from registration as national securities exchange so long as they comply with the specific rules[16] of Regulation ATS.

Regulation ATS establishes a regulatory framework for ATS, and mandates that all ATS need to be registered as a broker-dealer with the SEC and file a Form ATS detailing the functioning of the venue and then make this documentation available to the investing public.[17] As of January 2019, FINRA adopted an amendment to Reg ATS requiring the filing of a new Form ATS-N, addressing some of the shortcomings of the previous Form ATS, in particular around disclosure of potential conflict of interest and risks of information leakage arising from the ATS-related activities of the ATS's broker-dealer operator. The new form also discloses information pertaining to order types and market data used by the ATS, execution and priority procedures, and any process in place to segment orders in the ATS.

Three main types of venues fall under the definition of ATS:

- **Electronic Communication Network (ECN):** A computerized network allowing the trading of financial products outside traditional stock exchanges (even outside of traditional exchange hours), matching orders for a fee. As discussed

[12]https://www.theice.com/market-data/connectivity-and-feeds/wireless/chicago-to-new-jersey

[13]https://anovanetworks.com/

[14]https://en.wikipedia.org/wiki/Field-programmable_gate_array

[15]Under European regulations, ATS are designated as Multilateral Trading Facilities (MTF) and are governed by their own set of rules.

[16]Rules 300–304.

[17]http://www.FINRA.org/ATS

in Chapter 1 ECNs are historically the first setups created to compete against the monopoly of the exchanges with the first ECN, Instinet, dating back all the way to 1969.

- **Crossing Network:** Operated as broker-dealers' internal crossing network, matching orders in private using prices from public exchanges.
- **Dark Pool:** Private securities trading platform offering an undisplayed order book and matching engine. Dark pools also come in different flavors:
  - Independent Dark Pools: E.g., Instinet, Posit, Blockcross, Liquidnet.
  - Broker-Dealer Dark Pools: Where clients of the broker interact, such as CrossFinder, MSPOOL, SigmaX, JPMX, LX, ...
  - Hidden liquidity of public exchanges: E.g., BATS, NYSE, ASX CenterPoint in Australia. These, however do not fall under the ATS designation per se.

As private securities trading platforms, ATS which have an average trading volume of less than 5% of the aggregate daily volume in NMS stocks do not need to provide a feed of their top-of-book quotes to a national securities exchange, nor are they bound by fair access rules and thus allowing them to bar access to anyone they choose.[18] This barrier to entry allows owners to control the liquidity make up of their pool and is sometimes presented by lit exchange advocates as being detrimental to the price formation process. In this debate, it is interesting to note that, despite 5% being a rather high threshold, nothing prevents an operator from registering several ATS in order to remain under this value.

To address the increasing reliance on technology of the US market place, the SEC enacted regulation in 2015 designed to limit the occurrences of system-driven disturbances and speed up recovery time when they do occur. As part of this, above 1% of market share in NMS stocks (in four of the preceding six calendar months), ATS are subject to additional scrutiny, and in particular their infrastructure must be compliant with the far reaching requirements of Regulation SCI (Systems Compliance and Integrity). Reg SCI requires that in-scope entities establish written policies and procedures reasonably designed to ensure that their systems have levels of capacity, integrity, resiliency, availability, and security adequate to maintain their operational capability and promote the maintenance of fair and orderly markets, and that they operate in a manner that complies with the Exchange Act of 1933. Additionally, Regulation SCI requires that SCI entities conduct a review of their systems by objective, qualified personnel at least annually, submit quarterly reports regarding completed, ongoing, and planned material changes to their SCI systems to the SEC. It also mandates participation in scheduled testing of the operation of their business continuity and disaster recovery plans, including backup systems, and coordination of such testing on an industry or sector-wide basis with other SCI entities.[19]

---

[18]For instance, Luminex is a pool only accessible to Buy Side traders.

[19]Source: `www.sec.gov`

### 11.4.2 Matching Engine

In today's low latency trading landscape, the speed of the matching engine is also regarded as a significant competitive advantage among venues. Consequently, ATS matching engines and market data processing layers are often built leveraging technology of similar grade as the ones employed by main exchanges, using third party software.[20]

The matching logic, however remains relatively simple and—in its more basic form—similar to what happens on exchanges. In the next section we will provide some examples of specificities, but at its core, in its simplest form, the matching engine of an ATS or Dark Pool would adhere to the following logic for mid-point orders:

Buy order defined by: $BUY_{price}$, $BUY_{size}$
Sell order defined by: $SELL_{price}$, $SELL_{size}$
Prevailing mid-point defined as: MID

If $BUY_{price} \geq$ MID:
    If $SELL_{price} \leq$ MID:
        Executed order quantity = $\min(SELL_{size}, BUY_{size})$
        Executed order price = MID

Once trades happen, dark pools are required to "quickly report" them to the consolidated tape, but this process is not instantaneous, creating a temporary asymmetry between the two parties to the trade (who get immediate execution reports) and the rest of the market. More importantly, subscribers to the public tape also do not know which dark pool (or wholesaler/ELP) reported a given trade as it simply appears under the FINRA ADF Reporting Facility code.

To help shed some light on dark activity, FINRA Rule 4552 specifies that weekly dark pool volume be published per security and made public on a 2-week delayed basis for Tier 1 securities, and 1-month delayed basis for Tier 2 securities.[21] While the data is not high frequency enough to make direct trading decision, it does provide a helpful aggregate view of the non-displayed liquidity to allow market participants to adapt their liquidity sourcing strategies to an ever-evolving landscape (market share per security, average order size, ...).

### 11.4.3 Client Tiering and Other Rules

Registered ATS offer a range of differences over regular public exchanges, mainly in the way they handle order matching. While most exchanges match orders on a price-time priority, ATS might elect to have different structures such as matching orders

[20] www.cinnober.com, www.reconart.com

[21] https://otctransparency.finra.org/

with a price-size or price-tier-time priority, enforcing minimum trade sizes, etc. The common variations include:

- **Client Tiering:** Some ATS offer a counter-party tiering mechanism leveraging either a predetermined classification (Institutional Investors, Electronic Liquidity Providers (ELP), Broker Dealers, Principal flow,...) or a scoring of the order based on more arbitrary metrics (orders frequency, average order size, post-trade mark outs, ...). The tiering of the pool can allow participants to opt out of interacting with certain counterparties[22] or in certain cases alter the priority of order matching. Some pools have a matching engine that follows a price-tier-time priority, meaning that at an equivalent price level orders originating from certain institutions would be given priority (for instance, long only funds would transact first, even if they submitted their order after faster players like ELP, in an approach believed to level the playing field between natural investors and short term liquidity providers).

- **Price Improvement Methodology:** Most ATS matching engines rely heavily on the NBBO as a reference price for mid-point matching, allowing their participants to submit mid-point peg orders to increase the probability of matching undisplayed orders. However, near and far touch orders are also a common feature. In the scenario where a mid-point resting buy order transacts against an aggressive far touch sell order, the ATS can fill both parties at mid-point (as would happen on an exchange since the buy order was resting and was willing to transact at that price), or elect to fill both parties half way between the far touch and the mid-point (sharing the price improvement between both parties since the aggressive order was willing to transact all the way till the far touch).

- **Trade Frequency:** Earlier, we mentioned about the particular efforts made by ATS to operate as fast as possible in a highly competitive landscape. Some venues, however, offer a discrete time matching instead of a fully continuous one, focusing their latency reduction efforts on the processing of inbound data serving as reference prices rather than pure matching engine speed. For instance, these venues might be offering an 'average price matching' mechanism whereby two counterparties get locked into a matching period for a few minutes at the end of which both sides receive an execution at the period VWAP price (or any other agreed upon benchmark price). Despite the potential for temporary price inefficiencies, matching orders at discrete time intervals is perceived to mitigate the potential advantage enjoyed by the fastest market players. For similar reasons, IEX's dark pool slows down orders by 350 microseconds to protect resting orders against latency arbitrage by forcing incoming orders to go through a 38-mile optic fiber coil.[23]

---

[22]For instance, traditional long only funds might be concerned about adverse selection if they interact with ELP.

[23]`https://iextrading.com`

- **Minimum Execution Quantity:** Over time, ATS usage has mostly grown out of a desire by market participants to limit the information leakage associated with displaying their orders on publicly visible limit order books. While in theory the usage of dark pools would reduce such leakage, it is nonetheless necessary to protect orders against potential predatory behaviors such as "pinging."[24] A simple solution against abusive liquidity detection is to mandate a minimum execution quantity for all orders submitted. For instance, one might send a 20,000 shares order at mid-point but request executions to be of no less than 1,000 shares each. Another benefit of large minimum execution sizes is the fact that larger orders tend to originate from natural liquidity and, as such, present less short term adverse selection.

- **Order Aggregation:** To satisfy the minimum size requirements, ATS can either strictly compare the minimum size of each order in their book, or offer an aggregation solution. For instance, if only two buy orders (one for 200 shares with no minimum quantity, and one for 300 shares with a 200 minimum quantity) are resting at mid-point and a sell order for 1000 shares with a 500 shares minimum quantity enters the book, an ATS offering order aggregation would execute 500 shares. Without such aggregation, all three orders would remain on the book, unexecuted.

- **Firm vs. Conditional Matching:** Conditional orders allow the trader to receive a notification when the contra side is available to match and choose whether to commit to the trade or not (and for what quantity), as described in Figure 11.4. The benefit of conditional matching is the ability it gives to the trader to expose the trader's full order to multiple sources of liquidity at once. For instance, even if the true intention of Client A in Figure 11.4 is to purchase 60,000 shares, the trader can post 60,000 shares concurrently in three conditional venues, and start trading the 60,000 shares in a liquidity seeking algorithm at the same time. Doing so, the trader maximizes the probability of getting a block fill from a dark venue (compared to a sequential allocation which could miss transient liquidity), while not unduly incurring timing risk as the trader waits for such an opportunity to materialize. Once the contra side is matched and an invitation received, the trader can cancel the residual executions in the liquidity seeking algorithm and firm up with whatever quantity is left to execute. The request/response process known as firm-up mechanism can take the form of a fully automated machine-to-machine interaction (generally happening within milliseconds), or involve a human decision. Conditional venues targeting larger orders often offer a human firm-up component, which raises questions about the potential information leakage should one of the parties decide not to go through with the transaction.

---

[24]"Pinging" is a practice aimed at detecting the presence of large orders hidden in dark pools in order to gain insight about (or influence) the future price trajectory of a stock. To do so, a participant would usually send repeated small orders at mid-point, and use the executions received as a confirmation of the existence of a potential large order on the other side.

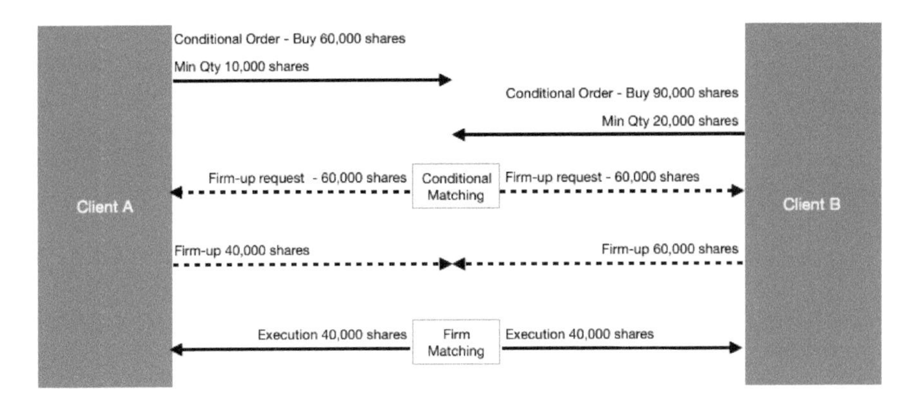


Figure 11.4: Conditional Order Matching Mechanism.

- **External Indication of Interest (IOIs):** Certain pools, in an attempt to improve fill rates for dark orders, might broadcast IOIs to liquidity providers and other destinations. This practice is generally quite controversial as it raises questions about undesirable information leakage, and most participants would opt out of the functionality if at all possible.

These examples are not meant to be exhaustive, but nonetheless highlight the flexibility that exists for ATS compared to traditional exchanges and help explain their rise in popularity over time.

# 12

# *The Research Stack*

Designing, testing, calibrating, measuring and improving a suite of algorithmic trading strategies is not for the faint of heart. As we have discussed, it is a large scale problem that requires deep investments and can take years to set up. One of the critical components of the overall stack, one that is often underfunded and thus underdeveloped, is the research environment. For an execution business this is usually an afterthought, a setup that is scrounged together from already existing pieces of the infrastructure they need to operate. A quantitative team often has to do with what is available: Historical market data, reference data, and historical transactional data in order to fulfill the transactions cost analysis (TCA) needs of the clients.

In this chapter, we attempt to provide some pointers of what a best-in-class research stack would look like. The basic components, approaches and considerations born out two decades of experience and the trials and tribulations (mostly tribulations!). While most of the chapter is reserved to the "standard" setup of an execution business we will reserve, where appropriate, a short paragraph to describe the setup most often found in an HFT outfit as it is quite different and has some interesting features at least worth mentioning.

The authors wish they could dedicate a full book (and maybe one day they will) to this topic, as there is so much to be said based on our practical experience.

## 12.1 Data Infrastructure

As we have seen throughout this book, without data, obviously, there is no research. It then goes without saying that the most important component of the research environment must be data infrastructure. There are two main parts of a typical data infrastructure: Historical Market Data, and Historical Transactional Data.

Additionally as observed from Chapter 1, we should have access to historical reference data as well. While most of the research and analytics will be intra-day and thus less dependent on historical data, it is sometimes necessary to look at prices/volumes over multiple days for consistency. This creates an added complexity of potentially having to adjust the data for corporate actions. This is not a simple thing and rarely reference data platforms store all history of ticker changes, symbology mapping and various corporate action multipliers that were used to adjust the data.

**Historical Ticker Plant:** Market data, just as the case was for the production tech stack can be the most challenging and expensive component of a data infrastructure. Even for the simplest usecase: Consolidated Level I quote and trade data TAQ data from NYSE[1] at present time can cost $3000 per month of history. It is more than 18 GB per day compressed. Any respectable operation needs at least one to two years so just the simplest setup requires an investment of at least $75,000 and storage and processing capacity for 8 TB of data. A more serious setup would have Level II data for the top exchanges with costs and sizes, several multiples of Level I data.

Once collected the data has to be processed, normalized and stored in an infrastructure that allows for analysis and for processing. While theoretically one could store this data in a relational database, realistically this never creates a usable infrastructure and historically only few vendor infrastructures have been successful in this space. Notably KDB from Kx System[2] has been the dominant player in this space, despite the steep cost and the terse and cryptic 'q' language that is hard to master and thus is a very much sought-after skill. In recent years this space has evolved significantly and there are now powerful open-source contenders like InfluxDB[3] and many others. Others have started exploring distributed systems on top of the Hadoop/Spark ecosystem. KDB continues to maintain its dominance in large scale trading operations but its reign is no longer as certain as it used to be.

The best setups also collect and provide, through a similar API, intraday real-time market data. This allows for an extensive analysis and tools to work both historically and intraday. Since it is quite hard to capture with 100% accuracy the data at all times, more often the intraday data is discarded overnight and replaced with the vendor provided official one that has been processed and adjusted for various erroneous trade events.

*HFT Setup*

Ultra-low latency HFT trading operations usually capture and store the network packets from raw production feeds for all major exchanges, not one but for each data center. This is to be able to train separately the strategy in each data center and to see the market data from other data centers with the actual delays as the strategy goes through production. Data is kept in a non-processed compressed format without the leverage of any time series database. The size of this data can be staggering, in the order of more than 50 GB compressed per day per data center.

**Historical Transactional Data:** This is another complicated and large infrastructure problem. Ideally all OMS orders and trading events should be stored in a time series database as they happen and then processed and stored in a consistent data structure.

---

[1] `https://www.nyse.com/market-data/historical#taqequities`

[2] `https://kx.com/`

[3] `https://www.influxdata.com/`

In Chapter 11, we mentioned the OMS often leverages a bus-based infrastructure. One of the important use cases for that to happen is to have a gateway process that can listen to all these events and insert them into the transactional database.

An important need for any execution setup is to be able to fully reconstruct the life-cycle of an order from its arrival to the inbound gateway, passing through the algorithm container generating orders for the SOR container to finally create child orders, to be sent to the market via the outbound gateway. Additionally if orders get amended by the clients it is necessary to be able to tie all this together within the order hierarchy. This is quite a subtle and complex data modeling problem. The buildout of a data warehousing solution that allows the researcher easy access is particularly important. This is something the authors feel very strongly about since it may otherwise cause countless hours of extra work and frustrations, grappling through the limitations for their own infrastructures.

With regards to data warehousing infrastructures we can make similar considerations around vendors and solutions as we made in the section on market data. While KDB is still a strong contender for these solutions it is not as dominant as in the market data domain.[4] Due to the complexity of the data modeling part, these approaches benefit more from the normalized approaches, typical of more standard data warehousing solutions built on relational databases.

As we have seen in Chapter 11, HFT and market-making setups are mostly built as single monolithic co-located processes. It is also a relatively simplified setup as trading initiation is completely endogenous to the trading strategy and thus does not have to be tied back to initiating order instructions coming down from the OMS. Typical solutions are more "low-tech" that come from processing raw log files and storing the instructions in compressed csv formatted files.

## 12.2 Calibration Infrastructure

From the discussion presented here, it is clear that a respectable trading execution operation requires a whole slew of models and analytics that have to be created, calibrated, and delivered into the production environment. Creating a solid, nimble and flexible calibration environment that provides a high degree of automation, has strong validation capabilities and one click productionization is not a simple exercise but one that we strongly encourage investing time and resources in. This, unless one is fine with countless hours spent each month keeping all these models up-to-date, fixing production breaks and answering embarrassing questions such as why one could not trade the recent IPO.

---

[4]This is somewhat surprising since, as we will see, a lot of the model calibration and TCA environments we will discuss in future sections require the joining and time alignment of both market data and transactional data (the so-called asof and windows joins) that are one of the absolute strengths of KDB.

The specific details of a solution for this is very much dependent on the current SDLC and productionization practices within a specific firm. So it would be hard to give more specific advice apart from a few general pointers:

- A frequent complication is that many of these processes are usually built within the research environment that is often different from the production installation setup. Ideally when the two setups match and the processes have been created, the code, not the data/analytics are pushed into production. The data is generated there for consumption of production applications.
- It is better to avoid a calibration of an entire process but opt for a pipeline of independent hierarchical components with intermediary data, available for various subsequent steps in the pipeline. This makes the process much easier to extend and parallelize.
- One should think carefully about segmentation of calibration, meaning when one should have per symbol analytics vs. per group. Different analytics might rely on different groupings. Should this complexity be supported by the production system or should one keep this as part of the calibration process but in the end can "explode" it to each individual symbol is not an easy one to answer. There are pros and cons.
- If possible, it is better to be robust around new issues/IPOs meaning that there should be a process in place that is forward looking and has a list of upcoming IPOs so that the reference data, and at minimum sensible default analytics, can be created. This is usually a very hard thing to do and IPOs are a frequent pain-point of this process.
- It is essential to think very carefully around default values. What happens if a symbol is not present? Because of the high reliance on normalization variables a mismatch between defaults and realized stock behavior could lead to strong under-performance of the algorithms.
- Maintain a history of the calibration data so that data can be used for historical simulation.
- There is generally an over-reliance on using flat files (e.g., csv) to deliver this content. A possible better approach is for the data to be stored into a database with a production tag primary key and the production process (possibly an intermediary process that actually creates flat files for the production environment) that generates these files based on the primary key which is controlled by production support.

## 12.3 Simulation Environment

This is a topic that deserves an extensive discussion but here we will provide only a cursory overview. In Chapter 11, we discussed how a strategy container is usually developed and how the main strategy loop contains most of the decision points on how the trading is done while other parts of the infrastructure keep the state of the executing order in sync. But even for very simple strategies how do we test that the logic is correct? The most intuitive approach is to build a simulation environment that allows the developer to ensure that all the pieces are working together and the strategy makes the "right" decisions based on the current state.

The most effective way to build a simulation environment is to start with the existing strategy container and to stub out its inputs and outputs so that it can run independently. The most important layers to replace are:

- **Inbound Order Layer:** Replace the injection of one or more parent orders reading from a flat file or command line.
- **Reference and Static Data Layers:** Replace the access to this data with csv files.
- **Market Data Layer:** Replace the market data, the real-time subscription with a component that injects historical market data from a flat file or a database.
- **Outbound Layer:** Replace the component with a substitute that contains some form of fill model.

The most complex layers of the above list are of course the Market Data and the Outbound layers which we will discuss later but there is a very important element that needs be considered as part of the overall architecture of a strategy container in order to be able to support effectively a simulation environment. This element is time management.

**Handling Time for Simulation:** When simulation of a whole-day VWAP order is desired, it is obvious that we may not want to have to wait many hours to find out the result. The simulation should execute faster, ideally as fast as events can be processed (the so-called "bullet time"), so the results can be quickly reviewed and the necessary changes can be made and repeated. The problem is that there are several components, part of the strategy container, that leverage actual time, e.g., trading sessions such as opening and closing auctions, order start and end time, etc., and there is usually a liberal use of timers to perform specific tasks such as timeouts, e.g., waiting for acknowledgement from the exchange after submitting an order, and delays, e.g., have the strategy wake up every '$n$' seconds or decide to sleep for a few seconds after a large block fill. Running the strategy simulation in accelerated time while maintaining realistic behavior requires that the time is set as part of the simulation process and the timers are accelerated in a consistent manner.

One approach often used in practice is to create an internal time service component within the infrastructure. Time is sourced and timers are scheduled so that they can then be replaced by a simulated version. It gets its time from one consistent source which is usually the time-stamps of the historical market data, being replayed and replicated within the Market Data layer. When time is requested by the strategy or container, the latest time-stamp processed is returned. If a timer is scheduled, the time service will look at the next event time-stamp and if that happens after the target expiry, the time event either calls back the component or, as it is more likely, will inject the time event in the event loop for processing.[5]

**Market Data and Order Submission for Simulation:** As discussed, this is where most of the complications lie and where there can be a great deal of differentiation between approaches. Some approaches are very simple and are only really useful as a form of compliance testing, ensuring that the strategy behaves correctly. Other approaches strive towards a more realistic behavior meant to be used to understand and fine-tune the expected behavior from a trading performance standpoint. We briefly look at some of the approaches used in practice.

**Data Replaying and Simple Fill Model:** Arguably, this is the simplest of approaches. The market data layer will simply replay Level I (or Level II depending on the implementation) data into the system. The outbound layer fills any market order at the price available at the far touch or, if using Level II data, it could provide the fills in a more realistic fashion using the available liquidity at each price level. For limit orders these are usually kept in a list and filled when either there is a trade at the limit price or the market moves toward the posted price so that now the price is marketable. The advantage of this approach is its simplicity but there are many disadvantages. Market data is not updated by the presence of our orders and thus does not incorporate the impact of any aggressive trades and completely ignores queue priority of the limit orders or any change, if the orders are meant to improve the near side (as if our orders were hidden). This makes the behavior of the simulator extremely optimistic and is not realistic if one wants to use it to fine-tune the order placement.

**Book Building, Matching Engine, and Small Order Assumption:** A more complex approach calls for order level data (or at least Level II data that can then be deconstructed into a simplified order level data) that can be used to reconstruct the order book. The order submission layer can be replaced with a simplified version of a matching engine where historical order-book is recreated via the order data feed. The same component is also used as the market data layers. Thus, the strategy receives the market data created by the matching engine. Now any strategy decision, to post or take liquidity, is overlaid on top of the historical data and is faithfully added to the matching engine and market data feed. This approach provides a more realistic environment as market data now changes because of the strategy actions. Queue positioning is now considered in some way and when liquidity is taken so that a level is removed, that effect will be visible at least until a new quote on the far side is inserted in the order book when the next market data tick happens.

While more realistic, this approach has the implicit assumption that the strategy actions do not change the state of the market, meaning that its decision does not affect the decision of other market participants and any impact on liquidity is strictly temporary in nature.

---

[5]The low level details of the threading model used by strategy containers is well beyond the scope of this book. We hope the reader can still grasp the underlying concepts.

**Market Impact and Relative Pricing:** An extension of the above approach is to try to incorporate some form of impact and order book dynamics, while still being connected to historical prices. One approach that has been tried in the past was to convert order book data into price relative data. That is, at the beginning, the bid and ask prices are stated based on historical data but new arrival of orders is converted as relative to the reference price. For example if a sell order at the best ask is entered it will be entered at the current (zero) level of the ask queue, regardless of the current price. This would allow for the impact of the strategy decision to have a more lasting effect (if a buying strategy takes away the whole level a new zero level ask order will be added at the new increased price).

In order for such an approach to avoid devolving into an unrealistic order book an additional, model driven, process is introduced. This injects orders into the engine in order to slowly drive the order book, over a certain "decay time," back to the historical setting. An approach like this will require some heuristics around liquidity replenishment but could incorporate some form of short term market impact component to add additional realism in the simulation.

**Model Based Order Book:** Finally, we turn to a purely model driven approach. To simulate a realistic order book is a challenging process and we refer you to the in-depth treatment of modeling LOB dynamics as discussed in Chapter 8. We can also point the interested reader to Huang, Lehalle, Rosenbaum (2015) [206]. These models provide some interesting features and might be of practical usage in HFT strategy simulation. But the authors are somewhat skeptical if the behavior of the models would be sufficiently nuanced to allow the researcher to use it as a tuning mechanism for execution strategies, in particular for liquidity seeking and aggressive strategies.

**Simulation of Multiple Venues for SOR:** The treatment we have made on simulation can expand to support multiple venues and some of the approaches might be relevant for the more complex scenarios. One would need to incorporate a model for simulating dark liquidity to account for trading in various dark pools. Additionally one would have to incorporate the facts that all these venues are connected by regulation and latency arbitrage to be effective in realistically replicating the complex interactions across the liquidity spectrum.

**Conclusion on Simulators:** Creating a realistic microstructure based simulation environment for execution strategies is extremely hard and probably out of reach of most practitioners at this point. This makes the joint simulation of several order books and dark pools necessary to train a Smart Order Router a completely unchartered area. There is still more research to be done in this space and we encourage researchers, academics and fintech firms to pursue this topic further. At the purely scheduling level, where order placement is essentially abstracted out and averaged across all liquidity sources, this seems a lot more promising in the short term. With some effort, one could create a somewhat more realistic environment on where to train some standard execution strategies.

## 12.4 TCA Environment

Once research and simulations are done there is only one way to really know how well the new algorithm works: Deploy the strategy in pilot testing, use it for trading, and then measure... and measure... and measure its performance. This is then a well architected Transaction Cost Analysis (TCA) infrastructure which becomes critical and possibly where most researchers will end up spending their time.

The shrewd reader could argue that this section would better fit in Chapter 11 since it is such an integral part of any large execution operation. We chose instead to make it part of this chapter on the research stack for a specific reason. As the authors have seen multiple times, this infrastructure is often built as an independent component that is not linked with rest of the research environment apart for maybe the data back-ends. We believe this to be a sub-optimal for multiple reasons:

- Many of the same analytics created for TCA are needed for general research as well, and bespoke analysis with the difference that the number of analytics and other various parameters are usually much larger and change more frequently. This implies that researchers end up building out their own infrastructure. This then diverges from the production leading to duplication of efforts, inconsistencies in the analytics and heavy reliance on bespoke, on-the-desk analysis of changes since the production environment always ends up lagging the research one.

- The focus of standard independent TCA infrastructure is usually to build standard, one-size-fits-all post-trade reports. This ignores the fact that, more and more, clients have specific needs and benchmarks and in some cases their own bespoke post-trade attribution frameworks. This fluid state fits poorly into the standard approaches which tend to be too inflexible for any meaningful development effort and thus results in long delays to implement client requested customizations.

- When evaluating the results of simulations we should ideally use the same post-trade infrastructure that is used for production trades. This is usually very hard, if not impossible, to do if one does not consider both research and production as one interdependent entity.

- More often than not the resources dedicated to build the TCA infrastructure are the same as the ones used for the overall data and research environments. This, combined with the tendency to put client deliverables always first, leads to usual under-investment of technical resources to build a flexible research and analytics environment which is the cornerstone of any successful operation.

Building a flexible, integrated environment is not an easy task. There are no well trodden solutions but it is a worthy effort to tackle and it pays off in spades if done correctly. It is challenging from many perspectives as the need for full access to multiple years of data for quantitative researchers and the flexibility and quick time to market of new analytics and benchmarks are often in contrast with the need for stability and repeatability necessary to manage the day-to-day client demands. This leads to difficult trade-offs and requires some creative thinking. Additionally any change in analytics and benchmarks often requires the backfilling of such analytics past in time so that they are available for historical reports. When we deal with many millions of orders over several years, with analytics often needing tick-by-tick data, backfilling itself can be a tough technical challenge.

Going into the full cumbersome details of such an infrastructure is unfortunately beyond the scope of the brief excursion. But we propose a few pointers that have worked well in the past as a guide for the reader who may be involved in building such a solution.

- Consider building a layered, componentized approach. Layered because most of the analytics are constructed using simpler lower level analytics. 'Componentized' implies subdividing the TCA process in logical components that can then be assembled into simple daily reports or full fledged quarterly review report. An approach that considers leveraging a micro-services middleware is growing in popularity in recent years.[6]

- Limit the number of times you pass through quotes and trades as this step is by far the slowest part of the process.[7]

- The ideal setup, rarely achieved in practice, should be able to generate most of the $T + 0$ analytics on the fly as the algorithms complete. This would allow for the post-trade data to be available at the end of the trading day and thus reports can be generated and sent to clients immediately. This is something that clients really appreciate and requires a strongly-integrated infrastructure that leverages all the same components at runtime as it does after the fact.

- Consider making the TCA development infrastructure as an extension of your research environment that is under continuous integration. This means building and running regression tests every time something is changed. This will allow for quick turn-around in changes and immediate promotion into production. Leverage the same enrichment and post-trade analytics and apply them to the output of any simulation that is run for a specific scenario and store the results separately for each scenario. This will simplify the analysis and comparison of different parametrization and functionality.

- Instead of backfilling all new analytics for all clients/time-frame, one may consider implementing forward demand backfilling. This means that each component that is requested and for which the analytics are unavailable will compute the analytics for the client/period requested. This same approach can be used overnight

---

[6]`https://en.wikipedia.org/wiki/Microservices`

[7]Many of the TCA analytics are based on the prevailing quotes and trades at specific times such as the start and the end of the order. This requires what are called asof and window joins. The excellent speed at which KDB can perform these joins is one of the main reason for the platform's success.

to trigger the backfill (or re-populate in case it had to be fixed) of analytics to accelerate the delivery of large scale requests.

- Unless necessary, avoid the temptation of investing into an infrastructure for a fully fledged custom report builder. Instead, focus on making customized reports that are easy to programmatically create and assemble with all branding and formatting already in place. More and more people in sales and coverage have (or should have) at least some basic programming skills that will enable them to build additional custom reports. Some of these reports can then be standardized for clients who may not have complex needs. Report builders' framework is in general a significant development effort, complicated and error prone to build. They are also usually inflexible and are hard to extend.

## 12.5 Conclusion

It has been the authors' experience that rarely enough time is ever spent in building a powerful and flexible research environment that provides quantitative researchers the tools to quickly access and analyze data, create supporting analytics, simulate behavior of algorithms and then to be able to analyze the behavior of the algorithms and generate post-trade analytics for their own and their clients' review. Done correctly though, this should become a real engine for innovation and performance improvement, and a real competitive advantage in a highly fierce field. We hope this brief but broad review will be useful to help the reader who is involved in such a buildout, and provide the tools to point in the right direction.

# *Bibliography*


[1] F. Abdi and A. Ranaldo. A simple estimation of bid-ask spreads from daily close, high and low prices. *The Review of Financial Studies*, 30:4437–4480, 2017.

[2] F. Abergel and A. Jedidi. A mathematical approach to order book modeling. *International Journal of Theoretical and Applied Finance*, 16:1–40, 2013.

[3] A.R. Admati and P. Pfleiderer. A theory of intraday patterns: Volume and price variability. *The Review of Financial Studies*, 1(1):3–40, 1988.

[4] Y. Aït-Sahalia, P.A. Mykland, and L. Zhang. How often to sample a continuous-time process in the presence of market microstructure noise. *The Review of Financial Studies*, 18(2):351–416, 2005.

[5] H. Akaike. A new look at the statistical model identification. *IEEE Transactions on Automatic Control*, AC–19:716–723, 1974.

[6] S.S. Alexander. Price movements in speculative markets: Trends of random walks. *Industrial Management Review*, pages 7–26, 1961.

[7] S.S. Alexander. Price movements in speculative markets: Trends of random walks, no 2. *Industrial Management Review*, pages 25–46, 1964.

[8] S. Alizadeh, M.W. Brandt, and F.X. Diebold. Range-based estimation of stochastic volatility models. *Journal of Finance*, 57:1047–1091, 2002.

[9] R. Almgren. Execution costs. *Encyclopedia of Quantitative Finance*, pages 1–5, 2008.

[10] R. Almgren and N. Chriss. Optimal execution of portfolio transactions. *The Journal of Risk*, 3:5–39, 2000.

[11] R. Almgren, C. Thum, E. Hauptmann, and H. Li. Equity market impact. *Risk*, 18(7):57–62, 2005.

[12] R.F. Almgren. Optimal execution with nonlinear impact functions and trading enhanced risk. *Applied Mathematical Finance*, 10:1–18, 2003.

[13] N. Amenc, F. Goltz, A. Lodh, and L. Martellini. Diversifying the diversifiers and tracking the tracking error: Outperforming cap-weighted indices with limited risk of underperformance. *The Journal of Portfolio Management*, 38(3):72–88, 2012.

[14] S. Anatolyev and A. Gospodinov. A trading approach to testing for predictability. *Journal of Business & Economic Statistics*, 23:455–461, 2005.

[15] S. Anatolyev and A. Gospodinov. Modeling financial return dynamics via decomposition. *Journal of Business & Economic Statistics*, 28:232–245, 2010.

[16] T. Andersen, I. Archakov, G. Cebiroglu, and N. Hautsch. Volatility information feedback and market microstructure noise: A tale of two regimes. *CFS Working Paper, Northwestern University*, 2017.

[17] T.G. Andersen. Return volatility and trading volume: An information flow interpretation of stochastic volatility. *Journal of Finance*, 51:116–204, 1996.

[18] T.G. Andersen and T. Bollerslev. Answering the skeptics: Yes, standard volatility models do provide accurate forecasts. *International Economic Review*, 39:885–905, 1998.

[19] T.W. Anderson. *An Introduction to Multivariate Statistical Analysis*. Second Edition. Wiley, New York, 1984.

[20] T.W. Anderson and A.M. Walker. On the asymptotic distribution of the autocorrelations of a sample from a linear stochastic process. *Annals of Mathematical Statistics*, 35:1296–1303, 1964.

[21] A. Ang and A. Timmermann. Regime changes and financial markets. *Annual Review of Finance and Economics*, 4:313–337, 2012.

[22] W. Antweiler and M.Z. Frank. Is all that talk just noise? The information content of internet stock message boards. *Journal of Finance*, 59(3):1259–1294, 2004.

[23] P. Asquith, R. Oman, and C. Safaya. Short sales and trade classification algorithms. *Journal of Financial Markets*, 13:157–173, 2010.

[24] M. Avellaneda and J.H. Lee. Statistical arbitrage in the US equities market. *Quantitative Finance*, 10:761–782, 2010.

[25] W. Bagehot. The only game in town. *Financial Analysis Journal*, 27:12–14, 1971.

[26] P. Bajgrowicz and O. Scaillet. Technical trading revisited: False discoveries, persistence tests, and transaction costs. *Journal of Financial Economics*, 106(3):473–491, 2012.

[27] M. Baker and J. Wurgler. Investor sentiment and the cross-section of stock returns. *Journal of Finance*, 61(4):1645–1680, 2006.

[28] M. Baker and J. Wurgler. Investor sentiment in the stock market. *Journal of Economic Perspectives*, 21(2):129–151, 2007.

[29] F.M. Bandi and J.R. Russell. Separating microstructure noise from volatility. *Journal of Financial Economics*, 79:655–692, 2006.

[30] N. Barberis, A. Shleifer, and R. Vishny. A model of investor sentiment. *Journal of Financial Economics*, 49:307–343, 1998.

[31] O.E. Barndorff-Nielsen and N. Shephard. Econometric analysis of realized volatility and its use in estimating stochastic volatility models. *Journal of the Royal Statistical Society: Series B (Statistical Methodology)*, 64(2):253–280, 2002.

[32] L. Barras, O. Scaillet, and R. Wermers. False discoveries in mutual fund performance: Measuring luck in estimated alphas. *Journal of Finance*, 65(1):179–216, 2010.

[33] R. Battalio, S.A. Corwin, and R. Jennings. Can brokers have it all? On the relation between make-take fees and limit order execution quality. *Journal of Finance*, 71:2193–2238, 2016.

[34] L. Bauwens, S. Laurent, and J.V.K. Rombouts. Multivariate GARCH models: A survey. *The Journal of Applied Econometrics*, 21:79–109, 2006.

[35] M. Bayraktar, I. Mashtaser, N. Meng, and S. Radchenko. Barra vs total market equity trading model, empirical notes. *MSCI Research*, 2015.

[36] P. Bertrand and C. Protopopescu. The statistics of the information ratio. *International Journal of Business*, 15:71–86, 2010.

[37] D. Bertsimas and A.W. Lo. Optimal control of executions costs. *Journal of Financial Markets*, 1:1–50, 1998.

[38] H. Bessembinder, M. Panayides, and K. Venkataraman. Hidden liquidity: An analysis of order exposure strategies in electronic stock markets. *Journal of Financial Economics*, 94:361–383, 2009.

[39] B. Biais, L. Glosten, and C. Spatt. Market microstructure; a survey of microfoundations, empirical results, and policy implications. *Journal of Financial Markets*, 8:217–264, 2005.

[40] B. Biais, P. Hillion, and C. Spatt. An empirical analysis of the limit order book and the order flow in the Paris bourse. *Journal of Finance*, 50:1655–1689, 1995.

[41] J.P. Bialkowski, S. Darolles, and Gaëlle G. Le Fol. Improving VWAP. strategies: A dynamical volume approach. *Journal of Banking and Finance*, 32, 2006.

[42] F. Black. Towards a fully automated exchange, Part I. *Financial Analysts Journal*, 27:29–34, 1971.

[43] F. Black. Capital market equilibrium with restricted borrowing. *The Journal of Business*, 45:444–454, 1972.

[44] F. Black and R. Litterman. Global portfolio optimization. *Financial Analysts Journal*, 48 No. 5:28–43, 1992.

[45] L. Blume, D. Easley, and M. O'Hara. Market statistics and technical analysis: The role of volume. *Journal of Finance*, 49:153–181, 1994.

[46] T. Bollerslev. Generalized autoregressive conditional heteroskedasticity. *Journal of Econometrics*, 31:307–327, 1986.

[47] M. Borkovec and H.G. Heidle. Building and evaluating a transaction cost model: A primer. *The Journal of Trading*, 5:57–77, 2010.

[48] P. Bossaerts. Common nonstationary components of asset prices. *The Journal of Economic Dynamics and Control*, 12(2):347–364, 1988.

[49] J.P. Bouchaud, J.D. Farmer, and F. Lillo. *How Markets Digest Supply and Demand and Slowly Incorporate Information into Prices*. Academic Press, 2009.

[50] J.P. Bouchaud, Y. Gefen, M. Potters, and M. Wyart. Fluctuations and response in financial markets: The subtle nature of "random" price changes. *Quantitative Finance*, 4:176–190, 2004.

[51] J.-P. Bouchaud, M. Mezard, and M. Potters. Statistical properties of stock order books: Empirical results and models. *Quantitative Finance*, 2:251–256, 2002.

[52] D. Bowen, M.C. Hutchinson, and N. O'Sullivan. High-frequency equity pairs trading: Transaction costs, speed of execution and patterns in returns. *The Journal of Trading, Summer*, pages 31–38, 2010.

[53] G.E.P. Box, G.M. Jenkins, G.C. Reinsel, and G.M. Ljung. *Time Series Analysis: Forecasting and Control, 5th edition*. Wiley, New York, 2015.

[54] G.E.P. Box and G.C. Tiao. A canonical analysis of multiple time series. *Biometrika*, 64:355–365, 1977.

[55] P. Boyle, L. Garlappi, R. Uppal, and T. Wang. Keynes meets Markowitz: The trade-off between familiarity and diversification. *Management Science*, 58:253–272, 2012.

[56] M.W. Brandt and P. Santa-Clara. Dynamic portfolio selection by augmenting the asset space. *Journal of Finance*, 61:2187–2217, 2006.

[57] L. Breiman. Bagging predictors. *Machine Learning*, 24(2):123–140, 1996.

[58] L. Breiman. Prediction games and arcing algorithms. *Neural Computation*, 11(7):1493–1517, 1999.

[59] D.R. Brillinger. *Time Series: Data Analysis and Theory*. Expanded edition. Holden-Day, San Francisco, 1981.

[60] W. Brock, J. Lakonishok, and B. LeBaron. Simple technical trading rules and the stochastic properties of stock returns. *Journal of Finance*, 47:1731–1764, 1992.

[61] J. Brodie, I. Daubechies, C. De Mol, D. Giannone, and I. Loris. Sparse and stable Markowitz portfolios. *Proceedings of the National Academy of Sciences*, 106:12267–12272, 2009.

[62] C. Brownlees, F. Cipollini, and G.M. Gallo. Intra-daily volume modeling and prediction for algorithmic trading. *The Journal of Financial Econometrics*, 9:489–518, 2011.

[63] B. Bruder, N. Gaussel, J.-C. Richard, and T. Roncalli. Regularization of portfolio allocation. *White Paper Issue #10*, 2013.

[64] E. Busseti and S. Boyd. *Volume Weighted Average Price Optimal Execution*. unpublished, Stanford University, 2015.

[65] J.Y. Campbell, S.J. Grossman, and J. Wang. Trading volume and serial correlation in stock returns. *The Quarterly Journal of Economics*, 108:905–939, 1993.

[66] J.Y. Campbell, A.W. Lo, and A.C. MacKinlay. *The Econometrics of Financial Markets*. Princeton University Press, New Jersey, 1996.

[67] C. Cao, O. Hansch, and X. Wang. The information content of an open limit order book. *The Journal of Futures Markets*, 29:16–41, 2009.

[68] M.M. Carhart. On persistence in mutual fund performance. *Journal of Finance*, 52(1):57–82, 1997.

[69] M. Centoni and G. Cubadda. Modeling co-movements of economic time series: A selective survey. *Statistica*, 71:267–293, 2011.

[70] A.P. Chaboud, B. Chiquoine, E. Hjalmarsson, and C. Vega. Rise of the machines: Algorithmic trading in the foreign exchange market. *Journal of Finance*, 69:2045–2084, 2014.

[71] B. Chakrabarty, P.C. Moulton, and A. Shkilko. Short sales, long sales, and the Lee-Ready trade classification algorithm revisited. *Journal of Financial Markets*, 15(4):467–491, 2012.

[72] K. Chan and W.-M. Fong. Trade size, order imbalance and the volatility-volume relation. *Journal of Financial Economics*, 57:247–273, 2000.

[73] L. Chan and J. Lakonishok. Institutional equity trading costs, NYSE versus Nasdaq. *Journal of Finance*, 52:713–735, 1997.

[74] L.K.C. Chan and J. Lakonishok. The behavior of stock prices around institutional trades. *Journal of Finance*, 50:1147–1174, 1995.

[75] L.K.C. Chan and J. Lakonishok. Institutional equity trading costs: NYSE versus Nasdaq. *Journal of Finance*, 52(2):176–190, 1997.

[76] N.F. Chen, R. Roll, and S.A. Ross. Economic forces and the stock market. *The Journal of Business*, 59:383–403, 1986.

[77] S. Chib. Estimation and comparison of multiple change-point models. *Journal of Econometrics*, 86:221–241, 1998.

[78] C. Chiyachantana, P.K. Jain, C. Jiang, and R.A. Wood. International evidence on institutional trading behavior and price impact. *Journal of Finance*, 59:869–898, 2004.

[79] T. Chordia, R. Roll, and A. Subrahmanyam. Commonality in liquidity. *Journal of Financial Economics*, 56:3–28, 2000.

[80] T. Chordia, R. Roll, and A. Subrahmanyam. Order imbalance, liquidity, and market returns. *Journal of Financial Economics*, 65:111–130, 2002.

[81] P.K. Clark. A subordinated stochastic process model with finite variance for speculative prices. *Econometrica*, 41:135–155, 1973.

[82] J. Conrad and G. Kaul. An anatomy of trading strategies. *The Review of Financial Studies*, 11:489–519, 1998.

[83] J. Conrad and S. Wahal. The term structure of liquidity provision. *Journal of Financial Economics*, 136:239–259, 2020.

[84] J.S. Conrad, A. Hameed, and C. Niden. Volume and autocovariances in short-horizon individual security returns. *Journal of Finance*, 49:1305–1329, 1994.

[85] A. Constantinos, J.A. Donkas, and A. Subrahmanyam. Cognitive dissonance, sentiment and momentum. *Journal of Financial and Quantitative Analysis*, 46:245–275, 2013.

[86] R. Cont and A. Kukanov. Optimal order placement in limit order markets. *Quantitative Finance*, 17:21–39, 2017.

[87] R. Cont, A. Kukanov, and S. Stoikov. The price impact of order book events. *The Journal of Financial Econometrics*, 12:47–88, 2014.

[88] R. Cont, S. Stoikov, and R. Talreja. A stochastic model for order book dynamics. *Operations Research*, 58:549–563, 2010.

[89] M. Cooper. Filter values based on price and volume in individual security overreaction. *The Review of Financial Studies*, 12:901–935, 1999.

[90] S.A. Corwin and P. Schultz. A simple way to estimate bid-ask spreads from daily high and low prices. *Journal of Finance*, 67:719–759, 2012.

[91] A. Cowles. Can stock market forecasters forecast. *Econometrica*, 1:309–324, 1933.

[92] A. Cowles. Stock market forecasting. *Econometrica*, 12:206–214, 1944.

[93] D.R. Cox and P.A.W. Lewis. *The Statistical Analysis of Series of Events*. Chapman and Hall, London, 1966.

[94] G. Creamer and Y. Freund. Automated trading with boosting and expert weighting. *Quantitative Finance*, 4:401–420, 2010.

[95] M. Cremers and D. Weinbaum. Deviations from put-call parity and stock return predictability. *Journal of Financial and Quantitative Analysis*, 45:335–367, 2010.

[96] D.M. Cutler, J.M. Poterba, and L.H. Summers. *What Moves Stock Prices?*, volume 15. National Bureau of Economic Research Cambridge, Mass., USA, 1989.

[97] S. Da, J. Engelberg, and P. Gao. In search of attention. *Journal of Finance*, 66(5):1461–1499, 2011.

[98] Z. Da, J. Engelberg, and P. Gao. The sum of all fears investor sentiment and asset prices. *The Review of Financial Studies*, 28(1):1–32, 2015.

[99] R. Dahlhaus and S. Subba Rao. Statistical inference for time-varying ARCH processes. *The Annals of Statistics*, 34:1075–1114, 2006.

[100] D.J. Daley and D. Vere-Jones. *An Introduction to the Theory of Point Processes, Volume I: Elementary Theory and Methods*. Springer, New York, 2003.

[101] H.E. Daniels. Autocorrelation between first differences of mid-ranges. *Econometrica*, pages 215–219, 1966.

[102] S.R. Das and M.Y. Chen. Yahoo! for Amazon: Sentiment extraction from small talk on the web. *Management Science*, 53:1375–1388, 2007.

[103] B.J. DeLong, A. Shleifer, L.H. Summers, and R.J. Waldmann. Noise trader risk in financial markets. *The Journal of Political Economy*, 98:703–738, 1990.

[104] V. DeMiguel, L. Garlappi, and R. Uppal. Optimal versus naïve diversification: How inefficient is the $1/N$ portfolio strategy? *The Review of Financial Studies*, 22:1915–1953, 2009.

[105] A.P. Dempster, N.-M. Laird, and D.B. Rubin. Maximum likelihood from incomplete data via the EM algorithm. *Journal of the Royal Statistical Society, Series B*, 39:1–38, 1977.

[106] B. Do and R. Faff. Does simple pairs trading still work? *The Financial Analysts Journal*, 66(4):83–95, 2010.

[107] I. Domowitz, J. Glen, and A. Madhavan. Liquidity, volatility and equity trading costs across countries and over time. *International Finance*, 4:221–255, 2001.

[108] I. Domowitz and H. Yegerman. The cost of algorithmic trading: A first look at comparative performance. *Algorithmic Trading: Precision, Control, Execution*, 2005.

[109] A. Dufour and R. Engle. Time and the price impact of a trade. *Journal of Finance*, 55(2):467–498, 2000.

[110] D. Easley, M.L. De Prado, and M. O'Hara. Flow toxicity and liquidity in a high frequency world. *The Review of Financial Studies*, 25:1457–1493, 2012.

[111] D. Easley, M.L. De Prado, and M. O'Hara. Optimal execution horizon. *Mathematical Finance*, 25:640–672, 2015.

[112] D. Easley and M. O' Hara. Price, trade size, and information in security markets. *Journal of Financial Economics*, 19:69–90, 1987.

[113] D. Easley and M. O' Hara. Time and the process of security price adjustment. *Journal of Finance*, 19:69–90, 1992.

[114] C. Eckart and G. Young. The approximation of one matrix by another of lower rank. *Psychometrika*, 1:211–218, 1936.

[115] B. Efron, T. Hastie, I. Johnstone, and R. Tibshirani. Least angle regression. *Annals of Statistics*, 32:407–499, 2004.

[116] J. Engelberg, P. Gao, and R. Jagannathan. An anatomy of pairs trading: the role of idiosyncratic news, common information and liquidity. In *Third Singapore International Conference on Finance*, 2009.

[117] R. Engle and R. Ferstenberg. Execution risk. *The Journal of Portfolio Management*, 33(4):34–44, 2007.

[118] R. Engle and F.K. Kroner. Multivariate simultaneous generalized ARCH. *Econometric Theory*, 11:122–150, 1995.

[119] R. Engle and A. Lunde. Trades and quotes: A bivariate point process. *The Journal of Financial Economics*, 1:159–188, 2003.

[120] R.F. Engle. Autoregressive conditional heteroscedasticity with estimates of the variance of United Kingdom inflations. *Econometrica*, 50:987–1007, 1982.

[121] R.F. Engle, R. Ferstenberg, and J.R. Russell. Measuring and modeling execution cost and risk. *The Journal of Portfolio Management*, 38(2):14–28, 2012.

[122] R.F. Engle and C.W.J. Granger. Co-integration and error correction: Representation, estimation, and testing. *Econometrica*, 55:251–276, 1987.

[123] R.F. Engle and S. Kozicki. Testing for common features. *Journal of Business & Economic Statistics*, 11(4):369–380, 1993.

[124] R.F. Engle and D. Kraft. *Multiperiod Forecast Error Variances of Inflation Estimated from ARCH Models*, in: A. Zellner, ed: *Applied Time Series Analysis of Economic Data*. Bureau of the Census, Washington, DC, 1983.

[125] R.F. Engle and J.R. Russell. Autoregressive conditioned duration: A new model for irregularly spaced transaction data. *Econometrica*, 66:1127–1162, 1998.

[126] R.F. Engle and R. Susmel. Common volatility in international equity markets. *Journal of Business & Economic Statistics*, 11(2):167–176, 1993.

[127] R.F. Engle and S. Kozicki. Testing for common features. *Journal of Business & Economic Statistics*, 11:369–380, 1993.

[128] R.F. Engle, V.K. Ng, and M. Rothschild. Asset pricing with a factor ARCH covariance structure: Empirical estimates for treasury bills. *Journal of Econometrics*, 45:213–218, 1990.

[129] T.W. Epps and M.L. Epps. The stochastic dependence of security price changes and transaction volumes: Implications for the mixture-of-distribution hypothesis. *Econometrica*, 44:305–321, 1976.

[130] T.W. Epps. Co-movements in stock prices in the very short run. *Journal of the American Statistical Association*, 74:291–298, 1979.

[131] F.J. Fabozzi, S.M. Focardi, and P.N. Kolm. *Quantitative Equity Investing: Techniques and Strategies*. Wiley and Sons, 2006.

[132] E.F. Fama and M.E. Blume. Filter rules and stock-market trading. *The Journal of Business*, 39:226–241, 1966.

[133] E.F. Fama and K.R. French. A five-factor after pricing model. *Journal of Financial Economics*, 116:1–22, 2015.

[134] E.F. Fama and K.R. French. International tests of a five-factor asset pricing model. *Fama-Miller Working Paper*, 2015.

[135] J. Fan, Y. Fan, and J. Lv. High dimensional covariance matrix estimation using a factor model. *Journal of Econometrics*, 147:187–197, 2008.

[136] J. Fan, F. Han, H. Liu, and B. Vickers. Robust inference of risks of large portfolios. *Journal of Econometrics*, 194:298–308, 2016.

[137] J. Fan, J. Zhang, H. Liu, and K. Yu. Vast portfolio selection with cross-exposure constraints. *Journal of the American Statistical Association*, 107:592–606, 2012.

[138] A. Farago and E. Hjalmarsson. Stock price co-movement and the foundations of pairs trading. *Journal of Financial and Quantitative Analysis*, 54:629–665, 2019.

[139] J.D. Farmer, A. Gerig, F. Lillo, and H. Waelbroeck. How efficiency shapes market impact. *Quantitative Finance*, 11:1743–1758, 2013.

[140] J.D. Farmer, L. Gillemot, F. Lillo, S. Mike, and A. Sen. What really causes large price changes? *Quantitative Finance*, 4:383–397, 2004.

[141] W.E. Ferson and A.F. Siegel. The use of conditioning information in portfolios. *Journal of Finance*, 56:967–982, 2001.

[142] T. Foucault and A.J. Menkveld. Competition for order flow and smart order routing systems. *Journal of Finance*, pages 119–157, 2008.

[143] A. Frazzini, R. Israel, and T.J. Moskowitz. Trading costs (unpublished), 2018.

[144] A. Frazzini and L.H. Pedersen. Betting against beta. *Journal of Financial Economics*, 111:1–25, 2014.

[145] Y. Freund and R.E. Schapire. A decision-theoretic generalization of on-line learning and an application to boosting. In *European Conference on Computational Learning Theory*, pages 23–37. Springer, 1995.

[146] Y. Freund and R.E. Schapire. Experiments with a new boosting algorithm. In *International Conference on Machine Learning*, volume 96, pages 148–156. Morgan Kaufmann Publishers Inc., San Francisco, CA, 1996.

[147] Y. Freund and R.E. Schapire. A decision-theoretic generalization of on-line learning and an application to boosting. In *Journal of Computer and System Sciences*, volume 55, pages 119–139, 1997.

[148] J. Friedman. Greedy function approximation: A gradient boosting machine. *Annals of Statistics*, pages 1189–1232, 2001.

[149] J. Friedman, T. Hastie, and R. Tibshirani. Additive logistic regression: A statistical view of boosting (with discussion and a rejoinder by the authors). *The Annals of Statistics*, 28(2):337–407, 2000.

[150] A. Frino, E. Jarnecic, and A. Lepone. The determinants of price impact of block trades: Further evidence. *Abacus*, 43(2):94–106, 2007.

[151] K.A. Froot and E.M. Dabora. How are stock prices affected by the location of trade? *Journal of Financial Economics*, 53(2):189–216, 1999.

[152] P. Fryzlewicz, T. Sapatinas, and S. Subba Rao. Normalized least-squared estimation in time-varying ARCH models. *The Annals of Statistics*, 36:742–786, 2008.

[153] W.A. Fuller. *Introduction to Statistical Time Series, Second Edition.* John Wiley, New York, 1996.

[154] L. Gagnon and G.A. Karolyi. Multi-market trading and arbitrage. *Journal of Financial Economics*, 97:53–80, 2010.

[155] A.R. Gallant, P.E. Rossi, and G. Tauchen. Nonlinear dynamic structures. *Econometrica*, 61:871–908, 1993.

[156] L. Gao, Y. Han, S.Z. Li, and G. Zhou. Market intraday momentum. *Journal of Financial Economics*, 129:394–414, 2018.

[157] N. Gârleanu and L.H. Pedersen. Dynamic trading with predictable returns and transaction costs. *The Journal of Finance*, 68:2309–2340, 2013.

[158] M.B. Garman and M.J. Klass. On the estimation of security price volatilities from historical data. *The Journal of Business*, 53:67–78, 1980.

[159] P.H. Garthwaite. An interpretation of partial least squares. *Journal of the American Statistical Association*, 89:122–127, 1994.

[160] E. Gatev, W.N. Goetzmann, and R.G. Rouwenhorst. Pairs trading: Performance of a relative value arbitrage rule. *The Review of Financial Studies*, 19:797–827, 2006.

[161] J. Gatheral. No-dynamic-arbitrage and market impact. *Quantitative Finance*, 10:769–759, 2010.

[162] R. Gencay. The predictability of security returns with simple technical trading rules. *Journal of Empirical Finance*, 5:347–359, 1998.

[163] S. Gervais, R. Kaniel, and D.H. Mingelgrin. The high-volume return premium. *Journal of Finance*, 56(3):877–919, 2001.

[164] S. Gervais, R. Kaniel, and D.H. Mingelgrin. The high-volume return premium. *Journal of Finance*, 56:877–919, 2001.

[165] E. Ghysels, C. Gourieroux, and J. Jasiak. Stochastic volatility duration models. *Journal of Econometrics*, 119:413–433, 2004.

[166] M.R. Gibbons, S.A. Ross, and J. Shanken. A test of the efficiency of a given portfolio. *Econometrica*, 57:1121–1152, 1989.

[167] L.R. Glosten and P.R. Milgrom. Bid, ask and transaction prices in a specialist market with heterogeneously informed traders. *Journal of Financial Economics*, 14:71–100, 1985.

[168] I. Goodfellow, Y. Bengio, and A. Courville. *Deep Learning*. M.I.T. Press, Boston, 2016.

[169] R.C. Grinold and R.N. Kahn. *Active Portfolio Management, Second Edition*. McGraw-Hill, 2000.

[170] A. Gross-Klussmann and N. Hautsch. When machines read the news: Using automated text analytics to quantify high frequency news-implied market reactions. *Journal of Empirical Finance*, 18:321–340, 2011.

[171] A.D. Hall and N. Hautsch. Order aggressiveness and order book dynamics. *Empirical Economics*, 30:973–1005, 2006.

[172] J.D. Hamilton. A new approach to the economic analysis of nonstationary time series and the business cycle. *Econometrica*, 57:357–384, 1989.

[173] J.D. Hamilton. Analysis of time series subject to changes in regime. *Journal of Econometrics*, 45:39–70, 1990.

[174] J.D. Hamiton. Macroeconomic regimes and regime shifts. In J.B. Taylor and H. Uhlig, editors, *Handbook of Macroeconomics*, chapter 3, pages 153–201. Elsevier, 2016.

[175] Y. Han, K. Yang, and G. Zhou. A new anomaly: The cross-sectional profitability of technical analysis. *Journal of Financial and Quantitative Analysis*, 48:1433–1461, 2013.

[176] P.R. Hansen. A test for superior predictive ability. *Journal of Business & Economic Statistics*, 23:365–380, 2005.

[177] M. O' Hara and M. Ye. Is market fragmentation harming market quality? *Journal of Financial Economics*, pages 454–474, 2011.

[178] L. Harris. *Trading and Exchanges: Market Microstructure for Practitioners*. Oxford University Press, New York, 2003.

[179] A.C. Harvey. *Forecasting, Structural Time Series Models and the Kalman Filter*. Cambridge University Press, Cambridge, 1989.

[180] J. Hasbrouck. Measuring the information content of stock trades. *Journal of Financial Economics*, 46:179–207, 1991.

[181] J. Hasbrouck. One security, many markets: Determining the contributions to price discovery. *Journal of Finance*, 50(4):1175–1199, 1995.

[182] J. Hasbrouck and G. Saar. Technology and liquidity provision: The blurring of traditional definitions. *Journal of Financial Markets*, 12:143–172, 2009.

[183] J. Hasbrouck and D.J. Seppi. Common factors in prices, order flows, and liquidity. *Journal of Financial Economics*, 59:383–411, 2001.

[184] T. Hastie, R. Tibshirani, and J. Friedman. *The Elements of Statistical Learning; Data Mining, Inference and Prediction, Second Edition*. Springer-Verlag, New York, 2009.

[185] N. Hautsch and R. Huang. The market impact of a limit order. *The Journal of Economic Dynamics and Control*, 36:501–522, 2012.

[186] A.G. Hawkes. Spectra of some self-exciting and mutually exciting point processes. *Biometrika*, 58:83–90, 1971.

[187] A.G. Hawkes. Hawkes processes and their applications to finance; a review. *Quantitative Finance*, 18:193–198, 2018.

[188] X. He and R. Velu. Volume and volatility in a common-factor mixture of distributions model. *Journal of Financial and Quantitative Analysis*, 49:33–49, 2014.

[189] I.S. Helland. On the structure of partial least squares regression. *Communications in Statistics, Simulation and Computation*, B17:581–607, 1988.

[190] I.S. Helland. Partial least squares regression and statistical models. *Scandinavian Journal of Statistics*, 17:97–114, 1990.

[191] T. Hendershott, J. Brogaard, and R. Riordon. High frequency trading and price discovery. *The Review of Financial Studies*, 27:2267–2306, 2014.

[192] T. Hendershott, C.M. Jones, and A.J. Menkveld. Does algorithmic trading improve liquidity? *Journal of Finance*, 66(1):1–33, 2011.

[193] T. Hendershott, C.M. Jones, and A.J. Menkveld. Does algorithmic trading improve liquidity? *Journal of Finance*, pages 1–3, 2011.

[194] T. Hendershott and R. Riordan. Algorithmic trading and information. *http://faculty.haas.berkeley.edu/hender/ATInformation.pdf*, 2011.

[195] T. Hendershott and M. Seasholes. Market maker inventories and stock prices. *American Economic Review*, 97:210–214, 2007.

[196] T. Ho and H.R. Stoll. Optimal dealer pricing under transactions and return uncertainty. *Journal of Financial Economics*, 9:47–73, 1981.

[197] T. Ho, R. Schwartz, and D. Whitcomb. The trading decision and market clearing under transaction price uncertainty. *The Journal of Finance*, 40:21–42, 1985.

[198] S. Hogan, R. Jarrow, M. Teo, and M. Warachka. Testing market efficiency using statistical arbitrage with application to momentum and value strategies. *Journal of Financial Economics*, 73:525–565, 2004.

[199] R.W. Holthausen, R.W. Leftwich, and D. Mayers. Large-block transactions, the speed of response, and temporary and permanent stock-price effect. *Journal of Financial Economics*, 26:71–95, 1990.

[200] H. Hotelling. Analysis of a complex of statistical variables into principal components. *Journal of Educational Psychology*, 4:417–441, 498–520, 1933.

[201] H. Hotelling. The most predictable criterion. *Journal of Educational Psychology*, 26:139–142, 1935.

[202] H. Hotelling. Relations between two sets of variables. *Biometrika*, 28:321–322, 1936.

[203] P.H. Hsu, Y.C. Hsu, and C.M. Kuan. Testing the Predictive Ability of Technical Analysis Using a New Stepwise Test Without Data Snooping Bias. *Journal of Empirical Finance*, 17:471–484, 2010.

[204] Y.-P. Hu and R.S. Tsay. Principal volatility component analysis. *Journal of Business & Economic Statistics*, 32:153–164, 2014.

[205] R. Huang and H. Stoll. Dealer versus auction markets: A paired comparison of execution costs on NASDAQ and the NYSE. *Journal of Financial Economics*, 41:313–357, 1996.

[206] W. Huang, C-A. Lehalle, and M. Rosenbaum. Simulating and analyzing order book data: The queue-reactive model. *Journal of the American Statistical Association*, 110:107–122, 2015.

[207] D. Huang, F. Jiang, J. Tu, and G. Zhou. Investor sentiment aligned: A powerful predictor of stock returns. *The Review of Financial Studies*, 28:791–837, 2015.

[208] G. Huberman. Familiarity breeds investment. *The Review of Financial Studies*, 14:659–680, 2001.

[209] G. Huberman and W. Stanzl. Price manipulation and quasi-arbitrage. *Econometrica*, 72:1247–1275, 2004.

[210] S. Hvidkjaer. A trade-based analysis of momentum. *The Review of Financial Studies*, 119:457–491, 2006.

[211] R. Israel, T. Moskowitz, A. Ross, and L. Serban. Implementing momentum: What have we learned. *NBER Working Paper*, 2017.

[212] R. Jagannathan and T. Ma. Risk reduction in large portfolios: Why imposing the wrong constraints helps. *Journal of Finance*, 58:1651–1683, 2003.

[213] C.M. Jarque and A.K. Bera. Efficient tests for normality, homoscedasticity and serial independence of regression residuals. *Economic Letters*, 6:255–259, 1980.

[214] N. Jegadeesh. Discussion of LMW (2000). *Journal of Finance*, pages 1765–1770, 2000.

[215] N. Jegadeesh and S. Titman. Returns to buying winners and selling losers: Implications for stock market efficiency. *Journal of Finance*, 48:65–91, 1993.

[216] N. Jegadeesh and S. Titman. Profitability of momentum strategies: An evaluation of alternative explanations. *Journal of Finance*, 56:699–720, 2001.

[217] N. Jegadeesh and S. Titman. Cross-sectional time series determinants of momentum returns. *The Review of Financial Studies*, 15:143–157, 2002.

[218] F. Jiang, J. Lee, X. Martin, and G. Zhou. Manager sentiment and stock returns. *Journal of Financial Economics*, 132:126–149, 2019.

[219] W. Jiang, L. Shu, and D.W. Apley. Adaptive CUSUM procedures with EWMA-based shift estimators. *IIE Transactions*, 40:992–1003, 2008.

[220] J.D. Jobson and B. Korkie. Estimation for markowitz efficient portfolios. *The Journal of American Statistical Association*, 75:544–554, 1980.

[221] S. Johansen. Statistical analysis of co-integration vectors. *The Journal of economic dynamics and control*, 12(2):231–254, 1988.

[222] S. Johansen. Estimation and hypothesis testing of co-integration vectors in Gaussian vector autoregressive models. *Econometrica: Journal of the Econometric Society*, pages 1551–1580, 1991.

[223] C. Jones, G. Kaul, and M. Lipson. Information, trading and volatility. *Journal of Financial Economics*, 36:127–154, 1994.

[224] L.P. Kaelbling, M.L. Littman, and A.W. Moore. Reinforcement learning: A survey. *Journal of Artificial Intelligence*, 4:237–285, 1996.

[225] R.N. Kahn and M. Lemmon. Smart beta: The owner's manual. *The Journal of Portfolio Management*, 41(2):76–83, 2015.

[226] H. Kawakatsu. Direct multiperiod forecasting for algorithmic trading. *Journal of Forecasting*, 37(1):83–101, 2018.

[227] D.B. Keim and A. Madhavan. The upstairs market for large-block transactions: Analysis and measurement of price effects. *The Review of Financial Studies*, 9:1–36, 1996.

[228] D.B Keim and A. Madhavan. Transaction costs and investment style: An inter-exchange analysis of institutional equity trades. *Journal of Financial Economics*, 46:265–292, 1997.

[229] J.L. Kelly. A new interpretation of information rate. *Bell System Technical Journal*, 35:917–926, 1956.

[230] J.M. Keynes. The general theory of employment. *The Quarterly Journal of Economics*, pages 209–223, 1937.

[231] P.D. Koch and T.W. Koch. Evolution in dynamic linkages across daily national stock indexes. *Journal of International Money and Finance*, 10.2:231–251, 1991.

[232] A. Kourtis. On the distribution and estimation of trading costs. *Journal of Empirical Finance*, 29:230–245, 2014.

[233] P. Kratz and T. Schöneborn. Optimal liquidity in dark pools. *Quantitative Finance*, 14:1519–1539, 2014.

[234] A. Kyle. Continuous time auctions and insider trading. *Econometrics*, 53:1315–1336, 1985.

[235] T.L. Lai and H. Xing. *Statistical Models and Methods for Financial Markets*. Springer, 2008.

[236] T.L. Lai and H. Xing. Stochastic change-point ARX GARCH models and their applications to econometric times series. *Statistica Sinica*, 23:1573–1594, 2013.

[237] T.L. Lai, H. Xing, and Z. Chen. Mean-variance portfolio optimization when means and covariances are unknown. *The Annals of Applied Statistics*, 5:798–823, 2011.

[238] J. Lakonishok, A. Shleifer, and R.W. Vishny. Contrarian investment, extrapolation, and risk. *Journal of Finance*, 49(5):1541–1578, 1994.

[239] O. Ledoit and M. Wolf. Improved estimation of the covariance matrix of stock returns with an application to portfolio selection. *Journal of Empirical Finance*, 10:603–621, 2003.

[240] C.M. Lee and B. Swaminathan. Price momentum and trading volume. *Journal of Finance*, LV:2017–2069, 2000.

[241] C.M. Lee and M. Ready. Inferring trade direction from intraday data. *Journal of Finance*, 46:733–746, 1991.

[242] J. Lewellen. Momentum and autocorrelation in stock returns. *The Review of Financial Studies*, 15:65–91, 2002.

[243] J.K.-S. Liew, S. Guo, and T. Zhang. Tweet sentiments and crowd-sourced earnings estimates as valuable sources of information around earnings releases. *The Journal of Alternative Investments*, Winter Issue:1–20, 2017.

[244] F. Lillo, J.D. Farmer, and R. Mantegna. Master curve for price impact function. *Nature*, 421:129–130, 2003.

[245] J. Lintner. The valuation of risky assets and the selection of risky investment in stock portfolios and capital budgets. *Review of Economics and Statistics*, 47:13–37, 1965.

[246] G. Llorente, R. Michaely, G. Saar, and J. Wang. Dynamic volume-return relation of individual stocks. *The Review of Financial Studies*, 15:1005–1047, 2002.

[247] A.W. Lo, H. Mamaysky, and J. Wang. Foundation of technical analysis: Computational algorithms, statistical inference and empirical implementation. *Journal of Finance*, LV(4):1705–1765, 2000.

[248] A.W. Lo. The statistics of Sharpe ratios. *Financial Analysis Journal*, pages 36–52, 2002.

[249] A.W. Lo and A.C. MacKinlay. When are contrarian profits due to stock market overreaction? *The Review of Financial Studies*, 3:175–205, 1990.

[250] A.W. Lo, A.C. MacKinlay, and J. Zhang. Econometric models of limit-order executions. *Journal of Financial Economics*, 65:31–71, 2002.

[251] T.F. Loeb. Trading cost: The critical link between investment information and results. *Financial Analyst Journal*, 39(3):39–44, 1983.

[252] M. Lopez de Prado. *Advances in Financial Machine Learning*. John Wiley & Sons, New Jersey, 2018.

[253] J. Lorenz and R. Almgren. Mean-variance optimal adaptive execution. *Applied Mathematical Finance*, 18(4):395–422, 2011.

[254] A. Madhavan. Market microstructure. *Journal of Financial Markets*, 3:205–258, 2000.

[255] A. Madhavan. VWAP strategies. *Trading*, 1:32–39, 2002.

[256] A. Madhavan, M. Richardson, and M. Roomans. Why do security prices change? A transaction-level analysis of NYSE stocks. *The Review of Financial Studies*, 10(4):1035–1064, 1997.

[257] C. Maglaras, C.C. Moallemi, and H. Zheng. Queueing dynamics and state space collapse in fragmented limit order book markets. *Operations Research* (to appear), 2019.

[258] B.G. Malkiel. *A Random Walk Down Wall Street: The Time-Tested Strategy for Successful Investing, 10th Edition*. Norton, 2012.

[259] B. Mandelbrot. The variation of certain speculative prices. *The Journal of Business*, 36:294–319, 1963.

[260] H. Markowitz. *Portfolio Selection: Efficient Diversification of Investments*. Wiley: New York, 1959.

[261] M. Martens and D. van Dijk. Measuring volatility with the realized range. *Journal of Econometrics*, 138:181–207, 2007.

[262] R. McCulloch and R. Tsay. Nonlinearity in high-frequency financial data and hierarchical models. *Studies in Nonlinear Dynamics and Econometrics*, 5:1067–1077, 2001.

[263] H. Mendelson. Market behavior in a clearing house. *Econometrica: Journal of the Econometric Society*, 1505–1524, 1982.

[264] L. Menkhoff, L. Sarno, M. Schmeling, and A. Schrimpf. Currency momentum strategies. *Journal of Financial Economics*, 106:660–684, 2012.

[265] L. Menkhoff and M.P. Taylor. The obstinate passion of foreign exchange professionals: Technical analysis. *The Journal of Economic Literature*, XLV:936–972, 2007.

[266] A.J. Menkveld, S.J. Koopman, and A. Lucas. Modeling around-the-clock price discovery for cross-listed stocks using state space methods. *Journal of Business & Economic Statistics*, 25:213–225, 2007.

[267] R.C. Merton. On estimating the expected return on the market. *Journal of Financial Economics*, 8:323–336, 1980.

[268] R.T. Merton. An intertemporal capital asset pricing model. *Econometrica*, 41:867–887, 1973.

[269] R.O. Michaud. *Efficient Asset Management*. Harvard Business School Press, Boston, 1989.

[270] G. Mitra and L. Mitra. *The Handbook of News Analytics in Finance (Ed)*. Wiley Finance, 2011.

[271] T. Moorman. An empirical investigation of methods to reduce transaction costs. *Journal of Empirical Finance*, 29:230–245, 2014.

[272] E. Moro, J. Vicente, L.G. Moyano, A. Gerig, J.D. Farmer, G. Vaglica, F. Lillo, and R.N. Mantegna. Market impact and trading profile of hidden orders in stock markets. *Physical Review E*, 80(6):066102, 2009.

[273] T.J. Moskowitz, Y.H. Ooi, and L.H. Pedersen. Time Series Momentum. *Journal of Financial Economics*, 104:228–250, 2012.

[274] A.A. Obizhaeva and J. Wang. Optimal trading strategy and supply/demand dynamics. *Journal of Financial Markets*, 16:1–32, 2013.

[275] M. O'Hara. Presidential address: Liquidity and price discovery. *Journal of Finance*, 58:1335–1354, 2003.

[276] M. O'Hara. High frequency market microstructure. *Journal of Financial Economics*, 116:257–270, 2015.

[277] J. Okunev and D. White. Do momentum - based strategies still work in foreign currency markets? *Journal of Financial and Quantitative Analysis*, 38:425–447, 2003.

[278] J.K. Ord, A.B. Koehler, and R.D. Snyder. Estimation and prediction for a class of dynamic nonlinear statistical models. *Journal of the American Statistical Association*, 92(440):1621–1629, 1997.

[279] D.A. Pachamanova and F.J. Fabozzi. Recent trends in equity portfolio construction analytics. *The Journal of Portfolio Management*, 40:137–151, 2014.

[280] A. Pardo and R. Pascual. On the hidden side of liquidity. *The European Journal of Finance*, 18:949–967, 2012.

[281] C. Parlour and D. Seppi. Liquidity-based competition for order flow. *The Review of Financial Studies*, 16:301–343, 2003.

[282] C.A. Parlour and D.J. Seppi. *Limit Order Markets – A Survey*, In *Handbook of Financial Intermediation and Banking*, Edited by A. Thakor and A. Boot. Elsevier, Amsterdam, 2008.

[283] L. Pastor and R.F. Stambough. The equity premium and structural breaks. *Journal of Finance*, 56:1207–1239, 2001.

[284] A.J. Patton and A. Timmermann. Monotonicity in asset returns: New tests with applications to the term structure, the CAPM, and the portfolio sorts. *Journal of Financial Economics*, 98:605–625, 2010.

[285] R.L. Peterson. *Trading on Sentiment: The Power of Minds over Markets*. Wiley Finance, 2016.

[286] M.J. Ready. Profits from technical trading rules. *Financial Management*, Autumn:43–61, 2002.

[287] G.C. Reinsel and S.K. Ahn. Vector autoregressive models with unit roots and reduced rank structure: estimation, likelihood ratio test, and forecasting. *The Journal of Time Series Analysis*, 13(4):353–375, 1992.

[288] G.C. Reinsel. *Element of Multivariate Time Series Analysis, Second Edition*. Springer-Verlag, New York, 2002.

[289] G.C. Reinsel and R. Velu. *Multivariate Reduced-Rank Regression, Theory and Application*. Springer-Verlag, New York, 1998.

[290] L.C.G. Rogers and S.E. Satchell. Estimating variance from high, low and closing prices. *Annals of Applied Probability*, 1:504–512, 1991.

[291] R. Roll. A simple model of the implicit bid-ask spread in an efficient market. *Journal of Finance*, 39:1127–1139, 1984.

[292] J.P. Romano and M. Wolf. Stepwise multiple testing as formalized data snooping. *Econometrica*, 73:1237–1282, 2005.

[293] S.A. Ross. The arbitrage theory of capital asset pricing. *The Journal of Economic Theory*, 13:341–360, 1976.

[294] I. Rosu. A dynamic model of the limit order book. *The Review of Financial Studies*, 22:4601–4641, 2009.

[295] T.H. Rydberg and N. Shephard. Dynamic trade-by-trade price movement: Decomposition and models. *Journal of Financial Econometrics*, 1:2–25, 2003.

[296] S. Satchell and A. Snowcraft. A demystification of the Black-Litterman model: Managing quantitative and traditional portfolio construction. *Journal of Asset Management*, 1:138–150, 2000.

[297] V. Satish, A. Saxena, and M. Palmer. Predicting intraday trading volume and volume percentages. *The Journal of Trading*, 9:15–25, 2014.

[298] M. Schneider and F. Lillo. Cross-impact and no-dynamics arbitrage. *Quantitative Finance*, 19:137–154, 2019.

[299] G. Schwarz. Estimating the dimension of a model. *Annals of Statistics*, 6:401–404, 1978.

[300] J.T. Scruggs. Noise trader risk: Evidence from the siamese twins. *Journal of Financial Markets*, 10:76–105, 2007.

[301] W.F. Sharpe. Capital asset prices: A theory of market equilibrium under conditions of risk. *Journal of Finance*, 19:425–442, 1964.

[302] R.H. Shumway and D.S. Stoffer. Arima models. In *Time Series Analysis and Its Applications*, pages 83–171. Springer, 2011.

[303] D. Smith, N. Wang, Y. Wang, and E.J. Zychowicz. Sentiment and the effectiveness of Technical Analysis: Evidence from the Hedge Fund Industry. *Journal of Financial and Quantitative Analysis*, 51:1991–2013, 2016.

[304] E. Smith, D.J. Farmer, L. Gillemot, and S. Krishnamurthy. Statistical theory of the continuous double auction. *Quantitative Finance*, 3:481–514, 2003.

[305] R. Stambaugh, J. Yu, and Y. Yuan. The short of it: Investor sentiment and anomalies. *Journal of Financial Economics*, 104:288–302, 2012.

[306] M. Statman, S. Thorley, and K. Vorkink. Investor overconfidence and trading volume. *The Review of Financial Studies*, 19(4):1531–1565, 2006.

[307] S. Stoikov. The micro-price: A high-frequency estimator of future prices. *Quantitative Finance*, 18:1959–1966, 2018.

[308] R. Sullivan, A. Timmermann, and H. White. Data snooping, technical trading rule performance, and the bootstrap. *Journal of Finance*, 54:1647–1691, 1999.

[309] R.S. Sulton and A.G. Barks. *Reinforcement Learning, An Introduction, Second Edition*. MIT Press, Boston, 2018.

[310] R.J. Sweeney. Some new filter rule tests: Methods and results. *Journal of Financial and Quantitative Analysis*, 20(3):285–300, 1988.

[311] G.E. Tauchen and M. Pitts. The price variability-volume relationship on speculative markets. *Econometrica*, 51:485–505, 1983.

[312] P.C. Tetlock. Giving content to investor sentiment: The role of media in the stock market. *Journal of Finance*, 62(3):1139–1168, 2007.

[313] I.M. Toke. An introduction to Hawkes processes with applications to finance, 2011. Lecture Notes from Ecole Centrale Paris, BNP, Paribas.

[314] B. Toth, Y. Lemperiere, C. Deremble, J. De Lataillade, J. Kockelkoren, and J.-P. Bouchaud. Anomalous price impact and the critical nature of liquidity in financial markets. *Physical Review X*, 1(2):021006, 2011.

[315] R. Tsay. *Analysis of Financial Time Series, Third Edition*. Wiley, 2010.

[316] G. Tsoukalas, J. Wang, and K. Giesecke. Dynamics portfolio execution. *Management Science*, 2019.

[317] J. Tu and G. Zhou. Markowitz meets talmud: A combination of sophisticated and naïve diversification. *Journal of Financial Economics*, 99:204–215, 2011.

[318] F. Vahid and R.F. Engle. Common trends and common cycles. *The Journal of Applied Econometrics*, 8:341–360, 1993.

[319] R. Velu, A. Gretchika, M. Benaroch, D. Nehren, and K. Kuber. Market Impact: To Trade Small or to Trade Seldom? Evidence from Algorithmic Execution Data. *Unpublished*, 2015.

[320] B. von Beschwitz, D.B. Keim, and M. Massa. First to "read" the news: News analytics and algorithmic trading. *Review of Financial Studies* (to appear), 2019.

[321] A.A. Weiss. ARMA models with ARCH errors. *The Journal of Time Series Analysis*, 5:129–143, 1984.

[322] M. West and J. Harrison. *Bayesian Forecasting and Dynamic Models, Second Edition*. Springer-Verlag, New York, 1997.

[323] H. White. A reality check for data snooping. *Econometrica*, 68:1097–1126, 1999.

[324] R. De Winne and C. D'Hondt. Hide-and-seek in the market: Placing and detecting hidden order. *Review of Finance*, 11:663–692, 2007.

[325] H. Wold. *PLS regression*, volume 6. Eds. N.L. Johnson and S. Kotz, 1984.

[326] H. Working. Note on the correlation of first differences of averages in a random chain. *Econometrica: Journal of the Econometric Society*, pages 916–918, 1960.

[327] D. Yang and Q. Zhang. Drift-independent volatility estimation based on high, low, open and close prices. *The Journal of Business*, 73:477–491, 2000.

[328] J.W. Yang. Transaction duration and asymmetric price impact of trades - Evidence from Australia. *Journal of Empirical Finance*, 18:91–102, 2011.

[329] J. Yu and Y. Yuan. Investor sentiment and the mean–variance relation. *Journal of Financial Economics*, 100:367–381, 2011.

[330] E. Zarinelli, M. Treccani, J.D. Farmer, and F. Lillo. Beyond the square root: Evidence for logarithmic dependence of market impact on size and participation rate. *Market Microstructure and Liquidity*, 1:1–31, 2015.

[331] G. Zhou. Measuring investor sentiment. *Annual Review of Financial Economics*, 10:239–259, 2018.

[332] Y. Zhu and G. Zhou. Technical analysis: An asset allocation perspective on the use of moving averages. *Journal of Financial Economics*, 92:519–544, 2009.

# *Subject Index*